CELL FREE SYNTHESIS OF BIOLOGICS

CELL FREE SYNTHESIS OF BIOLOGICS

A COMPREHENSIVE GUIDE

SARFARAZ NIAZI
College of Pharmaceutical Sciences, Washington State University, Spokane, WA, United States

ELSEVIER
ACADEMIC PRESS
An imprint of Elsevier

Academic Press is an imprint of Elsevier
125 London Wall, London EC2Y 5AS, United Kingdom
50 Hampshire Street, 5th Floor, Cambridge, MA 02139, United States

ISBN: 978-0-443-33271-5

For information on all Academic Press publications visit our website at https://www.elsevier.com/books-and-journals

Publisher: Andre Wolff
Editorial Project Manager: Himanshi Yadav
Production Project Manager: Sathyavani Deivasigamani
Cover Designer: Raman Kumar

Typeset by Integra Software Services

Contents

11 Diagnostics and biosensors

12 Industrial enzyme production

13 Synthetic biology integration

Part V

Specialized applications

14 Membrane protein synthesis

15 Complex biomolecule production

16 Emerging molecular targets

VI

Economic and regulatory framework

17 Economic analysis and business models

VIII

Practical implementation

24 Laboratory setup and implementation for cell-free protein synthesis

25 Protocols and standard operating procedures for cell-free protein synthesis

26 Case studies and success stories in cell-free protein synthesis

IX

Resources and references

27 Commercial suppliers and resources for cell-free systems

28 Databases and computational tools for cell-free protein synthesis

Preface

The contemporary era represents a fundamental transformation in our relationship with biological systems, marking a transition from passive observation to active engineering of living processes with unprecedented precision, intelligence, and responsiveness. Among the most promising technologies facilitating this paradigm shift is cell-free protein synthesis (CFPS), which, through its capacity to circumvent cellular constraints, accelerate production timelines, and enable decentralized manufacturing, transcends its origins as a laboratory convenience to emerge as a strategic instrument for addressing global health challenges, driving pharmaceutical innovation, and preparing for extraterrestrial life sciences applications.

This work emerges at a critical juncture in technological development. The convergence of synthetic biology, nanotechnology, machine learning, quantum computing, and additive manufacturing has rendered CFPS not merely relevant but imperatively necessary for addressing contemporary challenges. In operational environments where speed, adaptability, and portability constitute essential requirements, including pandemic response scenarios, battlefield therapeutic applications, and extraterrestrial habitat maintenance, traditional cell-based systems demonstrate fundamental limitations in terms of temporal constraints, structural fragility, and operational complexity. CFPS addresses these limitations directly.

The scope of this work extends beyond introducing CFPS as a scientific methodology to positioning it as a platform technology that serves as a foundational engine for molecular manufacturing. Through comprehensive visual frameworks, technological readiness matrices, and detailed architectural schematics presented throughout this volume, a systems-level perspective of CFPS is provided, encompassing its operational capabilities, integration opportunities with emerging technologies, and transformative role in reshaping global bioproduction infrastructure.

The intended audience for this work encompasses researchers, investors, engineers, and policymakers, all of whom are encouraged to conceptualize CFPS beyond its function as an experimental tool. Rather, it represents a comprehensive solution to next-generation challenges that demand rapid response capabilities, operational reliability, and global accessibility. As this book demonstrates, such applications are not speculative projections, but present realities prepared for immediate implementation and deployment.

This book represents the culmination of decades of research, innovation, and observation in biotechnology, protein therapeutics, regulatory science, and emerging life science platforms. The central thesis underlying this work is that cell-free protein synthesis (CFPS) has evolved beyond its origins as a laboratory curiosity to become a foundational pillar of future-ready biomanufacturing. Whether applied to biologics production in resource-limited environments, the development of space-based medicine, or the decentralization of therapeutic access, CFPS offers the most scalable, responsive, and intelligent pathway to molecular production.

The purpose of this volume extends beyond merely explaining the mechanisms of CFPS to provide a comprehensive framework that encompasses the full spectrum from underlying biochemistry to platform engineering, from regulatory strategy to commercial deployment, and from terrestrial systems to extraterrestrial architectures. This book serves simultaneously as a strategic document and a scientific treatise.

Motivation

The impetus for this work arose from a significant gap in the existing literature. No single source adequately integrated the molecular protocols of CFPS with the engineering logic required for scalable platforms, the quality systems necessary for reproducible operation, the market insights essential for translational opportunity, and the regulatory understanding needed to bridge science and compliance. This book represents an attempt to address this deficiency by bridging rigorous science with strategic application, thereby enabling a broader audience to understand and effectively utilize CFPS technology.

Scope and organization

This volume comprises 26 chapters, each focused yet interconnected, collectively building a comprehensive understanding of CFPS as a transformative technology. The opening Chapters 1 through 5 establish the technology foundations, defining CFPS within the context of modern synthetic biology and introducing analytical frameworks, such as the technology convergence readiness matrix and cross-technology value chains. These chapters outline key enabling technologies, including CRISPR–Cas systems, three-dimensional bioprinting, and quantum algorithms, examining their synergistic relationships with CFPS in driving platform convergence and readiness.

Chapters 6 through 9 examine system design and workflow integration, exploring the architectural and operational layers of CFPS, including lysate preparation, energy optimization, translation systems, and modular reaction configurations. These chapters provide detailed visual workflows, control checkpoints, and component validation strategies essential for consistent and high-yield production.

The quality and analytical systems that form the foundation of mature CFPS technology are addressed in Chapters 10 through 13. These chapters introduce comprehensive frameworks for lysate quality control workflows, process monitoring protocols, statistical control strategies, and data management systems, including real-time monitoring and predictive analytics capabilities. These systems support both research and development activities and Good Manufacturing Practice-compliant production environments.

Advanced engineering and deployment considerations are examined in Chapters 14 through 16, addressing distributed, automated, and embedded CFPS systems. These chapters present engineered use cases, including autonomous CFPS modules for space stations, point-of-care therapeutic manufacturing platforms, and miniaturized bioreactor systems for field deployment. The space-based CFPS architecture framework presented in Chapter 16 exemplifies the reimagining of these technologies for space medicine, disaster response, and defense applications.

Market analysis and investment mapping are comprehensively covered in Chapters 17 through 19, presenting quantitative and visual analyses of CFPS commercialization patterns. These chapters examine market segmentation representing over 5 billion dollars in value, funding patterns spanning from 2000 to 2025, and the complex ecosystem of investors and startup enterprises. Visual analytics such as the Startup Funding Evolution framework provide dynamic perspectives on how capital investment has shaped CFPS innovation trajectories.

The institutional frameworks necessary for scaling CFPS technology are explored in Chapters 20 through 22, which examine collaboration and translational ecosystems. These chapters analyze academic–industry collaboration models, intellectual property management and licensing strategies, and comprehensive success metrics and performance alignment frameworks. This section introduces the collaboration sustainability framework, offering a systematic approach to managing stakeholder balance across temporal dimensions.

Regulatory, compliance, and manufacturing strategy considerations are addressed in Chapters 23 through 25. Any technology intended for scale must demonstrate regulatory readiness, and these chapters provide detailed roadmaps for alignment with Food and Drug Administration and European Medicines Agency requirements, the design of process control and documentation systems, and the integration of statistical quality assurance with digital batch record systems. These chapters are specifically designed to serve quality assurance and regulatory affairs leaders, regulatory consultants, and biotechnology executives.

The concluding Chapter 26 maps the outlook and convergence horizon, examining personalized CFPS platforms, programmable CFPS systems with artificial intelligence integration, zero-gravity production capabilities, and convergence with quantum design tools for protein folding and expression optimization. The argument presented is that CFPS transcends its role as a mere tool to function as a platform for distributed biological intelligence.

Visual framework integration

Throughout this work, over 30 custom-designed figures, infographics, and process maps have been integrated, each constructed to serve not merely as illustration but as actionable frameworks. These visual elements include development timelines, readiness matrices, modular system maps, investment evolution charts, and regulatory compliance diagrams. They are designed to serve educators, investors, and engineers as practical tools for planning and implementation.

Tabulation advantages

Textbooks are primarily descriptive, but this book should be viewed more as a handbook that scientists can use for quick reference and systematic learning of the practical teachings. To achieve this goal, I have introduced tables wherever they would help reach the reference and direction quickly. References provided at the end of each chapter should help readers who wish to delve deeper into the subject. For example, Table 8.26 lists examples of recent success stories of the utility of CFPS; by the time the book goes to the press, there would have been more stories to quote but the purpose for such tables is to stimulate thinking of the reader to anticipate many new possibilities of the application of CFPS.

Intended audience

This book has been developed to serve multiple constituencies within the biotechnology ecosystem. Scientists will find validated workflows and systematic thinking frameworks, while biotechnology entrepreneurs can evaluate CFPS platforms for productization potential. Investors tracking the emergence of high-yield, short-cycle biological production will find comprehensive market analytics, and regulatory teams will discover quality system alignment and documentation frameworks. Educators and students exploring next-generation bioengineering curricula will benefit from the integrated pedagogical approach, while space and defense agencies will find frameworks for designing autonomous biological production systems under extreme conditions.

Final reflections

The development of this work has represented a deeply fulfilling intellectual journey, bringing together fundamental scientific principles, systems engineering methodologies, considerations of global equity, and visionary technological thinking. CFPS technology enables us to envision a future where therapeutic proteins are not confined to centralized manufacturing facilities but can be produced at the point of need, regardless of geographic or logistical constraints.

If this book contributes to the development of breakthrough systems, safer therapeutic modalities, or innovative business models, it will have fulfilled its intended purpose. I invite readers to utilize this work not merely as a reference text but as a platform for building, challenging, and advancing the future of biological systems.

Sarfaraz K. Niazi
Affiliate Professor, College of Pharmaceutical Sciences, Washington State University

List of abbreviations

ACS	American Chemical Society
ADC	antibody–drug conjugate
ADCC	antibody-dependent cellular cytotoxicity
ADP	adenosine diphosphate
AI	artificial intelligence
AI/ML	artificial intelligence/machine learning
AND	analog to digital
AP	alkaline phosphatase
APPLICATIONS	software applications
ASTM	American Society for Testing and Materials
ATP	adenosine triphosphate
ATP/ADP	adenosine triphosphate/adenosine diphosphate
ATP/GTP	Adenosine triphosphate/guanosine triphosphate
AUTOMATION	process automation
BGC	biosynthetic gene cluster
BLA	biologics license application
BME	beta-mercaptoethanol
BSA	bovine serum albumin
CAGR	compound annual growth rate
CAPEX	capital expenditure
CECF	continuous exchange cell-free
CFPS	cell-free protein synthesis
CFR	code of federal regulations
CFSP	cell-free synthesis platform
CH	heavy chain (antibody)
CHO	Chinese hamster ovary
CL	light chain (antibody)
CMC	Chemistry, Manufacturing, and Controls
COVID-19	Coronavirus disease 2019
CRISPR	clustered regularly interspaced short palindromic repeats
CRISPR-CFPS	CRISPR cell-free protein synthesis
CUSUM	cumulative sum (statistical process control)
CV	coefficient of variation
DBTL	design–build–test–learn
DIGITAL	digital technologies
DIRECTIONS	future directions
DNA	deoxyribonucleic acid
DOD	Department of Defense
DOE	design of experiments
DTT	dithiothreitol
ECONOMIC	economic analysis
EDTA	ethylenediaminetetraacetic acid
EM	electron microscopy
EMA	European Medicines Agency
EMA/CHMP	European Medicines Agency/Committee for Medicinal Products for Human Use
EMERGING	emerging technologies

EMSA	electrophoretic mobility shift assay
ETH	ethanol
FDA	Food and Drug Administration
FEBS	Federation of European Biochemical Societies
FOUNDATIONS	theoretical foundations
FRAMEWORK	regulatory framework
FUTURE	future perspectives
GFP	green fluorescent protein
GMP	good manufacturing practice
GPU	graphics processing unit
GTP	guanosine triphosphate
HEPES	4-(2-hydroxyethyl)-1-piperazineethanesulfonic acid
HLA	human leukocyte antigen
HPLC	high-performance liquid chromatography
HRP	horseradish peroxidase
HTS	high-throughput screening
HVAC	heating, ventilation, and air conditioning
ICH	International Council for Harmonisation
IDP	intrinsically disordered protein
IND	investigational new drug
INTEGRATION	system integration
IP	intellectual property
ISO	International Organization for Standardization
IEC	International Electrotechnical Commission
LAL	limulus amebocyte lysate
LAMP	loop-mediated isothermal amplification
LC–MS	liquid chromatography–mass spectrometry
LCA	life cycle assessment
LED	light-emitting diode
LIMS	Laboratory Information Management System
LOC	Lab-on-a-Chip
MIT	Massachusetts Institute of Technology
ML	machine learning
MPC	model predictive control
MS	mass spectrometry
MSP	managed service provider
NAD	nicotinamide adenine dinucleotide
NADH	nicotinamide adenine dinucleotide (reduced form)
NADPH	nicotinamide adenine dinucleotide phosphate (reduced form)
NIH	National Institutes of Health
NMR	nuclear magnetic resonance
NOT	negation operator
NPV	net present value
NRPS	nonribosomal peptide synthetase
NSF	National Science Foundation
OEE	overall equipment effectiveness
OPC-UA	open platform communications unified architecture
OPEX	operating expenditure
OR	operating room or operations research
PAGE	polyacrylamide gel electrophoresis
PAM	protospacer adjacent motif
PAT	process analytical technology
PCR	polymerase chain reaction
PDA	photodiode array

PDI	protein disulfide isomerase
PEG	polyethylene glycol
PEP	phosphoenolpyruvate
PKS	polyketide synthase
PLANT	plant-based systems
PLATFORMS	technology platforms
PMDA	Pharmaceuticals and Medical Devices Agency (Japan)
PMSF	phenylmethylsulfonyl fluoride
POC	proof of concept
PPE	personal protective equipment
PTM	posttranslational modification
PURE	protein synthesis using recombinant elements
QA	quality assurance
RBS	ribosome binding site
RL	ribosomal large subunit
RNA	ribonucleic acid
RNP	ribonucleoprotein
ROI	return on investment
RPA	robotic process automation
RRL	rabbit reticulocyte lysate
RT-CRISPR	real-time CRISPR
RTD	resistance temperature detector
SARS	severe acute respiratory syndrome
SBIR/STTR	small business innovation research/small business technology transfer
SDS–PAGE	sodium dodecyl sulfate–polyacrylamide gel electrophoresis
SiLA	standardization in lab automation
SPC	statistical process control
SRP	signal recognition particle
TLCK	N-α-tosyl-l-lysyl-chloromethylketone
TRENDS	current trends
TXTL	transcription–translation
UNIDO	United Nations Industrial Development Organization
UTR	untranslated region
VH	variable heavy chain
VL	variable light chain
VLP	virus-like particle
VOC	volatile organic compound
WGE	wheat germ extract
WHO	World Health Organization

PART I

Foundations and principles

CHAPTER

1

Introduction to cell-free protein synthesis

1.1 Historical perspective: From Nirenberg to modern platforms

The journey of cell-free protein synthesis (CFPS) began with one of the most fundamental questions in molecular biology: How is genetic information translated into functional proteins? In 1961 Marshall Nirenberg and Heinrich Matthaei conducted their groundbreaking experiments that would not only crack the genetic code but also establish the foundation for modern CFPS. Using a simple mixture of ribosomes, transfer RNAs, and other cellular components extracted from *Escherichia coli*, they demonstrated that synthetic RNA could direct protein synthesis in a test tube, a discovery that would earn Nirenberg the Nobel Prize and forever change our understanding of life itself.

This pivotal experiment represented the birth of cell-free synthesis as we know it today. However, the path from Nirenberg's basic system to today's sophisticated platforms has been marked by continuous innovation and technological breakthroughs. The 1970s saw the development of more robust systems, particularly the rabbit reticulocyte lysate system, which became a workhorse for eukaryotic protein synthesis research. The 1980s brought the pioneering work of Alexander Spirin, who developed continuous-exchange systems that could sustain protein synthesis for extended periods (Fig. 1.1) (Table 1.1).

The turn of the millennium marked a new era for cell-free synthesis. James Swartz and his colleagues at Stanford University revolutionized the field by developing optimized *E. coli*-based systems that could produce therapeutically relevant quantities of proteins. Simultaneously, the development of the Protein synthesis Using Recombinant Elements (PURE) system by Ueda and colleagues provided an entirely defined, reconstituted platform free from the complexity of cellular lysates.

The 2010s witnessed the transformation of CFPS from a research tool into a viable manufacturing platform. Commercial entities began adopting cell-free systems for rapid prototyping and small-scale production of therapeutic proteins. The development of freeze-dried systems enabled the creation of stable, portable diagnostic platforms that could operate without refrigeration, a breakthrough that would prove crucial during global health emergencies.

Cell Free Synthesis of Biologics
https://doi.org/10.1016/B978-0-443-33271-5.00022-3

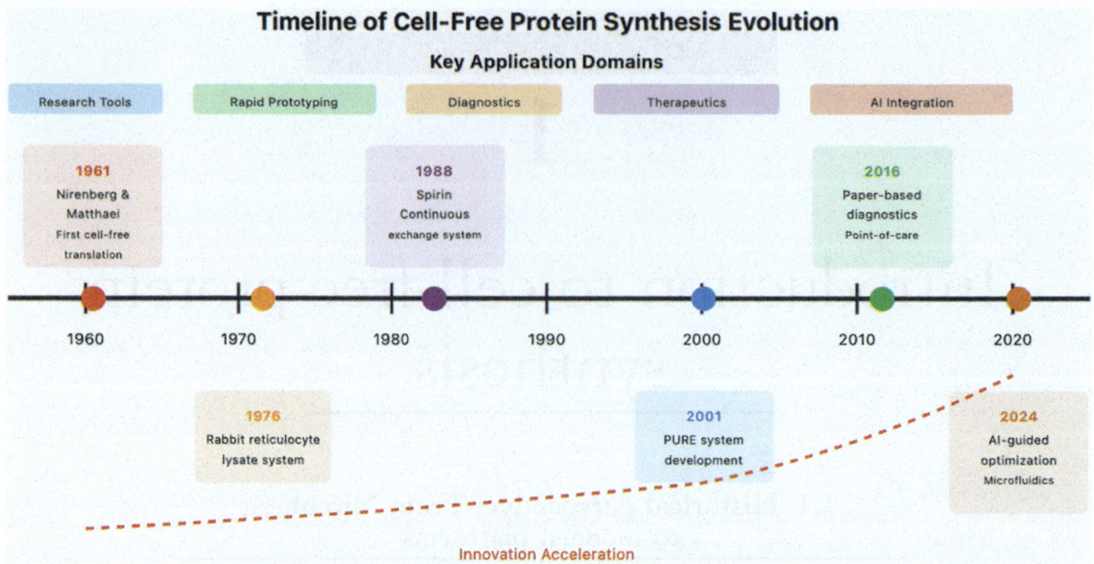

FIG. 1.1 Timeline of cell-free protein synthesis evolution. The development of CFPS from Nirenberg's initial experiments to modern AI-guided platforms. Major milestones are shown with corresponding technological capabilities and application domains. The timeline illustrates the acceleration of innovation, particularly in the 21st century with the convergence of biotechnology, automation, and artificial intelligence.

TABLE 1.1 Milestones in cell-free protein synthesis development.

Year	Milestone	Key contributors	Impact
1961	First cell-free translation system	Nirenberg and Matthaei	Genetic code discovery
1976	Rabbit reticulocyte lysate system	Pelham and Jackson	Eukaryotic protein synthesis
1988	Continuous-exchange system	Spirin et al.	Extended reaction duration
1999	*E. coli* S30 optimization	Kim and Swartz	Enhanced bacterial systems
2001	PURE system development	Shimizu et al.	Defined component system
2011	Commercial scale-up	Zawada et al.	Industrial applications
2016	Paper-based diagnostics	Pardee et al.	Point-of-care applications
2023	AI-guided optimization	Pandi et al.	Machine learning integration
2024	Microfluidic automation	Baranwal and Maerkl	High-throughput platforms

The most recent phase of CFPS evolution, spanning 2020–2025, has been characterized by the integration of artificial intelligence (AI) and advanced automation. Machine learning algorithms now guide system optimization, while microfluidic platforms enable unprecedented control and parallelization. The COVID-19 pandemic accelerated adoption by demonstrating the value of rapid, distributed manufacturing capabilities.

1.2 The central dogma in cell-free systems

To understand the power and potential of CFPS, one must first appreciate how it relates to the fundamental process of life itself, the flow of genetic information from DNA to proteins. Francis Crick's *Central Dogma of Molecular Biology*, formulated in 1957, describes this process as occurring in two key steps: transcription, where genetic information in DNA is copied into messenger RNA (mRNA), and translation, where the mRNA directs the synthesis of proteins at ribosomes.

In living cells, this process is tightly regulated and occurs within a complex, compartmentalized environment. The nucleus houses DNA and the transcription machinery, while translation occurs in the cytoplasm, with intricate feedback mechanisms controlling every step. While this complexity allows cells to respond to environmental changes and maintain homeostasis, it also creates constraints for protein production applications.

Cell-free systems liberate this molecular machinery from cellular constraints, creating what can be described as "life in a test tube." The essential components—ribosomes, transfer RNAs, aminoacyl-tRNA synthetases, translation factors, and energy regeneration systems—are preserved and reconstituted in a controlled environment. This liberation provides unprecedented access to and control over the protein synthesis process (Fig. 1.2).

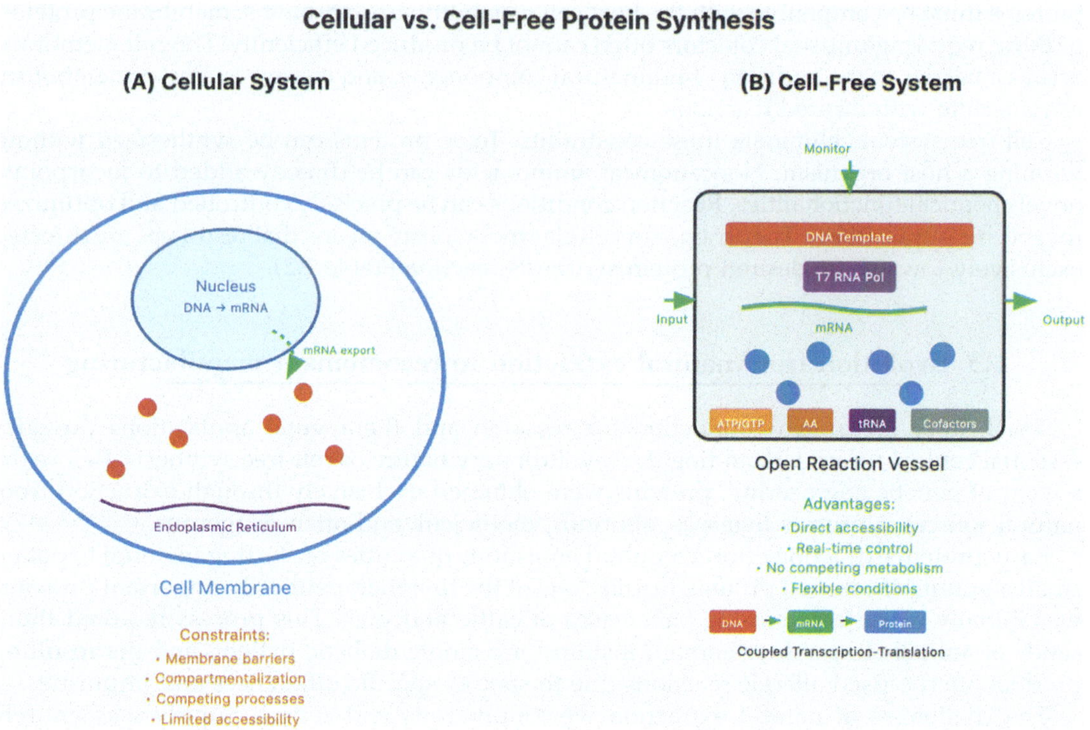

FIG. 1.2 Comparison of cellular vs. cell-free protein synthesis. (A) In vivo protein synthesis occurs within the complex, compartmentalized environment of living cells, with transcription in the nucleus and translation in the cytoplasm. (B) Cell-free systems reconstitute the essential molecular machinery in an open, accessible format, enabling direct manipulation and optimization of reaction conditions. The diagram illustrates key components, including ribosomes, tRNAs, energy systems, and regulatory elements.

TABLE 1.2 Advantages of cell-free vs. cellular protein synthesis.

Aspect	Cellular systems	Cell-free systems
Speed	Days to weeks	Hours to days
Toxicity tolerance	Limited by host viability	Unlimited
Environmental control	Constrained by cellular requirements	Complete control
Nonnatural components	Membrane barrier limitations	Direct addition possible
Resource allocation	Competing cellular processes	Dedicated to protein synthesis
Monitoring access	Limited by cellular barriers	Real-time analysis possible
Scalability	Complex bioreactor requirements	Simple reaction scaling
Contamination risk	High (living system)	Lower (controlled environment)

The advantages of this "open system" architecture become immediately apparent when considering the challenges of traditional recombinant protein production. In cellular systems, proteins must be compatible with the host cell's viability; toxic proteins, membrane proteins, or those requiring unusual cofactors often cannot be produced efficiently. The cell membrane acts as a barrier to the addition of nonnatural components, and complex cellular metabolism can interfere with desired reactions.

Cell-free systems eliminate these constraints. Toxic proteins can be synthesized without harming a host organism. Noncanonical amino acids can be directly added to incorporate novel chemical functionalities. Reaction conditions can be precisely controlled and optimized for specific applications. The absence of cellular metabolism means that resources are directed exclusively toward the desired protein synthesis reaction (Table 1.2).

1.3 Evolution from natural extraction to recombinant manufacturing

The history of protein production for research and therapeutic applications provides essential context for understanding the revolutionary nature of cell-free synthesis. Before the advent of genetic engineering, proteins were obtained exclusively through extraction from natural sources, a process that was laborious, inefficient, and often dangerous.

Early protein extraction efforts required enormous quantities of starting material to obtain small amounts of purified protein. Insulin, one of the first therapeutically important proteins, was initially extracted from the pancreases of cattle and pigs. This process required thousands of animals to produce enough insulin for a single diabetic patient, and the resulting product often caused allergic reactions due to species-specific differences and impurities.

The challenges of natural extraction were numerous and severe. Supply was entirely dependent on the availability of biological material, yields were typically very low, and purification was extremely difficult. The risk of pathogen transmission was constant, and for many important proteins, natural sources simply did not exist in sufficient quantities or appropriate forms (Fig. 1.3).

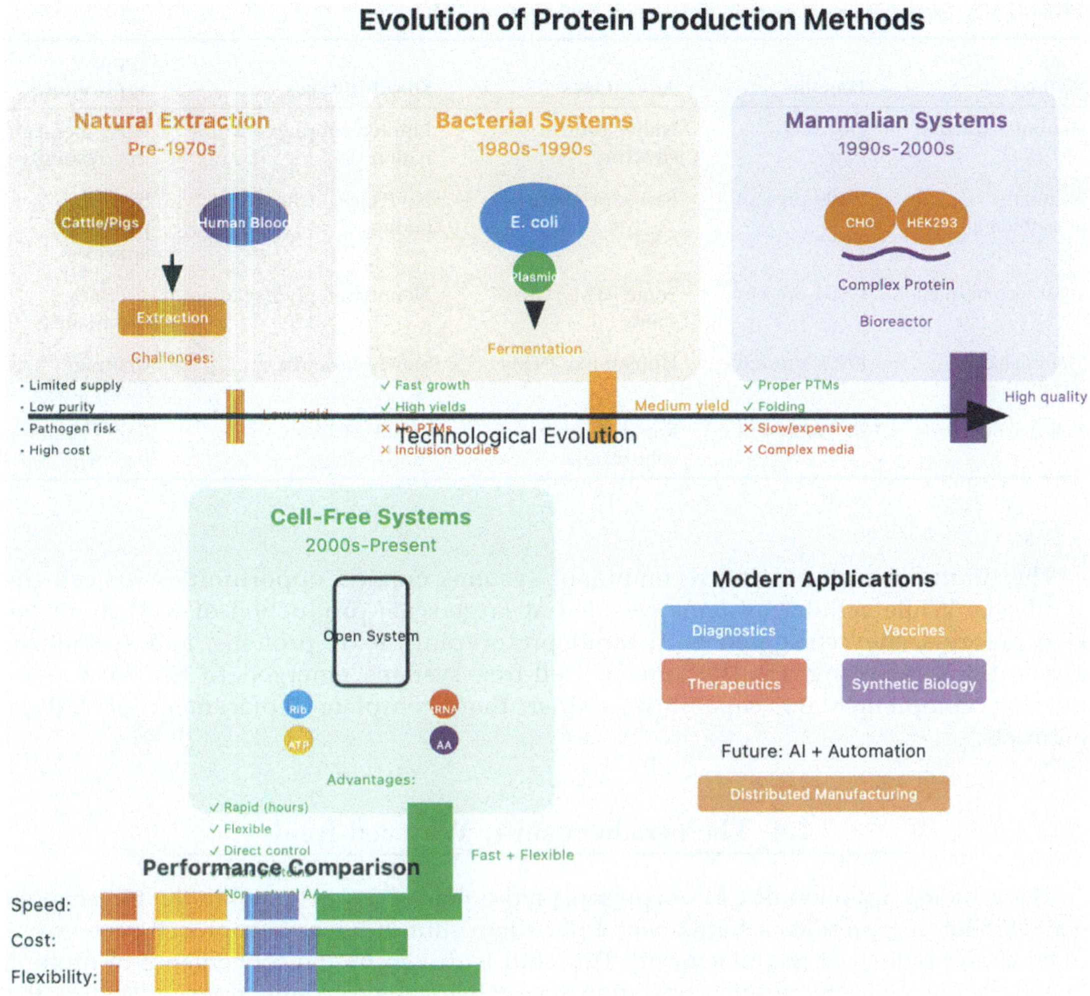

FIG. 1.3 Evolution of protein production methods. The progression from natural extraction through recombinant cellular production to modern cell-free synthesis. Each era is characterized by improvements in yield, purity, safety, and accessibility. Timeline shows key technological developments and their impact on protein availability and cost.

The development of recombinant DNA technology in the 1970s promised to revolutionize protein production. By inserting genes encoding desired proteins into bacterial, yeast, or mammalian cells, researchers could theoretically produce unlimited quantities of any protein. The first commercial success was human insulin produced in genetically modified *E. coli*, approved by the Food and Drug Administration (FDA) in 1982.

However, recombinant cellular production brought its own set of challenges. Different proteins required different host systems, each with specific advantages and limitations. Bacterial systems were fast and cost-effective but could not perform complex posttranslational modifications required for many eukaryotic proteins. Mammalian cell systems could produce properly modified proteins but were slow, expensive, and technically demanding (Table 1.3).

TABLE 1.3 Comparison of protein production paradigms.

Method	Timeline	Advantages	Disadvantages	Typical applications
Natural extraction	Pre-1970s	Native protein structure	Limited supply, contamination risk	Early therapeutics (insulin)
Bacterial recombinant	1980s–present	Fast, cost-effective	No PTMs, inclusion bodies	Simple proteins, enzymes
Yeast recombinant	1990s–present	Some PTMs, GRAS status	Nonhuman glycosylation	Vaccines, therapeutics
Mammalian recombinant	1990s–present	Human-like PTMs	Slow, expensive	Complex therapeutics
Cell-free synthesis	2000s–present	Rapid, flexible, controllable	Limited PTMs (improving)	Rapid prototyping, diagnostics

The limitations of existing recombinant systems created opportunities for cell-free synthesis. While cellular systems excelled at large-scale production of well-characterized proteins, they struggled with rapid prototyping, toxic proteins, and specialized applications requiring precise control. Cell-free systems emerged to fill these gaps, offering complementary capabilities rather than complete replacement of cellular methods.

1.4 The paradigm shift: Why cell-free?

The growing adoption of CFPS represents more than just a new tool in the biotechnologist's toolkit; it represents a fundamental paradigm shift in how we approach protein production and biological manufacturing. This shift is driven by the convergence of multiple factors: technological maturation, changing market demands, and emerging applications that cellular systems cannot address effectively.

The speed advantage of cell-free systems cannot be overstated. Traditional recombinant protein production requires weeks to months for cell line development, process optimization, and scale-up. Cell-free systems can produce functional proteins within hours of receiving genetic information. This speed advantage has proven crucial in applications ranging from pandemic response to personalized medicine, where time-to-product is often more important than absolute cost per unit.

Beyond speed, cell-free systems offer unprecedented flexibility. The open architecture allows for real-time monitoring and adjustment of reaction conditions. Nonnatural components can be incorporated directly into reactions. Multiple proteins can be cosynthesized in controlled ratios. These capabilities enable applications that are simply impossible with cellular systems (Fig. 1.4).

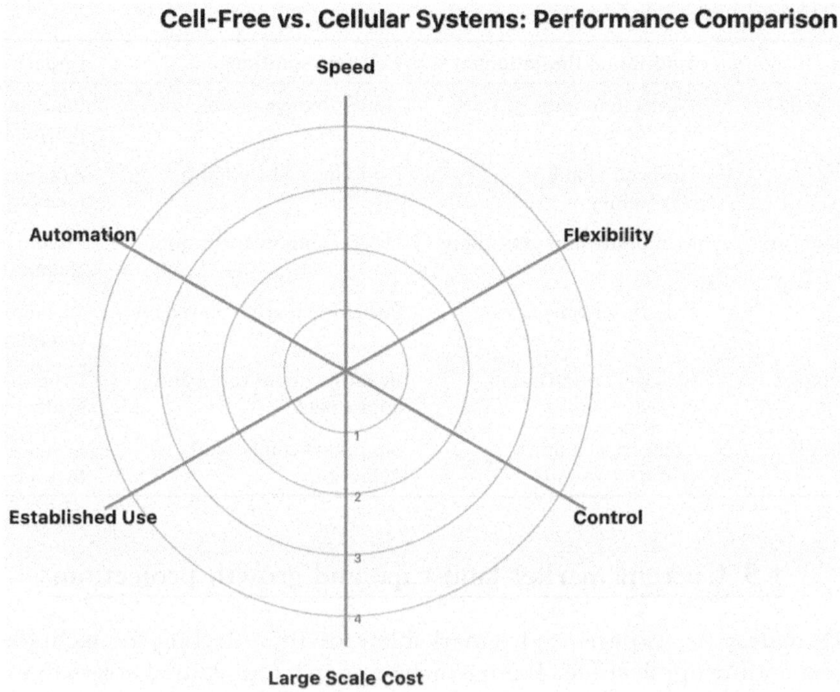

FIG. 1.4 Cell-free synthesis advantages radar chart. Comparative analysis of cell-free systems versus traditional cellular approaches across key performance metrics. The radar chart illustrates relative strengths in speed, flexibility, control, toxicity tolerance, and ease of automation. Cell-free systems (blue) show particular advantages in speed and flexibility, while cellular systems (red) excel in established scalability and cost for large-volume production.

The ability to produce toxic proteins represents another crucial advantage. Many proteins of therapeutic or research interest are inherently toxic to living cells, making their production in cellular systems difficult or impossible. Cell-free systems eliminate this constraint entirely, as there are no living cells to be harmed. This capability has enabled the production of antimicrobial peptides, membrane-disrupting proteins, and other challenging targets.

The rise of personalized medicine has created new demands that cellular systems struggle to meet. Patient-specific therapeutics require small batch sizes, rapid turnaround times, and the ability to produce proteins on demand. Cell-free systems excel in these applications, offering the potential for point-of-care protein production and distributed manufacturing models (Table 1.4).

The COVID-19 pandemic served as a dramatic demonstration of cell-free synthesis advantages. While traditional vaccine development required months, cell-free systems enabled the production of SARS-CoV-2 antigens within hours of sequence availability. Diagnostic platforms based on cell-free systems were deployed globally, providing rapid, accurate detection capabilities in resource-limited settings.

TABLE 1.4 Driving forces behind cell-free adoption.

Driving force	Traditional limitations	Cell-free solutions	Impact sectors
Speed requirements	Weeks to months development	Hours to days production	Pandemic response, personalized medicine
Toxic proteins	Host cell viability constraints	No living cells to harm	Antimicrobials, membrane proteins
Specialized chemistry	Membrane impermeability	Direct component addition	Protein engineering, chemical biology
Distributed manufacturing	Complex infrastructure needs	Portable, simplified systems	Global health, emergency response
Rapid prototyping	Lengthy optimization cycles	Real-time monitoring and adjustment	Drug discovery, synthetic biology
Regulatory simplicity	Complex cell line characterization	Simplified component definition	Diagnostics, therapeutics

1.5 Current market landscape and growth projections

The CFPS market has experienced remarkable growth, reflecting the technology's maturation and expanding applications. From a niche research tool valued at less than $50 million in 2010, the market has grown to approximately $300 million in 2024, with projections suggesting it will reach $585 million by 2034, representing a robust compound annual growth rate (CAGR) of 7.0%.

This growth is driven by multiple factors, including increasing adoption in pharmaceutical research and development, expanding diagnostic applications, and growing recognition of the technology's unique advantages. The pharmaceutical sector represents the largest market segment, accounting for approximately 60% of current revenues, followed by academic research (25%) and diagnostics (15%) (Table 1.5).

TABLE 1.5 Cell-free protein synthesis market analysis (2024).

Market segment	Revenue ($ million)	Growth rate (CAGR)	Key applications	Leading companies
Pharmaceutical R&D	180	8.2%	Drug discovery, antibody development	Thermo Fisher, Promega
Academic research	75	5.5%	Basic research, education	New England Biolabs, Takara
Diagnostics	45	12.1%	Point-of-care testing, biosensors	Arbor Biosciences, Custom vendors
Industrial biotechnology	30	15.3%	Enzyme production, biocatalysis	Synthetic Genomics, Zymergen
Total market	330	7.8%		

Regional analysis reveals North America as the dominant market, accounting for approximately 45% of global revenues, driven by strong pharmaceutical and biotechnology sectors. Europe follows with 30%, while Asia-Pacific shows the highest growth rate at 9.2% CAGR, reflecting increasing research investments and growing biotechnology capabilities in countries such as China, Japan, and India.

The competitive landscape is characterized by a mix of established life sciences companies and innovative startups. Major players include Thermo Fisher Scientific, Promega Corporation, New England Biolabs, and Takara Bio, which collectively control approximately 60% of the market. However, the field remains highly dynamic, with numerous startups developing specialized applications and novel platform technologies (Fig. 1.5).

Investment in cell-free technology has accelerated significantly, with venture capital funding exceeding $200 million in 2023–2024 alone. This investment is driving innovation in

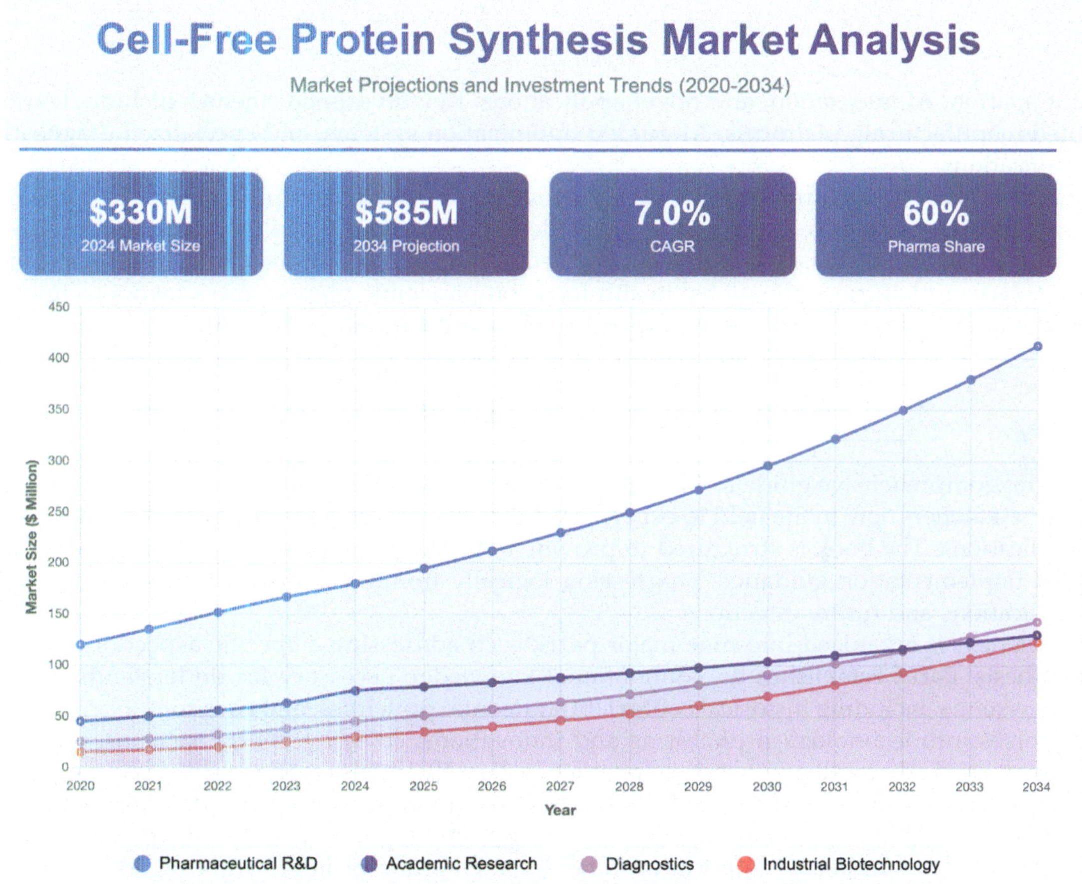

FIG. 1.5 Cell-free protein synthesis market projections (2020–2034). Historical growth and future projections for the global cell-free protein synthesis market, broken down by application sectors. The chart shows accelerating growth driven by pharmaceutical applications, diagnostic expansion, and emerging industrial biotechnology uses. Shaded areas represent confidence intervals for projections.

TABLE 1.6 Investment and innovation trends (2023–2024).

Investment theme	Funding ($ million)	Key companies	Technology focus
AI-guided optimization	75	Zymergen, Ginkgo Bioworks	Machine learning, automated design
Distributed manufacturing	60	Portable Genomics, Field Bio	Point-of-care production
Diagnostic applications	45	Mammoth Biosciences, Sherlock Bio	CRISPR-based detection
Automation platforms	35	Transcriptic, Emerald Cloud	Robotic integration
Novel applications	25	Various startups	Membrane proteins, natural products

automation, AI integration, and novel applications. Key investment themes include distributed manufacturing platforms, AI-guided optimization systems, and specialized diagnostic applications.

The market outlook remains strongly positive, supported by several trends. The growing emphasis on personalized medicine creates demand for rapid, small-batch protein production capabilities. Increasing interest in synthetic biology applications drives need for rapid prototyping platforms. The global health focus on pandemic preparedness favors technologies that enable rapid response and distributed manufacturing (Table 1.6).

1.6 Book structure and learning objectives

This comprehensive guide to CFPS is designed to serve multiple audiences, from students and researchers new to the field to experienced practitioners seeking to implement advanced applications. The book is structured to provide both fundamental understanding and practical implementation guidance, progressing logically from basic principles to cutting-edge applications and future directions.

The text is organized into nine major parts, each addressing a specific aspect of cell-free synthesis. Part I establishes the foundational knowledge necessary for understanding cell-free systems, including historical context, fundamental principles, and core components. Part II delves into technological platforms and innovations, covering energy systems, reaction formats, and advanced engineering approaches. Part III addresses the rapidly evolving integration of automation and digital technologies, including microfluidics, AI, and robotics.

Parts IV and V focus on applications, ranging from biopharmaceutical development and diagnostics to specialized applications such as membrane protein synthesis and complex biomolecule production. Part VI examines the economic and regulatory framework essential for commercial implementation, while Part VII explores future directions and emerging trends. Finally, Parts VIII and IX provide practical implementation guidance and comprehensive resources for practitioners (Table 1.7).

TABLE 1.7 Learning objectives by part.

Part	Focus area	Primary learning objectives	Target audience
I	Foundations	Understand principles, components, preparation methods	All readers
II	Technology platforms	Master reaction systems, scale-up, engineering	Technical practitioners
III	Digital integration	Apply automation, AI, microfluidics	Advanced researchers
IV	Core applications	Implement pharmaceutical, diagnostic applications	Industry professionals
V	Specialized applications	Address challenging targets, complex molecules	Specialized researchers
VI	Commercial framework	Navigate economics, regulations, quality systems	Business professionals
VII	Future directions	Anticipate trends, emerging technologies	Strategic planners
VIII	Implementation	Execute practical setup, protocols, troubleshooting	Laboratory managers
IX	Resources	Access suppliers, databases, tools	All practitioners

Each chapter includes specific learning objectives, key concepts, practical examples, and suggested exercises or experiments. Chapters conclude with comprehensive references and suggested further reading. Special emphasis is placed on recent developments and emerging trends, ensuring readers gain current and forward-looking perspectives.

The book incorporates multiple learning aids, including detailed tables comparing different approaches, comprehensive figures illustrating key concepts, case studies demonstrating real-world applications, and practical protocols for implementation. Troubleshooting guides and optimization strategies are integrated throughout to support practical implementation.

By the conclusion of this text, readers should be able to understand the fundamental principles underlying CFPS, evaluate different platform technologies for specific applications, design and optimize cell-free reactions for their particular needs, implement quality control and regulatory compliance strategies, assess economic and business considerations for commercial applications, and anticipate future developments and emerging opportunities in the field.

Bibliography

Anon Cell-free protein synthesis market analysis. Cell-free protein synthesis market deep analysis report 2025-2034. *Market Research Report* (2024).

Baranwal, A.K., Maerkl, S.J., 2024. A comprehensive review of microfluidic approaches in cell-free synthetic biology. Front. Synth. Biol. 2, 1397533.

Carlson, E.D., Gan, R., Hodgman, C.E., Jewett, M.C., 2012. Cell-free protein synthesis: applications come of age. Biotechnol. Adv. 30, 1185–1194.

Hunt, A.C., Rasor, B.J., et al., 2024. Cell-free gene expression: methods and applications. Chem. Rev. 125, 91–149.

Nirenberg, M.W., Matthaei, J.H., 1961. The dependence of cell-free protein synthesis in E. coli upon naturally occurring or synthetic polyribonucleotides. Proc. Natl. Acad. Sci. USA. 47, 1588–1602.

Pandi, A., Adam, D., Zare, A., et al., 2023. Cell-free biosynthesis combined with deep learning accelerates de novo-development of antimicrobial peptides. Nat. Commun. 14, 7197.

Pardee, K., Green, A.A., Takahashi, M.K., et al., 2016. Rapid, low-cost detection of Zika virus using programmable biomolecular components. Cell. 165, 1255–1266.

Shimizu, Y., Inoue, A., Tomari, Y., et al., 2001. Cell-free translation reconstituted with purified components. Nat. Biotechnol. 19, 751–755.

Walsh, G., 2018. Biopharmaceutical benchmarks 2018. Nat. Biotechnol. 36, 1136–1145.

Zawada, J.F., Yin, G., Steiner, A.R., et al., 2011. Microscale to manufacturing scale-up of cell-free cytokine production—a new approach for shortening protein production development timelines. Biotechnol. Bioeng. 108, 1570–1578.

CHAPTER

2

Fundamental principles and core components

2.1 The "open system" architecture

The defining characteristic that sets cell-free protein synthesis apart from all other biological manufacturing approaches is its "open system" architecture. Unlike traditional recombinant protein production, where the synthesis machinery operates within the confines of living cells surrounded by selectively permeable membranes, cell-free systems liberate this machinery into an accessible, controllable environment. This fundamental difference creates unprecedented opportunities for manipulation, optimization, and real-time monitoring that are simply impossible in cellular systems.

In living cells, the protein synthesis machinery operates within a highly compartmentalized environment. The nuclear membrane separates transcription from translation in eukaryotes, while the cell membrane controls the movement of nutrients, cofactors, and products. Regulatory networks maintain homeostasis, often at the expense of maximum protein production efficiency. The cell must balance protein synthesis with growth, reproduction, and survival, competing priorities that limit the resources available for heterologous protein production.

Cell-free systems eliminate these constraints by removing the cellular boundaries while preserving the essential molecular machinery. The result is a "biological factory" that can be directly accessed, monitored, and modified. Substrates can be added at optimal concentrations, products can be removed continuously, and reaction conditions can be precisely controlled. Nonnatural components that would be toxic to living cells or unable to cross membranes can be incorporated directly into the reaction mixture (Fig. 2.1).

This accessibility enables real-time process analytical technology (PAT) that provides unprecedented insight into protein synthesis kinetics. Researchers can monitor adenosine triphosphate (ATP)/ADP ratios, amino acid consumption, intermediate accumulation, and protein folding in real time. These data can be used to optimize conditions dynamically or to understand fundamental mechanisms of protein synthesis.

The open architecture also enables the incorporation of noncanonical amino acids (ncAAs) with remarkable efficiency. While cellular systems require complex engineering to transport ncAAs across membranes and compete with natural amino acid uptake systems, cell-free

https://doi.org/10.1016/B978-0-443-33271-5.00014-4

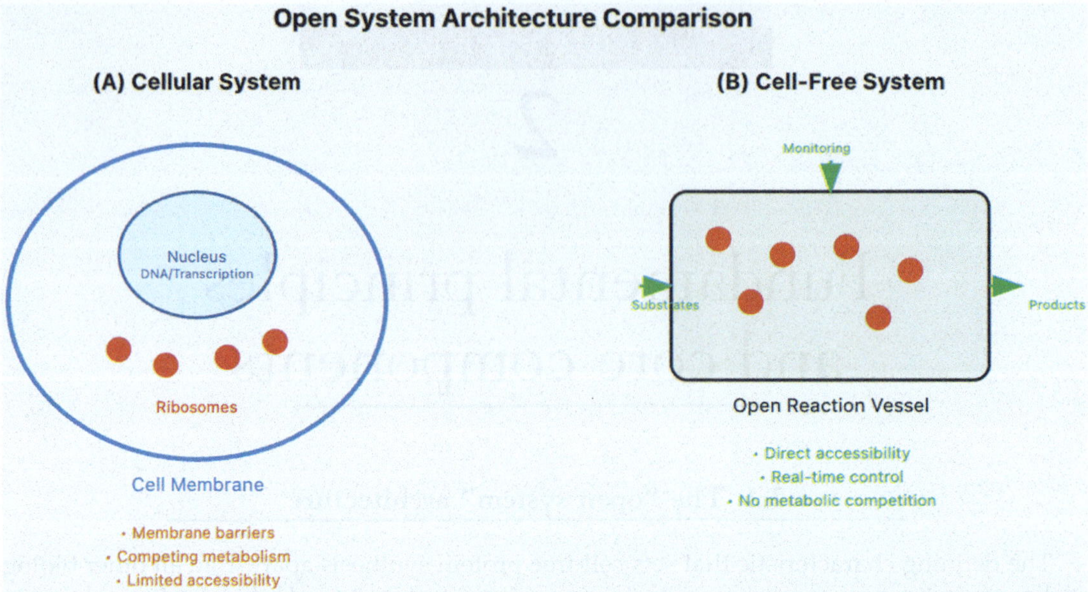

FIG. 2.1 Open system architecture comparison. (A) Cellular systems showing compartmentalization, membrane barriers, and competing metabolic processes. (B) Cell-free systems demonstrating direct accessibility, a controllable environment, and dedicated protein synthesis machinery. The diagram illustrates key advantages, including real-time monitoring capabilities, direct substrate addition, and the elimination of metabolic competition.

TABLE 2.1 Open system advantages in cell-free synthesis.

Capability	Cellular limitation	Cell-Free advantage	Applications
Direct access	Membrane barrier	Immediate substrate addition	ncAA incorporation, cofactor optimization
Real-time monitoring	Cellular interference	Continuous analysis	Process optimization, mechanistic studies
Toxic protein production	Host cell death	No viability constraints	Antimicrobials, membrane-disrupting proteins
Environmental control	Homeostatic regulation	Precise condition control	Temperature, pH, ionic strength optimization
Product recovery	Cellular retention	Direct access	Simplified purification, continuous removal
Nonnatural chemistry	Membrane Impermeability	Unlimited Component addition	Chemical biology, protein Modification

systems allow the direct addition of ncAAs at controlled concentrations. This capability has revolutionized protein engineering, enabling the incorporation of fluorescent labels, click chemistry handles, and novel functional groups (Table 2.1).

The implications of this open architecture extend beyond simple convenience. It enables entirely new approaches to protein engineering, synthetic biology, and biotechnology applications. Researchers can create "artificial cells" by encapsulating cell-free systems in synthetic membranes, engineer metabolic pathways without cellular constraints, and develop portable biotechnology platforms that operate without living organisms.

2.2 Essential molecular machinery

The remarkable capability of cell-free systems to synthesize proteins stems from the careful preservation and reconstitution of the essential molecular machinery that drives protein synthesis in living cells. Understanding these components and their interactions is crucial for anyone seeking to implement, optimize, or troubleshoot cell-free systems.

At the heart of every cell-free system lies the ribosome, the molecular machine responsible for translating genetic information into proteins. Ribosomes are complex ribonucleoprotein particles consisting of two subunits: the small subunit (30S in prokaryotes, 40S in eukaryotes) and the large subunit (50S in prokaryotes, 60S in eukaryotes). The small subunit binds mRNA and ensures accurate codon recognition, while the large subunit catalyzes peptide bond formation and contains the peptidyl transferase center where amino acids are joined together.

The preservation of ribosomal integrity during lysate preparation is critical for system performance. Ribosomes are sensitive to nuclease degradation, ionic conditions, and mechanical stress. Modern preparation protocols employ specific buffers, protease inhibitors, and gentle lysis methods to maintain ribosomal structure and function. The concentration and activity of ribosomes often represent the limiting factor in cell-free reactions, making their optimization a primary focus of system development (Fig. 2.2).

Transfer RNAs (tRNAs) serve as the critical adapters that decode genetic information by carrying specific amino acids to the ribosome. Each tRNA molecule has two crucial regions: the anticodon, which recognizes specific mRNA codons, and the 3′ end, where amino acids are attached. The cell contains multiple tRNA species for each amino acid, and the relative concentrations of different tRNAs can significantly impact protein synthesis efficiency and accuracy.

Aminoacyl-tRNA synthetases (aaRS) represent one of the most sophisticated components of the protein synthesis machinery. These enzymes perform the crucial task of "charging" tRNAs with their cognate amino acids, ensuring the fidelity of genetic code translation. Each synthetase recognizes both a specific amino acid and its corresponding tRNA(s), creating aminoacyl-tRNA molecules ready for protein synthesis. The precision of this recognition is remarkable; synthetases achieve error rates of <1 in 10,000, ensuring accurate protein synthesis (Table 2.2).

Translation factors orchestrate the complex process of protein synthesis through their roles in initiation, elongation, and termination. Initiation factors ensure that ribosomes assemble correctly at the start of mRNA molecules, positioning them to begin translation at the appropriate start codon. Elongation factors facilitate the movement of ribosomes along mRNA and the delivery of aminoacyl-tRNAs to the ribosomal A-site. Release factors recognize stop codons and promote the release of completed proteins from ribosomes.

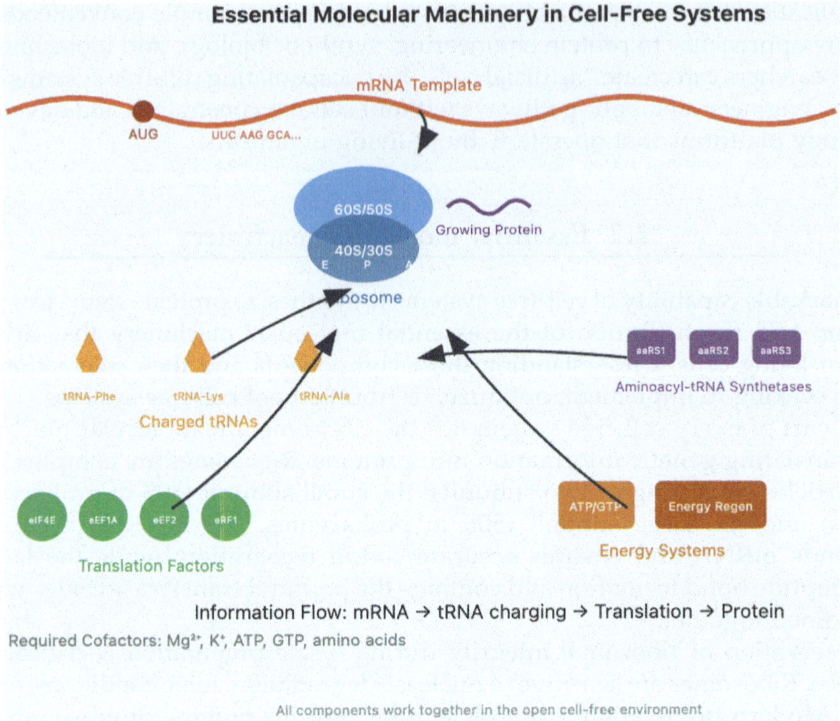

FIG. 2.2 Essential molecular machinery in cell-free systems. A schematic representation of the key components required for cell-free protein synthesis, including ribosomes, tRNAs, aminoacyl-tRNA synthetases, translation factors, and associated enzymes. The diagram shows the flow of information from mRNA through tRNA charging to protein synthesis, highlighting the interconnected nature of the molecular machinery.

TABLE 2.2 Core molecular components and their functions.

Component	Prokaryotic version	Eukaryotic version	Primary function	Optimization targets
Ribosomes	70S (30S + 50S)	80S (40S + 60S)	Protein synthesis	Concentration, integrity
tRNAs	~40 species	~60 species	Amino acid delivery	Charging levels, ratios
Aminoacyl-tRNA synthetases	20 enzymes	20 enzymes	tRNA charging	Activity, specificity
Initiation factors	IF1, IF2, IF3	eIF1, eIF1A, eIF2, eIF3, eIF4E, eIF4G, eIF5, eIF6	Translation initiation	Concentration balance
Elongation factors	EF–Tu, EF–G	eEF1A, eEF2	Translation elongation	Activity, regeneration
Release factors	RF1, RF2, RF3	eRF1, eRF3	Translation termination	Specificity, efficiency

TABLE 2.3 Supporting enzymes and cofactors.

Category	Components	Functions	Optimization strategies
Chaperones	GroEL/GroES, DnaK/DnaJ/GrpE	Protein folding assistance	Concentration optimization, coexpression
Proteases	Minimized through strain engineering	Protein degradation (to be avoided)	Genetic knockouts, inhibitors
Nucleases	Minimized through strain engineering	RNA/DNA degradation (to be avoided)	Genetic knockouts, inhibitors
Metabolic enzymes	Pyruvate kinase, creatine kinase	Energy regeneration	Activity enhancement, stability
Folding catalysts	Protein disulfide isomerase	Disulfide bond formation	Addition for eukaryotic proteins
Modification enzymes	Variable depending on application	Posttranslational modifications	Specific supplementation

The balance and activity of these factors critically influence cell-free system performance. Insufficient initiation factors can limit the number of ribosomes that begin translation, creating a bottleneck even when ribosomes are abundant. Inadequate elongation factors can slow ribosomal movement, reducing overall synthesis rates. The optimization of factor concentrations represents a key area for system improvement.

Molecular chaperones play increasingly recognized roles in cell-free protein synthesis. These proteins assist in proper protein folding, prevent aggregation, and ensure that newly synthesized proteins achieve their native conformations. In cell-free systems, chaperone activity can be particularly important because the open environment may not provide the optimal folding conditions found in living cells. The addition of exogenous chaperones or the engineering of chaperone activity into lysates represents an active area of research (Table 2.3).

2.3 Energy regeneration systems

Protein synthesis is an energy-intensive process that requires continuous supplies of ATP and guanosine triphosphate (GTP) to drive ribosomal function, tRNA charging, and various translation factors. In living cells, these energy requirements are met through complex metabolic networks that efficiently regenerate high-energy phosphate bonds. Cell-free systems must recreate this energy regeneration capability using simplified but effective systems that can sustain protein synthesis for extended periods.

The most widely used energy regeneration system in cell-free synthesis couples phosphoenolpyruvate (PEP) with pyruvate kinase to regenerate ATP from ADP. This system is highly effective because PEP represents one of the highest-energy phosphate compounds in biochemistry, providing a strong thermodynamic driving force for ATP regeneration. The reaction is essentially irreversible under typical cell-free conditions, ensuring efficient energy capture.

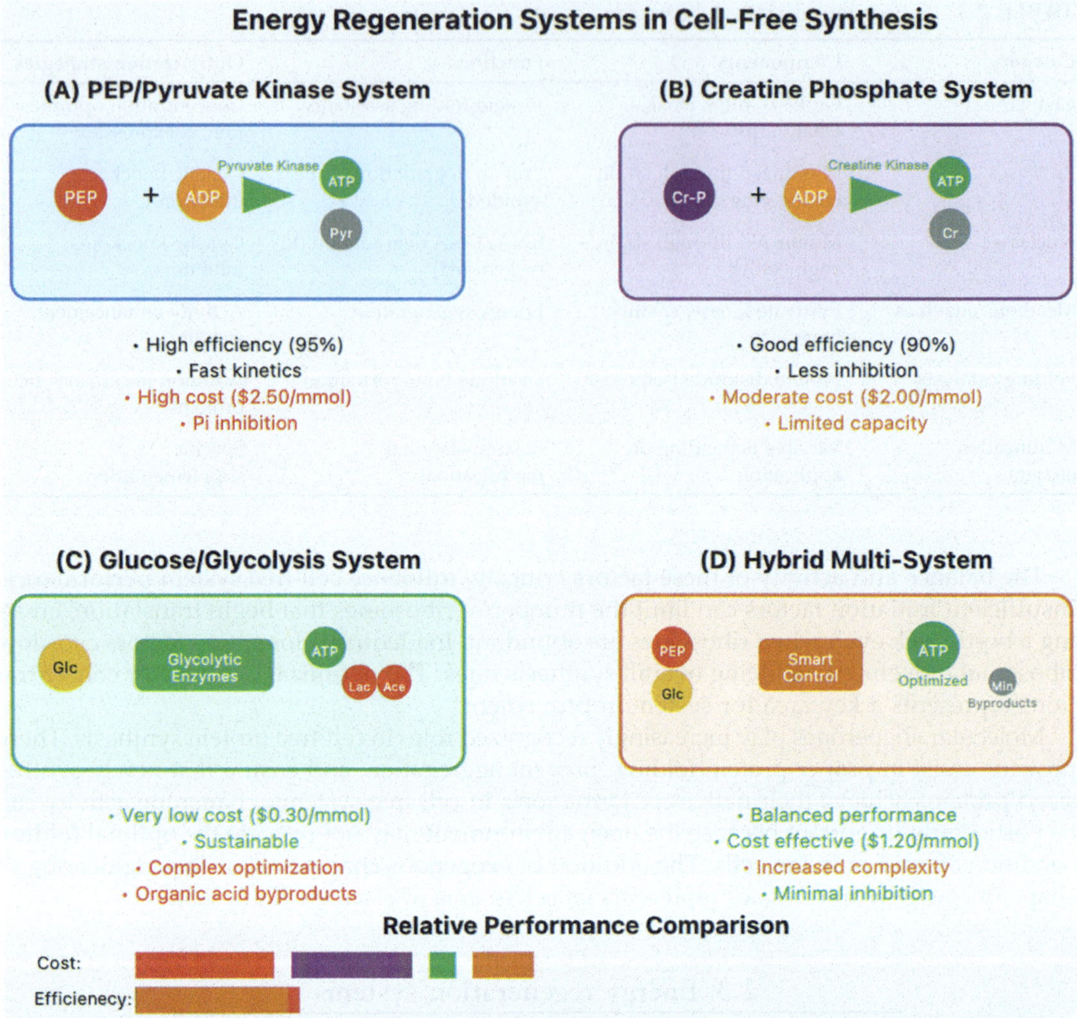

FIG. 2.3 Energy regeneration systems in cell-free synthesis. Comparison of major energy regeneration approaches showing (A) PEP/pyruvate kinase system, (B) creatine phosphate/creatine kinase system, (C) glucose-based glycolytic regeneration, and (D) hybrid multi-system approaches. Each system shows energy flow, by-product formation, and relative advantages for different applications.

The PEP/pyruvate kinase system offers several advantages that have made it the gold standard for cell-free applications. The reaction is simple, requiring only two components that are relatively stable and commercially available. The kinetics are favorable, with rapid regeneration rates that can keep pace with protein synthesis demands. The system is compatible with both prokaryotic and eukaryotic cell-free platforms, providing versatility across different applications (Fig. 2.3).

However, the PEP system also has limitations that have driven the development of alternative approaches. The primary concern is cost: PEP is expensive relative to other energy

TABLE 2.4 Comparison of energy regeneration systems.

System	Energy source	Enzyme	Advantages	Disadvantages	Typical cost ($/mmol ATP)
PEP/pyruvate kinase	Phosphoe-nolpyruvate	Pyruvate kinase	High efficiency, fast kinetics	Expensive, phosphate inhibition	$2.50–3.00
Creatine phosphate	Creatine phosphate	Creatine kinase	Less inhibition, physiological	Moderate cost, limited capacity	$1.80–2.20
Glucose/ glycolysis	Glucose	Multiple enzymes	Low cost, sustainable	Complex optimization, by-product issues	$0.20–0.40
Acetyl-CoA	Acetyl phosphate	Acetate kinase	Very low cost	Limited studies, optimization needed	$0.15–0.25
Hybrid systems	Multiple sources	Multiple enzymes	Optimized performance	Increased complexity	$0.80–1.50

sources, contributing significantly to overall reaction costs. In addition, the system generates inorganic phosphate as a by-product, which can accumulate and inhibit various enzymes involved in protein synthesis. This phosphate inhibition becomes particularly problematic in long-duration reactions or continuous-flow systems.

The creatine phosphate/creatine kinase system represents the most common alternative to PEP-based energy regeneration. This system is particularly favored in eukaryotic cell-free platforms because creatine kinase is naturally abundant in many eukaryotic cells, and the system closely mimics natural energy metabolism in muscle and nervous tissues. The creatine system produces creatine as its primary by-product, which is generally less inhibitory than inorganic phosphate (Table 2.4).

Glucose-based energy regeneration has emerged as an attractive alternative that addresses the cost limitations of phosphate-based systems. These systems harness the natural glycolytic pathway present in cell lysates to generate ATP from glucose through a series of enzymatic reactions. While more complex than single-enzyme systems, glucose-based regeneration can provide sustained energy production at a fraction of the cost of PEP or creatine phosphate.

The implementation of glucose-based systems requires careful optimization to balance ATP generation with by-product management. Glycolysis produces several organic acids, including lactate and acetate, which can accumulate and inhibit protein synthesis. Successful glucose systems often employ pH buffering, selective by-product removal, or metabolic engineering to redirect carbon flow toward less inhibitory products.

Recent advances in energy system design have focused on hybrid approaches that combine multiple regeneration mechanisms to optimize both performance and cost. These systems might use PEP for rapid initial ATP generation, while glucose provides sustained long-term energy production. Alternatively, they might employ different systems for different phases of the reaction or in different reactor compartments (Table 2.5).

The choice of energy regeneration system significantly impacts overall cell-free system performance and economics. For research applications where cost is less critical than

TABLE 2.5 Energy system performance metrics.

Metric	PEP system	Creatine system	Glucose system	Hybrid system
Initial ATP generation rate (mM/min)	15–25	10–20	5–15	12–30
Sustained generation capacity (hours)	3–6	2–4	8–24	6–48
Protein yield (typical, mg/mL)	1.5–3.0	1.0–2.5	0.8–2.0	1.8–4.0
Cost per gram protein ($)	45–60	35–50	8–15	20–35
By-product inhibition severity	High	Moderate	Variable	Low
System complexity (1–5 scale)	1	1	4	3

reliability and simplicity, PEP-based systems remain popular. For commercial applications or cost-sensitive research, glucose-based or hybrid systems offer compelling advantages. The development of novel energy systems continues to be an active area of research, with promising approaches, including photosynthetic ATP generation, enzymatic fuel cells, and synthetic metabolic cycles.

2.4 Cofactors and ionic requirements

The molecular machinery of protein synthesis requires precise ionic conditions and specific cofactors to function optimally. These requirements, evolved over billions of years of cellular evolution, must be carefully recreated in cell-free systems to achieve high levels of protein synthesis. Understanding and optimizing these conditions represents one of the most critical aspects of cell-free system design and operation.

Magnesium ions (Mg^{2+}) play perhaps the most crucial role in protein synthesis, serving multiple essential functions throughout the process. Ribosomes require Mg^{2+} for structural stability; the ion coordinates between ribosomal RNA molecules, maintaining the precise three-dimensional architecture necessary for function. Mg^{2+} also participates directly in the peptidyl transferase reaction, coordinating the positioning of substrates and facilitating peptide bond formation.

The concentration of free Mg^{2+} in cell-free reactions requires careful optimization because both deficiency and excess can severely impair protein synthesis. Insufficient Mg^{2+} leads to ribosomal instability and reduced translation efficiency, while excess Mg^{2+} can cause ribosomal aggregation and precipitation of nucleic acids. The optimal range typically falls between 8 and 15 mM total Mg^{2+}, but this can vary depending on other buffer components and the specific system being used (Fig. 2.4).

Potassium ions (K^+) serve complementary roles to magnesium, primarily maintaining optimal ribosomal conformation and supporting translation factor function. K^+ concentrations in the range of 100–150 mM typically provide optimal activity, though this optimum can shift depending on the specific proteins being synthesized and the source of the cell lysate. The K^+/Mg^{2+} ratio is particularly important, as these ions can have competing effects on ribosomal structure.

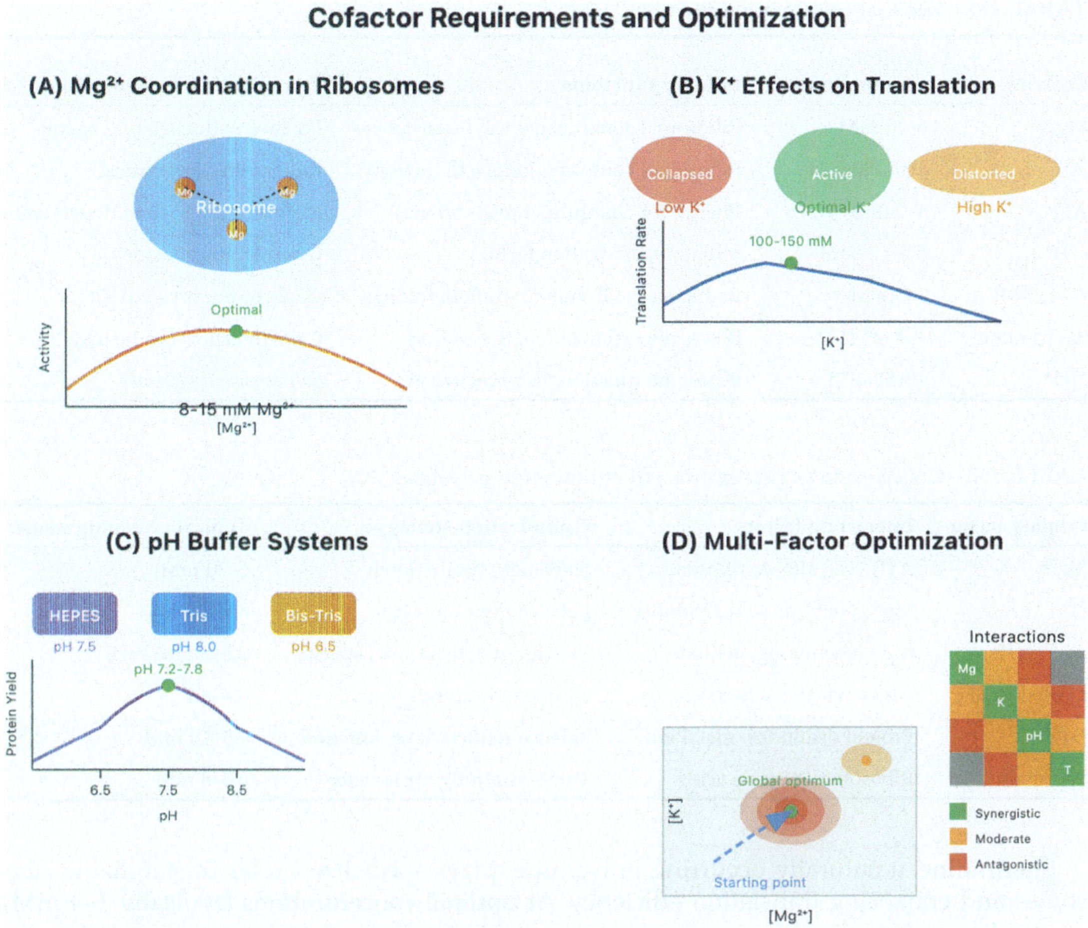

FIG. 2.4 Cofactor requirements and optimization. (A) Magnesium coordination in ribosomal structure and function. (B) Potassium effects on ribosomal conformation and translation efficiency. (C) pH buffer systems and their impact on enzyme activities. (D) Optimization landscape showing interaction effects between major cofactors and their impact on protein yield.

The buffering system used in cell-free reactions profoundly impacts protein synthesis efficiency. Most systems employ 4-(2-hydroxyethyl)-1-piperazineethanesulfonic acid (HEPES) as the primary buffer because of its excellent buffering capacity at physiological pH and minimal interference with enzymatic reactions. However, alternative buffers, such as Tris or phosphate, may be preferred for specific applications or when studying particular aspects of protein synthesis (Table 2.6).

Reducing agents, such as dithiothreitol (DTT) or β-mercaptoethanol (BME), are essential for maintaining protein thiols in their reduced state and supporting proper protein folding. These compounds prevent unwanted disulfide bond formation and protect cysteine residues from oxidation. However, excessive concentrations can interfere with desired disulfide bond formation in proteins that require them for stability or function.

TABLE 2.6 Essential cofactors and their functions.

Cofactor	Typical concentration	Primary functions	Optimization considerations
Mg^{2+}	8–15 mM	Ribosomal stability, peptidyl transferase	Balance with chelating agents
K^{+}	100–150 mM	Ribosomal conformation, factor activity	Ratio with Mg^{2+} critical
ATP	1–3 mM	Energy for charging, translocation	Regeneration system dependent
GTP	0.5–1.5 mM	Initiation, elongation factors	Often limiting factor
DTT/BME	1–5 mM	Reducing environment, protein folding	Concentration vs. toxicity
Spermidine	1–4 mM	Ribosome stabilization, translation	Growth factor, can be toxic
NH^{4+}	10–50 mM	Ribosome function, factor activity	*Salt sensitivity balance*

TABLE 2.7 Cofactor interaction matrix and optimization guidelines.

Primary factor	Interacting factors	Optimization strategy	Typical screening range
Mg^{2+}	ATP, GTP, tRNA, ribosomes	Start high, titrate down	5–20 mM
K^{+}	Mg^{2+}, NH^{4+}, protein solubility	Balance ionic strength	80–200 mM
ATP/GTP	Mg^{2+}, energy regeneration	Match to regeneration capacity	0.5–5 mM each
pH (HEPES)	All enzymatic activities	Optimize for lysate source	6.8–8.2
DTT	Protein disulfides, metal ions	Balance reduction vs. function	0.5–10 mM
Spermidine	Ribosomes, nucleic acids	Titrate carefully for toxicity	0.5–8 mM

Spermidine, a naturally occurring polyamine, plays specialized roles in stabilizing ribosomes and enhancing translation efficiency. At optimal concentrations (typically 1–4 mM), spermidine can significantly improve protein yields by promoting ribosomal assembly and reducing translation errors. However, polyamines can be toxic at higher concentrations, requiring careful optimization for each specific system.

The optimization of cofactor concentrations is complicated by their interactions with other system components. Many cofactors bind to proteins, nucleic acids, or other molecules in the reaction mixture, reducing their free concentration. EDTA or other chelating agents present in some buffer systems can sequester divalent cations, requiring adjustment of metal ion concentrations to compensate (Table 2.7).

Temperature represents another critical parameter that affects all aspects of cell-free protein synthesis. Most prokaryotic systems operate optimally at 30°C–37°C, while eukaryotic systems often prefer slightly lower temperatures (25°C–30°C) to maintain proper enzyme activities and prevent protein aggregation. Temperature optimization often requires balancing synthesis rate against protein stability and proper folding.

Recent advances in cofactor optimization have employed systematic approaches, such as design of experiments, to map interaction effects between multiple factors simultaneously. These approaches have revealed unexpected synergies and antagonisms between cofactors that would be missed by traditional one-factor-at-a-time optimization. Machine learning approaches are increasingly being applied to predict optimal cofactor combinations based on the specific proteins being synthesized and the particular lysate system being used.

2.5 Template design and optimization

The genetic template that encodes the target protein represents one of the most critical yet controllable variables in cell-free protein synthesis. Unlike cellular systems, where template optimization must account for complex regulatory networks and competing cellular processes, cell-free systems offer direct access to the translation machinery, enabling precise control over gene expression elements. This accessibility has led to sophisticated approaches for template design that can dramatically impact protein yields and quality.

The promoter system used for transcription in coupled transcription–translation reactions profoundly influences mRNA production rates and, consequently, protein synthesis levels. The bacteriophage T7 promoter system has become the gold standard for most bacterial cell-free applications due to its exceptional strength and specificity. T7 RNA polymerase, a single-subunit enzyme, maintains high activity in cell-free environments and can be easily added to reactions or preexpressed in lysate preparation strains.

T7 promoter variants offer different levels of transcriptional activity, allowing researchers to tune mRNA production to optimal levels for specific applications. Strong promoters may saturate the translation machinery, while weak promoters may limit protein production. The choice often depends on the specific protein being synthesized, the lysate system being used, and the desired balance between yield and quality (Fig. 2.5).

Alternative promoter systems, including SP6, T3, and hybrid promoters, offer different kinetic properties and regulatory capabilities. Some applications benefit from controllable promoters that can be activated or repressed during the reaction, enabling temporal control over protein synthesis. The development of orthogonal promoter systems allows for the controlled expression of multiple proteins in the same reaction mixture.

The ribosome binding site (RBS) represents perhaps the most crucial element for translation efficiency in prokaryotic systems. The RBS includes the Shine–Dalgarno sequence, which base-pairs with ribosomal RNA to position ribosomes correctly for translation initiation, and the spacing between this sequence and the start codon. Even small changes in the RBS sequence or spacing can result in dramatic differences in protein expression levels.

Computational tools for RBS design have revolutionized template optimization. Programs such as the RBS Calculator and RBS Designer can predict translation initiation rates based on thermodynamic models of ribosome binding. These tools enable researchers to design RBS sequences that provide desired expression levels or to create libraries of variants for systematic optimization (Table 2.8).

Codon optimization represents another powerful tool for enhancing protein synthesis in cell-free systems. Different organisms prefer different codons for the same amino acid, reflecting the abundance of corresponding tRNAs in their cells. Cell-free systems derived from *E. coli* perform best with genes optimized for *E. coli* codon usage, while eukaryotic systems benefit from species-appropriate codon optimization.

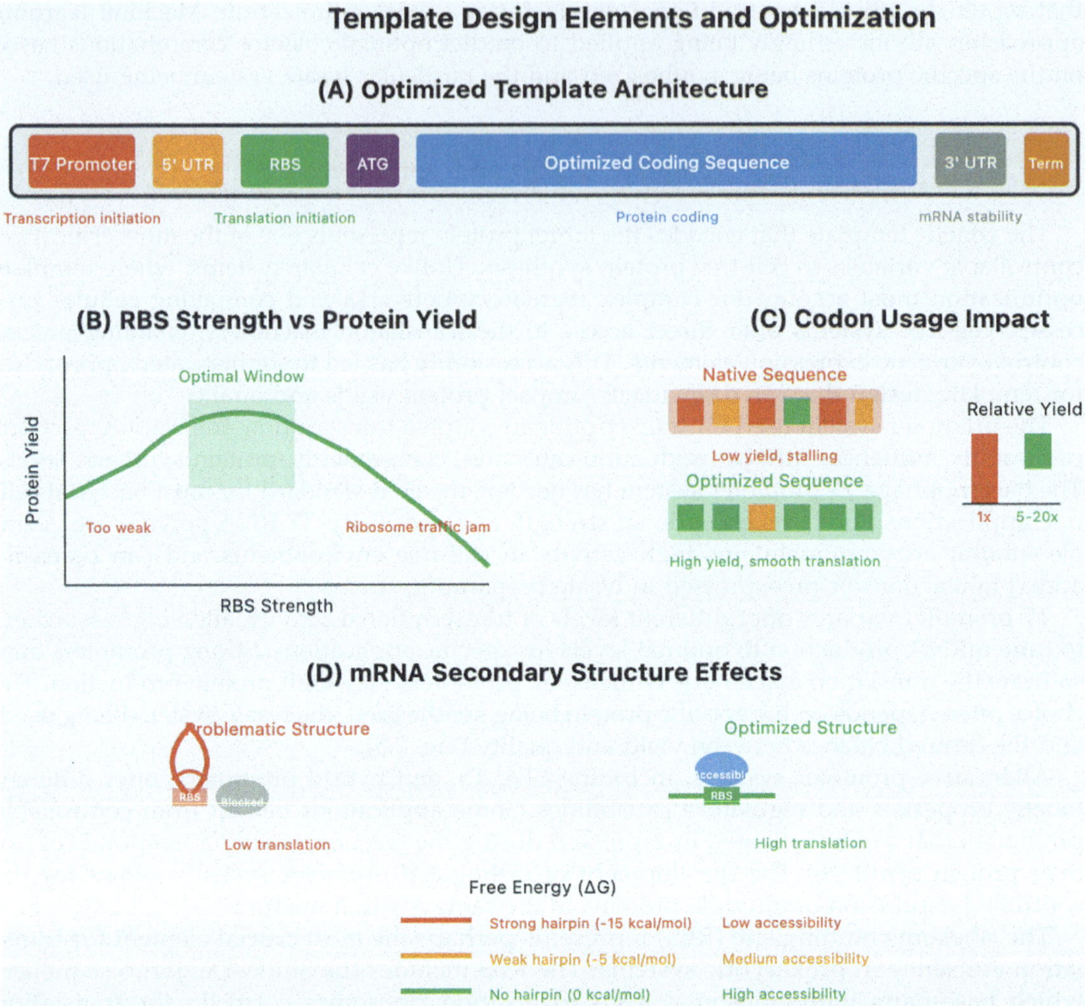

FIG. 2.5 Template design elements and optimization. (A) Schematic of optimized template showing promoter, 5′ UTR, ribosome binding site, coding sequence, and 3′ UTR elements. (B) Ribosome binding site strength vs. protein yield showing optimal expression windows. (C) Codon optimization impact on synthesis rates for different target proteins. (D) mRNA secondary structure effects on translation initiation and efficiency.

Modern codon optimization goes beyond simple codon frequency matching to consider factors such as tRNA availability, translation kinetics, and protein folding. Some codons may be translated more slowly, providing time for proper protein folding, while others enable rapid synthesis when speed is prioritized. The strategic use of rare codons can create translational pauses that facilitate proper folding of complex proteins.

mRNA secondary structure profoundly impacts translation efficiency by affecting ribosome access to the RBS and start codon. Stable hairpin structures near the translation initiation region can severely inhibit ribosome binding and reduce protein synthesis. Computational

TABLE 2.8 Template design elements and their impact.

Element	Function	Optimization parameters	Typical impact on yield
Promoter	Transcription initiation	Strength, specificity, regulation	5–50× variation
5′ UTR	mRNA stability, ribosome access	Length, structure, sequences	2–10× variation
RBS/Shine–Dalgarno	Translation initiation	Sequence, spacing, accessibility	10–1000× variation
Start codon	Translation start	ATG, GTG, TTG variants	1.5–3× variation
Coding sequence	Amino acid sequence	Codon usage, rare codons	2–20× variation
3′ UTR	mRNA stability	Length, structure, termination	1.5–5× variation

TABLE 2.9 Advanced template optimization strategies.

Strategy	Approach	Benefits	Applications
Systematic RBS libraries	Combinatorial RBS variants	Empirical optimization	Unknown proteins, complex systems
Codon harmonization	Match codon usage to tRNA pools	Improved folding, reduced stalling	Complex proteins, membrane proteins
UTR engineering	Optimize 5′ and 3′ regions	Enhanced stability, translation	Unstable mRNAs, long reactions
Synonymous variant libraries	Silent mutations affecting structure	Fine-tune expression levels	Protein engineering applications
Orthogonal systems	Multiple independent templates	Multiprotein expression	Pathway reconstruction, complexes
Computational design	AI-guided optimization	Reduced experimental burden	High-throughput applications

tools can predict mRNA folding and identify problematic structures that should be eliminated through sequence modification (Table 2.9).

The 5′ and 3′ untranslated regions (UTRs) significantly influence mRNA stability and translation efficiency. The 5′ UTR affects ribosome accessibility and can contain regulatory elements that modulate translation. The 3′ UTR influences mRNA degradation and can affect the overall mRNA lifetime in the reaction. Optimized UTRs can extend reaction duration and improve protein yields.

Recent advances in template design have incorporated machine learning approaches that can predict optimal designs based on target protein characteristics and lysate properties. These systems can analyze vast databases of successful constructs to identify patterns that correlate with high expression levels. Some approaches use iterative design-build-test cycles guided by AI to rapidly converge on optimal template designs.

Template preparation methods have also evolved to support rapid prototyping and high-throughput applications. PCR-based methods enable template generation directly from oligonucleotides, eliminating the need for cloning. Linear templates can be as effective as plasmid DNA for many applications while being much faster to prepare. These approaches have enabled same-day template design, synthesis, and testing workflows that dramatically accelerate protein engineering and optimization.

2.6 Quality control and system validation

The success of cell-free protein synthesis applications depends critically on robust quality control measures and systematic validation approaches. Unlike cellular systems where protein production occurs within the protected environment of living cells, cell-free systems operate in open environments where contamination, degradation, and inhibition can severely impact performance. Comprehensive quality control ensures reproducible results, enables troubleshooting when systems underperform, and provides the documentation necessary for regulatory compliance in commercial applications.

Quality control in cell-free systems begins with the characterization and validation of individual components before system assembly. Cell lysates, the most complex component, require extensive testing to ensure consistent activity, minimal inhibitory factors, and appropriate composition. Lysate quality can vary significantly between preparations due to differences in cell growth conditions, lysis procedures, and storage conditions.

Standard lysate qualification protocols typically include measurement of protein concentration, assessment of translation activity using control templates, evaluation of endogenous nuclease and protease activities, and quantification of key cofactors and metabolites. These measurements provide baseline performance metrics and enable batch-to-batch comparisons that are essential for maintaining system consistency (Fig. 2.6).

Energy regeneration components require validation to ensure adequate activity and stability. Enzymes such as pyruvate kinase or creatine kinase should be tested for specific activity, pH stability, and temperature tolerance. Energy substrates like PEP or creatine phosphate should be verified for purity and concentration, as impurities can significantly impact reaction performance.

Template validation ensures that genetic constructs are correctly designed and free from inhibitory elements. This includes verification of DNA concentration and integrity, confirmation of correct sequences through sequencing or restriction analysis, and assessment of transcription and translation efficiency using the actual cell-free system. Template storage conditions and stability should also be validated to ensure consistent performance over time (Table 2.10).

System validation involves testing complete cell-free reactions under standardized conditions to establish baseline performance and identify sources of variability. Control reactions using well-characterized reporter proteins (such as Green Fluorescent Protein, GFP, luciferase, or β-galactosidase) provide benchmarks for comparing system performance across different batches, conditions, or modifications.

Validation protocols should include the assessment of reproducibility through replicate reactions, evaluation of system stability over time, testing of sensitivity to environmental conditions, and documentation of typical performance ranges. These studies establish the

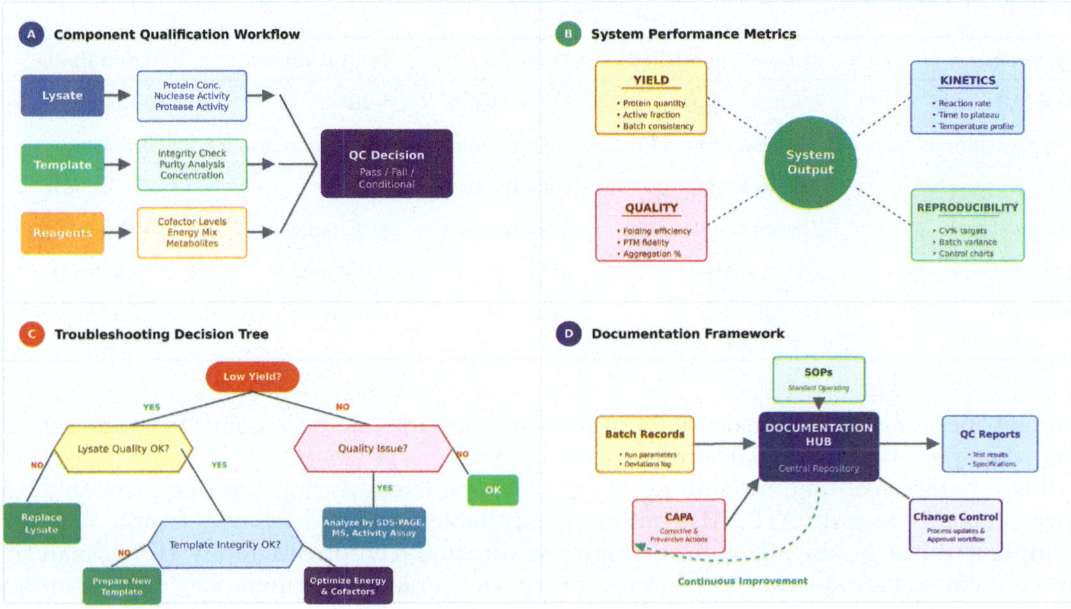

FIG. 2.6 Quality control and validation framework. (A) Component qualification workflow showing lysate, template, and reagent testing protocols. (B) System performance metrics, including yield, kinetics, and quality indicators. (C) Troubleshooting decision tree for common performance issues. (D) Documentation framework for regulatory compliance and process improvement.

TABLE 2.10 Critical quality control parameters.

Component category	Key parameters	Acceptance criteria	Test methods
Cell lysate	Protein concentration	12–20 mg/mL	Bradford assay, BCA
	Translation activity	>0.5 mg/mL GFP	Standard reporter assay
	Nuclease activity	<10% RNA degradation	RNA stability test
	Endotoxin levels	<10 EU/mL	LAL assay
Energy system	Enzyme activity	>80% of specification	Coupled enzyme assay
	Substrate purity	>95% by HPLC	Chromatographic analysis
	ATP regeneration rate	>5 mM/hour	Luciferase assay
Templates	DNA concentration	±10% of target	Spectrophotometry
	Sequence integrity	100% match	Sequencing confirmation
	Transcription efficiency	>80% of control	In vitro transcription
Final product	Protein yield	Within 2-fold of target	SDS-PAGE, quantification
	Protein quality	>90% correct size	Western blot, MS
	Functional activity	System-dependent	Activity assays

TABLE 2.11 Process monitoring and analytical methods.

Parameter	Monitoring method	Frequency	Typical values	Action limits
ATP/ADP ratio	Luciferase assay	Every 30 min	5–20	<3 (intervention)
Protein synthesis rate	Fluorescence (GFP)	Continuous	0.1–0.5 mg/mL/h	<0.05 (stop)
pH	pH electrode	Continuous	7.2–7.8	<7.0 or > 8.0
Temperature	Thermocouple	Continuous	±1°C target	±3°C (alarm)
Amino acid levels	HPLC analysis	Every 2 h	>50% initial	<20% (feed)
By-product accumulation	LC–MS	Every 4 h	<5 mM Pi	>15 mM (concern)

normal operating characteristics of the system and provide reference points for troubleshooting when performance deviates from expectations.

PAT enables real-time monitoring of critical parameters during cell-free reactions. Key measurements include ATP/ADP ratios (using luciferase-based assays), amino acid consumption (through derivatization and high pressure liquid chromatography, HPLC analysis), and protein accumulation (using fluorescent reporters or label-free methods). These real-time data can identify problems early in reactions and guide optimization efforts (Table 2.11).

Troubleshooting protocols provide systematic approaches for diagnosing and correcting common problems in cell-free systems. Poor protein yields may result from lysate degradation, suboptimal cofactor concentrations, template problems, or inhibitory contaminants. Systematic testing can isolate the source of problems and guide corrective actions.

Documentation and record-keeping are essential components of quality control, particularly for commercial applications that may require regulatory approval. Complete records should include component lot numbers and certificates of analysis, preparation procedures and environmental conditions, reaction setup details and any deviations, analytical results and quality assessments, and corrective actions taken when problems occur.

Advanced quality control approaches increasingly employ statistical process control methods to monitor system performance over time and detect trends that might indicate developing problems. Control charts can track key performance indicators and trigger investigations when values drift outside normal ranges, enabling proactive maintenance and optimization.

The integration of laboratory information management systems with cell-free platforms enables automated data collection, analysis, and reporting. These systems can track component genealogy, automatically flag out-of-specification results, generate compliance reports, and facilitate continuous improvement efforts through trend analysis and correlation studies.

Bibliography

Carlson, E.D., Gan, R., Hodgman, C.E., Jewett, M.C., 2012. Cell-free protein synthesis: applications come of age. Biotechnol. Adv. 30, 1185–1194.

Caschera, F., 2017. Bacterial cell-free expression technology to in vitro systems engineering and optimization. Synth. Biol. 2, 1–16.

Hunt, A.C., Rasor, B.J., et al., 2024. Cell-free gene expression: methods and applications. Chem. Rev. 125, 91–149.

Jewett, M.C., Swartz, J.R., 2004. Mimicking the *Escherichia coli* cytoplasmic environment activates long-lived and efficient cell-free protein synthesis. Biotechnol. Bioeng. 86, 19–26.

Karim, A.S., Jewett, M.C., 2016. A cell-free framework for rapid biosynthetic pathway prototyping and enzyme discovery. Metab. Eng. 36, 116–126.

Kim, D.M., Swartz, J.R., 2001. Regeneration of adenosine triphosphate from glycolytic intermediates for cell-free protein synthesis. Biotechnol. Bioeng. 74, 309–316.

Liu, G., Mack, S.G., Torres-Montemayor, R., Swartz, J.R., 2021. Engineering *Escherichia coli* for cell-free protein synthesis. Biotechnol. Prog. 37, e3131.

Marshall, R., Maxwell, C.S., Collins, S.P., et al., 2019. Rapid and scalable characterization and scoring of engineered cellular systems with flow cytometry. Nat. Protoc. 14, 2317–2328.

Salis, H.M., Mirsky, E.A., Voigt, C.A., 2009. Automated design of synthetic ribosome binding sites to control protein expression. Nat. Biotechnol. 27, 946–950.

Shimizu, Y., Inoue, A., Tomari, Y., et al., 2001. Cell-free translation reconstituted with purified components. Nat. Biotechnol. 19, 751–755.

CHAPTER

3

Cell lysate preparation and characterization

3.1 Introduction

Cell lysates represent the fundamental foundation of cell-free protein synthesis (CFPS) systems, serving as the primary source of the molecular machinery required for translation. The quality, composition, and activity of these lysates directly determine the efficiency, yield, and functionality of the resulting protein products. This chapter provides comprehensive coverage of the preparation methods, characterization techniques, and optimization strategies for various types of cell lysates used in CFPS applications.

The selection of an appropriate lysate system involves multiple critical factors that must be carefully evaluated for each application. Table 3.1 and Fig. 3.1 present a comprehensive overview of the key selection factors and their impact on system choice, providing a framework for decision-making in lysate selection.

3.2 Prokaryotic systems (*Escherichia coli, Bacillus subtilis*)

3.2.1 *Escherichia coli* cell-free systems

Escherichia coli remains the most widely used organism for CFPS due to its well-characterized biology, rapid growth, and robust translational machinery. The choice of *E. coli* strain significantly impacts the performance characteristics of the resulting lysate, as detailed in Table 3.2.

Optimal growth conditions are critical for producing high-quality lysates with maximum translation activity. Table 3.3 summarizes the key growth parameters that must be carefully controlled during cell cultivation.

The S30 extract preparation protocol involves multiple critical steps, each with specific requirements for maintaining lysate quality. Table 3.4 provides a detailed breakdown of each step with associated quality checkpoints and critical parameters.

Performance benchmarks are essential for evaluating lysate quality and consistency across batches. Table 3.5 establishes the criteria for classifying lysate performance levels.

Cell Free Synthesis of Biologics
https://doi.org/10.1016/B978-0-443-33271-5.00019-3

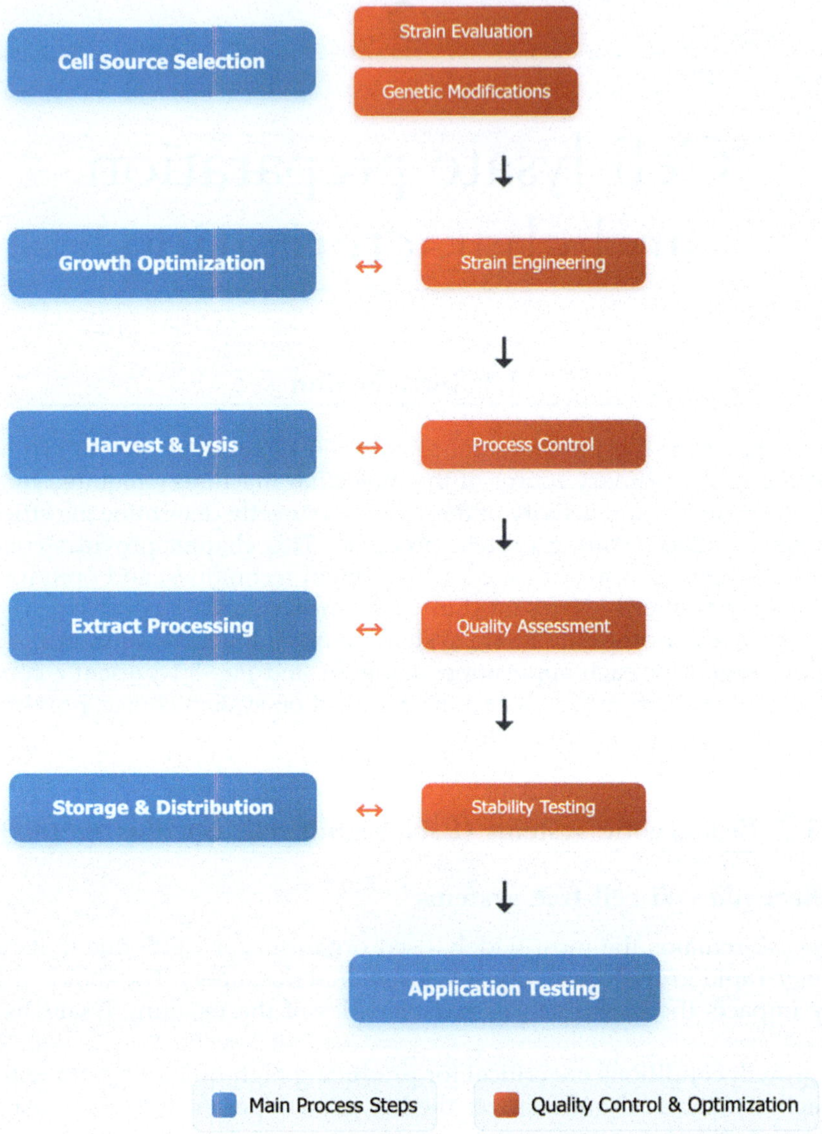

FIG. 3.1 General workflow for cell lysate preparation showing key decision points and quality control steps throughout the process.

TABLE 3.1 Key selection factors for lysate systems.

Factor	Considerations	Impact on selection
Target protein type	Prokaryotic vs. eukaryotic origin	Determines folding machinery compatibility
Posttranslational modifications	Glycosylation, phosphorylation, acetylation	Limits system choice to appropriate cellular machinery
Expression yield requirements	Research scale vs. industrial production	Affects cost–benefit analysis
Cost constraints	Research budget vs. commercial viability	Influences system complexity and source
Downstream applications	Research, diagnostics, therapeutics	Determines purity and regulatory requirements
Regulatory environment	Food and Drug Administration (FDA), European Medicines Agency (EMA) approval pathways	Affects documentation and validation needs

TABLE 3.2 *E. coli* strain selection criteria for cell-free systems.

Strain	Key features	Best applications	Limitations
BL21(DE3)	• T7 RNA polymerase • High expression capability • Well-characterized genetics	• General protein expression • Industrial applications • High-yield requirements	• Limited rare codon support • Standard folding machinery only
Rosetta(DE3)	• Enhanced rare codon usage • Supplemented transfer RNA (tRNA) pool • T7 system compatibility	• Eukaryotic protein expression • Complex codon usage • Heterologous proteins	• Higher cost • More complex preparation
C41(DE3)	• Membrane protein optimization • Reduced toxicity • Enhanced viability	• Membrane proteins • Toxic protein targets • Difficult-to-express proteins	• Lower general yields • Specialized applications only
A19	• CFPS-specific optimization • Enhanced extract activity • Reduced background	• Cell-free applications • Research purposes • High-quality extracts	• Limited commercial availability • Specialized protocols required

3.2.2 *Bacillus subtilis* systems

Bacillus subtilis cell-free systems offer unique advantages for specific applications, particularly those requiring different cellular machinery or lower endotoxin content. Table 3.6 outlines the specific advantages and their corresponding applications.

The preparation of *B. subtilis* lysates requires several modifications compared to standard *E. coli* protocols. Table 3.7 summarizes these key adaptations and their rationale.

TABLE 3.3 Optimal growth conditions for *E. coli* lysate preparation.

Parameter	Standard range	Optimal value	Critical notes
Temperature	35°C –39°C	37°C	• Higher temps reduce viability • Lower temps extend growth time
Medium type	LB, 2xYT, TB	2xYT + 0.8% glucose	• Glucose enhances energy reserves • Rich media improve yields
Harvest OD_{600}	1.5–3.0	1.8–2.5	• Mid-exponential phase optimal • Higher optical density (OD) reduces activity
Growth rate (μ)	0.3–0.8 h^{-1}	0.4–0.6 h^{-1}	• Moderate growth preserves quality • Fast growth causes stress
pH control	6.8–7.4	7.0–7.2	• Automatic pH control preferred • Avoid extreme fluctuations

TABLE 3.4 S30 extract preparation protocol with quality control points.

Step	Procedure	Critical parameters	Quality checkpoints
Cell harvesting	• Rapid cooling to 4°C • Centrifugation 5000 × g, 10 min • Immediate processing	• Temperature maintenance • Minimized harvest time • Gentle handling	• Cell pellet integrity • Minimal cell lysis • Color assessment
Cell washing	• 3× washes with S30 buffer • Resuspension and centrifugation • Final wash verification	• Buffer temperature (4°C) • Complete resuspension • Wash volume accuracy	• Supernatant clarity • pH stability • Conductivity measurement
Cell disruption	• French press (2–3 passes) • Sonication alternative • Cooling between passes	• Pressure optimization • Temperature control • Disruption efficiency	• Microscopic examination • Protein release • DNA release monitoring
Clarification	• High-speed centrifugation • 30,000 × g, 30 min, 4°C • Careful supernatant collection	• Temperature maintenance • Rotor balance • Pellet examination	• Supernatant clarity • Protein concentration • Ribosome integrity
Dialysis	• Overnight against S30 buffer • 3–4 buffer changes • 1:100 vol ratio minimum	• Buffer equilibration • Temperature control • Time optimization	• Conductivity measurement • Small-molecule removal • Volume recovery
Quality control	• Activity testing with GFP • Protein and RNA quantification • Aliquoting and storage	• Standard assay conditions • Reproducible testing • Proper documentation	• Activity benchmarks • Batch consistency • Storage verification

TABLE 3.5 Performance benchmarks for *E. coli* cell-free systems.

Parameter	Target range	Excellent performance	Acceptable performance	Poor performance
Protein yield	0.5–2.0 mg/mL	>1.5 mg/mL	0.8–1.5 mg/mL	<0.8 mg/mL
Reaction duration	2–8 h	6–8 h	4–6 h	<4 h
Template efficiency	10–50 nM optimal	<20 nM	20–40 nM	>40 nM
Temperature stability	30°C –37°C	35°C –37°C	30°C –35°C	<30°C
Storage stability	6–12 months	>12 months	6–12 months	<6 months

TABLE 3.6 Unique advantages of *B. subtilis* cell-free systems.

Advantage	Description	Applications	Benefits
Natural secretion	• Signal peptide processing • Secretory pathway machinery • Reduced contamination	• Secreted proteins • Enzyme production • Therapeutic proteins	• Simplified purification • Native processing • Industrial relevance
Alternative chaperones	• Different folding environment • Unique chaperone systems • Thermostable factors	• Difficult-to-fold proteins • Thermostable enzymes • Industrial biocatalysts	• Enhanced stability • Improved yields • Process robustness
Low endotoxin	• Gram-positive cell wall • Reduced lipopolysaccharide (LPS) content • Cleaner preparations	• Therapeutic applications • In vivo studies • Medical devices	• Regulatory advantages • Safety improvements • Quality benefits

TABLE 3.7 Modified preparation requirements for *B. subtilis* systems.

Modification area	Standard *E. coli*	*B. subtilis* adaptation	Rationale
Growth phase	Mid-exponential	Extended exponential	• Sporulation avoidance • Enzyme accumulation • Metabolic optimization
Buffer composition	Standard S30	Modified salt ratios	• Cell wall compatibility • Enzyme stability • Ionic requirements
Disruption method	French press	Lysozyme + mechanical	• Cell wall structure • Gentle release • Membrane preservation
Storage conditions	Standard −80°C	Enhanced stabilizers	• Enzyme protection • Activity preservation • Extended shelf life

3.3 Eukaryotic systems (wheat germ, rabbit reticulocyte)

3.3.1 Wheat germ extract systems

Wheat germ extracts (WGEs) represent one of the most successful eukaryotic cell-free systems, offering superior protein folding capabilities and extended reaction times. Table 3.8 details the key advantages and their corresponding applications.

The preparation of WGEs involves several specialized steps that differ significantly from prokaryotic systems. Table 3.9 provides detailed requirements for each preparation step.

Optimization of wheat germ systems requires careful attention to multiple parameters that affect both translation efficiency and protein folding. Table 3.10 presents the key optimization parameters and their recommended ranges.

TABLE 3.8 Wheat germ extract system advantages and applications.

Advantage	Technical basis	Primary applications	Performance benefits
Superior folding	• Eukaryotic chaperones • ER-like environment • Disulfide formation	• Complex eukaryotic proteins • Membrane proteins • Therapeutic targets	• Higher activity yields • Native conformations • Functional products
Extended reactions	• Reduced nuclease activity • Stable energy systems • Optimal temperature	• High-value proteins • Research applications • Structural studies	• Improved economics • Higher total yields • Reduced batch frequency
Minimal modifications	• Authentic processing • Eukaryotic machinery • Natural environment	• Pharmaceutical proteins • Vaccine antigens • Diagnostic proteins	• Product authenticity • Regulatory acceptance • Quality assurance

TABLE 3.9 Wheat germ extract preparation methodology.

Preparation Step	Specific requirements	Critical parameters	Quality indicators
Raw material selection	• Fresh wheat germ (<3 months) • High viability (>90%) • Contamination-free source	• Storage temperature • Moisture content • Microbial load	• Germination test • Visual inspection • Enzymatic activity
Grinding and sieving	• Controlled particle size • Temperature control • Yield optimization	• Mesh size (40–60) • Grinding speed • Cooling efficiency	• Particle distribution • Extract yield • Protein release
Extract preparation	• Buffer extraction ratio • Protease inhibitor cocktail • pH optimization	• 1:3 w/v buffer ratio • Inhibitor concentrations • Extraction time	• Protein concentration • Ribosome integrity • Nuclease activity
Purification steps	• mRNA removal • Ribosome enrichment • Small molecule dialysis	• Centrifugation speed • Wash efficiency • Dialysis duration	• Background reduction • Activity retention • Purity assessment

3.3.2 Rabbit reticulocyte lysate

Rabbit reticulocyte lysate (RRL) has been a cornerstone of eukaryotic CFPS, though its use involves specific challenges and limitations. Table 3.11 summarizes the system characteristics and associated considerations.

The use of RRL involves significant regulatory and ethical considerations that must be addressed. Table 3.12 outlines these requirements and potential alternatives.

TABLE 3.10 Wheat germ extract optimization parameters.

Parameter	Range	Optimal value	Optimization notes
Mg^{2+} concentration	1–6 mM	2–4 mM	• Ribosome stability • Translation efficiency • Folding assistance
K^+ concentration	60–150 mM	80–120 mM	• Ionic strength balance • Ribosome binding • Protein stability
Energy system	Various options	Creatine phosphate/kinase	• ATP regeneration • Long-term stability • Cost-effectiveness
Temperature	22°C–32°C	26°C–30°C	• Enzyme activity • Protein stability • Reaction kinetics
pH range	7.0–8.5	7.5–8.0	• Enzyme optimum • Protein stability • Buffer capacity

TABLE 3.11 Rabbit reticulocyte lysate system characteristics.

Aspect	Details	Advantages	Disadvantages
Preparation source	• Phenylhydrazine-treated rabbits • Reticulocyte enrichment • Hemoglobin removal	• High translation activity • Mammalian machinery • Established protocols	• Animal welfare concerns • Batch variability • Limited availability
Performance profile	• Moderate protein yields • Excellent folding quality • Authentic modifications	• Research applications • Mechanistic studies • Regulatory proteins	• High cost • Limited scalability • Ethical considerations
Applications	• Academic research • Proof-of-concept studies • Specialized proteins	• Well-characterized system • Literature precedent • Quality standards	• Commercial limitations • Supply constraints • Regulatory complexity

3.4 Mammalian cell-free systems

3.4.1 Chinese hamster ovary cell-based systems

Chinese hamster ovary (CHO) cell extracts offer authentic mammalian machinery for complex protein production, though they come with significant preparation challenges. Table 3.13 details the system advantages and their applications.

CHO cell-based systems face several preparation challenges that require specialized solutions. Table 3.14 outlines these challenges and proposed mitigation strategies.

TABLE 3.12 Regulatory and ethical considerations for rabbit reticulocyte lysate.

Consideration	Requirements	Documentation	Alternatives
Animal welfare	• Institutional Animal Care and Use Committee (IACUC) approval • Veterinary oversight • Humane protocols	• Protocol submissions • Training records • Monitoring reports	• Commercial sources • Synthetic alternatives • Other cell types
Batch consistency	• Standardized procedures • Quality control testing • Traceability systems	• Batch records • Test results • Chain of custody	• Pooled preparations • Characterization standards • Alternative systems

TABLE 3.13 CHO cell-based system advantages and applications.

Feature	Technical basis	Applications	Performance benefits
Authentic folding	• Mammalian chaperone systems • ER-like environment • Native cofactors	• Therapeutic proteins • Complex antibodies • Membrane receptors	• Native conformations • Proper assembly • Functional activity
Complex glycosyl-ation	• N-linked pathways • O-linked modifications • Sialic acid addition	• Glycoproteins • Hormones • Therapeutic mAbs	• Proper pharmacokinetics • Immunogenicity control • Efficacy enhancement
Disulfide formation	• Protein disulfide isomerase • Oxidizing environment • Quality control systems	• Antibodies • Growth factorsEnzyme cofactors	• Structural integrity • Biological activity • Stability improvement

TABLE 3.14 CHO cell system preparation challenges and solutions.

Challenge	Technical issues	Proposed solutions	Implementation
Cell culture require-ments	• Expensive media • Contamination risk • Scale limitations	• Serum-free media • Bioreactor systems • Automated monitoring	• Cost reduction strategies • Quality systems • Process optimization
Lower extract activity	• Fragile cellular machinery • Complex requirements • Inhibitory factors	• Gentle preparation • Optimized conditions • Activity enhancers	• Protocol refinement • Component supplementation • System engineering
Cost considera-tions	• High preparation costs • Limited throughput • Resource intensity	• Process efficiency • Yield optimization • Alternative approaches	• Economic modeling • Value engineering • Technology development

3.4.2 HeLa cell systems

HeLa cell extracts provide human-specific translational machinery for specialized research applications. Table 3.15 outlines the specific applications and associated considerations.

3.5 Insect cell systems

3.5.1 Sf9/Sf21 cell extracts

Spodoptera frugiperda cell extracts offer a compromise between prokaryotic simplicity and eukaryotic complexity. Table 3.16 compares the performance characteristics of the two commonly used cell lines.

The preparation of insect cell extracts follows a standardized protocol with specific critical control points. Table 3.17 outlines the key steps and quality metrics.

Insect cell systems excel in specific applications where their unique capabilities provide advantages. Table 3.18 summarizes these applications and their benefits.

TABLE 3.15 HeLa cell system specialized applications.

Application area	Specific uses	Technical advantages	Considerations
Human protein studies	• Native human machinery • Authentic processing • Disease relevance	• Physiological relevance • Translational value • Clinical applicability	• Ethical considerations • Safety protocols • Regulatory requirements
Viral protein expression	• Host-pathogen interactions • Viral assembly • Infection mechanisms	• Natural host environment • Authentic interactions • Research relevance	• Biosafety requirements • Containment protocols • Risk assessment
Cancer research	• Oncogene products • Tumor suppressors • Therapeutic targets	• Disease-relevant context • Pathway studies • Drug development	• Cell line characterization • Genetic stability • Reproducibility

TABLE 3.16 Sf9 vs. Sf21 cell extract performance characteristics.

Characteristic	Sf9 Cells	Sf21 cells	comparative notes
Growth properties	• Faster growth rate • Higher density cultures • Standard maintenance	• Slower growth • Lower final density • Enhanced protein quality	• Sf9 for high throughput • Sf21 for quality applications
Extract quality	• Good protein yields • Standard processing • Cost-effective	• Superior folding • Enhanced modifications • Premium applications	• Application-dependent choice • Cost-benefit analysis
Applications	• General expression • Screening studies • Process development	• High-value proteins • Structural studies • Therapeutic applications	• Selection based on requirements • Quality vs. cost trade-off

TABLE 3.17 Insect cell extract preparation protocol overview.

Step	Procedure	Critical factors	Quality metrics
Cell culture	• Serum-free medium • Suspension culture • Bioreactor scale-up	• Medium quality • Contamination control • Growth monitoring	• Cell viability > 95% • Density 2–4 × 10^6 cells/mL • Doubling time consistency
Harvest timing	• Exponential phase • High viability • Optimal density	• Growth curve monitoring • Viability assessment • Metabolic indicators	• Glucose consumption • Lactate production • pH stability
Lysis procedures	• Gentle disruption • Osmotic lysis • Mechanical methods	• Temperature control • Buffer optimization • Disruption efficiency	• Protein release • Organelle integrity • Enzyme activity
Processing	• Centrifugal clarification • Nuclease treatment • Concentration	• Speed optimization • Treatment conditions • Recovery efficiency	• Clarity assessment • Activity retention • Concentration accuracy

TABLE 3.18 Insect cell system applications and benefits.

Application	Technical benefits	Economic considerations	implementation notes
Virus-like particles	• Natural assembly • Proper morphology • Vaccine applications	• High-value products • Premium pricing • Market demand	• Specialized protocols • Quality requirements • Regulatory pathways
Membrane proteins	• Eukaryotic machinery • Proper insertion • Functional activity	• Research applications • Drug development • Structural studies	• Technical expertise • Specialized equipment • Quality assessment
Glycoproteins	• Moderate glycosylation • Eukaryotic patterns • Functional products	• Therapeutic relevance • Commercial potential • Quality premiums	• Process optimization • Analytical methods • Regulatory compliance

3.6 Protein synthesis Using Recombinant Elements and reconstituted systems

3.6.1 Protein synthesis Using Recombinant Elements system component analysis

The Protein synthesis Using Recombinant Elements (PURE) system represents the ultimate in defined cell-free systems. Table 3.19 provides a comprehensive breakdown of the core translation machinery components.

The aminoacylation system is critical for accurate translation and requires careful component balance. Table 3.20 details the aminoacylation system components and their quality control requirements.

Energy regeneration is essential for sustained protein synthesis in PURE systems. Table 3.21 outlines the energy system components and their performance metrics.

TABLE 3.19 PURE system core translation machinery components.

Component category	Individual components	Function	Quality requirements
Ribosomal components	• 70S ribosomes (purified)	• Protein synthesis platform • tRNA binding • Peptide bond formation	• High purity (>95%) • Functional integrity • Consistent activity
Translation initiation	• IF1, IF2, IF3	• Start codon recognition • Ribosome assembly • fMet–tRNA binding	• Stoichiometric ratios • Activity verification • Stability assessment
Translation elongation	• EF–Tu, EF–G, EF–P	• tRNA delivery • Translocation • Proline incorporation	• Concentration optimization • GTP binding activity • Functional testing
Translation termination	• RF1, RF2, RF3, ribosyme release factor (RRF)	• Stop codon recognition • Peptide release • Ribosome recycling	• Specificity testing • Release efficiency • Recycling activity

TABLE 3.20 PURE system aminoacylation components.

Component	Quantity	Specificity	Quality control
Synthetases (aaRS)	20 enzymes	One per amino acid	• Amino acid specificity • tRNA recognition • ATP consumption
tRNA species	40+ molecules	Codon coverage	• Aminoacylation efficiency • Codon reading • Structural integrity
Amino acids	20 standard	Purity requirements	• Chemical purity > 99% • Optical purity • Solubility verification

TABLE 3.21 PURE system energy regeneration components.

Component	Function	Optimal concentration	Performance metrics
Creatine kinase (CK)	ATP regeneration	0.1–0.5 mg/mL	• ATP/ADP ratio maintenance • Long-term stability • Temperature sensitivity
Myokinase (MK)	Nucleotide balance	0.05–0.2 mg/mL	• AMP elimination • Energy charge • Reaction coupling
Nucleoside diphosphate kinase (NDK)	GTP regeneration	0.02–0.1 mg/mL	• GTP supply • Translation factors • Energy efficiency
Pyrophosphatase (PPase)	Pyrophosphate removal	0.01–0.05 mg/mL	• Reaction driving • Product removal • Equilibrium shift

3.6.2 System advantages and limitations

PURE systems offer unique advantages for specific research and development applications. Table 3.22 summarizes these advantages and their corresponding benefits.

Despite their advantages, PURE systems face several limitations that restrict their widespread adoption. Table 3.23 outlines these limitations and potential mitigation strategies.

TABLE 3.22 PURE system advantages and applications.

Advantage	Technical basis	Applications	Benefits
Complete control	• Defined composition • Component manipulation • Systematic optimization	• Mechanistic studies • System engineering • Fundamental research	• Reproducible results • Predictable behavior • Scientific understanding
Minimal side reactions	• Purified components • Reduced complexity • Controlled environment	• Clean synthesis • Specific products • Research applications	• Higher purity • Reduced background • Easier analysis
Systematic optimization	• individual component control • Rational design • Predictive modeling	• Process development • Commercial applications • Technology transfer	• Faster optimization • Reduced screening • Predictable scaling

TABLE 3.23 PURE system limitations and mitigation strategies.

Limitation	Technical issues	Impact	Mitigation strategies
High component costs	• Protein purification • Quality requirements • Yield limitations	• Economic barriers • Limited applications • Research constraints	• Process improvement • Alternative sources • Cost reduction R&D
Complex preparation	• Multiple components • Quality control • Assembly requirements	• Technical barriers • Expertise requirements • Time consumption	• Standardized protocols • Commercial kits • Training programs
Limited modifications	• Minimal post-translational modifications (PTM) machinery • No cellular compartments • Reduced complexity	• Application limitations • Product constraints • Market restrictions	• System augmentation • Hybrid approaches • Specialized applications

3.7 Lysate engineering and optimization

3.7.1 Metabolic engineering strategies

Modern lysate optimization involves systematic genetic modifications to enhance performance. Table 3.24 outlines key knockout strategies for improved lysate performance.

Overexpression of specific components can significantly enhance lysate performance. Table 3.25 details key overexpression targets and their implementation strategies.

3.7.2 Chemical optimization matrices

Buffer system optimization is crucial for maximizing lysate performance across different conditions. Table 3.26 provides comprehensive optimization ranges for key buffer components.

Energy system optimization requires careful balance of multiple components for sustained ATP regeneration. Table 3.27 details the key energy system components and their optimization criteria.

TABLE 3.24 Knockout strategies for enhanced lysate performance.

Target pathway	Genes targeted	Rationale	Expected benefits
Nuclease removal	• endA (endonuclease I) • rne (RNase E) • rnc (RNase III)	• DNA/RNA protection • Template stability • Reaction longevity	• Extended reactions • Higher yields • Improved reproducibility
Toxin–antitoxin systems	• mazEF system • relBE system • higBA system	• Cell death prevention • Stress response reduction • Extract activity preservation	• Higher extract activity • Better stability • Consistent performance
Competing pathways	• Proteases (lon, clpP) • Phosphatases • Competing ATPases	• Resource conservation • Reduced side reactions • Energy efficiency	• Resource redirection • Cleaner reactions • Cost reduction

TABLE 3.25 Overexpression targets for lysate enhancement.

Component	Overexpression strategy	Performance impact	Implementation
Chaperones	• GroEL/GroES system • DnaK/DnaJ/GrpE • Trigger factor (tig)	• Enhanced folding • Reduced aggregation • Higher activity	• Plasmid-based • Chromosomal integration • Inducible systems
Translation factors	• EF–Tu, EF–G • Ribosome biogenesis • tRNA modifications	• Translation efficiency • Accuracy improvement • Speed enhancement	• Balanced expression • Stoichiometric control • Quality monitoring
Energy metabolism	• Glycolytic enzymes • ATP synthase • Nucleotide kinases	• Energy supply • Regeneration capacity • Reaction longevity	• Pathway engineering • Flux optimization • Balance maintenance

TABLE 3.26 Buffer system optimization parameters.

Parameter	Low range	Optimal range	High range	Optimization notes
pH	7.0–7.5	8.0–8.2	8.3–8.5	• Temperature compensation • Enzyme pH optima • Protein stability
Mg^{2+} (mM)	8–12	12–16	16–20	• Ribosome assembly • RNA structure • Enzyme cofactor
K^+ (mM)	50–100	100–150	150–200	• Ionic strength • Protein interactions • Membrane potential
DTT (mM)	1–2	2–3	3–5	• Reducing environment • Cysteine protection • Enzyme activity
Polyamines (mM)	0.5–1.0	1.0–1.5	1.5–2.0	• Ribosome stabilization • RNA protection • Translation enhancement

TABLE 3.27 Energy system component optimization.

Component	Concentration	Function	Optimization criteria
Phosphoenolpyruvate (PEP)	10–30 mM	Primary energy source	• Cost-effectiveness • Stability requirements • Reaction longevity
Pyruvate kinase	0.1–0.5 mg/mL	ATP regeneration	• Activity maintenance • Temperature stability • Cofactor requirements
Creatine phosphate	10–20 mM	Alternative energy	• Higher cost • Superior performance • Specialized applications
Creatine kinase	0.1–0.3 mg/mL	ATP regeneration	• Enzyme stability • Activity optimization • Cost considerations

3.7.3 Physical parameter optimization

Physical conditions significantly impact lysate performance and must be optimized for each specific system. Table 3.28 outlines the key physical parameters and their system dependencies (Fig. 3.2).

TABLE 3.28 Physical parameter optimization guidelines.

Parameter	Range	Optimal conditions	System dependencies
Temperature	25°C–37°C	System-specific optimum	• *E. coli*: 30°C–37°C • Wheat germ: 26°C–30°C • Mammalian: 37°C
Agitation	0–300 rpm	Gentle mixing (50–100 rpm)	• Oxygen transfer • Mixing efficiency • Shear sensitivity
Atmosphere	Air to anaerobic	Reduced oxygen preferred	• Oxidation prevention • Enzyme protection • Redox balance
Pressure	Atmospheric	Standard conditions	• Specialized applications • Gas solubility • Equipment complexity

3.8 Storage, stability, and quality assessment

3.8.1 Storage condition optimization

Short-term storage strategies must balance convenience with activity preservation. Table 3.29 outlines various short-term storage options and their applications.

Long-term storage requires specialized protocols and stabilizing agents to maintain lysate activity over extended periods. Table 3.30 compares different long-term storage methods.

3.8.2 Comprehensive quality assessment

Quality assessment involves multiple analytical approaches to ensure lysate performance and consistency. Table 3.31 details the primary quality metrics and their acceptance criteria.

Advanced characterization methods provide deeper insights into lysate composition and quality. Table 3.32 outlines these methods and their applications.

3.8.3 Stability assessment protocols

Accelerated stability testing provides rapid assessment of lysate degradation patterns. Table 3.33 outlines the key stability testing conditions and acceptance criteria.

Degradation monitoring helps identify specific failure modes and guide optimization efforts. Table 3.34 details different types of degradation and their detection methods.

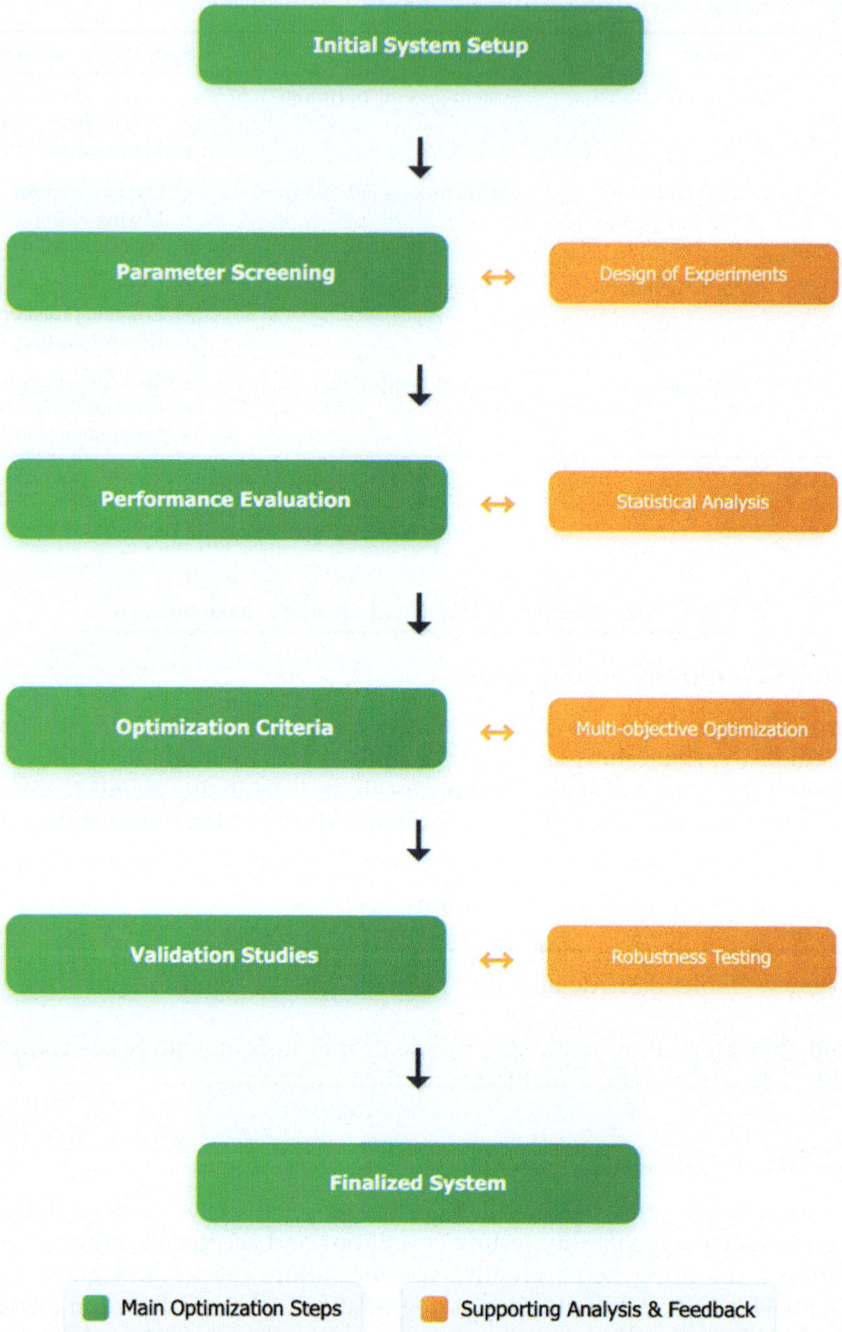

FIG. 3.2 Systematic optimization workflow for cell-free protein synthesis systems showing the iterative process of parameter screening, evaluation, and validation.

TABLE 3.29 Short-term storage strategies for cell lysates.

Storage method	Temperature	Duration	Applications	Quality retention
Ice storage	0°C–4°C	2–8 h	• Immediate use • Transport • Processing delays	• 95%–100% activity • Minimal degradation • Full functionality
Refrigeration	4°C	1–7 days	• Laboratory use • Multiple experiments • Quality control	• 80%–95% activity • Gradual decline • Acceptable performance
Controlled atmosphere	4°C + N_2/Ar	3–14 days	• Extended studies • Shipping • Quality preservation	• 85%–98% activity • Oxidation prevention • Enhanced stability

TABLE 3.30 Long-term storage protocols for cell lysates.

Method	Temperature	Stabilizers	Duration	Recovery efficiency
Standard freezing	−80°C	10–20% glycerol	6–12 months	• 7%–90% activity • Standard protocol • Acceptable losses
Cryoprotected	−80°C	Glycerol + trehalose	12–24 months	• 80%–95% activity • Enhanced protection • Premium storage
Liquid nitrogen	−196°C	Optimized cocktail	>24 months	• 85%–98% activity • Maximum stability • Research applications
Lyophilized	Room temp	Complex formulation	12–36 months	• 60%–85% activity • Shipping advantages • Specialized applications

TABLE 3.31 Primary quality metrics for cell lysate assessment.

Test category	Specific assays	Acceptance criteria	Testing frequency
Activity assays	• GFP expression test • Luciferase reporter • Standard protein panel	• >80% of reference • Reproducible kinetics • Dose–response linearity	• Every batch • Storage intervals • Preuse verification
Biochemical analysis	• Total protein (Bradford/BCA) • RNA content (A_{260}) • Ribosome concentration	• Protein: 15–25 mg/mL • A_{260}/A_{280}: 1.8–2.0 • Ribosomes: 2–4 μM	• Batch characterization • Stability studies • Troubleshooting
Contamination testing	• Endotoxin (Limulus Amebocyte Lysate, LAL assay) • Sterility (USP < 71 >) • Nuclease activity	• <1.0 EU/mL • No growth • <10% RNA degradation	• Batch release • Regulatory compliance • Safety verification

TABLE 3.32 Advanced characterization methods for cell lysates.

Analysis type	Methods	Information obtained	Applications
Proteomic analysis	• Liquid chromatography-mass spectra (LC–MS)/MS • 2D gel electrophoresis • Western blotting	• Protein composition • Degradation products • Contaminant identification	• System optimization • Batch comparison • Quality improvement
Ribosome profiling	• Sucrose gradient • Dynamic light scattering • Cryo-EM analysis	• Ribosome integrity • Subunit ratios • Structural integrity	• Fundamental studies • Quality verification • Method development
Metabolite analysis	• Nuclear magnetic resonance (NMR) spectroscopy • High pressure liquid chromatography (HPLC) analysis • Enzymatic assays	• Energy charge • Cofactor levels • Inhibitor presence	• System optimization • Troubleshooting • Process development

TABLE 3.33 Accelerated stability testing protocols.

Condition	Temperature	Duration	Sampling points	Acceptance criteria
Stress testing	37°C	1–4 weeks	Daily/weekly	• Activity decline <20% • No major degradation • Maintained functionality
Freeze–thaw	−80°C to RT	5–10 cycles	Each cycle	• Activity retention >70% • No precipitation • Consistent performance
Real-time	Storage temp	6–24 months	Monthly/quarterly	• Stability trending • Shelf-life determination • Quality maintenance

TABLE 3.34 Degradation monitoring and mitigation strategies.

degradation type	Detection methods	Critical indicators	Mitigation strategies
Protein degradation	• SDS–PAGE • Activity assays • Western blotting	• Band smearing • Activity loss • Fragment formation	• Protease inhibitors • Temperature control • pH optimization
RNA degradation	• Gel electrophoresis • Bioanalyzer • Activity testing	• rRNA fragmentation • A_{260}/A_{280} changes • Translation decline	• RNase inhibitors • Proper handling • Storage optimization
Cofactor loss	• Enzymatic assays • HPLC analysis • Activity testing	• ATP depletion • Mg^{2+} precipitation • Energy charge decline	• Cofactor supplementation • pH control • Ionic strength management

3.9 Comparative analysis of lysate systems

3.9.1 Performance comparison matrix

A comprehensive comparison of different lysate systems helps guide selection for specific applications. Table 3.35 provides a detailed performance comparison using a standardized rating system.

3.9.2 Selection decision matrix

Application-specific selection criteria help guide the choice of lysate system based on project requirements. Table 3.36 provides guidance for different application types.

Economic evaluation is crucial for system selection, particularly for commercial applications. Table 3.37 provides a comprehensive cost analysis framework.

TABLE 3.35 Comprehensive performance comparison of lysate systems.

System	Protein yield	Folding quality	PTM capability	Cost level	Complexity	Best applications
E. coli S30	☆☆☆☆☆	☆☆☆	☆	☆☆☆☆☆	☆☆	• High-volume production • Research applications • Cost-sensitive projects
Wheat germ	☆☆☆	☆☆☆☆☆	☆☆☆	☆☆☆	☆☆☆	• Eukaryotic proteins • Research grade • Structural studies
Rabbit reticulocyte	☆☆	☆☆☆☆☆	☆☆☆☆	☆	☆☆☆☆	• Mammalian research • Regulatory studies • Specialized applications
CHO cells	☆☆☆	☆☆☆☆☆	☆☆☆☆☆	☆	☆☆☆☆☆	• Therapeutic proteins • Complex glycoproteins • Pharmaceutical applications
PURE system	☆☆	☆☆☆	–	☆	☆☆☆☆	• Mechanistic studies • System engineering • Fundamental research
Insect cells	☆☆☆☆	☆☆☆☆	☆☆☆	☆☆	☆☆☆☆	• Virus-like particles • Membrane proteins • Vaccine production

TABLE 3.36 Application-specific lysate selection guidelines.

Application type	Primary considerations	Recommended systems	Alternative options
Academic research	• Cost-effectiveness • Ease of use • Literature precedent	• *E. coli* S30 • Wheat germ • PURE (specialized)	• Rabbit reticulocyte • Insect cells
Pharmaceutical development	• Regulatory acceptance • Product quality • Scalability	• CHO cells • Insect cells • Wheat germ	• *E. coli* (simple proteins) • Specialized systems
Industrial production	• Cost per unit • Scalability • Yield requirements	• *E. coli* S30 • Engineered systems • Process-optimized	• Wheat germ (premium) • Hybrid approaches
Diagnostic applications	• Stability • Reproducibility • Field deployment	• *E. coli* S30 • Freeze-dried systems • Simplified formats	• Wheat germ • Stabilized preparations

TABLE 3.37 Economic evaluation framework for lysate systems.

Cost component	*E. coli*	Wheat germ	Mammalian	PURE	Relative impact
Raw materials	Low	Medium	High	Very high	30%–40% of total cost
Preparation labor	Low	Medium	High	Very high	20%–30% of total cost
Quality control	Low	Medium	High	High	10%–20% of total cost
Equipment/infrastructure	Low	Medium	Very high	High	15%–25% of total cost
Regulatory compliance	Low	Medium	Very high	Medium	5%–15% of total cost
Total cost index	1.0×	3–5×	10–20×	15–25×	Baseline comparison

3.10 Future directions and emerging trends

3.10.1 Novel lysate sources and engineering

Extremophile-based systems represent an emerging area with potential for specialized applications. Table 3.38 outlines the unique properties and potential applications of these systems.

Synthetic biology integration offers new possibilities for customized lysate systems. Table 3.39 outlines key approaches and their implementation timelines.

3.10.2 Advanced optimization technologies

Machine learning integration promises to revolutionize lysate optimization approaches. Table 3.40 details various ML applications and their expected benefits.

TABLE 3.38 Extremophile-based cell-free systems.

Organism Type	Unique properties	Potential applications	Development status
Thermophiles	• High-temperature stability • Thermostable enzymes • Robust machinery	• Industrial biocatalysts • Harsh environment applications • Process intensification	• Early research • Proof-of-concept • Limited availability
Psychrophiles	• Cold-active enzymes • Flexible proteins • Cold adaptation	• Low-temperature processes • Energy-saving applications • Specialty enzymes	• Research stage • Academic interest • Technical challenges
Halophiles	• Salt tolerance • Unique biochemistry • Stress resistance	• High-salt environments • Specialized applications • Extremophile studies	• Early development • Limited scope • Niche applications

TABLE 3.39 Synthetic biology integration approaches.

Approach	Technology	Applications	Implementation timeline
Chassis engineering	• CRISPR–Cas systems • Metabolic engineering • Pathway optimization	• Custom lysate properties • Enhanced performance • Specialized functions	• 2–5 years • Active development • Commercial interest
Modular systems	• Standardized components • Mix-and-match design • Predictable behavior	• Customizable platforms • Rapid optimization • Standardized products	• 3–7 years • Technology development • Industry collaboration
Artificial cells	• Minimal cell systems • Defined components • Controllable properties	• Research tools • Therapeutic applications • Biosensor platforms	• 5–10 years • Fundamental research • Long-term potential

TABLE 3.40 Machine learning integration in lysate optimization.

Application Area	ML techniques	Implementation	Expected benefits
Process optimization	• Neural networks • Genetic algorithms • Response surface methodology	• Automated screening • Predictive modeling • Real-time control	• Faster optimization • Reduced experiments • Better performance
Quality prediction	• Classification algorithms • Time series analysis • Pattern recognition	• Batch quality prediction • Failure prevention • Process monitoring	• Improved reliability • Reduced waste • Quality assurance
System design	• Deep learning • Reinforcement learning • Multiobjective optimization	• Novel system design • Component selection • Performance prediction	• Innovation acceleration • Systematic development • Optimal solutions

High-throughput technologies are enabling more efficient lysate development and optimization. Table 3.41 outlines key technologies and their current development status.

3.10.3 Standardization and regulatory evolution

Industry standardization efforts are crucial for the widespread adoption of cell-free systems. Table 3.42 outlines current standardization initiatives and their implementation timelines.

TABLE 3.41 High-throughput technologies for lysate development.

Technology	Capabilities	Applications	Development status
Microfluidics	• Miniaturization • Automation • Parallel processing	• Rapid screening • Condition optimization • System characterization	• Commercial systems • Active research • Growing adoption
Robotics	• Automated preparation • Consistent processing • High throughput	• Large-scale screening • Process standardization • Quality improvement	• Available systems • Integration challenges • Cost considerations
AI-driven design	• Predictive algorithms • Optimization engines • Decision support	• System design • Process development • Performance prediction	• Early adoption • Promising results • Expanding capabilities

TABLE 3.42 Industry standardization initiatives.

Standardization area	Current status	Proposed standards	Implementation timeline
Reference materials	• Limited availability • Variable quality • No consensus	• Certified reference standards • Quality benchmarks • Performance metrics	• 2–4 years • Industry collaboration • Regulatory support
Test methods	• Laboratory-specific • Limited comparability • Variable protocols	• Standardized assays • Validation protocols • Harmonized methods	• 3–5 years • Technical committees • Validation studies
Quality systems	• Basic guidelines • Limited enforcement • Variable implementation	• Good Manufacturing Practice (GMP) guidelines • Quality standards • Compliance frameworks	• 5–7 years • Regulatory development • Industry adoption

3.11 Summary tables and decision tools

3.11.1 Quick selection guide

For rapid system selection, Table 3.43 provides a simplified decision matrix based on primary application requirements.

3.12 Troubleshooting matrix

When lysate systems underperform, systematic troubleshooting is essential. Table 3.44 provides a comprehensive troubleshooting matrix for common problems.

TABLE 3.43 Quick selection guide for cell-free lysate systems.

Application requirement	Recommended system	Alternative options	Key considerations
High yield, low cost	*E. coli* S30	Engineered *E. coli*	• Yield vs. quality trade-off • Scalability requirements
Eukaryotic protein folding	Wheat germ	Insect cells	• Complexity vs. cost • PTM requirements
Mammalian PTMs	CHO cells	HeLa cells	• Regulatory pathway • Cost justification
Mechanistic studies	PURE system	*E. coli* minimal	• Control vs. complexity • Budget constraints
Membrane proteins	Insect cells	Wheat germ	• Expression level • Functional requirements
Therapeutic proteins	CHO cells	Insect/Wheat germ	• Regulatory requirements • Quality specifications

TABLE 3.44 Comprehensive troubleshooting matrix for cell-free systems.

Symptom	Possible causes	Diagnostic tests	Solutions
Low protein yield	• Poor lysate activity • Suboptimal conditions • Template problems	• GFP control assay • Mg^{2+} titration • Template verification	• Fresh lysate preparation • Condition optimization • Template redesign
High back-ground	• Endogenous mRNA • Template contamination • Poor nuclease treatment	• No-template control • RNase treatment • Gel analysis	• Nuclease optimization • Template purification • Preparation improvement
Poor reproduc-ibility	• Batch variation • Storage degradation • Protocol inconsistency	• Batch comparison • Stability testing • Protocol audit	• Standardization • Storage optimization • Training improvement

Continued

TABLE 3.44 Comprehensive troubleshooting matrix for cell-free systems—cont'd

Symptom	Possible causes	Diagnostic tests	Solutions
Protein aggregation	• Folding problems • Concentration effects • Temperature stress	• Native gel analysis • Dynamic light scattering (DLS) measurement • Microscopy	• Chaperone addition • Dilution optimization • Temperature adjustment
Short reaction time	• Energy depletion • Nuclease activity • Inhibitor accumulation	• ATP monitoring • RNA stability • Time course analysis	• Energy optimization • Inhibitor addition • Condition adjustment

Bibliography

Bundy, B.C., Swartz, J.R., 2010. Site-specific incorporation of p-propargyloxyphenylalanine in a cell-free environment for direct protein– protein click conjugation. Bioconjug. Chem. 21 (2), 255–263.

Carlson, E.D., Gan, R., Hodgman, C.E., Jewett, M.C., 2012. Cell-free protein synthesis: applications come of age. Biotechnol. Adv. 30 (5), 1185–1194.

Endo, Y., Sawasaki, T., 2006. Cell-free expression systems for eukaryotic protein production. Curr. Opin. Biotechnol. 17 (4), 373–380.

Failmezger, J., Nitschel, R., Sánchez-Kopper, A., Kraml, M., Siemann-Herzberg, M., 2018. Site-specific cleavage of ribosomal RNA in Escherichia coli-based cell-free protein synthesis systems. PLoS One 13 (11), e0207994.

Jackson, R.J., Hunt, T., 1983. Preparation and use of nuclease-treated rabbit reticulocyte lysates for the translation of eukaryotic mRNA. Methods Enzymol. 96, 50–74.

Katzen, F., Chang, G., Kudlicki, W., 2005. The past, present and future of cell-free protein synthesis. Trends Biotechnol. 23 (3), 150–156.

Marshall, R., Maxwell, C.S., Collins, S.P., Jacobsen, T., Luo, M.L., Begemann, M.B., ... Noireaux, V., 2018. Rapid and scalable characterization of CRISPR technologies using an *E. coli* cell-free transcription-translation system. Mol. Cell 69 (1), 146–157.

Mureev, S., Kovtun, O., Nguyen, U.T., Alexandrov, K., 2009. Species-independent translational leaders facilitate cell-free expression. Nat. Biotechnol. 27 (8), 747–752.

Nirenberg, M.W., Matthaei, J.H., 1961. The dependence of cell-free protein synthesis in E. coli upon naturally occurring or synthetic polyribonucleotides. Proc. Natl. Acad. Sci. U.S.A. 47 (10), 1588–1602.

Schwarz, D., Junge, F., Durst, F., Frölich, N., Schneider, B., Reckel, S., Bernhard, F., 2007. Preparative scale expression of membrane proteins in Escherichia coli-based continuous exchange cell-free systems. Nat. Protoc. 2 (11), 2945–2957.

Shimizu, Y., Inoue, A., Tomari, Y., Suzuki, T., Yokogawa, T., Nishikawa, K., Ueda, T., 2001. Cell-free translation reconstituted with purified components. Nat. Biotechnol. 19 (8), 751–755.

Vilkhovoy, M., Dai, D., Mehta, S., Adhikari, A., Varner, J.D., Linhardt, R.J., 2018. Sequence specific modeling of *E. coli* cell-free protein synthesis. ACS Synth. Biol. 7 (8), 1844–1857.

Voytsekh, I., Prokhorova, I., Martynov, A., Konevega, A.L., Rodnina, M.V., 2021. Optimization of bacterial cell-free protein synthesis systems: recent developments in *E. coli*-based systems. Front. Bioeng. Biotechnol. 9, 717090.

Zawada, J.F., Yin, G., Steiner, A.R., Yang, J., Naresh, A., Roy, S.M., Swartz, J.R., 2011. Microscale to manufacturing scale-up of cell-free cytotoxin production—an *E. coli* cell-free protein synthesis case study. Biotechnol. Bioeng. 108 (7), 1570–1578.

Zubay, G., 1973. In vitro synthesis of protein in microbial systems. Annu. Rev. Genet. 7 (1), 267–287.

PART II

Technological platforms and innovations

CHAPTER

4

Energy systems and metabolic engineering

4.1 Introduction

Energy systems represent the metabolic heart of cell-free protein synthesis (CFPS), providing the essential ATP and GTP required for translation, maintaining optimal energy charge, and supporting sustained protein production. Unlike living cells that can dynamically adjust their metabolism, cell-free systems require carefully engineered energy regeneration systems that can operate efficiently in the absence of cellular regulatory mechanisms. This chapter provides comprehensive coverage of energy system design, metabolic engineering approaches, and cost optimization strategies for various CFPS platforms.

The design of effective energy systems involves balancing multiple competing factors, including energy yield, regeneration kinetics, cost considerations, and system complexity. Table 4.1 outlines the fundamental requirements that must be addressed in any energy system design.

The evolution of energy systems for CFPS has progressed from simple ATP supplementation to sophisticated metabolic networks capable of sustained energy regeneration. Fig. 4.1 illustrates the major energy system architectures and their relative complexity and performance characteristics.

4.2 Phosphoenolpyruvate-based energy regeneration

4.2.1 System architecture and components

Phosphoenolpyruvate (PEP)-based energy regeneration represents the most widely used and well-characterized energy system for CFPS. The system relies on the high-energy phosphate bond in PEP to drive ATP regeneration through pyruvate kinase, providing a thermodynamically favorable and kinetically efficient energy source.

Cell Free Synthesis of Biologics
https://doi.org/10.1016/B978-0-443-33182-4.00004-1

TABLE 4.1 Fundamental requirements for cell-free energy systems.

Requirement category	Specific needs	Design implications	Performance metrics
Energy supply	• Sustained ATP regeneration • GTP availability • Energy charge maintenance	• High-capacity systems • Rapid regeneration • Balanced nucleotide pools	• ATP/ADP ratio > 10 • Energy charge > 0.8 • GTP concentration > 1 mM
Temporal stability	• Extended reaction duration • Consistent energy levels • Minimal degradation	• Stable enzyme systems • Protected cofactors • Regenerative capacity	• Activity retention > 80% at 8 h • Linear energy supply • Predictable kinetics
Economic viability	• Cost-effective substrates • Efficient utilization • Scalable production	• Low-cost energy sources • High conversion efficiency • Minimal waste generation	• <$5/g protein produced • > 90% substrate utilization • Competitive economics
System compatibility	• pH stability • Ionic strength control • Minimal interference	• Buffered systems • Compatible cofactors • Clean byproducts	• pH variation < 0.2 units • No inhibitory metabolites • System integration

The core components of PEP-based systems include multiple enzymes and cofactors that work in concert to maintain energy homeostasis. Table 4.2 details the essential components and their specific functions within the energy regeneration network.

4.2.2 Thermodynamic and kinetic analysis

The thermodynamic driving force of PEP-based energy regeneration stems from the high-energy phosphate bond in PEP ($\Delta G^{\circ\prime} = -61.9$ kJ/mol), making ATP regeneration highly favorable. Table 4.3 provides a comprehensive thermodynamic analysis of the key reactions in PEP-based systems.

Kinetic parameters are equally important for maintaining rapid energy regeneration during active protein synthesis. Table 4.4 summarizes the key kinetic parameters for the major energy regeneration enzymes.

4.2.3 System optimization strategies

Optimization of PEP-based energy systems requires careful attention to multiple parameters that affect both energy regeneration efficiency and overall system performance. Table 4.5 outlines the key optimization parameters and their recommended ranges.

4.2.4 Performance characteristics and limitations

PEP-based energy systems offer excellent performance characteristics but also have inherent limitations that must be considered in system design. Table 4.6 provides a comprehensive analysis of system performance metrics and associated limitations.

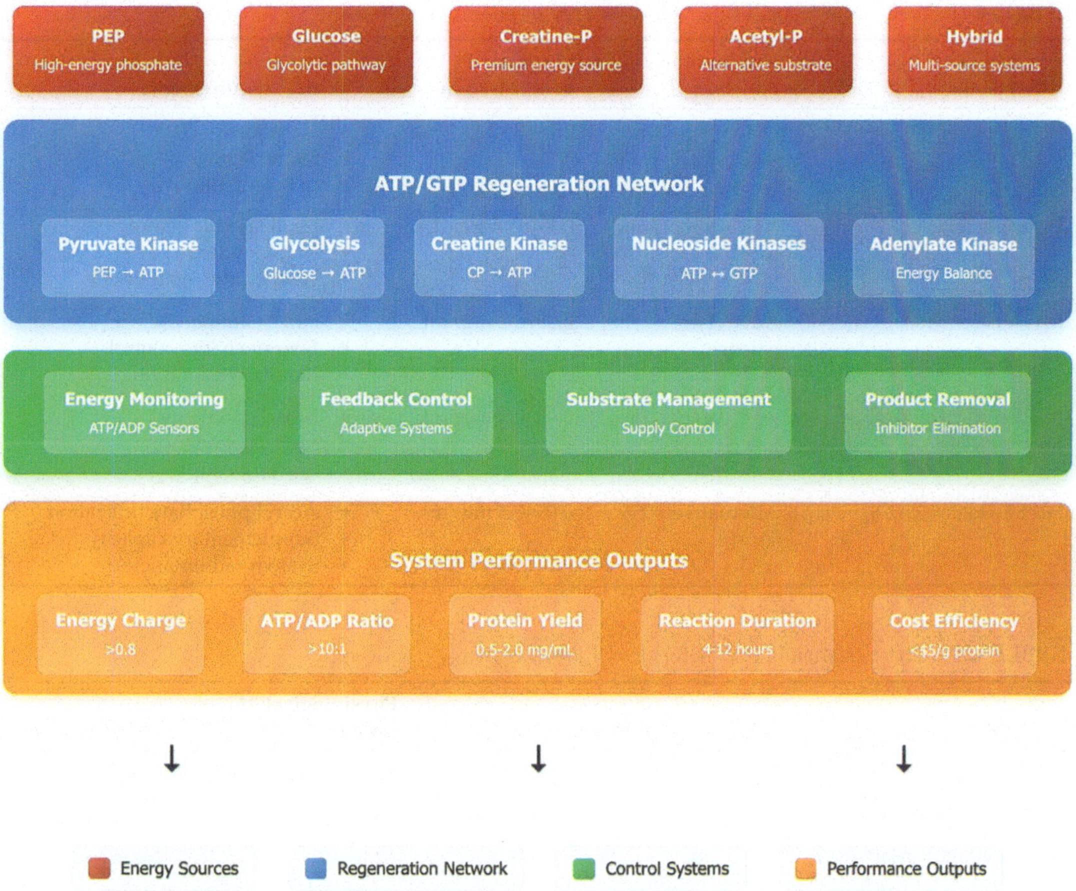

FIG. 4.1 Comprehensive energy system architecture for cell-free protein synthesis showing the integration of multiple energy sources (red), ATP/GTP regeneration network (blue), control and monitoring system (green), and performance outputs (orange). The system demonstrates the complexity and integration required for sustained energy supply in cell-free systems.

4.3 Glucose-based systems

4.3.1 System architecture and metabolic pathways

Glucose-based energy systems leverage the natural glycolytic pathway to provide sustained ATP regeneration through multiple enzymatic steps. These systems offer the advantage of utilizing a low-cost, readily available substrate while providing multiple points for energy extraction and regulation.

The glucose-based energy regeneration pathway involves multiple enzymatic steps that collectively convert glucose to pyruvate while generating ATP. Table 4.7 details the key enzymatic steps and their contributions to overall energy generation.

TABLE 4.2 Core components of PEP-based energy regeneration systems.

Component	Function	Optimal concentration	Critical requirements
Phosphoenolpyruvate (PEP)	Primary energy substrate	20–40 mM	• High purity (>95%) • Stable storage • pH compatibility
Pyruvate kinase (PK)	ATP regeneration catalyst	0.1–0.5 mg/mL	• High specific activity • K^+ dependence • Temperature stability
Myokinase (MK)	Adenylate balance	0.05–0.2 mg/mL	• AMP elimination • Energy charge control • Rapid equilibration
Nucleoside diphosphate kinase (NDK)	GTP regeneration	0.02–0.1 mg/mL	• Broad specificity • Mg^{2+} dependence • Efficient coupling
Acetate kinase (AK)	Alternative ATP source	0.05–0.3 mg/mL	• Acetyl phosphate utilization • Supplementary capacity • System redundancy

TABLE 4.3 Thermodynamic parameters for PEP-based energy reactions.

Reaction	ΔG°′ (kJ/mol)	Keq	Physiological ΔG (kJ/mol)	Driving force
PEP + ADP → Pyruvate + ATP	−31.4	2.5×10^5	−25 to −35	Very strong
ATP + AMP → 2 ADP	0	1.0	−5 to +5	Equilibrium
ATP + GDP → ADP + GTP	0	1.0	−2 to +2	Near equilibrium
Acetyl-P + ADP → Acetate + ATP	−43.1	1.8×10^7	−30 to −40	Very strong
Overall energy charge	Variable	>100	−20 to −30	System dependent

TABLE 4.4 Kinetic parameters for PEP-based energy regeneration enzymes.

Enzyme	Km (substrate)	Vmax (μmol/min/mg)	kcat (s^{-1})	Efficiency (kcat/Km)
Pyruvate kinase	• PEP: 0.2 mM • ADP: 0.3 mM	800–1200	400–600	• PEP: 2000 $M^{-1}s^{-1}$ • ADP: 1500 $M^{-1}s^{-1}$
Myokinase	• ATP: 0.8 mM • AMP: 0.1 mM	1500–2000	800–1000	• ATP: 1000 $M^{-1}s^{-1}$ • AMP: 8000 $M^{-1}s^{-1}$
NDK	• ATP: 0.5 mM • GDP: 0.2 mM	500–800	300–500	• ATP: 600 $M^{-1}s^{-1}$ • GDP: 2000 $M^{-1}s^{-1}$
Acetate kinase	• Acetyl-P: 1.0 mM • ADP: 0.4 mM	300–500	200–350	• Acetyl-P: 200 $M^{-1}s^{-1}$ • ADP: 750 $M^{-1}s^{-1}$

TABLE 4.5 PEP-based energy system optimization parameters.

Parameter	Range	Optimal value	Optimization strategy	Performance impact
PEP concentration	10–50 mM	20–30 mM	• Cost-benefit analysis • Stability assessment • Ionic strength control	• Energy capacity • Reaction duration • Economic efficiency
Pyruvate kinase activity	0.05–1.0 mg/mL	0.2–0.4 mg/mL	• Activity titration • Stability testing • Cofactor optimization	• ATP regeneration rate • System responsiveness • Energy efficiency
K^+ concentration	50–200 mM	100–150 mM	• PK activation • Ionic balance • Protein stability	• Enzyme activation • Translation efficiency • System compatibility
Mg^{2+} concentration	5–20 mM	8–12 mM	• Cofactor balance • Enzyme activity • RNA stability	• Multiple enzyme functions • Nucleotide binding • Overall performance
pH control	7.0–8.5	7.8–8.2	• Buffer capacity • Enzyme stability • System integration	• Enzyme activity • Protein folding • Reaction kinetics

TABLE 4.6 PEP-based energy system performance analysis.

Performance metrics	Typical values	Excellent performance	Limitations	Mitigation strategies
Energy regeneration rate	5–15 μmol ATP/min/mL	>12 μmol ATP/min/mL	• Enzyme saturation • Substrate depletion • Product inhibition	• Enzyme optimization • Substrate excess • Product removal
System duration	4–12 h	>10 h	• Enzyme degradation • Substrate instability • Accumulation effects	• Stabilizer addition • Fresh substrate • Continuous exchange
Energy efficiency	85%–95%	>92%	• Side reactions • Uncoupled hydrolysis • Futile cycling	• Pure enzymes • Optimized ratios • System monitoring
Economic performance	$3–8/g protein	<$4/g protein	• PEP cost • Enzyme costs • System complexity	• Alternative substrates • Enzyme recycling • Process optimization
Scalability	Lab to pilot	Full commercial scale	• Enzyme availability • Quality control • Process complexity	• Enzyme production • Standardization • Automation

TABLE 4.7 Glucose-based energy regeneration pathway components.

Enzyme	Reaction	ATP yield	Km values	Regulatory role
Hexokinase (HK)	Glucose + ATP → G6P + ADP	−1	• Glucose: 0.1 mM • ATP: 0.4 mM	• Entry point control • Glucose phosphorylation • Initial commitment
Phosphoglucose isomerase (PGI)	G6P ↔ F6P	0	• G6P: 0.3 mM • F6P: 0.8 mM	• Pathway flow • Equilibrium control • Branch point
Phosphofructokinase (PFK)	F6P + ATP → F1,6BP + ADP	−1	• F6P: 0.1 mM • ATP: 0.5 mM	• Rate-limiting step • Allosteric control • Energy charge sensing
Aldolase (ALD)	F1,6BP → DHAP + G3P	0	• F1,6BP: 0.02 mM	• Carbon splitting • Flux amplification • Pathway commitment
Glyceraldehyde-3-P dehydrogenase (GAPDH)	G3P + NAD^+ + Pi → 1,3BPG + NADH	0	• G3P: 0.05 mM • NAD^+: 0.2 mM	• Energy coupling • Reducing power • Phosphorylation
Phosphoglycerate kinase (PGK)	1,3BPG + ADP → 3PG + ATP	+2	• 1,3BPG: 0.01 mM • ADP: 0.2 mM	• ATP generation • Energy harvest • Coupling control
Pyruvate kinase (PK)	PEP + ADP → Pyruvate + ATP	+2	• PEP: 0.2 mM • ADP: 0.3 mM	• Final ATP generation • Energy completion • Product formation

4.3.2 Cofactor requirements and regeneration

Glucose-based systems require careful management of multiple cofactors, including NAD^+/NADH, inorganic phosphate, and various metal ions. Table 4.8 outlines the essential cofactors and their regeneration requirements.

4.3.3 System engineering and optimization

Engineering glucose-based energy systems requires balancing multiple competing factors, including enzyme ratios, cofactor availability, and pathway regulation. Table 4.9 provides optimization guidelines for key system parameters.

4.3.4 Advantages and challenges

Glucose-based energy systems offer unique advantages but also present specific challenges that must be addressed through careful system design. Table 4.10 provides a comprehensive analysis of system advantages and associated challenges.

TABLE 4.8 Cofactor requirements for glucose-based energy systems.

Cofactor	Function	Concentration range	Regeneration method	Critical considerations
NAD^+/NADH	Electron carrier	• NAD^+: 2–5 mM • NADH: 0.1–0.5 mM	• NADH oxidation • Enzymatic cycling • Chemical regeneration	• Redox balance • Ratio maintenance • Stability issues
Inorganic phosphate (Pi)	Phosphorylation substrate	10–30 mM	• Buffer component • Substrate addition • Recycling	• pH effects • Precipitation • Availability
Mg^{2+}	Enzyme cofactor	8–15 mM	• Salt addition • Buffer component • System integration	• Enzyme activation • Nucleotide binding • Precipitation
K^+	Enzyme activation	100–200 mM	• Salt addition • Buffer system • Ionic strength	• PK activation • Protein stability • Osmotic effects
Mn^{2+}	Specific enzymes	0.1–0.5 mM	• Salt addition • Trace supplements	• Enzyme specificity • Metal competition • Toxicity

TABLE 4.9 Glucose-based energy system engineering parameters.

Engineering aspect	Optimization target	Implementation strategy	Performance metrics
Enzyme ratios	Balanced flux distribution	• Stoichiometric analysis • Kinetic modeling • Experimental optimization	• Steady-state flux • Minimal accumulation • Efficient conversion
Cofactor balance	Sustained regeneration	• Cofactor titration • Regeneration systems • Ratio optimization	• NAD^+/NADH ratio • Pi availability • Energy charge
pH buffering	Stable reaction environment	• Buffer selection • Capacity optimization • pH monitoring	• pH stability ± 0.2 • Buffer capacity • System compatibility
Temperature control	Optimal enzyme activity	• Temperature optimization • Stability assessment • Kinetic analysis	• Activity retention • Reaction rate • System longevity
Inhibitor management	Minimal product inhibition	• Product removal • Dilution strategies • Alternative pathways	• Reduced inhibition • Sustained activity • Extended duration

TABLE 4.10 Glucose-based energy systems: advantages and challenges.

Aspect	Advantages	Challenges	Mitigation strategies
Economic	• Low substrate cost • Readily available • High energy yield	• Complex enzyme requirements • Cofactor costs • System complexity	• Enzyme optimization • Cofactor recycling • Process simplification
Perfor-mance	• Sustained energy supply • Multiple ATP generation • Natural pathway	• Slower kinetics • Complex regulation • Side reactions	• Enzyme engineering • Pathway optimization • Selective inhibition
Scalability	• Abundant substrate • Established production • Industrial relevance	• Enzyme production • Quality control • Process complexity	• Recombinant enzymes • Standardization • Automation
Technical	• Well-understood biochemistry • Regulatory mechanisms • Metabolic flexibility	• Pathway complexity • Cofactor management • Product inhibition	• Systems engineering • Cofactor regeneration • Product removal

4.4 Alternative energy sources

4.4.1 Creatine phosphate systems

Creatine phosphate represents a premium energy source offering superior performance characteristics but at higher cost. The system provides rapid ATP regeneration through creatine kinase, offering excellent kinetics and minimal side reactions.

Table 4.11 compares creatine phosphate systems with other energy sources, highlighting the performance advantages and economic considerations.

4.4.2 Polyphosphate-based systems

Polyphosphate represents an emerging energy source with unique properties, including pH buffering, metal chelation, and sustained energy release. Table 4.12 details the characteristics and applications of polyphosphate energy systems.

4.4.3 Acetyl phosphate systems

Acetyl phosphate offers an alternative high-energy phosphate source with unique metabolic coupling possibilities. Table 4.13 outlines the properties and applications of acetyl phosphate energy systems.

4.4.4 Hybrid energy systems

Hybrid energy systems combine multiple energy sources to optimize performance, cost, and operational characteristics. Table 4.14 analyzes various hybrid energy system configurations and their advantages.

TABLE 4.11 Creatine phosphate energy system characteristics.

Parameter	Creatine phosphate	PEP system	Glucose system	Comparative analysis
Energy density	32 mM ATP equivalent	30 mM ATP equivalent	38 mM ATP equivalent	• CP: High density, clean • PEP: Moderate, efficient • Glucose: Highest yield
regeneration rate	20–50 μmol/min/mL	10–25 μmol/min/mL	5–15 μmol/min/mL	• CP: Fastest kinetics • PEP: Good performance • Glucose: Slower but sustained
System complexity	Low (1 enzyme)	Low (2–3 enzymes)	High (7+ enzymes)	• CP: Simplest system • PEP: Moderate complexity • Glucose: Most complex
Cost per ATP	$0.15–0.25	$0.05–0.10	$0.02–0.05	• CP: Premium pricing • PEP: Moderate cost • Glucose: Most economical
Byproducts	Creatine (inert)	Pyruvate (compatible)	Multiple metabolites	• CP: Clean, no inhibition • PEP: Minimal interference • Glucose: Complex products

TABLE 4.12 Polyphosphate-based energy system properties.

Property	Characteristics	System benefits	Implementation considerations
Energy capacity	• Variable chain length • 10–100 Pi units • Sustained release	• Long-duration energy • Controlled kinetics • Buffering capacity	• Chain length optimization • Release rate control • Enzyme requirements
pH buffering	• Multiple pKa values • Buffering range 6–8 • Intrinsic capacity	• pH stabilization • Reduced buffer needs • System integration	• Buffer calculation • pH monitoring • Compatibility testing
Metal chelation	• Mg^{2+}, Ca^{2+} binding • Controlled release • Metal homeostasis	• Cofactor regulation • Precipitation prevention • Metal buffering	• Metal balance • Chelation effects • Enzyme activity
Enzymatic requirements	• Polyphosphatase • Nucleoside kinases • Specialized enzymes	• Controlled hydrolysis • ATP regeneration • Pathway coupling	• Enzyme availability • Activity optimization • System integration

TABLE 4.13 Acetyl phosphate energy system analysis.

Aspect	System properties	Performance characteristics	Application considerations
Energetics	• $\Delta G°' = -43.1$ kJ/mol • High-energy phosphate • Favorable thermodynamics	• Rapid ATP regeneration • Strong driving force • Efficient coupling	• Thermodynamic advantage • Kinetic optimization • System integration
Stability	• pH dependent • Hydrolysis susceptible • Temperature sensitive	• Limited shelf life • Fresh preparation • Controlled conditions	• Storage optimization • Stabilizer addition • Usage protocols

Continued

TABLE 4.13 Acetyl phosphate energy system analysis—cont'd

Aspect	System properties	Performance characteristics	Application considerations
Metabolic coupling	• Acetate production • CoA metabolism • Central metabolism	• Pathway integration • Metabolic flexibility • Biosynthetic coupling	• Product utilization • Pathway engineering • System design
Economic factors	• Moderate cost • Synthetic accessibility • Production scalability	• Cost-effective option • Industrial feasibility • Commercial potential	• Cost optimization • Scale-up considerations • Market development

TABLE 4.14 Hybrid energy system configurations.

Hybrid configuration	Component sources	System advantages	Design considerations
PEP + creatine phosphate	• Primary: PEP/PK • Secondary: CP/CK • Balanced ratio	• Cost optimization • Performance enhancement • Risk mitigation	• Ratio optimization • Kinetic matching • Cost balance
Glucose + PEP	• Primary: Glucose pathway • Boost: PEP system • Complementary kinetics	• Extended duration • High energy capacity • Economic efficiency	• Enzyme compatibility • Cofactor management • Complexity control
Multisubstrate cascade	• Sequential utilization • Staged activation • Temporal control	• Sustained energy supply • Optimized kinetics • Flexible operation	• Sequential design • Timing control • System complexity
Compartmentalized systems	• Spatial separation • Specialized zones • Controlled mixing	• Optimized environments • Reduced interference • Enhanced control	• Compartment design • Mass transfer • System integration

4.5 Metabolic pathway integration

4.5.1 Central metabolism coupling

Integration of energy systems with central metabolic pathways provides opportunities for enhanced efficiency and reduced costs through metabolic coupling and pathway engineering. Table 4.15 outlines key metabolic integration strategies and their implementation approaches.

4.5.2 Cofactor regeneration networks

Effective cofactor regeneration is essential for sustained metabolic activity in cell-free systems. Table 4.16 details the major cofactor regeneration networks and their optimization strategies.

4.5.3 Metabolic flux control

Control of metabolic flux is crucial for optimizing energy system performance and preventing metabolic bottlenecks. Table 4.17 outlines key flux control strategies and their implementation methods.

TABLE 4.15 Central metabolism integration strategies.

Integration strategy	Metabolic pathways	Implementation approach	Expected benefits
Glycolysis coupling	• Glucose to pyruvate • ATP generation • NADH production	• Complete pathway reconstruction • Enzyme optimization • Cofactor management	• Sustained energy supply • High ATP yield • Cost reduction
Pentose phosphate pathway	• NADPH generation • Ribose production • Antioxidant systems	• Selective pathway inclusion • Reducing power supply • Biosynthetic support	• Reducing equivalents • Nucleotide synthesis • System protection
TCA cycle elements	• Acetyl-CoA metabolism • NADH generation • Biosynthetic precursors	• Key enzyme addition • Cofactor regeneration • Metabolic flexibility	• Enhanced energy yield • Metabolic versatility • Pathway efficiency
Amino acid metabolism	• Amino acid biosynthesis • Nitrogen metabolism • Energy coupling	• Biosynthetic pathways • Nitrogen assimilation • Energy integration	• Amino acid supply • Nitrogen utilization • Cost reduction

TABLE 4.16 Cofactor regeneration network design.

Cofactor system	Regeneration pathways	Key enzymes	Optimization strategies
NAD^+/NADH	• Enzymatic cycling • Chemical regeneration • Coupled reactions	• Lactate dehydrogenase • Alcohol dehydrogenase • Malate dehydrogenase	• Ratio optimization • Reaction coupling • Stability enhancement
$NADP^+$/NADPH	• Glucose-6-P dehydrogenase • Transhydrogenase • Isocitrate dehydrogenase	• G6PDH • Transhydrogenase • ICDH	• Reducing power balance • Pathway integration • Cofactor cycling
CoA/Acetyl-CoA	• Acetyl-CoA synthesis • CoA recycling • Thioester metabolism	• Acetyl-CoA synthetase • Phosphotransacetylase • Thioesterase	• CoA pool maintenance • Acetyl unit supply • Metabolic coupling
ATP/GTP balance	• Nucleoside kinases • Adenylate kinase • Cross-phosphorylation	• NDK • Adenylate kinase • Guanylate kinase	• Energy charge control • Nucleotide balance • Translation support

4.5.4 Pathway engineering approaches

Systematic pathway engineering enables the development of customized energy systems with optimized performance characteristics. Table 4.18 details various pathway engineering approaches and their applications.

TABLE 4.17 Metabolic flux control strategies.

Control mechanism	Implementation method	Target processes	Control objectives
Enzyme concentration	• Enzyme titration • Activity optimization • Ratio balancing	• Rate-limiting steps • Branch points • Regulatory enzymes	• Flux optimization • Bottleneck elimination • Balanced pathways
Allosteric regulation	• Effector addition • Regulatory design • Feedback control	• Key regulatory enzymes • Pathway control points • Metabolic switches	• Dynamic control • Responsive regulation • System stability
Competitive inhibition	• Inhibitor design • Selective blocking • Pathway redirection	• Side reactions • Competing pathways • Wasteful processes	• Flux redirection • Efficiency improvement • Selectivity enhancement
Compartmentalization	• Spatial organization • Microenvironments • Controlled mixing	• Pathway separation • Specialized zones • Optimized conditions	• Reduced interference • Enhanced efficiency • Improved control

TABLE 4.18 Pathway engineering approaches for energy systems.

Engineering approach	Technical strategy	Implementation tools	Expected outcomes
Rational design	• Known biochemistry • Thermodynamic analysis • Kinetic modeling	• Computational tools • Database mining • Literature analysis	• Predictable performance • Optimized efficiency • Reduced development time
Directed evolution	• Random mutagenesis • Screening/selection • Iterative improvement	• Mutagenesis methods • High-throughput screening • Selection pressure	• Enhanced enzyme properties • Novel activities • Improved performance
Synthetic biology	• Standardized parts • Modular design • Predictable assembly	• BioBrick standards • Genetic circuits • Chassis organisms	• Standardized components • Predictable behavior • Rapid prototyping
Systems engineering	• Network analysis • Multiobjective optimization • Integrated design	• Mathematical modeling • Optimization algorithms • Systems integration	• System-level optimization • Trade-off analysis • Integrated solutions

4.6 Cost-effectiveness analysis

4.6.1 Economic framework and metrics

Cost-effectiveness analysis of energy systems requires comprehensive evaluation of multiple cost components and their impact on overall system economics. Table 4.19 establishes the framework for economic analysis of different energy systems.

TABLE 4.19 Economic analysis framework for energy systems.

Cost component	Analysis parameters	Calculation methods	Impact factors
Raw material costs	• Substrate prices • Cofactor costs • Enzyme expenses	• Market price analysis • Vendor quotations • Bulk pricing models	• Market volatility • Scale economies • Supply availability
Processing costs	• Preparation labor • Equipment amortization • Utility consumption	• Time-motion studies • Equipment costing • Energy analysis	• Process complexity • Automation level • Scale factors
Quality control costs	• Testing expenses • Analytical costs • Validation activities	• Laboratory costing • Analytical methods • Regulatory requirements	• Testing frequency • Method complexity • Regulatory burden
Opportunity costs	• Alternative systems • Development time • Risk factors	• Comparative analysis • Time value calculations • Risk assessment	• Technology alternatives • Development risks • Market timing

TABLE 4.20 Comparative cost analysis of energy systems.

Energy system	Substrate cost ($/mol ATP)	Enzyme cost ($/mol ATP)	Total cost ($/mol ATP)	Cost drivers	Optimization potential
PEP-based	$0.025	$0.015	$0.040	• PEP substrate cost • Pyruvate kinase • Supporting enzymes	• PEP production optimization • Enzyme engineering • System integration
Glucose-based	$0.008	$0.025	$0.033	• Multiple enzymes • Cofactor costs • System complexity	• Enzyme production • Cofactor recycling • Process simplification
Creatine phosphate	$0.080	$0.008	$0.088	• Premium substrate • Specialized production • Limited suppliers	• Alternative synthesis • Scale economies • Supply diversification
Acetyl phosphate	$0.035	$0.012	$0.047	• Synthesis costs • Stability issues • Specialized handling	• Synthesis optimization • Stabilization • Process development
Hybrid systems	$0.020–0.050	$0.015–0.030	$0.035–0.080	• Component balance • System complexity • Integration costs	• Ratio optimization • Component selection • Process integration

4.6.2 Comparative cost analysis

Direct comparison of energy system costs provides crucial decision-making information for system selection and optimization. Table 4.20 presents a comprehensive cost comparison of major energy systems.

4.6.3 Scale-dependent cost models

Energy system costs vary significantly with production scale, requiring detailed analysis of scale-dependent factors. Table 4.21 analyzes cost scaling relationships for different energy systems.

4.6.4 Economic optimization strategies

Systematic approaches to economic optimization can significantly improve the cost-effectiveness of energy systems. Table 4.22 outlines key optimization strategies and their implementation approaches.

TABLE 4.21 Scale-dependent cost models for energy systems.

Scale category	Production volume	Cost scaling factors	Critical cost elements	Economies of scale
Laboratory scale	1–100 mL batches	• Fixed cost dominance • High unit costs • Manual processing	• Labor intensity • Small-scale reagents • Setup costs	Limited economies
Pilot scale	1–10 L batches	• Semi-automated • Bulk reagent discounts • Process optimization	• Equipment costs • Process development • Quality systems	Moderate economies
Manufacturing scale	100–1000 L batches	• Full automation • Bulk supply contracts • Optimized processes	• Capital equipment • Regulatory compliance • Quality assurance	Significant economies
Commercial scale	>1000 L continuous	• Maximum efficiency • Lowest unit costs • Integrated operations	• Infrastructure • Market development • Technology licensing	Maximum economies

TABLE 4.22 Economic optimization strategies for energy systems.

Optimization strategy	Implementation approach	Target areas	Expected benefits
Substrate optimization	• Alternative sources • Production optimization • Supply chain efficiency	• Raw material costs • Supply reliability • Quality consistency	• 20%–40% cost reduction • Improved supply security • Quality enhancement
Enzyme engineering	• Activity enhancement • Stability improvement • Production optimization	• Enzyme costs • Performance enhancement • Operational efficiency	• 30%–50% cost reduction • Performance improvement • Extended system life
Process integration	• System consolidation • Shared infrastructure • Operational efficiency	• Processing costs • Equipment utilization • Labor efficiency	• 15%–25% cost reduction • Improved productivity • Simplified operations
Scale optimization	• Production scaling • Capacity utilization • Market development	• Fixed cost amortization • Unit cost reduction • Market positioning	• 40%–60% cost reduction • Market competitiveness • Growth enablement

4.7 Continuous energy supply strategies

4.7.1 Continuous-exchange systems

Continuous-exchange cell-free systems enable sustained energy supply through continuous substrate addition and product removal. Table 4.23 details the design principles and operational characteristics of continuous energy supply systems.

4.7.2 Fed-batch strategies

Fed-batch operation provides a compromise between batch simplicity and continuous operation complexity while enabling extended energy supply. Table 4.24 analyzes different fed-batch strategies and their applications.

TABLE 4.23 Continuous energy supply system design.

System component	Design principles	Implementation methods	Performance characteristics
Substrate supply	• Continuous addition • Concentration control • Flow rate optimization	• Peristaltic pumps • Concentration monitoring • Feedback control	• Steady-state concentrations • Minimized fluctuations • Extended operation
Product removal	• Selective extraction • Membrane separation • Continuous processing	• Dialysis membranes • Ultrafiltration • Continuous reactors	• Reduced inhibition • Maintained activity • Extended duration
Energy monitoring	• Real-time analysis • Energy charge tracking • Performance optimization	• ATP/ADP sensors • Energy charge meters • Automated control	• Consistent energy levels • Optimized performance • Predictable operation
System control	• Automated operation • Feedback regulation • Optimization algorithms	• Process controllers • Sensor integration • Control algorithms	• Stable operation • Optimized efficiency • Reduced variability

TABLE 4.24 Fed-batch energy supply strategies.

Fed-batch strategy	Feeding protocol	Advantages	Implementation considerations
Constant rate feeding	• Fixed flow rate • Predetermined schedule • Simple control	• Easy implementation • Predictable operation • Minimal complexity	• Suboptimal efficiency • Limited flexibility • Fixed parameters
Exponential feeding	• Growth-matched rate • Exponential profile • Dynamic adjustment	• Optimized efficiency • Maintained activity • Enhanced performance	• Complex control • Model requirements • Parameter estimation
Feedback-controlled feeding	• Sensor-based control • Real-time adjustment • Optimal conditions	• Maximum efficiency • Adaptive operation • Optimal performance	• Sensor requirements • Control complexity • System integration
Multistage feeding	• Sequential additions • Stage-specific optimization • Temporal control	• Optimized stages • Flexible operation • Enhanced control	• Complex scheduling • Multiple preparations • Coordination requirements

4.7.3 Energy storage and buffering

Energy storage and buffering systems provide temporal decoupling between energy generation and consumption, enabling more stable and efficient operation. Table 4.25 outlines various energy storage strategies and their characteristics.

4.7.4 System integration and control

Effective integration and control of continuous energy supply systems requires sophisticated monitoring and control strategies. Table 4.26 details key integration and control approaches.

Fig. 4.2 illustrates the comprehensive energy system architecture showing the integration of different energy sources, regeneration pathways, and control systems.

TABLE 4.25 Energy storage and buffering strategies.

Storage strategy	Storage mechanism	Capacity characteristics	Release kinetics
High-energy phosphates	• Chemical bond energy • Phosphorylated compounds • Direct storage	• High energy density • Rapid release • Limited capacity	• Enzymatic release • Fast kinetics • Controllable rate
Polyphosphate chains	• Multiple phosphate bonds • Variable chain length • Sustained release	• Variable capacity • Extended duration • Controlled release	• Sequential hydrolysis • Moderate kinetics • Sustained supply
Metabolic intermediates	• Glycolytic intermediates • Energy-rich metabolites • Pathway coupling	• Moderate density • Pathway integration • Flexible utilization	• Enzymatic conversion • Variable kinetics • Pathway dependent
Electrochemical storage	• Redox couples • Electron carriers • Chemical batteries	• High efficiency • Reversible storage • Clean conversion	• Electrode kinetics • Moderate rate • System integration

TABLE 4.26 Energy system integration and control strategies.

Control aspect	Monitoring parameters	Control methods	Integration requirements
Energy balance	• ATP/ADP ratios • Energy charge • Consumption rates	• Feedback control • Predictive algorithms • Adaptive systems	• Real-time sensors • Control systems • Process integration
Substrate management	• Concentration levels • Consumption rates • Supply status	• Flow control • Concentration regulation • Inventory management	• Supply systems • Monitoring networks • Control integration
Product management	• Inhibitor levels • Product concentrations • Removal efficiency	• Separation control • Removal optimization • Product handling	• Separation systems • Product monitoring • Process integration
System optimization	• Overall efficiency • Performance metrics • Economic indicators	• Multiobjective optimization • Advanced control • Machine learning	• Data integration • Control algorithms • Decision support

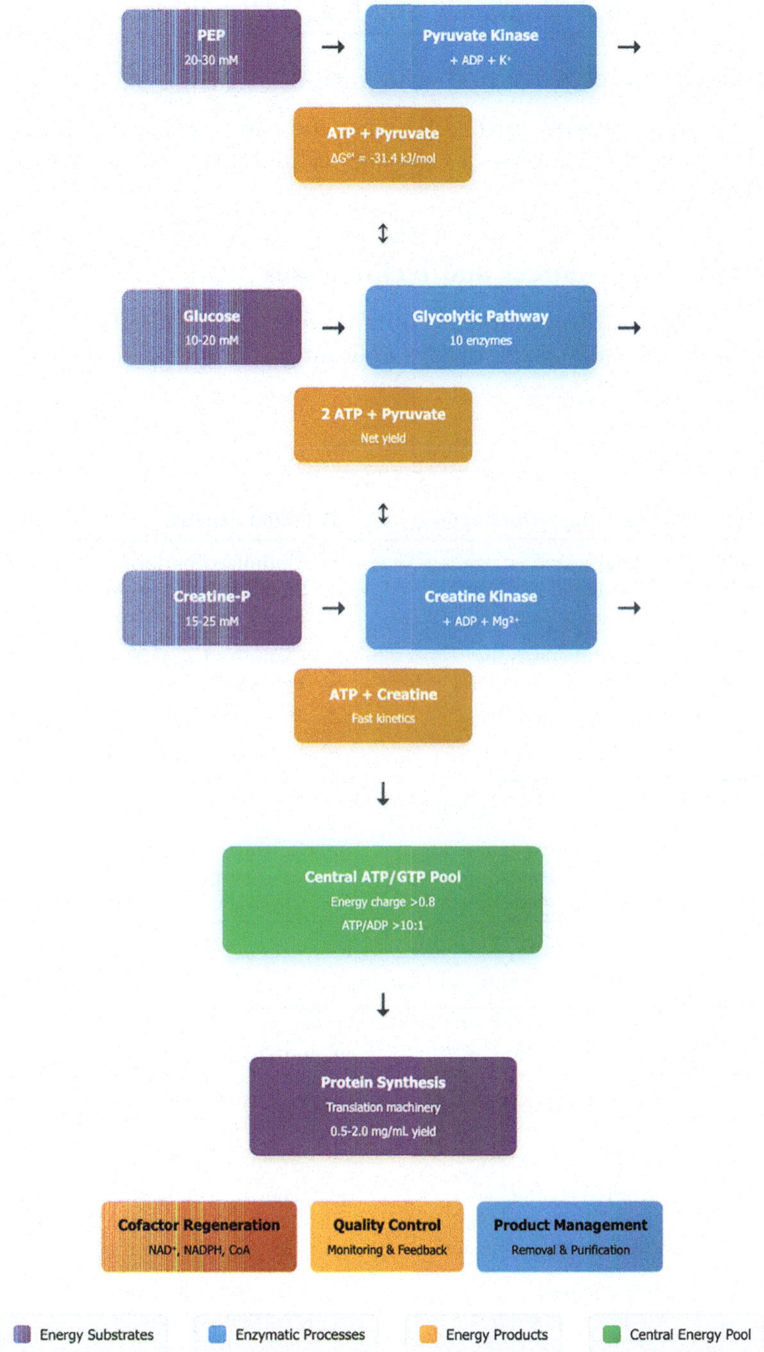

FIG. 4.2 Energy flow and conversion pathways in CFPS. The diagram shows the major energy sources (purple), energy products (orange), and the central ATP/GTP pool (green) that drives protein synthesis. Supporting processes for cofactor regeneration, quality control, and product management are shown at the bottom.

4.8 Summary and future directions

4.8.1 Technology convergence and integration

The future of energy systems for CFPS lies in the convergence of multiple technologies and approaches. Table 4.27 outlines emerging technology convergence trends and their potential impacts.

4.8.2 Emerging energy sources and technologies

New energy sources and technologies continue to emerge, offering potential advantages for specific applications. Table 4.28 surveys emerging energy technologies and their development status.

TABLE 4.27 Energy system technology convergence trends.

Convergence area	Integrating technologies	Potential benefits	Development timeline
Synthetic biology + energy systems	• Engineered enzymes • Designed pathways • Modular systems	• Optimized performance • Standardized components • Predictable behavior	2–5 years
AI/ML + process control	• Machine learning • Predictive modeling • Adaptive control	• Optimized operation • Predictive maintenance • Self-optimization	3–7 years
Microfluidics + energy systems	• Miniaturization • Precise control • High-throughput testing	• Reduced costs • Rapid optimization • Parallel processing	2–4 years
Biotechnology + chemical engineering	• Bioprocess engineering • Reactor design • Process intensification	• Enhanced efficiency • Scalable processes • Commercial viability	5–10 years

TABLE 4.28 Emerging energy sources and technologies.

Technology	Technical approach	Potential advantages	Development status
Enzymatic fuel cells	• Biocatalytic electrodes • Direct electron transfer • Continuous operation	• High efficiency • Clean operation • Sustainable energy	Early research stage
Photosynthetic systems	• Light harvesting • Artificial photosynthesis • Solar energy conversion	• Renewable energy • Sustainable operation • Low operating costs	Proof-of-concept
Thermochemical systems	• Heat engines • Temperature gradients • Thermal conversion	• Waste heat utilization • Process integration • Energy recovery	Research phase
Electrochemical regeneration	• Electrolytic systems • Redox cycling • Electrochemical control	• Precise control • Reversible operation • Clean technology	Development phase

Bibliography

Calhoun, K.A., Swartz, J.R., 2005. Energizing cell-free protein synthesis with glucose metabolism. Biotechnol. Bioeng. 90 (5), 606–613.

Caschera, F., Noireaux, V., 2014. Synthesis of 2.3 mg/ml of protein with an all *Escherichia coli* cell-free transcription–translation system. Biochimie 99, 162–168.

Cole, S.D., Beabout, K., Turner, K.B., Smith, Z.K., Funk, V.L., Harbaugh, S.V., … Lux, M.W., 2019. Quantification of interlaboratory cell-free protein synthesis variability. ACS Synth. Biol. 8 (9), 2080–2091.

Des Soye, B.J., Gerbasi, V.R., Thomas, P.M., Kelleher, N.L., Jewett, M.C., 2019. A highly productive, one-pot cell-free protein synthesis platform based on genomically recoded *Escherichia coli*. Cell Chem. Biol. 26 (12), 1743–1754.

Dudley, Q.M., Karim, A.S., Jewett, M.C., 2015. Cell-free metabolic engineering: biomanufacturing beyond the cell. Biotechnol. J. 10 (1), 69–82.

Garamella, J., Marshall, R., Rustad, M., Noireaux, V., 2016. The all E. coli TX-TL toolbox 2.0: a platform for cell-free synthetic biology. ACS Synth. Biol. 5 (4), 344–355.

Hunt, J.P., Yang, S.O., Wilding, K.M., Bundy, B.C., 2017. The growing impact of lyophilized cell-free protein synthesis. Bioengineering 4 (2), 39.

Jewett, M.C., Calhoun, K.A., Voloshin, A., Wuu, J.J., Swartz, J.R., 2008. An integrated cell-free metabolic platform for protein production and synthetic biology. Mol. Syst. Biol. 4 (1), 220.

Jewett, M.C., Swartz, J.R., 2004. Mimicking the *Escherichia coli* cytoplasmic environment activates long-lived and efficient cell-free protein synthesis. Biotechnol. Bioeng. 86 (1), 19–26.

Kim, D.M., Swartz, J.R., 2001. Regeneration of adenosine triphosphate from glycolytic intermediates for cell-free protein synthesis. Biotechnol. Bioeng. 74 (4), 309–316.

Kim, D.M., Swartz, J.R., 2004. Efficient production of a bioactive, multiple disulfide-bonded protein using modified extracts of Escherichia coli. Biotechnol. Bioeng. 85 (2), 122–129.

Kwon, Y.C., Jewett, M.C., 2015. High-throughput preparation methods of crude extract for robust cell-free protein synthesis. Sci. Rep. 5 (1), 8663.

Moore, S.J., MacDonald, J.T., Wienecke, S., Ishwarbhati, A., Tsipa, A., Aw, R., … Freemont, P.S., 2018. Rapid acquisition and model-based analysis of cell-free transcription–translation reactions from nonmodel bacteria. Proc. Natl. Acad. Sci. U. S. A. 115 (19), E4340–E4349.

Swartz, J.R., 2012. Transforming biochemical engineering with cell-free biology. AIChE J. 58 (1), 5–13.

Vilkhovoy, M., Dai, D., Mehta, S., Adhikari, A., Varner, J.D., Linhardt, R.J., 2018. Sequence specific modeling of *E. coli* cell-free protein synthesis. ACS Synth. Biol. 7 (8), 1844–1857.

CHAPTER

5

Reaction formats and scale-up technologies

5.1 Introduction

The transition from laboratory-scale cell-free protein synthesis to commercial production requires sophisticated reactor design and scale-up strategies that maintain high productivity while ensuring economic viability. Reaction format selection critically impacts protein yield, product quality, operational complexity, and manufacturing costs. This chapter provides comprehensive coverage of reactor technologies, scale-up methodologies, and process intensification strategies for cell-free protein synthesis systems.

The selection of appropriate reaction formats depends on multiple interconnected factors, including production scale, product characteristics, economic constraints, and operational requirements. Table 5.1 outlines the fundamental considerations that guide reactor selection and design decisions.

Fig. 5.1 illustrates the reactor technology landscape showing the relationship between different reaction formats, their complexity levels, and typical applications across various production scales.

5.2 Batch reaction systems

5.2.1 Standard batch reactor design

Standard batch reactors represent the foundation of cell-free protein synthesis, offering simplicity, flexibility, and well-understood operation principles. These systems provide complete mixing, uniform conditions, and straightforward scale-up characteristics that make them suitable for a wide range of applications (Fig. 5.2).

The design of effective batch reactors requires careful attention to multiple engineering parameters that affect both mass transfer and reaction kinetics. Table 5.2 details the critical design parameters and their optimization targets for batch reactor systems.

Cell Free Synthesis of Biologics
https://doi.org/10.1016/B978-0-443-33271-5.00017-X

TABLE 5.1 Fundamental considerations for reaction format selection.

Selection criteria	Key factors	Design implications	Performance impact
Production scale	• Batch size requirements • Throughput demands • Market volume projections	• Reactor volume design • Equipment selection • Infrastructure planning	• Capital investment • Operating efficiency • Unit cost economics
Product characteristics	• Protein stability • Aggregation tendency • Purification requirements	• Residence time control • Mixing optimization • Product handling	• Product quality • Recovery efficiency • Downstream complexity
Economic constraints	• Capital availability • Operating cost targets • Market price pressure	• Technology selection • Automation level • Process simplification	• Investment payback • Competitive positioning • Profitability margins
Operational requirements	• Process flexibility • Campaign length • Product changeover	• Modular design • Cleaning protocols • Validation requirements	• Operational efficiency • Product mix capability • Regulatory compliance

TABLE 5.2 Batch reactor design parameters and optimization targets.

Design parameter	Typical range	Optimal target	Engineering considerations	Performance impact
Aspect ratio (H/D)	0.5–3.0	1.0–2.0	• Mixing efficiency • Heat transfer • Foam control	• Mass transfer rates • Temperature uniformity • Surface area effects
Working volume	50%–80% of total	65%–75%	• Expansion room • Mixing clearance • Safety margins	• Volumetric productivity • Process safety • Equipment utilization
Agitation rate	50–300 rpm	100–200 rpm	• Shear sensitivity • Oxygen transfer • Power consumption	• Mixing quality • Protein integrity • Energy costs
Temperature control	±0.5°C	±0.2°C	• Heat transfer area • Control system design • Sensor placement	• Enzyme activity • Protein stability • Process consistency
pH control	±0.2 units	±0.1 units	• Buffer capacity • Acid/base addition • Sensor accuracy	• Enzyme function • Protein folding • System stability

5.2.2 Mixing and mass transfer optimization

Effective mixing is crucial for batch reactor performance, ensuring uniform distribution of substrates, products, and temperature while minimizing protein damage from excessive shear. Table 5.3 analyzes different mixing strategies and their characteristics.

5.2.3 Temperature and environmental control

Precise environmental control is essential for maintaining optimal reaction conditions throughout the batch cycle. Table 5.4 outlines the key environmental control systems and their performance characteristics.

5.2.4 Performance characteristics and limitations

Batch reactors offer specific advantages but also face inherent limitations that must be considered in system design. Table 5.5 provides a comprehensive analysis of batch reactor performance characteristics.

FIG. 5.1 The comprehensive reactor technology landscape showing the relationships between different reaction formats, their complexity, and scale applications.

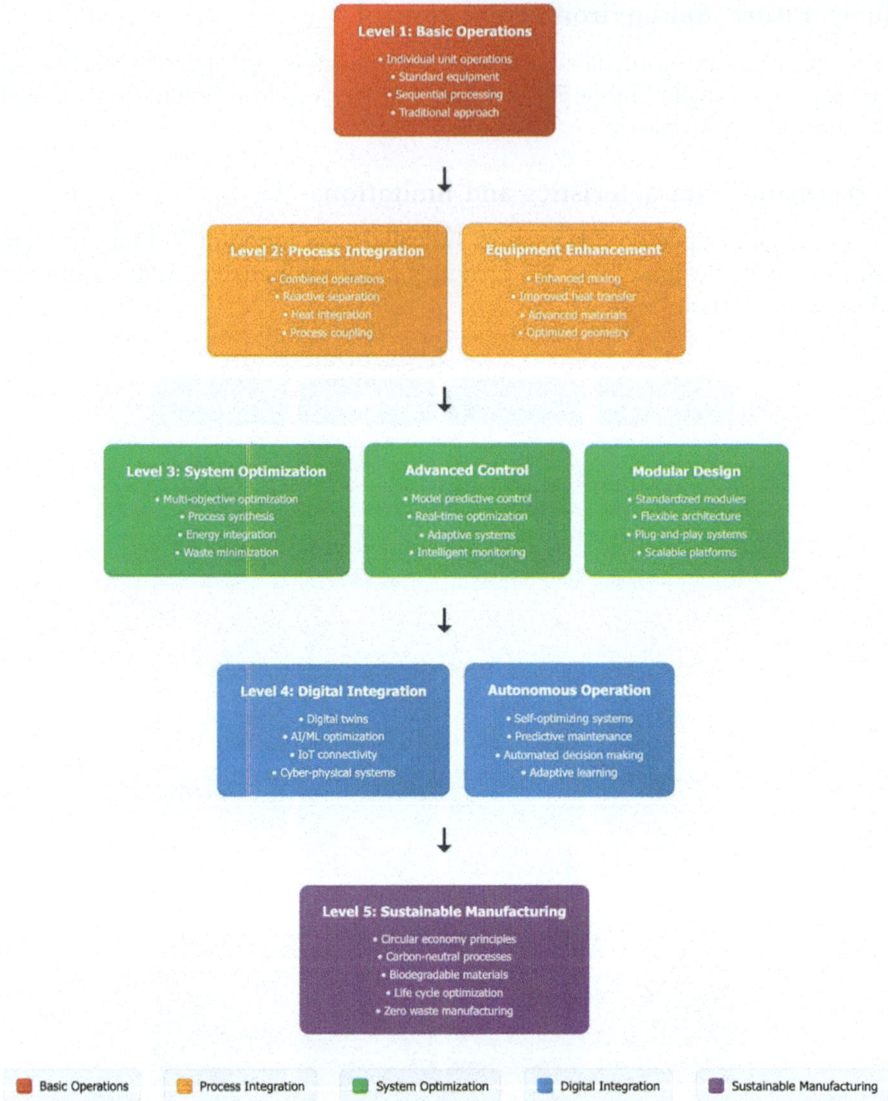

FIG. 5.2 The process intensification hierarchy showing how different technologies can be integrated to create more efficient and sustainable systems.

5.3 Continuous exchange cell-free systems

5.3.1 Continuous exchange cell-free system architecture and design

Continuous exchange cell-free (CECF) systems represent a significant advancement in reaction technology, enabling sustained protein synthesis through continuous substrate supply and product removal. These systems overcome the temporal limitations of batch reactions while maintaining the benefits of cell-free synthesis.

The design of CECF systems requires integration of multiple subsystems, including reaction chambers, separation membranes, fluid handling, and control systems. Table 5.6 details the core components and their design specifications.

TABLE 5.3 Mixing strategies for batch cell-free reactors.

Mixing strategy	Equipment type	Advantages	Limitations	Optimal applications
Mechanical stirring	• Pitched blade turbines • Marine propellers • Rushton turbines	• Predictable performance • Scalable design • Good heat transfer	• Shear damage potential • Mechanical complexity • Maintenance requirements	• Large-scale production • Standard proteins • Established processes
Magnetic stirring	• Magnetic stir bars • External drive systems • Orbital shakers	• Gentle mixing • Contamination prevention • Simple operation	• Scale limitations • Limited power • Positioning constraints	• Laboratory scale • Sensitive proteins • Research applications
Pneumatic mixing	• Air sparging • Gas recirculation • Bubble columns	• Gentle operation • No mechanical parts • Self-cleaning	• Limited control • Foam generation • Gas requirements	• Foam-tolerant systems • Simple operations • Maintenance-free
Wave mixing	• Rocking platforms • Wave bags • Orbital motion	• Very gentle • Disposable systems • Contamination-free	• Scale limitations • Equipment costs • Limited applicability	• Small–medium scale • High-value proteins • Pilot studies

TABLE 5.4 Environmental control systems for batch reactors.

Control system	Technology options	Performance specifications	Critical considerations
Temperature control	• Jacket cooling/heating • Internal coils • External heat exchangers	• Response time: < 5 min • Accuracy: ± 0.1°C • Uniformity: < 0.5°C variation	• Heat transfer coefficient • Control algorithm tuning • Sensor placement strategy
pH control	• Automatic titration • Buffer systems • Real-time monitoring	• Response time: <2 min • Accuracy: ±0.05 pH • Stability: <0.1 pH drift	• Buffer capacity design • Acid/base selection • Control loop optimization
Dissolved oxygen	• Sparging systems • Membrane contactors • Oxygen-enriched air	• Control range: 0%–100% • Response time: <3 min • Accuracy: ±2%	• Mass transfer design • Foam control strategies • Gas flow optimization
Pressure control	• Relief valves • Back-pressure regulators • Vacuum systems	• Accuracy: ±0.1 bar • Safety margin: 2x working • Response: < 1 min	• Safety system design • Process requirements • Equipment compatibility

TABLE 5.5 Batch reactor performance analysis.

Performance aspect	Characteristics	Advantages	Limitations	Mitigation strategies
Productivity	• 0.5–2.0 mg/mL protein • 4–12 h cycles • 70%–85% equipment utilization	• High concentrations • Flexible operation • Simple scheduling	• Batch-to-batch variation • Downtime between batches • Equipment limitations	• Process standardization • Automated operations • Parallel reactors
Product quality	• Consistent within batch • Variable between batches • Standard specifications	• Uniform conditions • Complete control • Quality assurance	• Batch variations • End-of-batch effects • Storage degradation	• Process control • Real-time monitoring • Optimized timing
Economic performance	• Moderate capital costs • Flexible capacity • Standard equipment	• Proven technology • Lower risk • Established supply chain	• Labor intensive • Equipment idle time • Batch overhead costs	• Automation systems • Multiproduct capability • Process intensification
Scalability	• Linear scale-up • Proven methodologies • Predictable performance	• Established principles • Risk mitigation • Regulatory precedent	• Geometric constraints • Mixing limitations • Heat transfer challenges	• Advanced modeling • Pilot-scale validation • Modular approaches

5.3.2 Membrane technology and selection

Membrane selection is critical for CECF system performance, determining separation efficiency, system longevity, and operational costs. Table 5.7 compares different membrane technologies and their applications.

5.3.3 Process control and optimization

CECF systems require sophisticated process control to maintain optimal conditions and maximize productivity. Table 5.8 outlines the key process variables and their control strategies.

5.3.4 Performance advantages and challenges

CECF systems offer significant performance advantages over batch systems but also present unique challenges. Table 5.9 provides a comprehensive comparison of CECF system characteristics.

TABLE 5.6 CECF system component design specifications.

System component	Design elements	Specifications	Performance requirements
Reaction chamber	• Volume: 10–1000 mL • Geometry: cylindrical/ rectangular • Materials: glass, stainless steel	• Working volume efficiency > 80% • Mixing time < 30 s • Temperature uniformity ± 0.2°C	• Uniform mixing • Minimal dead zones • Easy cleaning/ sterilization
Membrane separator	• Molecular weight cutoff: 10–100 kDa • Surface area: 0.1–10 m^2 • Configuration: flat sheet, hollow fiber	• Protein retention > 99% • Small-molecule passage > 90% • Flux rate > 10 $L/m^2/h$	• Selective permeability • Minimal fouling • Chemical compatibility
Feed system	• Pumps: peristaltic, diaphragm • Flow rates: 0.1–50 mL/min • Control: flow, pressure, composition	• Flow accuracy ± 2% • Pressure stability ± 0.1 bar • Composition control ± 1%	• Precise delivery • Contamination prevention • System integration
Control system	• Sensors: pH, temperature, flow • Controllers: Proportional-Integral-Derivative (PID), advanced control • Data logging: real-time monitoring	• Response time < 1 min • Accuracy: sensor dependent • Data storage: 1 Hz minimum	• Process stability • Trend analysis • Regulatory compliance

TABLE 5.7 Membrane technology comparison for CECF systems.

Membrane type	Material composition	Molecular weight cutoff	Advantages	Limitations
Regenerated cellulose	• Cellulose polymer • Hydrophilic surface • Low protein binding	3–100 kDa	• Low fouling • Good biocompatibility • Stable performance	• pH limitations (2–12) • Temperature sensitivity • Chemical resistance
Polyethersulfone (PES)	• Synthetic polymer • Asymmetric structure • High flux rates	1–1000 kDa	• Chemical resistance • High temperature stability • Mechanical strength	• Higher protein binding • Hydrophobic surface • Potential fouling
Polysulfone (PS)	• Aromatic polymer • Tight pore structure • Excellent selectivity	1-300 kDa	• Sharp molecular cutoff • Chemical stability • Industrial proven	• Protein adsorption • Hydrophobic character • Cleaning requirements

Continued

TABLE 5.7 Membrane technology comparison for CECF systems—cont'd

Membrane type	Material composition	Molecular weight cutoff	Advantages	Limitations
Ceramic membranes	• Inorganic materials • Rigid structure • Extreme stability	1–200 kDa	• Chemical inertness • High-temperature resistance • Long lifetime	• High cost • Brittle nature • Limited flexibility
Composite membranes	• Multilayer structure • Tailored properties • Optimized performance	0.1–500 kDa	• Customized characteristics • Enhanced performance • Application-specific	• Complex manufacturing • Higher costs • Limited availability

TABLE 5.8 CECF process control variables and strategies.

Process variable	Control range	Control strategy	Monitoring method	Optimization targets
Residence time	2–24 h	• Flow rate control • Volume adjustment • Cascade control	• Flow sensors • Volume measurement • Tracer studies	• Maximum productivity • Optimal conversion • Quality maintenance
Substrate concentration	10–100 mM	• Feed composition control • Flow ratio regulation • Feedback adjustment	• Online analyzers • Sampling analysis • Mass balance	• Saturation kinetics • Cost optimization • Waste minimization
Product removal rate	0.1–10 mL/min	• Permeate flow control • Pressure management • Membrane performance	• Flow measurement • Pressure monitoring • Concentration analysis	• Inhibition prevention • Product recovery • System stability
Energy supply	Continuous	• ATP monitoring • Energy regeneration • Cofactor management	• Energy charge sensors • ATP/ADP ratios • Enzymatic assays	• Sustained activity • Energy efficiency • Cost control
Temperature profile	±0.1°C	• Feedback control • Cascade systems • Predictive control	• Multiple sensors • Thermal imaging • Heat balance	• Enzyme activity • Protein stability • Energy efficiency

TABLE 5.9 CECF system performance analysis.

Performance metric	CECF performance	Batch comparison	Advantages	Challenges
Productivity	2–10 mg/mL/day	0.5–2.0 mg/mL/batch	• Continuous operation • Extended duration • Higher space-time yield	• System complexity • Membrane fouling • Control requirements
Product quality	Consistent over time	Variable between batches	• Steady-state conditions • Continuous monitoring • Real-time adjustment	• Membrane selectivity • System aging • Contamination risk

Performance metric	CECF performance	Batch comparison	Advantages	Challenges
Resource utilization	85%–95% efficiency	70%–80% efficiency	• Continuous substrate use • Minimal waste • Optimized kinetics	• Membrane replacement • System maintenance • Energy consumption
Operational flexibility	Moderate	High	• Steady production • Predictable output • Automated operation	• Limited product changes • System reconfiguration • Startup complexity
Economic impact	Higher CAPEX, lower OPEX	Lower CAPEX, higher OPEX	• Reduced labor • Higher productivity • Better economics at scale	• Initial investment • Technical expertise • Maintenance costs

5.4 Flow-through reactors

5.4.1 Reactor design and configuration

Flow-through reactors provide an alternative continuous operation mode where the cell-free reaction mixture flows through a reactor system without recirculation. These systems offer unique advantages for specific applications and can be optimized for different residence times and mixing patterns.

The design of flow-through reactors requires careful consideration of residence time distribution (RTD), mixing patterns, and mass transfer characteristics. Table 5.10 details the key design configurations and their applications.

5.4.2 RTD and mixing

Understanding and controlling RTD is crucial for flow-through reactor performance. Table 5.11 analyzes different mixing patterns and their impact on reactor performance.

5.4.3 Heat and mass transfer optimization

Effective heat and mass transfer are essential for maintaining optimal reaction conditions in flow-through reactors. Table 5.12 outlines the key transport phenomena and optimization strategies.

5.4.4 Application-specific designs

Flow-through reactors can be optimized for specific applications and product requirements. Table 5.13 details different application-specific reactor designs and their characteristics.

TABLE 5.10 Flow-through reactor design configurations.

Reactor configuration	Design characteristics	Typical applications	Advantages	Design considerations
Tubular reactors	• Length: 1–100 m • Diameter: 1–50 mm • L/D ratio: 50-2000	• Continuous production • Simple proteins • High-volume applications	• Plug flow behavior • Predictable residence time • Easy scale-up	• Pressure drop • Mixing limitations • Temperature control
Packed bed reactors	• Packing: beads, supports • Void fraction: 0.3–0.7 • Flow: upflow, downflow	• Immobilized systems • Enzyme reactors • Continuous processing	• High surface area • Enzyme immobilization • Efficient contact	• Pressure drop • Channeling • Fouling issues
Microreactors	• Channel width: 10–1000 μm • Length: 1–100 cm • Surface/volume: 10,000–100,000 m^{-1}	• Small-scale production • Process development • Parallel processing	• Excellent mixing • Precise control • Minimal reagent use	• Scale limitations • Fabrication costs • Clogging risks
Spiral reactors	• Spiral geometry • Dean flow mixing • Compact design	• Enhanced mixing • Moderate scale • Space-efficient	• Secondary flow mixing • Compact footprint • Good heat transfer	• Complex geometry • Manufacturing challenges • Cleaning difficulties

TABLE 5.11 Residence time distribution patterns in flow-through reactors.

Flow pattern	RTD characteristics	Mixing quality	Performance impact	Control strategies
Plug flow	• Narrow RTD • Minimal dispersion • Uniform residence time	• Minimal axial mixing • Radial mixing only • Predictable profiles	• Optimal conversion • Predictable quality • Efficient utilization	• Proper L/D ratio • Flow conditioning • Velocity uniformity
Laminar flow	• Parabolic velocity profile • Wide RTD • Variable residence times	• Poor radial mixing • Axial dispersion • Nonuniform conditions	• Reduced efficiency • Quality variations • Incomplete conversion	• Static mixers • Flow enhancers • Geometry optimization
Turbulent flow	• Flat velocity profile • Good mixing • Moderate RTD spread	• Excellent radial mixing • Some axial dispersion • Uniform conditions	• Good performance • Uniform quality • Predictable behavior	• Reynolds number control • Turbulence promoters • Flow rate optimization
Mixed flow	• Exponential RTD • Complete mixing • Wide distribution	• Perfect mixing • Uniform composition • No gradients	• Reduced driving force • Lower conversion • Consistent quality	• Mixing enhancement • Recycle streams • Staging strategies

TABLE 5.12 Heat and mass transfer optimization in flow-through reactors.

Transport phenomenon	Governing parameters	Enhancement methods	Performance metrics	Design targets
Heat transfer	• Overall heat transfer coefficient • Temperature difference • Heat transfer area	• Extended surfaces • Enhanced mixing • Flow optimization	• Nusselt number > 50 • Temperature uniformity ± 0.5°C • Response time < 5 min	• Optimal enzyme activity • Protein stability • Energy efficiency
Mass transfer	• Mass transfer coefficient • Concentration gradients • Interfacial area	• Mixing enhancement • Surface area increase • Flow pattern optimization	• Sherwood number > 20 • Concentration uniformity > 95% • Mass transfer rate > 1 $g/m^2/s$	• Uniform substrate supply • Product removal • Reaction efficiency
Momentum transfer	• Pressure drop • Velocity distribution • Flow resistance	• Streamlined design • Surface smoothness • Geometry optimization	• Friction factor < 0.02 • Pressure drop < 1 bar/m • Pumping power < 100 W/m^3	• Energy efficiency • Uniform flow • Operational stability

TABLE 5.13 Application-specific flow-through reactor designs.

Application	Reactor design	Specific features	Performance advantages	Implementation considerations
High-volume production	• Large-diameter tubes • Multiple parallel units • Continuous operation	• Economy of scale • Standardized design • Automated control	• Low unit costs • High reliability • Consistent quality	• Capital investment • Market demand • Regulatory approval
Sensitive proteins	• Gentle flow conditions • Temperature control • Minimal shear design	• Low Reynolds numbers • Enhanced heat transfer • Protective environments	• Maintained activity • Reduced degradation • High-quality products	• Complex control • Higher costs • Specialized equipment
Rapid prototyping	• Microreactor systems • Modular design • Flexible configuration	• Small volumes • Quick changeover • Parallel testing	• Fast development • Minimal material use • Rapid optimization	• Scale-up challenges • Limited throughput • Fabrication complexity
Multiproduct systems	• Flexible geometry • Easy cleaning • Quick reconfiguration	• Changeover protocols • Cleaning validation • Cross-contamination prevention	• Product flexibility • Efficient utilization • Market responsiveness	• Validation requirements • Cleaning complexity • Product scheduling

5.5 Membrane-based systems

5.5.1 Membrane reactor configurations

Membrane-based systems integrate reaction and separation processes, enabling selective retention of synthesis machinery while allowing continuous substrate supply and product removal. These systems offer unique advantages for cell-free protein synthesis applications.

Different membrane reactor configurations provide varying levels of integration and performance characteristics. Table 5.14 compares the major membrane reactor configurations and their applications.

5.5.2 Membrane performance and optimization

Membrane performance is critical for the success of membrane-based reactor systems. Table 5.15 details the key performance parameters and optimization strategies for membrane operations.

TABLE 5.14 Membrane reactor configuration comparison.

Configuration	System architecture	Operating principles	Advantages	Optimal applications
Stirred tank with UF	• Reactor with membrane unit • External circulation • Separate components	• Batch reaction • Continuous separation • Independent optimization	• Flexible operation • Easy maintenance • Proven technology	• Research applications • Pilot studies • Multiproduct facilities
Immersed membrane	• Membrane submerged in reactor • Direct contact • Integrated system	• In situ separation • Minimal dead volume • Direct mass transfer	• Compact design • Reduced complexity • Lower costs	• Small–medium scale • Single products • Continuous operation
Cross-flow membrane	• Tangential flow • High shear rates • Continuous circulation	• Fouling prevention • High flux rates • Steady performance	• Reduced fouling • High productivity • Predictable operation	• Large-scale production • Fouling-prone systems • Commercial applications
Hollow fiber bioreactor	• Fiber bundle design • Shell-and-tube configuration • High surface area	• Excellent mass transfer • High surface/ volume ratio • Compartmentalized design	• High efficiency • Compact footprint • Scalable design	• High-density culture • Continuous production • Space-constrained facilities

TABLE 5.15 Membrane performance parameters and optimization.

Performance parameter	Measurement methods	Typical values	Optimization strategies	Impact on system performance
Permeate flux	• Volume/area/time measurement • Pressure monitoring • Temperature correction	• Initial: 50–200 L/m²/h • Steady-state: 20–100 L/m²/h • Decline rate: < 5%/day	• Transmembrane pressure optimization • Cross-flow velocity control • Membrane surface modification	• Productivity levels • Operating costs • System efficiency
Selectivity	• Protein retention analysis • Small-molecule passage • Molecular weight cutoff	• Protein retention: > 99% • Salt passage: > 90% • Substrate passage: > 85%	• Membrane selection • Operating condition optimization • Pore size control	• Product purity • Separation efficiency • Process effectiveness
Fouling resistance	• Flux decline monitoring • Resistance analysis • Cleaning efficiency	• Fouling rate: < 10%/day • Recovery: > 95% • Cleaning cycles: < daily	• Pretreatment • Operating optimization • Cleaning protocols	• Operating stability • Maintenance requirements • Equipment lifetime
Chemical stability	• Long-term testing • Performance monitoring • Material analysis	• pH range: 2–12 • Temperature: < 60°C • Chemical resistance: > 1000 cycles	• Material selection • Operating condition control • Protective measures	• Membrane lifetime • Replacement costs • System reliability

5.5.3 Fouling prevention and control

Membrane fouling represents a major challenge in membrane-based reactor systems. Table 5.16 outlines comprehensive fouling prevention and control strategies.

5.5.4 Economic and operational considerations

Membrane-based systems require careful economic and operational analysis to ensure commercial viability. Table 5.17 provides a comprehensive framework for evaluating membrane system economics.

TABLE 5.16 Membrane fouling prevention and control strategies.

Fouling Type	Fouling mechanisms	Prevention strategies	Control methods	Cleaning protocols
Protein fouling	• Hydrophobic interactions • Electrostatic binding • Pore blocking	• Surface modification • Operating pH optimization • Ionic strength control	• Cross-flow velocity increase • Pulsed operation • Backwashing	• Alkaline cleaning • Enzymatic treatment • Chaotropic agents
Concentration polarization	• Mass transfer limitations • Boundary layer formation • Concentration gradients	• High cross-flow velocity • Turbulence promotion • Flow optimization	• Dynamic control • Pulsed flow • Flow reversal	• Physical cleaning • Flow optimization • Membrane replacement
Particulate fouling	• Physical blocking • Cake layer formation • Pore plugging	• Prefiltration • Particle removal • Feed treatment	• Backflushing • Air sparging • Mechanical cleaning	• Physical removal • High-flow cleaning • Ultrasonic treatment
Biological fouling	• Microbial growth • Biofilm formation • Metabolic products	• Sterile operation • Biocide addition • UV treatment	• Continuous monitoring • Preventive treatment • Early intervention	• Biocidal cleaning • Hot water treatment • Chemical disinfection

TABLE 5.17 Economic and operational analysis framework for membrane systems.

Analysis category	Cost components	Evaluation metrics	Optimization opportunities	Decision factors
Capital costs	• Membrane modules: \$100–1000/m^2 • Housing systems: \$50–500/m^2 • Pumps and controls: \$10–100/m^2	• Cost per unit capacity • Payback period • Return on investment	• Technology selection • Supplier negotiation • System integration	• Production scale • Technology maturity • Risk tolerance
Operating costs	• Membrane replacement: \$50–500/m^2/year • Energy consumption: \$5–50/m^2/year • Cleaning chemicals: \$10–100/m^2/year	• Cost per unit product • Annual operating expenses • Cost trend analysis	• Membrane lifetime extension • Energy efficiency • Cleaning optimization	• Product value • Market competition • Operational efficiency

Analysis category	Cost components	Evaluation metrics	Optimization opportunities	Decision factors
Performance metrics	• Productivity: 1–50 g/ m^2/day • Selectivity: > 95% • Uptime: > 90%	• Product quality • System reliability • Process efficiency	• Process optimization • Technology upgrade • Operational excellence	• Quality requirements • Reliability demands • Competitive position
Risk factors	• Technology risk • Market risk • Operational risk	• Risk assessment • Mitigation strategies • Contingency planning	• Technology diversification • Risk reduction • Insurance coverage	• Risk tolerance • Business strategy • Market dynamics

5.6 Scale-up challenges and solutions

5.6.1 Fundamental scale-up principles

Scale-up of cell-free protein synthesis systems requires understanding and managing the complex relationships between physical phenomena, reaction kinetics, and equipment design. Successful scale-up depends on maintaining critical performance parameters while accommodating the geometric and operational changes associated with larger systems.

The fundamental principles governing scale-up must address multiple engineering disciplines and their interactions. Table 5.18 outlines the key scale-up principles and their implementation strategies.

5.6.2 Common scale-up challenges

Scale-up operations face numerous challenges that can significantly impact system performance and economics. Table 5.19 identifies the major scale-up challenges and their potential solutions.

5.6.3 Scale-up methodologies

Systematic scale-up methodologies provide structured approaches for successful technology transfer from laboratory to commercial scale. Table 5.20 compares different scale-up methodologies and their applications.

5.6.4 Validation and risk management

Scale-up validation and risk management are essential for successful commercial implementation. Table 5.21 outlines comprehensive validation strategies and risk management approaches.

TABLE 5.18 Fundamental scale-up principles for cell-free systems.

Scale-Up principle	Technical basis	Implementation strategy	Critical parameters	Validation methods
Geometric similarity	• Dimensional analysis • Scaling laws • Equipment proportions	• Constant aspect ratios • Proportional dimensions • Similar geometries	• L/D ratios • Surface/volume ratios • Flow patterns	• CFD modeling • Flow visualization • Performance comparison
Dynamic similarity	• Reynolds number • Schmidt number • Mixing parameters	• Constant dimensionless groups • Similar flow regimes • Equivalent mixing	• Re, Sc, Pe numbers • Mixing time • Power consumption	• Mixing studies • Tracer experiments • Power measurements
Kinetic similarity	• Reaction kinetics • Mass transfer rates • Heat transfer rates	• Constant kinetic parameters • Equivalent driving forces • Similar rate processes	• Reaction rates • Transfer coefficients • Conversion levels	• Kinetic studies • Performance testing • Product analysis
Thermal similarity	• Heat generation • Heat removal • Temperature control	• Constant heat transfer • Equivalent cooling • Similar temperature profiles	• Heat transfer coefficients • Temperature gradients • Cooling capacity	• Temperature mapping • Heat balance • Thermal modeling

TABLE 5.19 Major scale-up challenges and solutions.

Challenge category	Specific issues	Root causes	Solution strategies	Implementation approaches
Mixing and mass transfer	• Nonuniform conditions • Poor mass transfer • Dead zones formation	• Scale-dependent mixing • Geometric constraints • Flow pattern changes	• Advanced impeller design • Multiple mixing zones • Computational optimization	• CFD modeling • Pilot-scale testing • Mixing validation
Heat transfer	• Temperature gradients • Hot spots • Cooling limitations	• Surface/volume scaling • Heat generation increases • Transfer area limitations	• Enhanced heat transfer • Distributed cooling • Temperature control	• Heat transfer enhancement • Advanced cooling systems • Temperature monitoring

Challenge category	Specific issues	Root causes	Solution strategies	Implementation approaches
Process control	• Longer response times • Measurement delays • Control instability	• Increased thermal mass • Sensor limitations • Control loop design	• Advanced control systems • Multiple sensor arrays • Predictive control	• Control system upgrade • Sensor optimization • Model-based control
Economic factors	• Higher capital costs • Increased complexity • Risk factors	• Scale economics • Technology limitations • Market uncertainties	• Modular approaches • Phased implementation • Risk mitigation	• Economic modeling • Staged development • Technology assessment

TABLE 5.20 Scale-up methodology comparison.

Methodology	Approach	Advantages	Limitations	Best applications
Constant power per volume	• Maintain P/V ratio • Scale power input • Similar energy dissipation	• Simple calculation • Proven approach • Good mixing similarity	• May not maintain all parameters • Limited applicability • Geometric constraints	• Well-mixed systems • Standard geometries • Proven processes
Constant tip speed	• Maintain impeller tip velocity • Scale rotational speed • Similar shear rates	• Maintains shear environment • Good for shear-sensitive • Established correlations	• Power increases significantly • May cause excessive mixing • Heat generation issues	• Shear-sensitive systems • Protein applications • Quality-critical processes
Constant mixing time	• Maintain blend time • Optimize mixing efficiency • Scale mixing parameters	• Ensures adequate mixing • Quality assurance • Performance-based	• Complex calculations • Multiple variables • Validation requirements	• Critical mixing • Uniform requirements • Quality-focused
Dimensional analysis	• Dimensionless groups • Scaling laws • First principles	• Fundamental approach • Broadly applicable • Scientific basis	• Complex analysis • Multiple parameters • Validation needed	• Novel systems • Research applications • Complex processes
Pilot-scale testing	• Intermediate scale • Experimental validation • Risk reduction	• Reduces scale-up risk • Validates assumptions • Provides data	• Time consuming • Expensive • Limited flexibility	• High-value products • Novel processes • Risk-averse projects

TABLE 5.21 Scale-up validation and risk management framework.

Validation aspect	Validation methods	Success criteria	Risk factors	Mitigation strategies
Process performance	• Yield comparison • Quality assessment • Kinetic analysis	• Yield within 90%–110% • Quality specifications met • Kinetics maintained	• Performance degradation • Quality issues • Kinetic changes	• Process optimization • Quality systems • Kinetic modeling
Equipment function	• Mechanical testing • Control validation • Safety assessment	• Design specifications • Control performance • Safety compliance	• Equipment failure • Control issues • Safety concerns	• Equipment qualification • Control validation • Safety systems
Economic viability	• Cost analysis • Profitability assessment • Market evaluation	• Target cost achievement • Positive ROI • Market acceptance	• Cost overruns • Market changes • Competition	• Cost management • Market analysis • Competitive strategy
Regulatory compliance	• Good manufacturing production (GMP) assessment • Documentation review • Inspection readiness	• GMP compliance • Complete documentation • Inspection approval	• Regulatory issues • Documentation gaps • Compliance failures	• Regulatory strategy • Documentation systems • Compliance training

5.7 Process intensification strategies

5.7.1 Technology integration approaches

Process intensification in cell-free protein synthesis involves integrating multiple unit operations and technologies to achieve enhanced performance, reduced footprint, and improved economics. These approaches leverage synergies between different technologies to create more efficient and compact systems.

Modern process intensification strategies combine reactor technology, separation processes, and control systems to create integrated solutions. Table 5.22 details various technology integration approaches and their benefits.

5.7.2 Advanced control and automation

Advanced control and automation strategies are essential for process intensification, enabling precise control of complex integrated systems and optimizing overall performance. Table 5.23 outlines key control and automation technologies.

TABLE 5.22 Technology integration approaches for process intensification.

Integration strategy	Combined technologies	System benefits	Implementation challenges	Success factors
Reactive separation	• Reaction + membrane separation • In situ product removal • Continuous processing	• Reduced inhibition • Continuous operation • Compact design	• System complexity • Fouling issues • Control challenges	• Membrane selection • Process optimization • Advanced control
Multistage systems	• Sequential reactors • Staged feeding • Progressive optimization	• Optimized conditions • Higher conversion • Better control	• Increased complexity • Higher costs • Coordination requirements	• Stage optimization • Control integration • System design
Microreactor networks	• Parallel microreactors • Distributed processing • Modular scaling	• Excellent control • Rapid optimization • Flexible capacity	• Manufacturing complexity • Connection challenges • Scale limitations	• Fabrication technology • Modular design • System integration
Hybrid reactor systems	• Multiple reactor types • Complementary strengths • Optimized performance	• Best of both worlds • Application-specific • Enhanced flexibility	• Design complexity • Integration challenges • Higher costs	• Careful design • System optimization • Technology selection

TABLE 5.23 Advanced control and automation technologies.

Control technology	Technical capabilities	Application areas	Performance benefits	Implementation requirements
Model predictive control (MPC)	• Multivariable control • Constraint handling • Predictive optimization	• Complex systems • Multiple objectives • Constrained operations	• Optimal performance • Constraint satisfaction • Predictive capability	• Process models • Computing power • Expert knowledge
Artificial intelligence	• Machine learning • Pattern recognition • Adaptive optimization	• Process optimization • Fault detection • Quality prediction	• Self-learning • Adaptive control • Intelligent operation	• Data infrastructure • AI expertise • Training data

Continued

TABLE 5.23 Advanced control and automation technologies—cont'd

Control technology	Technical capabilities	Application areas	Performance benefits	Implementation requirements
Digital twins	• Real-time modeling • Virtual simulation • Predictive analysis	• Process optimization • Predictive maintenance • Scenario analysis	• Enhanced understanding • Risk reduction • Optimization insights	• Modeling capability • Real-time data • Computing resources
Internet of things (IoT)	• Sensor networks • Data connectivity • Remote monitoring	• Process monitoring • Data collection • Remote operation	• Enhanced visibility • Data-driven decisions • Remote capabilities	• Sensor infrastructure • Connectivity systems • Cybersecurity

TABLE 5.24 Modular and flexible design strategies.

Design approach	Modular elements	Flexibility features	Advantages	Design considerations
Standardized modules	• Standard reactor sizes • Interchangeable components • Common interfaces	• Easy reconfiguration • Component swapping • Rapid deployment	• Reduced costs • Faster implementation • Simplified maintenance	• Interface standardization • Component compatibility • System integration
Scalable architecture	• Parallel operation • Capacity multiplication • Distributed processing	• Linear scale-up • Capacity flexibility • Risk distribution	• Predictable scaling • Flexible capacity • Reduced risk	• Load balancing • Control coordination • System synchronization
Multiproduct capability	• Changeover protocols • Cleaning systems • Product-specific modules	• Product flexibility • Market responsiveness • Asset utilization	• Market adaptability • Economic efficiency • Competitive advantage	• Contamination prevention • Validation requirements • Changeover efficiency
Plug-and-play integration	• Modular connections • Automatic recognition • Self-configuration	• Easy expansion • Technology upgrade • System evolution	• Future proofing • Technology adoption • Simplified operation	• Interface standards • Autoconfiguration • Compatibility protocols

5.7.3 Modular and flexible design

Modular and flexible design approaches enable rapid reconfiguration and optimization of cell-free synthesis systems for different products and scales. Table 5.24 details modular design strategies and their advantages.

TABLE 5.25 Sustainability and green processing strategies.

Sustainability aspect	Implementation strategies	Environmental benefits	Economic benefits	Measurement metrics
Energy efficiency	• Heat integration • Energy recovery • Efficient equipment	• Reduced CO_2 emissions • Lower energy consumption • Resource conservation	• Lower operating costs • Energy savings • Cost competitiveness	• Energy per unit product • Carbon footprint • Energy efficiency ratio
Waste minimization	• Process optimization • By-product utilization • Recycling systems	• Reduced waste generation • Minimal environmental impact • Resource efficiency	• Lower disposal costs • Material savings • Revenue generation	• Waste generation rate • Recycling efficiency • Material utilization
Water conservation	• Water recycling • Closed-loop systems • Efficient cleaning	• Reduced water consumption • Minimal wastewater • Water resource protection	• Lower water costs • Reduced treatment costs • Regulatory compliance	• Water consumption • Wastewater generation • Water reuse rate
Green chemistry	• Biodegradable materials • Nontoxic solvents • Renewable feedstocks	• Reduced toxicity • Environmental compatibility • Sustainable sourcing	• Risk reduction • Regulatory advantages • Market acceptance	• Toxicity indicators • Biodegradability • Renewable content

5.7.4 Sustainability and green processing

Sustainability considerations are increasingly important in process intensification, driving the development of environmentally friendly and resource-efficient systems. Table 5.25 outlines key sustainability strategies and their implementation.

5.8 Summary and future directions

5.8.1 Technology evolution and trends

The field of reaction formats and scale-up technologies for cell-free protein synthesis continues to evolve rapidly. Table 5.26 outlines the major technology trends and their potential impacts on future developments.

TABLE 5.26 Technology evolution trends and future outlook.

Technology trend	Current status	Expected developments	Timeline	Impact potential
Continuous processing	• Pilot-scale demonstrations • Limited commercial use • Technology development	• Commercial implementation • Process optimization • Regulatory acceptance	2–5 years	• Improved economics • Higher productivity • Better quality control
Microreactor technology	• Laboratory applications • Process development • Limited scale-up	• Manufacturing scale-up • Parallel processing • Commercial viability	3–7 years	• Precise control • Rapid development • Flexible production
Digitalization	• Basic automation • Data collection • Simple control	• AI/ML integration • Digital twins • Autonomous operation	2–5 years	• Optimized performance • Reduced costs • Predictive capabilities
Sustainable processing	• Green chemistry adoption • Waste reduction • Energy efficiency	• Circular economy • Carbon neutrality • Sustainable materials	5–10 years	• Environmental benefits • Regulatory compliance • Market advantages

TABLE 5.27 Emerging technologies and applications.

Emerging technology	Technical approach	Potential advantages	Development status	Commercial prospects
3D printed reactors	• Additive manufacturing • Custom geometries • Rapid prototyping	• Design flexibility • Rapid development • Customization	Early development	Niche applications
Electrochemical reactors	• Electrochemical control • Redox regulation • Electrical activation	• Precise control • Energy coupling • Novel chemistry	Research stage	Specialized applications
Plasma-enhanced systems	• Plasma activation • Surface modification • Sterilization	• Enhanced performance • Surface properties • Sterility assurance	Research phase	Limited applications
Magnetic field systems	• Magnetic mixing • Field effects • Remote control	• Noncontact operation • Precise control • Sterility maintenance	Development stage	Specific applications

5.8.2 Emerging technologies and applications

New technologies and applications continue to emerge, offering potential advantages for specific cell-free synthesis applications. Table 5.27 surveys emerging technologies and their development status.

Bibliography

Carlson, E.D., Gan, R., Hodgman, C.E., Jewett, M.C., 2012. Cell-free protein synthesis: applications come of age. Biotechnol. Adv. 30 (5), 1185–1194.

Caschera, F., Noireaux, V., 2014. Synthesis of 2.3 mg/ml of protein with an all *Escherichia coli* cell-free transcription–translation system. Biochimie 99, 162–168.

Cole, S.D., Beabout, K., Turner, K.B., Smith, Z.K., Funk, V.L., Harbaugh, S.V., … Lux, M.W., 2019. Quantification of interlaboratory cell-free protein synthesis variability. ACS Synth. Biol. 8 (9), 2080–2091.

Des Soye, B.J., Gerbasi, V.R., Thomas, P.M., Kelleher, N.L., Jewett, M.C., 2019. A highly productive, one-pot cell-free protein synthesis platform based on genomically recoded *Escherichia coli*. Cell Chem. Biol. 26 (12), 1743–1754.

Gagoski, D., Polinkovsky, M.E., Mureev, S., Kunert, A., Johnston, W., Gambin, Y., Alexandrov, K., 2016. Performance benchmarking of four cell-free protein expression systems. Biotechnol. Bioeng. 113 (2), 292–300.

Garamella, J., Marshall, R., Rustad, M., Noireaux, V., 2016. The all *E. coli* TX-TL toolbox 2.0: a platform for cell-free synthetic biology. ACS Synth. Biol. 5 (4), 344–355.

Hunt, J.P., Yang, S.O., Wilding, K.M., Bundy, B.C., 2017. The growing impact of lyophilized cell-free protein synthesis. Bioengineering 4 (2), 39.

Jewett, M.C., Swartz, J.R., 2004. Mimicking the Escherichia coli cytoplasmic environment activates long-lived and efficient cell-free protein synthesis. Biotechnol. Bioeng. 86 (1), 19–26.

Kim, D.M., Swartz, J.R., 2004. Efficient production of a bioactive, multiple disulfide-bonded protein using modified extracts of Escherichia coli. Biotechnol. Bioeng. 85 (2), 122–129.

Krinsky, N., Kaduri, M., Shainsky-Roitman, J., Goldfeder, M., Ivanir, E., Benhar, I., … Schroeder, A., 2018. A simple and rapid method for preparing a cell-free bacterial lysate for protein synthesis. PLoS One 13 (2), e0193056.

Moore, S.J., MacDonald, J.T., Wienecke, S., Ishwarbhati, A., Tsipa, A., Aw, R., … Freemont, P.S., 2018. Rapid acquisition and model-based analysis of cell-free transcription–translation reactions from nonmodel bacteria. Proc. Natl. Acad. Sci. U.S.A. 115 (19), E4340–E4349.

Schwarz, D., Junge, F., Durst, F., Frölich, N., Schneider, B., Reckel, S., … Bernhard, F., 2007. Preparative scale expression of membrane proteins in *Escherichia coli*-based continuous exchange cell-free systems. Nat. Prot. 2 (11), 2945–2957.

Spirin, A.S., Baranov, V.I., Ryabova, L.A., Ovodov, S.Y., Alakhov, Y.B., 1988. A continuous cell-free translation system capable of producing polypeptides in high yield. Science 242 (4882), 1162–1164.

Swartz, J.R., 2012. Transforming biochemical engineering with cell-free biology. AIChE J. 58 (1), 5–13.

Zawada, J.F., Yin, G., Steiner, A.R., Yang, J., Naresh, A., Roy, S.M., … Swartz, J.R., 2011. Microscale to manufacturing scale-up of cell-free cytotoxin production—an *E. coli* cell-free protein synthesis case study. Biotechnol. Bioeng. 108 (7), 1570–1578.

CHAPTER

6

Advanced engineering approaches

6.1 Introduction

Advanced engineering approaches in cell-free protein synthesis (CFPS) represent the cutting edge of biotechnology, enabling the production of complex proteins with nonnatural modifications, enhanced folding characteristics, and novel functionalities. These approaches transcend the limitations of natural biological systems by incorporating engineered components, synthetic amino acids, and sophisticated posttranslational modifications. This chapter provides comprehensive coverage of advanced engineering strategies that expand the capabilities of cell-free systems beyond conventional protein synthesis.

The evolution of CFPS technology has enabled increasingly sophisticated engineering approaches that were previously impossible or impractical. Table 6.1 outlines the fundamental categories of advanced engineering approaches and their transformative potential for protein production.

Fig. 6.1 illustrates the advanced engineering landscape showing the integration of different engineering approaches and their impact on protein functionality and applications.

6.2 Noncanonical amino acid incorporation

6.2.1 Expanded genetic code systems

Noncanonical amino acid (ncAA) incorporation represents one of the most powerful tools for protein engineering, enabling the site-specific introduction of amino acids with novel chemical properties. These systems require orthogonal translation machinery that can recognize amber stop codons while maintaining high fidelity for natural amino acids.

The design and optimization of expanded genetic code systems involve multiple components that must work together seamlessly. Table 6.2 details the essential components and their optimization requirements for successful ncAA incorporation.

6.2.2 Chemical diversity and applications

The chemical diversity available through ncAA incorporation enables a wide range of applications from basic research to therapeutic development. Table 6.3 categorizes different types of ncAAs and their specific applications.

Cell Free Synthesis of Biologics
https://doi.org/10.1016/B978-0-443-33271-5.00007-7

TABLE 6.1 Categories of advanced engineering approaches in CFPS.

Engineering category	Technical scope	Key innovations	Applications	Complexity level
Noncanonical amino acid incorporation	• Expanded genetic code • Synthetic amino acids • Orthogonal systems	• Click chemistry handles • Fluorescent labels • Cross-linking residues	• Protein labeling • Drug conjugation • Structural studies	Medium–high
Posttranslational modification engineering	• Glycosylation systems • Phosphorylation networks • Ubiquitination machinery	• Mammalian-like modifications • Synthetic PTM systems • Controlled modification	• Therapeutic proteins • Functional studies • Drug development	High
Protein folding enhancement	• Chaperone engineering • Folding environments • Aggregation prevention	• Enhanced solubility • Improved stability • Native conformations	• Difficult proteins • Membrane proteins • Therapeutic applications	Medium–high
Disulfide bond formation	• Oxidative environments • Catalytic systems • Controlled oxidation	• Correct connectivity • Enhanced stability • Functional activity	• Antibodies • Enzymes • Structural proteins	Medium
Glycosylation systems	• Enzymatic machinery • Sugar donors • Quality control	• Human-like patterns • Controlled structures • Site-specific modification	• Biopharmaceuticals • Vaccine antigens • Diagnostic proteins	High
Cotranslational processing	• Ribosome engineering • Signal sequences • Processing machinery	• Membrane insertion • Signal peptide cleavage • Compartmentalization	• Membrane proteins • Secreted proteins • Complex assemblies	High

TABLE 6.2 Essential components for noncanonical amino acid incorporation systems.

System component	Function	Design requirements	Optimization parameters	Performance metrics
Orthogonal aminoacyl-tRNA synthetase (aaRS)	• Amino acid activation • tRNA aminoacylation • Specificity control	• High ncAA specificity • No natural AA recognition • Active site engineering	• Km for ncAA: $< 100\ \mu M$ • Selectivity ratio: > 100:1 • Expression level optimization	• Incorporation efficiency > 80% • Fidelity > 95% • Activity maintenance
Orthogonal tRNA (tRNAort)	• Amber codon recognition • Ribosome compatibility • aaRS recognition	• Amber anticodon • Orthogonal identity • Translation efficiency	• Tm optimization • Structure validation • Modification patterns	• Suppression efficiency > 60% • Orthogonality > 99% • Stability > 24 h

System component	Function	Design requirements	Optimization parameters	Performance metrics
Noncanonical amino acid	• Novel chemical properties • Functional groups • Bioorthogonal handles	• Cell permeability • Chemical stability • Biocompatibility	• Solubility > 10 mM • Stability > 48 h • Toxicity assessment	• Incorporation yield > 70% • Functional retention • Chemical reactivity
Expression system	• Protein production • Component integration • Quality control	• Component balancing • Optimal conditions • Monitoring systems	• Component ratios • Reaction conditions • Time course optimization	• Overall yield > 50% • Product quality • Reproducibility

FIG. 6.1 Advances engineering for CFSP showing the six core engineering domains and their integration. Each domain represents a distinct area of engineering capability with varying complexity levels and maturity. The progress metrics demonstrate current development status and capabilities in each area, while the integration domain shows the potential for combining multiple approaches to create enhanced protein functionalities.

TABLE 6.3 Noncanonical amino acid categories and applications.

ncAA category	Chemical properties	Representative examples	Primary applications	Technical advantages
Click chemistry handles	• Azide groups • Alkyne groups • Strained cycloalkynes	• p-Azido-L-phenylalanine • p-Propargyloxy-L-phenylalanine • Cyclooctyne lysine	• Protein labeling • Drug conjugation • Surface immobilization	• Bioorthogonal chemistry • High specificity • Mild conditions
Fluorescent amino acids	• Intrinsic fluorescence • Environmental sensitivity • FRET capability	• 7-Hydroxycoumarin alanine • BODIPY lysine • Naphthylalanine	• Protein dynamics • Structural studies • Live cell imaging	• Site-specific labeling • Minimal perturbation • Real-time monitoring
Cross-linking residues	• Photoactivatable groups • Chemical cross-linkers • Proximity detection	• p-Benzoyl-L-phenylalanine • Diazirine lysine • Benzophenone alanine	• Protein interactions • Conformational studies • Complex mapping	• Temporal control • Specific activation • Structural information
Metal-binding amino acids	• Coordination sites • Redox activity • Catalytic centers	• Bipyridyl alanine • Histidine analogs • Cysteine variants	• Artificial enzymes • Metal sensors • Catalytic studies	• Defined coordination • Tunable properties • Enhanced stability
Posttranslational mimics	• Phosphorylation mimics • Acetylation analogs • Methylation variants	• Phosphoserine analogs • N-acetyl lysine • Methyl lysine variants	• Signaling studies • Epigenetic research • Therapeutic targets	• Stable modifications • Controlled stoichiometry • Functional mimicry

6.2.3 Optimization strategies and challenges

Successful ncAA incorporation requires systematic optimization of multiple parameters to achieve high efficiency and fidelity. Table 6.4 outlines key optimization strategies and common challenges encountered in ncAA systems.

6.2.4 Advanced applications and case studies

The ncAA incorporation enables sophisticated applications that demonstrate the power of expanded genetic codes. Table 6.5 presents advanced applications and their implementation strategies.

TABLE 6.4 Optimization strategies for noncanonical amino acid incorporation.

Optimization aspect	Strategies	Common challenges	Solutions	Performance targets
Synthetase engineering	• Active site mutagenesis • Directed evolution • Rational design	• Maintaining specificity • Avoiding natural AAs • Activity preservation	• Multiround selection • Negative selection • Activity screening	• >100-fold selectivity • >50% wild-type activity • Stable expression
tRNA optimization	• Identity element tuning • Structural modifications • Expression optimization	• Orthogonality maintenance • Suppression efficiency • Stability issues	• Systematic mutagenesis • Structure-based design • Expression balancing	• > 99% orthogonality • > 60% suppression • > 24 h stability
System integration	• Component balancing • Condition optimization • Monitoring implementation	• Component competition • System complexity • Quality control	• Stoichiometric optimization • Systematic screening • Real-time monitoring	• > 70% incorporation • < 5% misincorporation • Reproducible results
Scale-up considerations	• Cost optimization • Supply chain • Process robustness	• ncAA availability • Economic viability • Process complexity	• Synthesis optimization • Alternative suppliers • Process simplification	• <$100/g ncAA • Reliable supply • Scalable processes

TABLE 6.5 Advanced applications of noncanonical amino acid incorporation.

Application	Technical approach	Innovation elements	Results achieved	Commercial potential
Antibody–drug conjugates	• Site-specific conjugation • Click chemistry • Controlled stoichiometry	• Homogeneous products • Defined drug ratios • Enhanced stability	• > 95% conjugation efficiency • Improved therapeutic index • Reduced side effects	High—multiple products in development
Protein–protein cross-linking	• Photoactivatable residues • Proximity detection • Structural mapping	• Temporal control • Specific activation • Complex analysis	• Detailed interaction maps • Dynamic studies • Functional insights	Medium—research applications

Continued

TABLE 6.5 Advanced applications of noncanonical amino acid incorporation—cont'd

Application	Technical approach	Innovation elements	Results achieved	Commercial potential
Fluorescent protein variants	• Intrinsic fluorophores • Environmental sensitivity • FRET pairs	• Minimal perturbation • Site-specific labeling • Ratiometric sensing	• Single-molecule studies • Live cell imaging • Conformational dynamics	Medium—specialized markets
Artificial metalloenzymes	• Metal-binding sites • Designed active sites • Novel catalysis	• Nonnatural reactions • Tunable properties • Enhanced stability	• Novel reaction types • Improved selectivity • Industrial applications	High—green chemistry potential

6.3 Posttranslational modification engineering

6.3.1 Glycosylation system design

Glycosylation represents one of the most complex and important posttranslational modifications, affecting protein folding, stability, immunogenicity, and biological activity. Engineering glycosylation systems in cell-free environments requires careful reconstruction of the enzymatic machinery and substrate supply networks.

The design of effective glycosylation systems involves multiple enzymatic steps and careful substrate management. Table 6.6 details the essential components and design considerations for cell-free glycosylation systems.

6.3.2 Phosphorylation network engineering

Protein phosphorylation represents a crucial regulatory mechanism that can be engineered in cell-free systems to create proteins with defined modification states. Table 6.7 outlines the components and strategies for engineering phosphorylation networks.

6.3.3 Ubiquitination and protein degradation

Engineering controlled protein degradation through ubiquitination systems enables temporal control of protein levels and functional studies. Table 6.8 details the components and applications of engineered ubiquitination systems.

6.3.4 Advanced post-translational modifications combinations

The combination of multiple posttranslational modifications creates sophisticated regulatory networks and enhanced protein functionalities. Table 6.9 analyzes strategies for implementing complex post-translational modifications (PTM) combinations.

TABLE 6.6 Essential components for cell-free glycosylation systems.

System component	Function	Key enzymes	Substrate requirements	Design considerations
N-linked glycosylation	• Oligosaccharyl transfer • Core structure formation • Processing reactions	• Oligosaccharyl transferase • Glucosidases I/II • Mannosidases	• Dolichol-PP-oligosaccharide • UDP-glucose • Acceptor proteins	• Membrane environment • Lipid requirements • Quality control
O-linked glycosylation	• Initial sugar transfer • Chain elongation • Branching reactions	• GalNAc-transferases • Galactosyltransferases • Sialyltransferases	• UDP-GalNAc • UDP-galactose • CMP-sialic acid	• Enzyme specificity • Sequential addition • Site selection
Sugar nucleotide synthesis	• Activated sugar production • Substrate regeneration • Metabolic coupling	• UDP-glucose pyrophosphorylase • GDP-mannose pyrophosphorylase • CMP-sialic acid synthetase	• Sugar phosphates • Nucleotide triphosphates • Cofactors	• Pathway reconstruction • Energy coupling • Product inhibition
Quality control systems	• Folding assessment • Glycan processing • Error correction	• Calnexin/calreticulin • UGGT1 • ERAD machinery	• Lectin cofactors • Chaperone systems • Degradation pathways	• Folding environment • Quality sensors • Disposal mechanisms

TABLE 6.7 Phosphorylation network engineering components.

Network component	Enzymatic elements	Regulation mechanisms	Implementation strategies	Control parameters
Kinase systems	• Serine/threonine kinases • Tyrosine kinases • Dual-specificity kinases	• Allosteric regulation • Compartmentalization • Substrate specificity	• Recombinant expression • Activity optimization • Specificity engineering	• Kinase concentration • ATP levels • Cofactor availability
Phosphatase networks	• Serine/threonine phosphatases • Tyrosine phosphatases • Dual-specificity phosphatases	• Inhibitor proteins • Subcellular localization • Substrate targeting	• Balanced expression • Inhibitor control • Temporal regulation	• Phosphatase activity • Inhibitor ratios • Reaction kinetics

Continued

TABLE 6.7 Phosphorylation network engineering components—cont'd

Network component	Enzymatic elements	Regulation mechanisms	Implementation strategies	Control parameters
Scaffold proteins	• Kinase anchoring proteins • Adaptor molecules • Organizing centers	• Proximity effects • Specificity enhancement • Signal amplification	• Modular design • Interaction engineering • Spatial organization	• Scaffold concentration • Binding affinities • Complex assembly
Regulatory circuits	• Feedback loops • Feedforward control • Bistable switches	• Positive feedback • Negative regulation • Ultrasensitivity	• Circuit design • Parameter tuning • Dynamic control	• Circuit parameters • Response time • Stability margins

TABLE 6.8 Ubiquitination system engineering for controlled protein degradation.

System element	Molecular components	Function	Engineering approaches	Applications
E1 activating enzymes	• UBA1, UBA6 • ATP-dependent activation • Ubiquitin charging	• Ubiquitin activation • Energy coupling • System initiation	• Expression optimization • Activity enhancement • Substrate supply	• Proteolysis control • Temporal regulation • Functional studies
E2 conjugating enzymes	• UBC family enzymes • Thioester intermediates • Specificity factors	• Ubiquitin transfer • Chain initiation • Specificity control	• Enzyme selection • Activity tuning • Specificity engineering	• Selective targeting • Chain type control • Degradation signals
E3 ligase systems	• RING, HECT, RBR domains • Substrate recognition • Catalytic mechanisms	• Substrate specificity • Ubiquitin ligation • Regulation integration	• Substrate engineering • Degron design • Conditional systems	• Targeted degradation • Conditional control • Synthetic circuits
Proteasome systems	• 26S proteasome • Recognition signals • Degradation machinery	• Protein degradation • Peptide production • Quality control	• Reconstitution strategies • Activity optimization • Substrate processing	• Controlled proteolysis • Peptide generation • Protein turnover

TABLE 6.9 Strategies for complex posttranslational modification combinations.

Combination strategy	PTM types	Implementation challenges	Solutions	Functional outcomes
Sequential processing	• Glycosylation → Phosphorylation • Acetylation → Methylation • Ubiquitination → Proteolysis	• Temporal coordination • Enzyme competition • Substrate availability	• Staged reactions • Enzyme ratios • Condition optimization	• Natural-like processing • Ordered modifications • Functional integration
Parallel modifications	• Multiple glycosylation sites • Phosphorylation patterns • Acetylation landscapes	• Site competition • Crosstalk effects • System complexity	• Site prioritization • Enzyme engineering • System modeling	• Complex modification patterns • Enhanced functionality • Regulatory diversity
Conditional systems	• pH-dependent modifications • Temperature-sensitive PTMs • Substrate-controlled processing	• Environmental sensitivity • System stability • Control precision	• Robust enzyme selection • Environmental optimization • Feedback control	• Dynamic regulation • Responsive systems • Controllable function
Synthetic PTM networks	• Artificial modifications • Nonnatural linkages • Designer regulatory circuits	• Novel chemistry • System design • Validation challenges	• Chemical biology tools • Synthetic enzymes • Systematic validation	• Novel functionalities • Engineered regulation • Synthetic biology applications

6.4 Protein folding enhancement

6.4.1 Chaperone system engineering

Protein folding in cell-free systems can be significantly enhanced through the engineering of chaperone networks that provide appropriate folding environments and prevent aggregation. Modern chaperone engineering approaches enable the production of properly folded proteins that would otherwise be impossible to obtain.

The design of effective chaperone systems requires understanding of folding pathways and chaperone mechanisms. Table 6.10 details different classes of chaperones and their engineering applications.

TABLE 6.10 Chaperone system engineering for enhanced protein folding.

Chaperone class	Mechanism of action	Engineering targets	Optimization strategies	Target applications
Hsp70 systems	• Cotranslational folding • Aggregation prevention • Folding assistance	• DnaK/DnaJ/GrpE • BiP/Grp94/ protein disulfide isomerase (PDI) • Cytosolic variants	• Cofactor optimization • Expression balancing • Substrate specificity	• Soluble proteins • Aggregation-prone targets • Cotranslational folding
Hsp60 chaperonins	• Encapsulated folding • Isolated environment • ATP-driven cycling	• GroEL/GroES • TRiC/CCT • Thermosomes	• Cavity engineering • Cycling optimization • Substrate accommodation	• Large proteins • Complex topologies • Aggregation prevention
Hsp90 systems	• Conformational maturation • Client stabilization • Regulatory integration	• Hsp90α/β • Cochaperones • Client specificity	• Cochaperone networks • ATP hydrolysis tuning • Client engineering	• Signaling proteins • Kinases • Transcription factors
Protein disulfide isomerases	• Disulfide formation • Isomerization • Quality control	• PDI family enzymes • Oxidative environment • Redox systems	• Redox potential control • Enzyme specificity • Substrate recognition	• Disulfide-bonded proteins • Secreted proteins • Therapeutic antibodies
Specialized chaperones	• Membrane insertion • Metal incorporation • Cofactor assembly	• SecYEG translocon • Metallochaperones • Assembly factors	• Environment reconstruction • Cofactor supply • Assembly coordination	• Membrane proteins • Metalloproteins • Multisubunit complexes

6.4.2 Folding environment optimization

Creating optimal folding environments in cell-free systems requires careful control of multiple physicochemical parameters. Table 6.11 outlines key environmental factors and their optimization strategies.

6.4.3 Antiaggregation strategies

Protein aggregation represents a major challenge in CFPS that can be addressed through various engineering strategies. Table 6.12 details comprehensive antiaggregation approaches.

6.4.4 Quality assessment and validation

Comprehensive quality assessment of folded proteins requires multiple analytical approaches to verify correct folding and functionality. Table 6.13 outlines quality assessment strategies and validation methods.

TABLE 6.11 Folding environment optimization parameters.

Environmental factor	Parameter range	Optimization targets	Control strategies	Impact on folding
Redox potential	−280 to +50 mV	Protein-specific optima	• Reduced glutathione (GSH)/oxidized glutathione (GSSG) ratios • Redox buffers • Enzymatic systems	• Disulfide formation • Metal coordination • Cofactor binding
Molecular crowding	100–400 g/L	Physiological levels	• Macromolecular crowders • PEG, Ficoll, dextran • Protein concentration	• Folding kinetics • Stability enhancement • Aggregation modulation
Osmolyte concentrations	0.1–1.0 M	Stabilizing conditions	• Glycine betaine • Trehalose • Proline derivatives	• Thermodynamic stability • Kinetic folding • Stress protection
pH microenvironments	6.5–8.5	Protein-specific ranges	• Local buffering • Compartmentalization • pH gradients	• Ionization states • Electrostatic interactions • Enzyme activity
Temperature profiles	15°C–45°C	Folding optima	• Temperature ramping • Controlled cooling • Thermal cycling	• Folding kinetics • Aggregation prevention • Native structure
Cofactor availability	µM to mM	Saturation levels	• Cofactor addition • Regeneration systems • Competitive binding	• Holoprotein formation • Functional activity • Structural integrity

TABLE 6.12 Antiaggregation strategies for cell-free protein synthesis.

Strategy category	Technical approaches	Mechanism of action	Implementation methods	Effectiveness metrics
Chemical additives	• Aggregation inhibitors • Stabilizing agents • Solubilizing compounds	• Hydrophobic interactions • Electrostatic shielding • Preferential binding	• Additive screening • Concentration optimization • Synergistic combinations	• Soluble fraction > 80% • Activity retention > 70% • Aggregation reduction > 90%
Cotranslational strategies	• Chaperone coexpression • Folding enhancers • Translation modulation	• Nascent chain protection • Folding assistance • Kinetic control	• Expression optimization • Timing coordination • Stoichiometric balancing	• Native folding > 75% • Reduced inclusion bodies • Enhanced yields

Continued

TABLE 6.12 Antiaggregation strategies for cell-free protein synthesis—cont'd

Strategy category	Technical approaches	Mechanism of action	Implementation methods	Effectiveness metrics
Environmental control	• Temperature management • pH optimization • Ionic strength control	• Aggregation kinetics • Stability enhancement • Solubility improvement	• Parameter screening • Multifactorial optimization • Real-time monitoring	• Optimized conditions • Reproducible results • Stable preparations
Protein engineering	• Surface modifications • Stability enhancement • Aggrega-tion-prone removal	• Reduced aggregation propensity • Enhanced stability • Improved solubility	• Computational design • Directed evolution • Rational engineering	• Designed improvements • Maintained function • Enhanced properties

TABLE 6.13 Quality assessment and validation methods for protein folding.

Assessment category	Analytical methods	Information obtained	Quality criteria	Validation standards
Structural analysis	• CD spectroscopy • NMR analysis • X-ray crystallography	• Secondary structure • Tertiary structure • Conformational dynamics	• Native-like spectra • Correct fold • Structural integrity	• Literature comparison • Reference standards • Database validation
Functional assessment	• Enzymatic activity • Binding assays • Biological activity	• Catalytic function • Binding specificity • Physiological activity	• >80% native activity • Correct Km values • Biological relevance	• Standard assays • Comparative analysis • Functional validation
Stability analysis	• Thermal unfolding • Chemical denaturation • Long-term storage	• Thermodynamic stability • Kinetic stability • Storage stability	• Tm within 5°C of native • >72 h stability • Minimal aggregation	• Stability protocols • Accelerated testing • Real-time studies
Aggregation assessment	• Dynamic light scattering • Size exclusion chromatography • Analytical ultracentrifugation	• Size distribution • Aggregation state • Homogeneity	• >95% monomer • Low polydispersity • Stable preparations	• Sizing standards • Aggregation controls • Quality specifications

6.5 Disulfide bond formation

6.5.1 Oxidative environment design

The formation of correct disulfide bonds is crucial for the proper folding and stability of many proteins, particularly those destined for secretion or therapeutic applications. Engineering appropriate oxidative environments in cell-free systems requires careful control of redox conditions and catalytic systems.

Designing effective oxidative environments involves balancing multiple redox systems and catalytic components. Table 6.14 details the key components and design principles for disulfide bond formation systems.

6.5.2 Catalytic disulfide formation systems

Advanced catalytic systems can significantly enhance the efficiency and fidelity of disulfide bond formation. Table 6.15 compares different catalytic approaches and their characteristics.

TABLE 6.14 Components and design principles for disulfide bond formation systems.

System component	Function	Design parameters	Optimization targets	Performance indicators
Redox buffer systems	• Redox potential control • Electron balance • Oxidation driving force	• GSH/GSSG ratios • Redox potential range • Buffer capacity	• Optimal redox potential • Stable conditions • Reproducible environment	• Correct disulfide formation • Minimal incorrect bonds • High yield folding
Protein disulfide isomerase (PDI)	• Disulfide formation • Isomerization • Quality control	• Enzyme concentration • Activity optimization • Substrate specificity	• Catalytic efficiency • Substrate recognition • Error correction	• >90% correct connectivity • Fast kinetics • Error reduction
Oxidizing agents	• Electron acceptors • Oxidation catalysts • Redox recycling	• Concentration ranges • Reaction kinetics • Selectivity factors	• Mild conditions • Controlled oxidation • Minimal side reactions	• Selective oxidation • Controlled kinetics • Clean reactions
Catalytic systems	• Metal catalysts • Enzymatic systems • Synthetic catalysts	• Catalyst selection • Activity modulation • Substrate compatibility	• High selectivity • Mild conditions • Recyclable systems	• Efficient catalysis • No side products • Stable performance

TABLE 6.15 Catalytic systems for enhanced disulfide bond formation.

Catalytic system	Mechanism	Advantages	Limitations	Optimal applications
Protein disulfide isomerase family	• Thiol-disulfide exchange • Conformational sampling • Quality control	• High fidelity • Error correction • Natural system	• Protein requirement • System complexity • Cost factors	• Complex proteins • Multiple disulfides • Therapeutic proteins
Small molecule catalysts	• Metal-catalyzed oxidation • Chemical redox cycling • Direct oxidation	• Simple systems • Low cost • Scalable	• Limited selectivity • Side reactions • Harsh conditions	• Simple proteins • Single disulfides • Cost-sensitive applications
Engineered redox systems	• Designed electron transfer • Controlled potentials • Synthetic biology	• Tunable properties • Designed specificity • Synthetic control	• Development complexity • Validation requirements • Limited availability	• Designer proteins • Novel applications • Research tools
Hybrid systems	• Multiple mechanisms • Synergistic effects • Complementary functions	• Enhanced performance • Robust systems • Flexible application	• System complexity • Optimization challenges • Higher costs	• Challenging proteins • High-value products • Custom applications

TABLE 6.16 Quality control strategies for disulfide-bonded proteins.

Quality control mechanism	Detection methods	Correction strategies	Implementation approaches	Success metrics
Conformational sensors	• Hydrophobic exposure • Structural probes • Folding reporters	• Chaperone recruitment • Refolding cycles • Degradation signals	• Sensor protein design • Reporter systems • Monitoring networks	• >95% correct detection • Fast response time • Low false positives
Disulfide connectivity analysis	• Mass spectrometry • Chemical mapping • Proteolytic analysis	• PDI-mediated correction • Reduction/refolding • Selective breaking	• Analytical methods • Correction protocols • Automation systems	• Correct connectivity mapping • Efficient correction • High fidelity
Functional assessment	• Activity measurements • Binding assays • Stability tests	• Refolding protocols • Condition optimization • Process adjustment	• Functional screens • Real-time monitoring • Feedback control	• Native-like activity • Stable function • Reproducible results

Quality control mechanism	Detection methods	Correction strategies	Implementation approaches	Success metrics
Aggregation prevention	• Size analysis • Solubility assessment • Aggregation kinetics	• Disaggregation systems • Solubilization agents • Refolding assistance	• Monitoring systems • Intervention protocols • Prevention strategies	• Low aggregation • High solubility • Stable preparations

TABLE 6.17 Disulfide bond formation in therapeutic protein applications.

Therapeutic category	Disulfide requirements	Critical factors	Quality standards	Regulatory considerations
Monoclonal antibodies	• Heavy-light chain linking • Intrachain disulfides • Structural integrity	• Correct pairing • Chain assembly • Aggregation prevention	• >99% correct structure • Minimal variants • Stable preparations	• ICH guidelines • Process validation • Comparability studies
Insulin and analogs	• Interchain disulfides • Intrachain bridges • Biological activity	• Chain association • Activity retention • Stability maintenance	• United States Pharmacopoeia (USP) specifications • Bioactivity requirements • Purity standards	• Food and Drug Administration (FDA) approval pathways • Clinical requirements • Manufacturing controls
Growth factors	• Structural disulfides • Receptor binding • Signaling activity	• Correct folding • Binding affinity • Biological potency	• Biological activity • Structural integrity • Safety profiles	• Regulatory guidelines • Safety requirements • Efficacy standards
Enzyme therapeutics	• Active site formation • Structural stability • Catalytic activity	• Enzyme activity • Substrate specificity • Stability profiles	• Catalytic efficiency • Selectivity requirements • Stability specifications	• Regulatory approval • Manufacturing standards • Quality requirements

6.5.3 Quality control and error correction

Ensuring correct disulfide connectivity requires sophisticated quality control mechanisms that can detect and correct misfolded proteins. Table 6.16 outlines quality control strategies for disulfide-bonded proteins.

6.5.4 Applications in therapeutic protein production

Disulfide bond formation is particularly critical for therapeutic protein production where correct folding directly impacts safety and efficacy. Table 6.17 details applications in therapeutic protein manufacturing.

6.6 Glycosylation systems

6.6.1 N-linked glycosylation engineering

N-linked glycosylation represents one of the most complex posttranslational modifications, requiring sophisticated enzymatic machinery and careful substrate management. Engineering effective N-linked glycosylation systems in cell-free environments enables the production of properly glycosylated therapeutic proteins.

The reconstruction of N-linked glycosylation pathways requires integration of multiple enzymatic steps and membrane systems. Table 6.18 details the essential components and engineering considerations for N-linked glycosylation systems.

TABLE 6.18 N-linked glycosylation system components and engineering.

Pathway component	Enzymatic elements	Substrate requirements	Engineering challenges	Optimization strategies
Oligosaccharyl transfer	• OST complex (STT3, OST4, etc.) • Membrane integration • Quality control	• Dolichol-PP-oligosaccharide • Nascent polypeptide • ER membrane environment	• Membrane reconstitution • Complex assembly • Substrate availability	• Proteoliposome systems • Complex optimization • Substrate synthesis
Core processing	• Glucosidase I/II • Mannosidase I • Quality control lectins	• Glucose trimming • Mannose processing • Calnexin/calreticulin cycle	• Sequential processing • Quality control integration • Temporal coordination	• Enzyme ratios • Timing control • Quality sensors
Complex formation	• GnT-I, GnT-II • Mannosidase II • Fucosyltransferases	• UDP-GlcNAc • GDP-fucose • Processing sequence	• Branch formation • Substrate competition • Pathway coordination	• Enzyme engineering • Substrate management • Pathway reconstruction
Terminal processing	• Galactosyltransferases • Sialyltransferases • Sulfotransferases	• UDP-galactose • CMP-sialic acid • PAPS	• Final modifications • Microheterogeneity • Quality control	• Enzyme specificity • Substrate supply • Product analysis

6.6.2 O-linked glycosylation systems

O-linked glycosylation provides additional complexity and functionality to glycoproteins through mucin-type and other glycosylation patterns. Table 6.19 outlines the components and engineering approaches for O-linked glycosylation systems.

6.6.3 Glycan analysis and quality control

Comprehensive analysis of glycosylation patterns is essential for quality control and product characterization. Table 6.20 details analytical methods and quality control strategies for glycosylated proteins.

6.6.4 Therapeutic applications and regulatory considerations

Glycosylation plays a critical role in therapeutic protein development and requires careful consideration of regulatory requirements. Table 6.21 outlines therapeutic applications and regulatory considerations for glycosylated proteins.

TABLE 6.19 O-linked glycosylation system design and implementation.

Glycosylation type	Key enzymes	Substrate specificity	Engineering approaches	Applications
Mucin-type O-glycosylation	• GalNAc-T family • Core ½ synthases • Extension enzymes	• Serine/threonine residues • Peptide context • Acceptor specificity	• Enzyme selection • Substrate optimization • Site engineering	• Mucins • Membrane proteins • Immunoglob-ulins
O-GlcNAc modification	• OGT (O-GlcNAc transferase) • OGA (O-GlcNAcase) • Regulatory proteins	• Nuclear/cytoplas-mic proteins • Dynamic modification • Stress response	• Dynamic control • Substrate recognition • Regulatory integration	• Transcription factors • Kinases • Stress proteins
O-linked mannose	• POMT1/POMT2 • Extension enzymes • Branching enzymes	• Dystroglycan • Basement membrane • Structural proteins	• Membrane reconstitution • Complex assembly • Function validation	• Dystroglycan • Structural proteins • Disease models
Other O-glycans	• Various transferases • Specific pathways • Regulatory enzymes	• Protein-specific • Tissue-specific • Function-specific	• Pathway reconstruction • Enzyme engineering • Validation studies	• Specialized functions • Research applications • Disease studies

TABLE 6.20 Glycan analysis and quality control methods.

Analysis category	Analytical techniques	Information obtained	Quality parameters	Applications
Structural analysis	• MALDI-TOF MS • LC–MS/MS • NMR spectroscopy	• Glycan composition • Linkage analysis • Structural details	• Correct structures • Expected patterns • Minimal variants	• Structure verification • Method development • Research applications
Site occupancy	• Peptide mapping • ETD-MS/MS • Site-specific analysis	• Glycosylation sites • Occupancy levels • Microheterogeneity	• >90% occupancy • Site consistency • Minimal variation	• Therapeutic proteins • Quality control • Process optimization
Functional assessment	• Binding assays • Activity measurements • Cell-based assays	• Biological activity • Receptor binding • Cellular functions	• Native-like activity • Functional consistency • Biological relevance	• Biopharmaceuticals • Functional studies • Bioactivity assessment
Comparative analysis	• Reference standards • Batch comparison • Trend analysis	• Consistency assessment • Batch-to-batch variation • Process stability	• Acceptable variation • Consistent patterns • Process control	• Manufacturing QC • Process monitoring • Regulatory compliance

TABLE 6.21 Therapeutic glycoprotein applications and regulatory framework.

Therapeutic category	Glycosylation requirements	Regulatory considerations	Quality standards	Development challenges
Monoclonal antibodies	• Fc glycosylation • Effector functions • ADCC/CDC activity	• ICH Q6B guidelines • Comparability studies • Process validation	• Defined glycan profiles • Functional activity • Batch consistency	• Glycan heterogeneity • Functional impact • Process control
Erythropoietin	• N-linked glycans • Sialic acid content • Biological activity	• Complex approval pathway • Bioequivalence studies • Safety requirements	• Specific glycan patterns • Bioactivity maintenance • Safety profiles	• Glycan complexity • Activity relationships • Manufacturing consistency

Therapeutic category	Glycosylation requirements	Regulatory considerations	Quality standards	Development challenges
Coagulation factors	• Multiple glycosylation • Activity requirements • Stability needs	• Blood product regulations • Safety requirements • Efficacy standards	• Hemostatic activity • Stability specifications • Safety profiles	• Complex structures • Activity preservation • Safety validation
Enzyme replacement therapies	• Lysosomal targeting • Mannose-6-phosphate • Uptake mechanisms	• Orphan drug pathways • Specialized requirements • Patient safety	• Targeting efficiency • Cellular uptake • Therapeutic efficacy	• Targeting specificity • Uptake optimization • Clinical efficacy

6.7 Cotranslational processing

6.7.1 Signal recognition and targeting

Cotranslational processing enables the proper targeting and modification of proteins during synthesis, particularly important for membrane proteins and secreted proteins. Engineering effective signal recognition and targeting systems requires reconstruction of cellular machinery in cell-free environments.

The design of signal recognition and targeting systems involves multiple components working in coordination. Table 6.22 details the essential components and their engineering requirements.

6.7.2 Membrane protein insertion and folding

Membrane protein synthesis requires specialized machinery for proper insertion and folding within lipid bilayers. Table 6.23 outlines the components and strategies for membrane protein cotranslational processing.

6.7.3 Protein complex assembly

The assembly of multisubunit protein complexes during translation requires sophisticated coordination mechanisms. Table 6.24 details strategies for engineering protein complex assembly systems.

6.7.4 Advanced cotranslational engineering

Modern cotranslational engineering approaches enable sophisticated protein modifications and assembly processes. Table 6.25 outlines advanced engineering strategies and their applications.

TABLE 6.22 Signal recognition and targeting system components.

System component	Function	Molecular elements	Engineering requirements	Performance targets
Signal recognition particle (SRP)	• Signal sequence recognition • Translation pausing • Ribosome targeting	• SRP54, SRP68/72 • 7SL RNA • SRP receptor	• Complex assembly • GTP cycling • Membrane targeting	• >90% signal recognition • Efficient targeting • Minimal mis-targeting
Translocon complex	• Protein translocation • Membrane insertion • Quality control	• Sec61$\alpha\beta\gamma$ complex • Associated proteins • Membrane environment	• Membrane reconstitution • Complex integration • Channel function	• Efficient translocation • Correct topology • Quality control
Signal peptidase	• Signal sequence cleavage • Protein maturation • Processing fidelity	• Signal peptidase complex • Cleavage specificity • Membrane association	• Enzyme activity • Specificity optimization • Membrane integration	• >95% cleavage efficiency • Correct processing • Minimal errors
Membrane environ-ment	• Lipid bilayer • Protein insertion • Functional support	• Phospholipid composition • Membrane fluidity • Protein-lipid interactions	• Membrane preparation • Composition optimization • Stability maintenance	• Functional membranes • Protein compatibility • System stability

TABLE 6.23 Membrane protein cotranslational processing systems.

Processing aspect	Molecular machinery	Implementation strategies	Critical factors	Success metrics
Transmembrane insertion	• Sec61 translocon • YidC insertase • Membrane integration	• Membrane reconstitution • Translocon optimization • Lipid composition	• Hydrophobic segments • Topology determination • Membrane orientation	• Correct topology • Stable insertion • Functional activity
Membrane protein folding	• Membrane chaperones • Lipid environment • Folding assistance	• Chaperone supplementation • Membrane optimization • Folding conditions	• Lipid-protein interactions • Membrane thickness • Lateral pressure	• Native structure • Functional activity • Stability in membranes
Quality control	• Membrane quality control • Misfolded protein detection • Degradation systems	• QC system reconstruction • Error detection • Correction mechanisms	• Folding assessment • Error recognition • Disposal pathways	• Quality maintenance • Error correction • System efficiency

Processing aspect	Molecular machinery	Implementation strategies	Critical factors	Success metrics
Functional integration	• Cofactor insertion • Subunit assembly • Activity maturation	• Cofactor supply • Assembly factors • Activity testing	• Cofactor availability • Assembly coordination • Functional validation	• Complete assembly • Full activity • Stable complexes

TABLE 6.24 Protein complex assembly engineering strategies.

Assembly strategy	Coordination mechanisms	Implementation approaches	Challenges	Solutions
Sequential assembly	• Ordered subunit addition • Assembly checkpoints • Quality control	• Staged synthesis • Temporal control • Assembly monitoring	• Timing coordination • Stoichiometry control • Intermediate stability	• Controlled expression • Assembly factors • Stabilization systems
Concurrent assembly	• Simultaneous synthesis • Cotranslational assembly • Chaperone assistance	• Balanced expression • Assembly chaperones • Scaffold proteins	• Expression balancing • Assembly efficiency • Complex stability	• Stoichiometric optimization • Chaperone engineering • Stability enhancement
Hierarchical assembly	• Sub-complex formation • Progressive assembly • Structural organization	• Modular synthesis • Assembly intermediates • Structural validation	• Interface optimization • Assembly pathways • Quality control	• Interface engineering • Pathway optimization • Validation methods
Template-assisted assembly	• Assembly templates • Spatial organization • Cooperative binding	• Scaffold design • Template optimization • Assembly enhancement	• Template design • Binding optimization • Assembly efficiency	• Rational design • Binding enhancement • Assembly optimization

Fig. 6.1 illustrates the advanced engineering landscape showing the integration of different engineering approaches and their synergistic effects on protein functionality.

Fig. 6.2 demonstrates the multilevel engineering hierarchy from molecular components to integrated systems and their applications.

TABLE 6.25 Advanced cotranslational engineering approaches.

Engineering approach	Technical strategy	Innovation elements	Applications	Development status
Ribosome engineering	• Modified ribosomes • Enhanced functions • Expanded capabilities	• Nonnatural substrates • Modified peptide bonds • Enhanced fidelity	• Peptide engineering • Novel chemistries • Synthetic biology	Research/ development
Orthogonal systems	• Independent translation • Specialized ribosomes • Isolated processing	• System orthogonality • Specialized functions • Controlled synthesis	• Synthetic proteins • Novel modifications • Research tools	Early development
Programmed processing	• Conditional processing • Trigger-based modification • Dynamic control	• Environmental control • Temporal regulation • Spatial organization	• Controlled synthesis • Dynamic systems • Smart materials	Research phase
Multifunctional integration	• Combined modifications • Integrated processing • Comprehensive engineering	• Multiple modifications • System integration • Enhanced functionality	• Complex proteins • Advanced therapeutics • Synthetic biology	Development stage

6.8 Summary and future directions

6.8.1 Technology integration and synergies

The integration of multiple advanced engineering approaches creates synergistic effects that enable unprecedented capabilities in protein design and production. Table 6.26 analyzes technology integration opportunities and their potential impact.

6.8.2 Emerging technologies and future directions

Advanced engineering approaches continue to evolve with emerging technologies offering new possibilities. Table 6.27 surveys emerging technologies and their development trajectories.

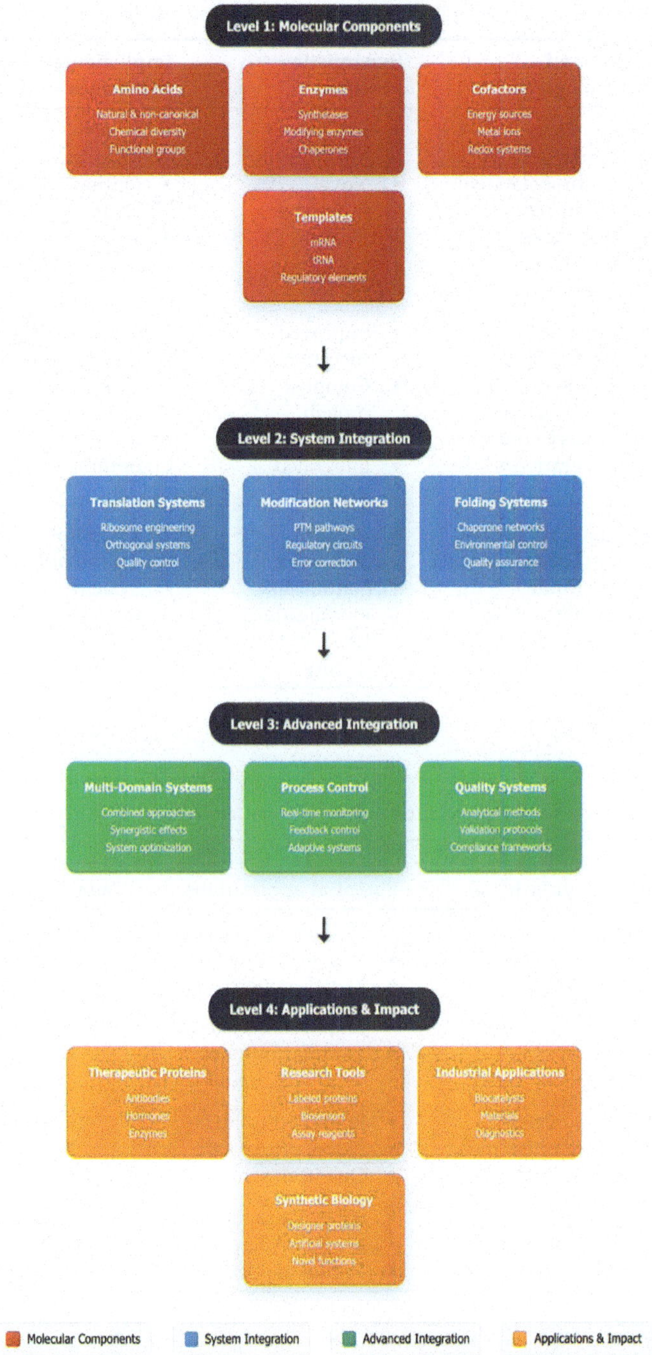

FIG. 6.2 The multilevel engineering hierarchy showing the progression from molecular components to real-world applications. Each level builds upon the previous one, with increasing complexity and integration. The hierarchy demonstrates how individual molecular components are assembled into integrated systems, which are then combined into advanced multidomain that enable diverse applications across therapeutics, research, industry and synthetic biology.

TABLE 6.26 Technology integration opportunities and synergistic effects.

Integration strategy	Combined technologies	Synergistic benefits	Implementation challenges	Future potential
ncAA + PTM engineering	• Noncanonical amino acids • Posttranslational modifications • Chemical-biological hybrids	• Enhanced functionality • Novel properties • Designer proteins	• System complexity • Optimization challenges • Validation requirements	• Therapeutic break-throughs • Novel materials • Synthetic biology
Folding + modification integration	• Enhanced folding systems • Simultaneous modifications • Quality control integration	• Improved success rates • Complex protein production • Integrated processing	• Process coordination • Quality assurance • System optimization	• Therapeutic proteins • Complex biologics • Industrial enzymes
Cotranslational + PTM systems	• Translation-coupled processing • Real-time modifications • Integrated quality control	• Efficient processing • Natural-like modifications • Enhanced fidelity	• Temporal coordination • System integration • Process control	• Native-like proteins • Enhanced therapeutics • Biological authenticity

TABLE 6.27 Emerging technologies and future development trajectories.

Emerging technology	Technical approach	Potential applications	Development timeline	Commercial impact
AI-guided protein design	• Machine learning • Protein prediction • Rational design	• Designer proteins • Enhanced functions • Novel properties	2–5 years	High impact expected
Synthetic biology integration	• Standardized parts • Modular systems • Predictable design	• Engineered systems • Modular proteins • Synthetic circuits	3–7 years	Transformative potential
Quantum Biology Applications	• Quantum effects • Coherence phenomena • Enhanced functions	• Enhanced catalysis • Novel properties • Quantum sensors	5–10 years	Speculative but promising
Digital–biological interfaces	• Digital control • Biological responses • Hybrid systems	• Smart therapeutics • Responsive systems • Digital biology	3–8 years	Revolutionary potential

Bibliography

Albayrak, C., Swartz, J.R., 2013. Cell-free co-production of an orthogonal transfer RNA activates efficient site-specific non-natural amino acid incorporation. Nucleic Acids Res. 41 (11), 5949–5963.

Bundy, B.C., Swartz, J.R., 2010. Site-specific incorporation of p-propargyloxyphenylalanine in a cell-free environment for direct protein– protein click conjugation. Bioconjug. Chem. 21 (2), 255–263.

Chin, J.W., 2017. Expanding and reprogramming the genetic code. Nature 550 (7674), 53–60.

Dondapati, S.K., Kreir, M., Quast, R.B., Wüstenhagen, D.A., Brüggemann, A., Fertig, N., Kubick, S., 2014. Membrane assembly of aquaporin-4 autoantibodies regulates classical complement activation in neuromyelitis optica. J. Clin. Invest. 124 (9), 3981–3993.

Guarino, C., DeLisa, M.P., 2012. A prokaryote-based cell-free translation system that efficiently synthesizes glycoproteins. Glycobiology 22 (5), 596–601.

Hong, S.H., Ntai, I., Haimovich, A.D., Kelleher, N.L., Isaacs, F.J., Jewett, M.C., 2014. Cell-free protein synthesis from a release factor 1 deficient Escherichia coli activates efficient and multiple site-specific nonstandard amino acid incorporation. ACS Synth. Biol. 3 (6), 398–409.

Jaroentomeechai, T., Stark, J.C., Natarajan, A., Glasscock, C.J., Yingling, L.E., Hsu, K.J., Mrksich, M., Jewett, M. C., DeLisa, M.P., 2018. Single-pot glycoprotein biosynthesis using a cell-free transcription-translation system enriched with glycosylation machinery. Nat. Commun. 9 (1), 2686.

Kightlinger, W., Duncker, K.E., Ramesh, A., Thames, A.H., Natarajan, A., Stark, J.C., Yang, A., Lin, L., Mrksich, M., DeLisa, M.P., Jewett, M.C., 2019. A cell-free biosynthesis platform for modular construction of protein glycosylation pathways. Nat. Commun. 10 (1), 5404.

Liu, C.C., Schultz, P.G., 2010. Adding new chemistries to the genetic code. Annu. Rev. Biochem. 79, 413–444.

Martin, R.W., Majewska, N.I., Chen, C.X., Albanetti, T.E., Jimenez, R.B., Schmelzer, A.E., Jewett, M.C., Roy, V., 2017. Development of a CHO-based cell-free platform for synthesis of active monoclonal antibodies. ACS Synth. Biol. 6 (7), 1370–1379.

Quast, R.B., Mrusek, D., Hoffmeisterc, C., Sonnabend, A., Kubick, S., 2015. Cotranslational incorporation of non-standard amino acids using cell-free protein synthesis. FEBS Lett. 589 (15), 1703–1712.

Schoborg, J.A., Clark, L.G., Cline, P.C., Gustafson, W.C., DeLisa, M.P., 2018. Escherichia coli-based cell-free protein synthesis for the generation of secretory and membrane proteins. Methods 147, 61–66.

Schoborg, J.A., Hodgman, C.E., Anderson, M.J., Jewett, M.C., 2014. Substrate replenishment and byproduct removal improve yeast cell-free protein synthesis. Biotechnol. J. 9 (5), 630–640.

Thoring, L., Dondapati, S.K., Stech, M., Wüstenhagen, D.A., Kubick, S., 2017. High-yield production of "difficult-to-express" proteins in a continuous exchange cell-free system based on CHO cell lysates. Sci. Rep. 7 (1), 11710.

Woodrow, K.A., Airen, I.O., Swartz, J.R., 2006. Enzymatic production of biotin sulfoximine and biotinylated proteins in cell-free extracts. Biotechnol. Bioeng. 95 (4), 633–643.

PART III

Automation and digital integration

CHAPTER

7

Microfluidics and miniaturization

7.1 Introduction

Microfluidics represents a transformative technology for cell-free protein synthesis (CFPS), enabling precise control over reaction conditions, dramatic reduction in reagent consumption, and unprecedented opportunities for parallelization and automation. The integration of microfluidic systems with CFPS platforms creates powerful tools for protein production, screening, and analysis at scales ranging from single molecules to industrial applications. This chapter provides comprehensive coverage of microfluidic technologies, design principles, and applications specific to CFPS.

The miniaturization of cell-free systems offers unique advantages that extend far beyond simple scale reduction. Table 7.1 outlines the fundamental advantages and considerations for microfluidic implementation in CFPS applications.

Fig. 7.1 illustrates the microfluidic technology landscape showing different device architectures and their applications in CFPS.

7.2 Droplet-based cell-free systems

7.2.1 Droplet generation and control

Droplet-based microfluidics enables the compartmentalization of cell-free reactions into discrete picoliter to nanoliter volumes, creating isolated reaction environments that prevent cross-contamination while enabling massive parallelization. The precise control of droplet formation and composition is crucial for reproducible protein synthesis.

The design and operation of droplet generation systems requires careful consideration of multiple factors affecting droplet formation and stability. Table 7.2 details the key parameters and design considerations for droplet-based CFPS systems.

Cell Free Synthesis of Biologics
https://doi.org/10.1016/B978-0-443-33271-5.00003-X

TABLE 7.1 Fundamental advantages and considerations for microfluidic CFPS.

Advantage category	Specific benefits	Technical basis	Implementation considerations	Impact metrics
Reduced reagent consumption	• 100–10,000× volume reduction • Cost savings • Rare reagent conservation	• High surface-to-volume ratios • Precise fluid control • Minimal dead volumes	• Channel design optimization • Dead volume minimization • Sample introduction methods	• Reagent cost reduction > 90% • Volume scaling: nL–μL • Cost per assay <$0.01
Enhanced mixing and control	• Rapid mixing • Precise concentration control • Temporal resolution	• Laminar flow patterns • Diffusion-based mixing • Geometric mixing enhancement	• Mixer design • Flow rate optimization • Residence time control	• Mixing time < 1 s • Concentration accuracy ± 1% • Temporal resolution < 100 ms
Parallelization capabilities	• High-throughput screening • Multiple conditions • Statistical analysis	• Parallel channel arrays • Automated fluid handling • Integrated detection	• Channel uniformity • Flow distribution • Cross-contamination prevention	• >1000 parallel reactions • Throughput > 10,000 assays/day • CV < 5% between channels
Real-time monitoring	• Kinetic measurements • Process optimization • Quality control	• Integrated sensors • Optical access • Real-time analysis	• Sensor integration • Signal processing • Data management	• Temporal resolution < 1 min • Signal-to-noise > 20 • Real-time feedback

7.2.2 Surfactant systems and biocompatibility

The selection and optimization of surfactant systems is critical for maintaining droplet stability while preserving the activity of cell-free components. Table 7.3 compares different surfactant systems and their compatibility with CFPS reactions.

7.2.3 Droplet manipulation and analysis

Advanced droplet manipulation techniques enable sorting, fusion, splitting, and analysis of individual droplets containing different reaction conditions or products. Table 7.4 outlines key manipulation techniques and their applications.

7.2.4 High-throughput screening applications

Droplet-based systems excel in high-throughput screening (HTS) applications where large numbers of conditions or variants must be tested rapidly. Table 7.5 details screening applications and their implementation strategies.

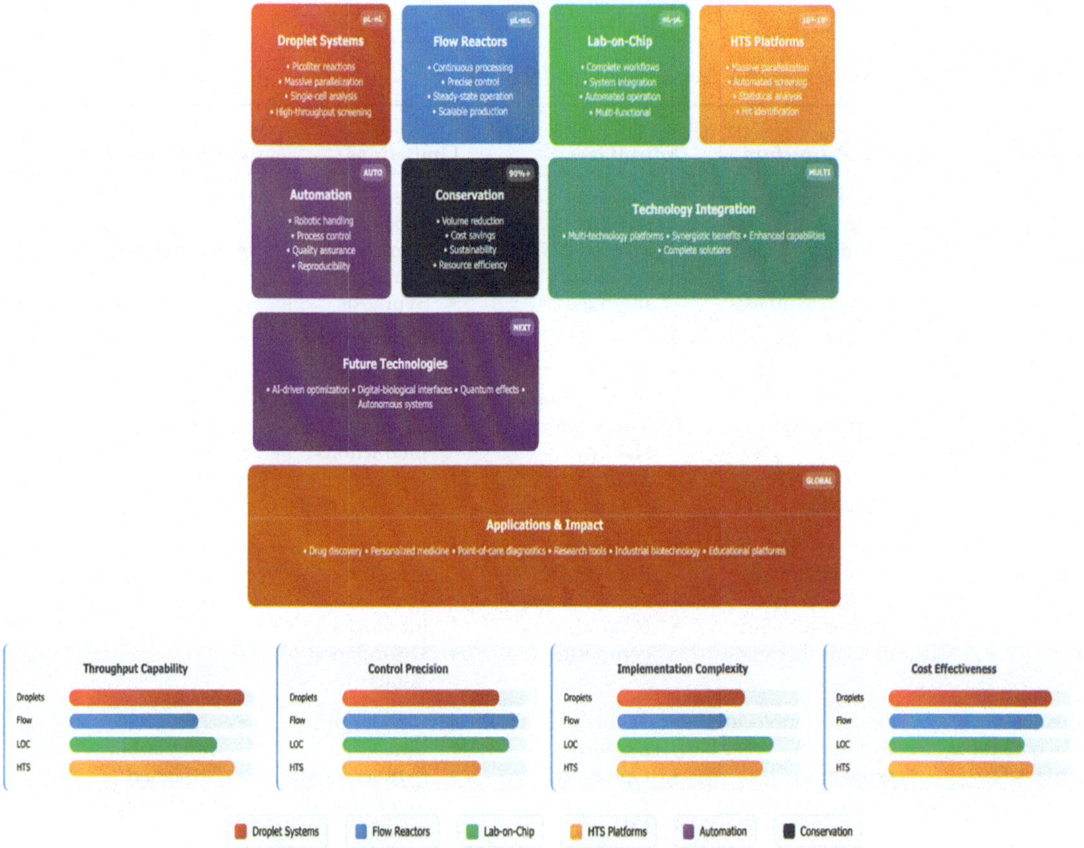

FIG. 7.1 Microfluidic technology landscape for CFPS showing the major technology categories and their characteristics. Each technology offers distinct advantages in terms of scale, throughout, control precision, and cost-effectiveness. The performance comparison charts demonstrate the relative strength of each approach across key metrics, enabling informed technology selection for specific applications.

TABLE 7.2 Droplet generation parameters and design considerations.

Parameter	Typical range	Optimal values	Control methods	Impact on performance
Droplet size	1–1000 μm diameter	10–100 μm	• Flow rate ratios • Channel geometry • Fluid properties	• Reaction volume • Surface effects • Incubation time
Generation frequency	1–10,000 Hz	100–1000 Hz	• Flow rates • Pressure control • Channel design	• Throughput • Uniformity • Stability
Size uniformity	CV: 1%–10%	CV < 3%	• Flow stability • Temperature control • Pressure regulation	• Reproducibility • Statistical analysis • Quality control
Droplet composition	Variable ratios	Application-specific	• Multiinput mixing • Concentration gradients • Temporal control	• Reaction screening capabilities • Functional outcomes
Stability	Minutes to days	>24 h	• Surfactant selection • Temperature control • Storage conditions	• Reaction completion • Long-term studies • Product preservation

TABLE 7.3 Surfactant systems for droplet-based CFPS.

Surfactant type	Chemical structure	Advantages	Limitations	CFPS compatibility
Fluorinated surfactants	• Perfluorinated chains • Hydrophilic head groups • Custom synthesis	• Excellent stability • Low solubility • Chemical inertness	• Cost considerations • Limited availability • Synthesis complexity	• High compatibility • Minimal interference • Long-term stability
Silicone-based surfactants	• Polydimethylsiloxane backbone • Functional end groups • Commercial availability	• Good biocompatibility • Temperature stability • Easy procurement	• Moderate stability • Potential interactions • Limited customization	• Good compatibility • Some protein interactions • Moderate performance
Protein-based surfactants	• Bovine serum albumin (BSA), casein • Natural proteins • Biological origin	• Excellent biocompatibility • No toxicity • Natural composition	• Limited stability • Protein interactions • Degradation issues	• Excellent biocompatibility • Potential competition • Short-term use
Custom biocompatible surfactants	• Designed molecules • Optimized properties • Application-specific	• Tailored properties • Optimized performance • Minimal interference	• Development time • Synthesis costs • Validation requirements	• Optimized compatibility • Designed performance • High cost

TABLE 7.4 Droplet manipulation techniques and applications.

Manipulation method	Technical approach	Capabilities	Applications	Performance metrics
Electrocoalescence	• Electric field application • Controlled droplet fusion • Selective merging	• Reagent addition • Reaction initiation • Multistep reactions	• Kinetic studies • Sequential reactions • Product analysis	• Fusion efficiency > 95% • Timing precision ± 10 ms • Volume accuracy ± 2%
Dielectrophoretic sorting	• Frequency-dependent forces • Dielectric property differences • Active sorting	• Product-based sorting • Quality selection • Enrichment	• Product purification • Quality control • Rare event detection	• Sorting efficiency > 90% • Throughput > 1000 Hz • Purity enhancement > 10 ×

Manipulation method	Technical approach	Capabilities	Applications	Performance metrics
Fluorescence-activated sorting	• Optical detection • Fluorescence-based decisions • Real-time analysis	• Activity-based sorting • Product identification • Functional screening	• Enzyme screening • Activity studies • Product development	• Detection sensitivity < 1 nM • Sorting accuracy > 95% • Processing rate > 1000 Hz
Pneumatic valving	• Pressure-controlled routing • Mechanical switching • Precise control	• Droplet routing • Storage • Sequential processing	• Reaction control • Storage systems • Complex workflows	• Switching time < 10 ms • Reliability > 99% • Pressure range 0–100 kPa

TABLE 7.5 High-throughput screening applications in droplet CFPS.

Screening application	Target parameters	Implementation strategy	Throughput capabilities	Success metrics
Protein variant screening	• Activity levels • Stability properties • Binding affinities	• Library expression • Activity detection • Sorting/collection	• $>10^6$ variants/day • Single-cell resolution • Quantitative readouts	• Hit identification > 0.1% • Dynamic range > 100× • False-positive rate < 1%
Condition optimization	• Buffer compositions • Temperature ranges • Additive effects	• Gradient generation • Combinatorial mixing • Parallel testing	• $>10^4$ conditions/day • Multiparameter space • Statistical analysis	• Optimal condition ID • Parameter interactions • Reproducible results
Inhibitor discovery	• Compound libraries • Dose–response curves • Selectivity profiles	• Compound addition • Concentration series • Activity measurement	• $>10^5$ compounds/day • Dose–response analysis • Multitarget screening	• Hit rate > 0.01% • IC_{50} determination • Selectivity ratios
Directed evolution	• Mutation libraries • Selection pressure • Evolutionary pathways	• Random mutagenesis • Selection systems • Iterative screening	• $>10^7$ mutants/round • Multiple generations • Fitness tracking	• Improvement rate > 2× • Generation time < 24 h • Diversity maintenance

7.3 Continuous-flow microreactors

7.3.1 Reactor design and flow dynamics

Continuous-flow microreactors enable steady-state protein synthesis with precise control over residence time, mixing, and environmental conditions. The design of effective microreactors requires understanding of fluid dynamics and mass transfer at the microscale.

The optimization of microreactor design involves balancing multiple competing factors to achieve optimal performance. Table 7.6 details key design parameters and their optimization strategies.

7.3.2 Mass transfer and mixing optimization

Effective mixing and mass transfer are crucial for microreactor performance, particularly for reactions involving multiple components or phases. Table 7.7 analyzes different mixing strategies and their effectiveness.

7.3.3 Multiphase systems and interfaces

Many advanced CFPS applications require multiphase systems for membrane protein synthesis, product separation, or environmental control. Table 7.8 details multiphase system designs and their applications.

TABLE 7.6 Microreactor design parameters and optimization strategies.

Design parameter	Typical range	Optimization target	Design strategy	Performance impact
Channel dimensions	• Width: 10–1000 μm • Height: 10–200 μm • Length: 1–100 cm	• Optimal mixing • Controlled residence time • Minimal pressure drop	• Aspect ratio optimization • Hydraulic diameter control • Scaling relationships	• Mixing efficiency • Reaction time • Pumping requirements
Flow rates	• 0.1–1000 μL/min	• Application-specific	• Reynolds number control • Mixing time optimization • Residence time targeting	• Mixing quality • Reaction kinetics • System stability
Mixing elements	• Passive mixers • Active mixers • Chaotic advection	• Complete mixing • Minimal residence time • Low-pressure drop	• Geometry optimization • Multiscale mixing • Flow conditioning	• Mixing uniformity • Reaction efficiency • Energy consumption
Temperature control	• Heating/cooling zones • Temperature gradients • Thermal cycling	• Precise control • Uniform distribution • Rapid response	• High surface area • Thermal mass minimization • Control algorithms	• Temperature uniformity • Response time • Energy efficiency

TABLE 7.7 Mixing strategies and mass transfer enhancement.

Mixing strategy	Mechanism	Advantages	Limitations	Optimal applications
Passive mixing	• Geometric features • Split-and-recombine • Chaotic advection	• No external power • Simple fabrication • Reliable operation	• Limited control • Fixed mixing ratio • Pressure drop	• Simple reactions • Fixed compositions • High-throughput systems
Active mixing	• External energy input • Ultrasonic agitation • Magnetic stirring	• Enhanced mixing • Variable control • Adaptive operation	• Complexity increase • Power requirements • Potential interference	• Complex reactions • Variable conditions • Precision applications
Electrokinetic mixing	• Electric field effects • Electroosmotic flow • Electrophoretic forces	• Rapid mixing • No moving parts • Precise control	• Buffer requi • rements • Electrode fouling • Limited materials	• Ionic solutions • Precise control needs • Research applications
Acoustic mixing	• Ultrasonic waves • Acoustic streaming • Cavitation effects	• Gentle mixing • No contamination • Broad applicability	• Equipment complexity • Power requirements • Frequency optimization	• Sensitive biological systems • Gentle processing • High-quality products

TABLE 7.8 Multiphase microfluidic systems for CFPS.

System type	Phase configuration	Interface control	Applications	Design challenges
Segmented flow	• Aqueous plugs • Oil/gas separation • Moving boundaries	• Surface tension • Flow rate ratios • Channel wetting	• Reaction compartments • Heat/mass transfer • Product separation	• Interface stability • Plug uniformity • Wall interactions
Parallel flow	• Side-by-side phases • Stable interfaces • Controlled contact	• Flow rate balance • Pressure matching • Surface properties	• Extraction processes • Gradient formation • Membrane processes	• Interface position • Flow stability • Mass transfer control
Membrane integration	• Selective barriers • Controlled permeation • Separation processes	• Membrane properties • Pressure differences • Selective transport	• Product purification • Substrate supply • Waste removal	• Membrane fouling • Pressure management • Long-term stability
Emulsion systems	• Dispersed phases • Controlled droplets • Stabilized interfaces	• Surfactant systems • Emulsion stability • Size control	• Encapsulation • Controlled release • Reaction isolation	• Stability maintenance • Size uniformity • Biocompatibility

TABLE 7.9 Process intensification strategies in microfluidic CFPS.

Intensification strategy	Technical approach	Performance enhancement	Implementation methods	Quantitative benefits
Reaction coupling	• Sequential reactions • Product consumption • Equilibrium shifting	• Higher conversions • Reduced byproducts • Simplified purification	• Multistage reactors • Residence time control • Product removal	• Conversion > 95% • Selectivity > 99% • Purification 10 × easier
Heat integration	• Heat exchange networks • Temperature cascades • Energy recovery	• Reduced energy consumption • Improved control • Enhanced stability	• Integrated heat exchangers • Temperature profiling • Thermal management	• Energy savings > 50% • Control improvement • Stability enhancement
Multifunctional integration	• Combined operations • Simultaneous processes • Integrated functionality	• Space efficiency • Reduced complexity • Enhanced performance	• Multifunctional devices • Parallel processing • System integration	• Footprint reduction 10× • Complexity reduction • Performance gains
Continuous processing	• Steady-state operation • Continuous product removal • Process stability	• Consistent quality • Higher productivity • Reduced batch effects	• Flow control systems • Product separation • Process monitoring	• Quality CV < 5% • Productivity 5× higher • Reduced variation

7.3.4 Process intensification strategies

Microfluidic systems enable various process intensification strategies that enhance productivity and efficiency. Table 7.9 outlines key intensification approaches and their implementation.

7.4 Lab-on-a-chip integration

7.4.1 Integrated system architecture

Lab-on-a-chip (LOC) systems integrate multiple laboratory functions onto a single microfluidic platform, enabling complete workflows from sample preparation to analysis. The design of integrated CFPS systems requires careful coordination of multiple subsystems.

The architecture of integrated LOC systems involves multiple functional modules that must work together seamlessly. Table 7.10 details the key functional modules and their integration requirements.

TABLE 7.10 Functional modules for integrated CFPS-LOC systems.

Functional module	Core functions	Integration requirements	Performance specifications	Interface considerations
Sample preparation	• Reagent storage • Mixing systems • Conditioning	• Fluid routing • Volume control • Cross-contamination prevention	• Volume accuracy ± 1% • Mixing uniformity > 95% • Contamination < 0.1%	• Standardized connections • Automated handling • Quality control
Reaction chambers	• Protein synthesis • Environmental control • Real-time monitoring	• Temperature control • Atmosphere management • Sensor integration	• Temperature stability ± 0.1°C • Atmosphere control • Sensor sensitivity	• Thermal isolation • Gas connections • Optical access
Product processing	• Purification • Concentration • Modification	• Separation systems • Recovery efficiency • Product handling	• Recovery > 90% • Purity > 95% • Processing time < 1 h	• Flow control • Collection systems • Storage integration
Analysis systems	• Detection • Quantification • Characterization	• Optical • systems • Sensor arrays • Data processing	• Detection limit < 1 nM • Dynamic range > 1000 × • Analysis time < 10 min	• Light coupling • Signal processing • Data management
Control systems	• Process control • Automation • Data management	• Sensor networks • Actuator control • Communication	• Response time < 1 s • Accuracy > 99% • Reliability > 95%	• Communication protocols • Power management • User interfaces

7.4.2 Automation and control integration

Advanced automation and control systems are essential for LOC operation, enabling complex workflows while maintaining reproducibility and reliability. Table 7.11 outlines automation strategies and their implementation.

7.4.3 Sensor integration and real-time monitoring

Real-time monitoring capabilities are crucial for LOC systems, enabling process control, quality assurance (QA), and optimization. Table 7.12 details sensor integration strategies and monitoring capabilities.

7.4.4 Data management and analysis

Integrated LOC systems generate large amounts of data that require sophisticated management and analysis systems. Table 7.13 outlines data management strategies and analytical approaches.

TABLE 7.11 Automation and control strategies for LOC-CFPS systems.

Automation level	Control elements	Implementation approach	Capabilities	Complexity management
Basic automation	• Pump control • Valve switching • Temperature regulation	• Programmable controllers • Simple algorithms • Manual supervision	• Reproducible operations • Reduced operator error • Basic quality control	• Modular programming • Standard interfaces • Simple diagnostics
Advanced automation	• Multiparameter control • Feedback systems • Adaptive algorithms	• PLC/computer control • Sensor networks • Real-time optimization	• Optimized performance • Adaptive operation • Quality assurance	• Hierarchical control • Modular architecture • Diagnostic systems
Intelligent automation	• AI-driven control • Machine learning • Predictive systems	• Advanced algorithms • Data analytics • Autonomous operation	• Self-optimization • Predictive maintenance • Autonomous operation	• Model-based control • Exception handling • Learning systems
Full integration	• End-to-end automation • System coordination • Workflow management	• Enterprise integration • Database systems • Remote monitoring	• Complete automation • Data integration • Remote operation	• System architecture • Interface standards • Security protocols

TABLE 7.12 Sensor integration and monitoring systems for LOC-CFPS.

Sensor type	Measurement parameters	Integration method	Performance specifications	Applications
Optical sensors	• Fluorescence • Absorbance • Scattering	• Fiber coupling • Integrated optics • Microscopy systems	• Sensitivity < 1 nM • Response time < 1 s • Dynamic range > 1000×	• Protein quantification • Activity monitoring • Quality assessment
Electrochemical sensors	• pH, conductivity • Redox potential • Ion concentrations	• Microelectrodes • Thin-film sensors • Reference systems	• Accuracy ± 0.01 pH • Response time < 10 s • Stability > 24 h	• Environment monitoring • Process control • Quality indicators
Physical sensors	• Temperature • Pressure • Flow rates	• Integrated transducers • MEMS devices • External coupling	• Accuracy ± 0.1°C, ± 0.01 bar • Response time < 1 s • Long-term stability	• Process conditions • Safety monitoring • Performance tracking

TABLE 7.12 Sensor integration and monitoring systems for LOC-CFPS—cont'd

Sensor type	Measurement parameters	Integration method	Performance specifications	Applications
Biological sensors	• Enzyme activity • Protein binding • Cellular responses	• Biosensor integration • Reporter systems • Cell-based assays	• Specificity > 90% • Sensitivity < 10 pM • Response time < 5 min	• Functional assessment • Bioactivity testing • Quality validation

TABLE 7.13 Data management and analysis systems for LOC-CFPS.

Data category	Information content	Management strategy	Analysis methods	Output applications
Process data	• Sensor readings • Control parameters • System status	• Real-time databases • Time-series storage • Automated logging	• Statistical analysis • Trend analysis • Process optimization	• Process control • Quality assurance • Performance optimization
Product data	• Yield measurements • Quality parameters • Characterization results	• Structured databases • Metadata management • Version control	• Comparative analysis • Quality assessment • Batch tracking	• Product release • Quality documentation • Regulatory compliance
Experimental data	• Screening results • Optimization studies • Method development	• Experimental databases • Data standardization • Cross-referencing	• Machine learning • Predictive modeling • Knowledge extraction	• Method improvement • Predictive capabilities • Knowledge management
System data	• Performance metrics • Maintenance records • Error logs	• Maintenance databases • Performance tracking • Diagnostic systems	• Reliability analysis • Predictive maintenance • Failure analysis	• System optimization • Maintenance planning • Reliability improvement

7.5 HTS platforms

7.5.1 Platform design and architecture

HTS platforms for CFPS applications require sophisticated design to handle thousands of samples while maintaining quality and reproducibility. The architecture must balance throughput, flexibility, and operational simplicity.

The design of effective HTS platforms involves integration of multiple subsystems optimized for high-throughput operation. Table 7.14 details key design considerations and performance targets.

TABLE 7.14 High-throughput screening platform design considerations.

Design aspect	Technical requirements	Implementation strategy	Performance targets	Validation methods
Sample handling	• Automated dispensing • Cross-contamination prevention • Volume accuracy	• Robotic systems • Disposable tips • Positive displacement	• Throughput > 10,000 samples/day • CV < 2% • Contamination < 0.01%	• Accuracy testing • Precision validation • Contamination studies
Reaction processing	• Parallel incubation • Environmental control • Timing precision	• Multiwell systems • Thermal control • Atmosphere management	• Parallel capacity > 1000 • Temperature uniformity ± 0.2°C • Timing accuracy ± 1%	• Temperature mapping • Uniformity testing • Timing validation
Detection systems	• Multiparameter detection • High sensitivity • Rapid analysis	• Plate readers • Imaging systems • Flow cytometry	• Detection limit < 1 nM • Analysis time < 1 min/sample • Dynamic range > 1000 ×	• Sensitivity testing • Linearity validation • Speed benchmarking
Data processing	• Real-time analysis • Quality control • Hit identification	• Automated algorithms • Statistical analysis • Database integration	• Processing time < 1 min/plate • Hit detection > 95% • False positive < 1%	• Algorithm validation • Statistical testing • Benchmark studies

7.5.2 Screening methodologies and applications

Different screening methodologies are suited for different types of investigations, from basic parameter optimization to complex evolutionary studies. Table 7.15 compares various screening approaches and their applications.

7.5.3 Quality control and validation

Robust quality control systems are essential for HTS platforms to ensure data reliability and reproducibility. Table 7.16 outlines quality control strategies and validation approaches.

7.5.4 Data analysis and hit identification

Sophisticated data analysis methods are required to extract meaningful information from HTS datasets and identify promising hits for further development. Table 7.17 details analysis approaches and hit identification strategies.

TABLE 7.15 Screening methodologies for microfluidic CFPS applications.

Screening method	Approach	Target applications	Throughput capability	Data quality
Parameter optimization	• Grid-based screening • Factorial designs • Response surfaces	• Buffer optimization • Temperature studies • Concentration effects	• 10^2–10^4 conditions • Systematic coverage • Statistical analysis	• High precision • Quantitative results • Reproducible data
Library screening	• Random libraries • Systematic variants • Combinatorial approaches	• Protein engineering • Catalyst discovery • Function optimization	• 10^4–10^6 variants • Diversity sampling • Hit identification	• Qualitative screening • Ranking systems • Hit validation
Evolutionary screening	• Iterative selection • Mutagenesis rounds • Fitness landscapes	• Directed evolution • Adaptation studies • Optimization paths	• 10^6–10^8 variants/round • Multiple generations • Evolutionary tracking	• Fitness measurements • Population dynamics • Selection analysis
Functional screening	• Activity assays • Binding studies • Phenotypic analysis	• Drug discovery • Biomarker identification • Diagnostic development	• 10^3–10^5 compounds • Multiparameter analysis • Mechanistic studies	• Functional readouts • Mechanistic data • Validation studies

TABLE 7.16 Quality control and validation strategies for HTS platforms.

Quality aspect	Control methods	Validation approach	Acceptance criteria	Corrective actions
System performance	• Positive controls • Negative controls • Reference standards	• Daily validation • Trend analysis • Performance tracking	• Control CV < 5% • Signal-to-noise > 20 • Detection consistency	• System recalibration • Component replacement • Method adjustment
Data Quality	• Statistical analysis • Outlier detection • Replicate analysis	• Automated QC algorithms • Manual review • Statistical validation	• Z-score < 3 • Replicate CV < 10% • Pass rate > 95%	• Data flagging • Repeat analysis • Method refinement
Cross-contamination	• Spatial controls • Temporal monitoring • Tracer studies	• Regular testing • Method validation • System qualification	• Contamination • < 0.1% • Spatial independence • Temporal stability	• Cleaning optimization • Protocol revision • Hardware modification
Reproducibility	• Interplate controls • Operator validation • Long-term tracking	• Multioperator studies • Interlaboratory comparison • Method transfer	• Plate-to-plate CV < 15% • Operator agreement > 90% • Lab-to-lab correlation > 0.9	• Training programs • Protocol standardization • Method improvement

TABLE 7.17 Data analysis and hit identification strategies.

Analysis category	Methods	Applications	Performance metrics	Output quality
Statistical analysis	• Descriptive statistics • Hypothesis testing • Confidence intervals	• Quality assessment • Significance testing • Uncertainty quantification	• Power > 80% • α-level < 0.05 • Confidence > 95%	• Statistical rigor • Quantified uncertainty • Validated conclusions
Hit identification	• Threshold methods • Ranking algorithms • Machine learning	• Active compound ID • Lead discovery • Optimization starting points	• Hit rate 0.1–1% • False positive < 5% • Enrichment > 10 ×	• Validated hits • Ranked priorities • Confirmed activities
Pattern recognition	• Clustering analysis • Principal components • Network analysis	• Structure-activity relationships • Mechanism identification • Target classification	• Cluster purity > 80% • Variance explained > 70% • Network modularity > 0.3	• Mechanistic insights • Grouping strategies • Predictive models
Predictive modeling	• Regression analysis • Machine learning • Neural networks	• Activity prediction • Property estimation • Optimization guidance	• $R^2 > 0.8$ • Cross-validation > 0.7 • Prediction accuracy > 80%	• Predictive capability • Model robustness • Optimization guidance

7.6 Automated sample processing

7.6.1 Automation system architecture

Automated sample processing systems integrate multiple components to handle the complex workflows required for microfluidic CFPS applications. The architecture must balance flexibility, reliability, and throughput while maintaining precise control over all process parameters.

The design of comprehensive automation systems requires careful integration of multiple subsystems. Table 7.18 details the key components and their integration requirements.

7.6.2 Workflow design and optimization

Efficient workflow design is crucial for maximizing automation system productivity while maintaining quality and flexibility. Table 7.19 outlines workflow optimization strategies and their implementation.

7.6.3 Quality assurance and validation

Automated systems require comprehensive QA programs to ensure consistent performance and regulatory compliance. Table 7.20 details QA strategies and validation approaches.

TABLE 7.18 Automation system components and integration requirements.

System component	Core functions	Integration requirements	Performance specifications	Reliability targets
Liquid handling systems	• Reagent dispensing • Sample preparation • Dilution series	• Volume control • Contamination prevention • Multichannel operation	• Volume range: 1 nL–1 mL • Accuracy: ± 1% • Precision CV < 2%	• Mean time between failures (MTBF) > 1000 h • Contamination < 0.01% • Calibration stability
Thermal management	• Temperature control • Heating/cooling cycles • Gradient generation	• Precise control • Uniform distribution • Rapid response	• Range: 4°C–95°C • Accuracy: ± 0.1°C • Ramp rate: 5°C/min	• Temperature stability • Long-term accuracy • Thermal uniformity
Detection systems	• Multimode detection • Real-time monitoring • Data acquisition	• Optical integration • Signal processing • Data management	• Sensitivity < 1 nM • Dynamic range > 1000× • Response time < 1 s	• Signal stability • Detector lifetime • Measurement reproducibility
Control software	• Workflow management • Process control • Data analysis	• Hardware integration • User interface • Database connectivity	• Response time < 100 ms • Uptime > 99% • Data integrity 100%	• Software reliability • Error handling • Update compatibility
Safety systems	• Emergency shutdown • Containment • Monitoring	• Sensor integration • Automatic response • Alert systems	• Response time < 1 s • Detection coverage 100% • False alarm rate < 1%	• Safety compliance • System integrity • Emergency response

TABLE 7.19 Workflow design and optimization strategies.

Workflow aspect	Design principles	Optimization methods	Performance benefits	Implementation challenges
Process sequencing	• Parallel operations • Resource optimization • Bottleneck elimination	• Critical path analysis • Resource allocation • Load balancing	• Throughput increase > 50% • Resource utilization > 80% • Cycle time reduction	• Complexity management • Synchronization • Error propagation
Sample scheduling	• Batch optimization • Priority management • Resource conflicts	• Scheduling algorithms • Queue management • Dynamic allocation	• Wait time reduction • Resource efficiency • Flexibility enhancement	• Algorithm complexity • Real-time decisions • Constraint handling

Continued

TABLE 7.19 Workflow design and optimization strategies—cont'd

Workflow aspect	Design principles	Optimization methods	Performance benefits	Implementation challenges
Quality integration	• In-process monitoring • Real-time QC • Adaptive control	• Sensor integration • Statistical control • Feedback systems	• Quality improvement • Waste reduction • Process robustness	• Sensor complexity • Control algorithms • System integration
Error handling	• Error detection • Recovery protocols • System resilience	• Redundancy design • Graceful degradation • Automatic recovery	• System reliability • Uptime improvement • Data integrity	• Error complexity • Recovery design • Validation requirements

TABLE 7.20 Quality assurance and validation frameworks for automated systems.

QA category	Validation elements	Testing protocols	Documentation requirements	Compliance standards
Installation qualification (IQ)	• Hardware installation • Software deployment • Utility connections	• Installation verification • Component testing • System integration	• Installation records • Component certificates • Integration testing	• GMP compliance • Equipment standards • Safety regulations
Operational qualification (OQ)	• Function testing • Performance verification • Operating ranges	• Function validation • Performance testing • Limit testing	• Test protocols • Results documentation • Deviation reports	• Performance standards • Operating procedures • Quality requirements
Performance qualification (PQ)	• Process validation • Product quality • Routine operation	• Process studies • Quality assessment • Long-term testing	• Validation reports • Statistical analysis • Trend data	• Process validation • Product specifications • Regulatory requirements
Continuous verification	• Performance monitoring • Drift detection • Preventive maintenance	• Control charts • Trend analysis • Maintenance schedules	• Performance data • Maintenance records • Change control	• Ongoing compliance • Performance maintenance • Continuous improvement

TABLE 7.21 Maintenance and support system requirements.

Maintenance category	Activities	Scheduling	Performance impact	Resource requirements
Preventive maintenance	• Component replacement • Calibration • Cleaning cycles	• Time-based • Usage-based • Condi-tion-based	• Reliability improvement • Performance stability • Cost optimization	• Maintenance staff • Spare parts inventory • Calibration standards
Predictive maintenance	• Condition monitoring • Trend analysis • Failure prediction	• Data-driven • Algorithm-based • Real-time monitoring	• Unplanned downtime reduction • Maintenance optimization • Cost savings	• Monitoring systems • Analysis software • Technical expertise
Corrective maintenance	• Failure diagnosis • Repair procedures • System restoration	• On-demand • Emergency response • Rapid resolution	• Downtime minimization • System recovery • Performance restoration	• Technical support • Repair capabilities • Emergency procedures
System upgrades	• Software updates • Hardware improvements • Capability enhancement	• Planned upgrades • Performance needs • Technology evolution	• Capability enhancement • Performance improvement • Technology currency	• Upgrade planning • Validation resources • Change management

7.6.4 Maintenance and support systems

Comprehensive maintenance and support systems are essential for sustained automation system performance and availability. Table 7.21 outlines maintenance strategies and support requirements.

7.7 Reagent conservation strategies

7.7.1 Miniaturization benefits and challenges

Reagent conservation through miniaturization offers significant economic and practical advantages, particularly for expensive or rare components. However, miniaturization also introduces unique challenges that must be addressed through careful system design.

The optimization of reagent conservation requires balancing multiple competing factors while maintaining system performance. Table 7.22 details conservation strategies and their implementation considerations.

TABLE 7.22 Reagent conservation strategies and implementation considerations.

Conservation strategy	Technical approach	Conservation benefits	Implementation challenges	Success metrics
Volume miniaturization	• Microscale reactions • High-density arrays • Droplet systems	• 100–10,000 × reduction • Cost savings > 90% • Rare reagent enablement	• Surface effects • Evaporation control • Handling precision	• Volume scaling factor • Cost reduction achieved • Performance maintenance
Reagent recycling	• Product separation • Component recovery • Reuse systems	• Effective utilization > 80% • Waste reduction • Resource efficiency	• Separation efficiency • Purity maintenance • System complexity	• Recovery efficiency • Reuse cycles • Quality maintenance
Gradient systems	• Concentration gradients • Systematic screening • Efficient sampling	• Systematic exploration • Reduced experiments • Optimized usage	• Gradient stability • Mixing control • Reproducibility	• Gradient quality • Screening efficiency • Result reproducibility
Just-in-time mixing	• On-demand preparation • Fresh reagents • Minimal waste	• Stability preservation • Waste elimination • Quality assurance	• Mixing precision • Timing control • System complexity	• Mixing accuracy • Timing precision • Waste reduction

7.7.2 Surface effects and mitigation

At microscale dimensions, surface effects become increasingly important and can significantly impact reaction performance. Table 7.23 analyzes surface effects and mitigation strategies.

7.7.3 Cost–benefit analysis

Comprehensive cost–benefit analysis is essential for evaluating the economic impact of reagent conservation strategies. Table 7.24 provides a framework for economic evaluation.

7.7.4 Sustainability considerations

Reagent conservation contributes to sustainability goals through reduced resource consumption and waste generation. Table 7.25 outlines sustainability metrics and assessment methods.

TABLE 7.23 Surface effects and mitigation strategies in microscale CFPS.

Surface effect	Mechanism	Impact on CFPS	Mitigation strategies	Effectiveness
Protein adsorption	• Hydrophobic interactions • Electrostatic binding • Van der Waals forces	• Component depletion • Activity loss • System fouling	• Surface passivation • Blocking agents • Surface modification	• Adsorption reduction > 90% • Activity preservation • System longevity
Evaporation effects	• High surface-to-volume ratio • Concentration changes • Osmotic stress	• Concentration drift • System imbalance • Reaction failure	• Vapor barriers • Humidity control • Sealed systems	• Evaporation control • Concentration stability • Reaction consistency
Wall catalysis	• Surface-mediated reactions • Unwanted catalysis • Side reactions	• Product contamination • Yield reduction • Selectivity loss	• Inert surfaces • Surface treatments • Passivation layers	• Catalysis elimination • Purity improvement • Selectivity enhancement
Mass transfer limitations	• Diffusion distances • Concentration gradients • Transport resistance	• Reaction limitations • Kinetic changes • Performance reduction	• Enhanced mixing • Flow optimization • Geometry design	• Transport enhancement • Kinetic improvement • Performance recovery

TABLE 7.24 Cost–benefit analysis framework for reagent conservation.

Cost category	Cost components	Benefit categories	Quantification methods	Return on investment (ROI) calculation
Implementation costs	• Equipment costs • Development time • Validation expenses	• Direct cost savings • Reagent reduction • Waste elimination	• Capital expenditure • Operating expenses • Time-value analysis	• Payback period • Net present value • Internal rate of return
Operating costs	• Maintenance • Consumables • Labor	• Efficiency gains • Throughput improvement • Quality enhancement	• Cost per experiment • Productivity metrics • Quality costs	• Operating ratio • Efficiency improvement • Quality benefits

Continued

TABLE 7.24 Cost–benefit analysis framework for reagent conservation—cont'd

Cost category	Cost components	Benefit categories	Quantification methods	Return on investment (ROI) calculation
Opportunity costs	• Alternative approaches • Delayed implementation • Risk factors	• Market advantages • Competitive position • Strategic benefits	• Market analysis • Competitive assessment • Strategic value	• Strategic value • Market position • Long-term benefits
Risk factors	• Technical risks • Implementation risks • Market risks	• Risk mitigation • Insurance value • Flexibility benefits	• Risk assessment • Probability analysis • Impact evaluation	• Risk-adjusted returns • Value at risk • Risk premium

TABLE 7.25 Sustainability assessment for reagent conservation systems.

Sustainability aspect	Metrics	Assessment methods	Improvement targets	Reporting standards
Resource efficiency	• Material utilization • Energy consumption • Water usage	• Life cycle assessment • Material flow analysis • Energy audits	• Resource reduction > 50% • Efficiency improvement • Circular economy principles	• Global Reporting Initiative (GRI) standards • ISO 14001 • Sustainability reports
Waste reduction	• Waste generation • Hazardous materials • Disposal requirements	• Waste audits • Material tracking • Disposal analysis	• Waste reduction > 75% • Hazard elimination • Zero waste goals	• Waste management • Environmental compliance • Sustainability metrics
Environmental impact	• Carbon footprint • Toxicity reduction • Ecosystem effects	• Environmental assessment • Impact analysis • Toxicity evaluation	• Carbon neutrality • Toxicity elimination • Ecosystem protection	• Environmental standards • Impact assessment • Sustainability certification
Economic sustainability	• Cost-effectiveness • Economic viability • Long-term sustainability	• Economic analysis • Cost-benefit assessment • Viability studies	• Economic optimization • Long-term viability • Sustainable business models	• Economic reporting • Financial sustainability • Business case development

Fig. 7.1 illustrates the comprehensive microfluidic technology landscape showing different approaches and their integration possibilities.

Fig. 7.2 demonstrates the reagent conservation hierarchy and its impact on sustainability and economics.

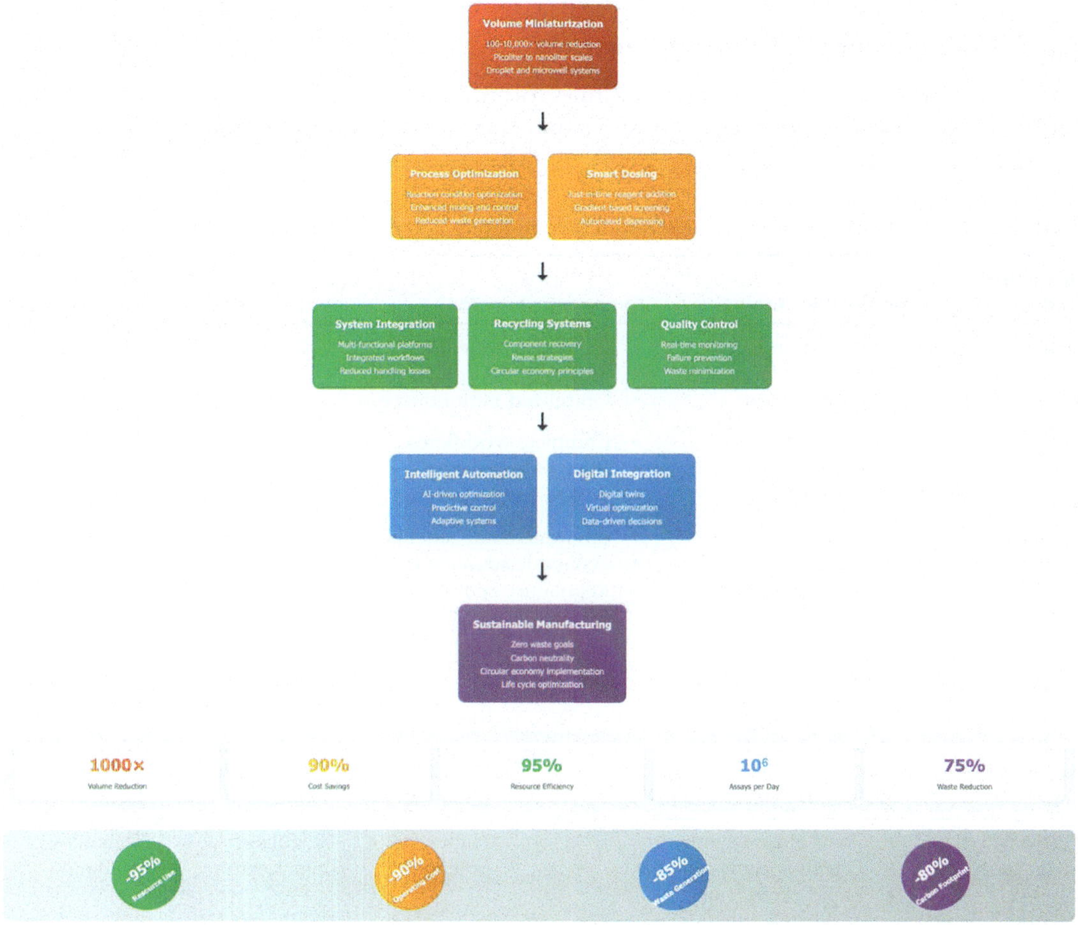

FIG. 7.2 Reagent conservation hierarchy showing the progression from basic volume miniaturization to comprehensive sustainability goals. Each level builds upon the previous one, creating cumulative benefits in resource efficiency, cost reduction, and environmental impact. The metrics demonstrates the quantitative benefits achievable through systemic implementation of conservation strategies, while the sustainability impact circles show the environmental benefits across multiple dimensions.

7.8 Summary and future directions

7.8.1 Technology integration and convergence

The future of microfluidic CFPS lies in the integration and convergence of multiple technologies to create comprehensive platforms. Table 7.26 outlines integration opportunities and their potential impact.

7.8.2 Emerging applications and markets

New applications and markets continue to emerge for microfluidic CFPS technology. Table 7.27 surveys emerging opportunities and their development potential.

TABLE 7.26 Technology integration opportunities and future potential.

Integration strategy	Combined technologies	Synergistic benefits	Development timeline	Commercial impact
Digital microfluidic integration	• Digital control • Microfluidic precision • AI optimization	• Intelligent systems • Autonomous operation • Optimized performance	2–5 years	Revolutionary potential
Multiscale integration	• Molecular to macro • Hierarchical systems • Scale bridging	• Complete workflows • Seamless operation • End-to-end solutions	3–7 years	Transformative impact
Sustainable technology	• Green chemistry • Resource efficiency • Circular economy	• Environmental benefits • Economic advantages • Regulatory compliance	2–6 years	Market driver

TABLE 7.27 Emerging applications and market opportunities.

Application area	Technology requirements	Market potential	Development challenges	Success factors
Personalized medicine	• Point-of-care synthesis • Patient-specific proteins • Rapid production	High growth potential	Regulatory complexity	Technology validation
Space biotechnology	• Resource efficiency • Autonomous operation • Extreme conditions	Niche but valuable	Environmental challenges	Robust systems
Field diagnostics	• Portable systems • Rapid analysis • Minimal infrastructure	Growing demand	Simplicity requirements	User-friendly design
Educational applications	• Teaching tools • Research platforms • Skill development	Expanding market	Cost constraints	Affordable solutions

Bibliography

Agresti, J.J., Antipov, E., Abate, A.R., Ahn, K., Rowat, A.C., Baret, J.C., Marquez, M., Klibanov, A.M., Griffiths, A.D., Weitz, D.A., 2010. Ultrahigh-throughput screening in drop-based microfluidics for directed evolution. Proc. Natl. Acad. Sci. U.S.A. 107 (9), 4004–4009.

Cecchini, M.P., Hong, J., Lim, C., Choo, J., Albrecht, T., deMello, A.J., Edel, J.B., 2011. Ultrafast surface enhanced resonance Raman scattering detection in droplet-based microfluidic systems. Anal. Chem. 83 (8), 3076–3081.

Clausell-Tormos, J., Lieber, D., Baret, J.C., El-Harrak, A., Miller, O.J., Frenz, L., Blouwolff, J., Humphry, K.J., Köster, S., Duan, H., Holtze, C., Weitz, D.A., Griffiths, A.D., Merten, C.A., 2008. Droplet-based microfluidic platforms for the encapsulation and screening of mammalian cells and multicellular organisms. Chem. Biol. 15 (5), 427–437.

Courtois, F., Olguin, L.F., Whyte, G., Theberge, A.B., Huck, W.T., Hollfelder, F., Abell, C., 2009. Controlling the retention of small molecules in emulsion microdroplets for use in cell-based assays. Anal. Chem. 81 (8), 3008–3016.

Fallah-Araghi, A., Meguellati, K., Baret, J.C., El Harrak, A., Mangeat, T., Karplus, M., Ladame, S., Marques, C.M., Griffiths, A.D., 2014. Enhanced chemical synthesis at soft interfaces: a universal reaction-adsorption mechanism in microcompartments. Phys. Rev. Lett. 112 (2), 028301.

Guo, M.T., Rotem, A., Heyman, J.A., Weitz, D.A., 2012. Droplet microfluidics for high-throughput biological assays. Lab Chip 12 (12), 2146–2155.

Huebner, A., Sharma, S., Srisa-Art, M., Hollfelder, F., Edel, J.B., deMello, A.J., 2008. Microdroplets: a sea of applications? Lab Chip 8 (8), 1244–1254.

Kaminski, T.S., Scheler, O., Garstecki, P., 2016. Droplet microfluidics for microbiology: techniques, applications and challenges. Lab Chip 16 (12), 2168–2187.

Mazutis, L., Gilbert, J., Ung, W.L., Weitz, D.A., Griffiths, A.D., Heyman, J.A., 2013. Single-cell analysis and sorting using droplet-based microfluidics. Nat. Protoc. 8 (5), 870–891.

Pekin, D., Skhiri, Y., Baret, J.C., Le Corre, D., Mazutis, L., Salem, C.B., Millot, F., El Harrak, A., Hutchison, J.B., Larson, J.W., Link, D.R., Laurent-Puig, P., Griffiths, A.D., Taly, V., 2011. Quantitative and sensitive detection of rare mutations using droplet-based microfluidics. Lab Chip 11 (13), 2156–2166.

Sciambi, A., Abate, A.R., 2015. Accurate microfluidic sorting of droplets at 30 kHz. Lab Chip 15 (1), 47–51.

Shim, J.U., Cristobal, G., Link, D.R., Thorsen, T., Jia, Y., Piattelli, K., Fraden, S., 2007. Control and measurement of the phase behavior of aqueous solutions using microfluidics. J. Am. Chem. Soc. 129 (28), 8825–8835.

Teh, S.Y., Lin, R., Hung, L.H., Lee, A.P., 2008. Droplet microfluidics. Lab Chip 8 (2), 198–220.

Theberge, A.B., Courtois, F., Schaerli, Y., Fischlechner, M., Abell, C., Hollfelder, F., Huck, W.T., 2010. Microdroplets in microfluidics: an evolving platform for discoveries in chemistry and biology. Angew. Chem. Int. Ed. 49 (34), 5846–5868.

Zhu, P., Wang, L., 2017. Passive and active droplet generation with microfluidics: a review. Lab Chip 17 (1), 34–75.

CHAPTER

8

Artificial intelligence and machine learning

8.1 Introduction

The integration of artificial intelligence (AI) and machine learning (ML) technologies with cell-free protein synthesis (CFPS) represents a paradigm shift that is transforming how we design, optimize, and control biological manufacturing processes. AI-driven approaches enable the analysis of complex, multidimensional datasets to identify optimal conditions, predict system performance, and automate decision-making processes that previously required extensive human expertise and time-consuming experimentation. This chapter provides comprehensive coverage of AI and ML applications in CFPS, from fundamental optimization algorithms to advanced deep learning architectures for protein engineering.

The complexity of cell-free systems, with their numerous interdependent variables and nonlinear relationships, makes them ideal candidates for AI-driven optimization and control. Table 8.1 outlines the fundamental areas where AI and ML technologies are making significant impact in CFPS applications.

Fig. 8.1 illustrates the AI/ML technology landscape in CFPS, showing the interconnections between different approaches and their applications.

8.2 AI-guided system optimization

8.2.1 Multiparameter optimization strategies

CFPS systems involve numerous interdependent parameters that collectively determine system performance. Traditional optimization approaches often struggle with the high-dimensional parameter space and complex interactions between variables. AI-guided optimization strategies provide systematic approaches for navigating these complex landscapes efficiently.

The implementation of effective multiparameter optimization requires careful selection of algorithms and problem formulation. Table 8.2 details various optimization algorithms and their suitability for different CFPS optimization challenges.

Cell Free Synthesis of Biologics
https://doi.org/10.1016/B978-0-443-33271-5.00008-9

TABLE 8.1 AI and ML impact areas in cell-free protein synthesis.

Application area	AI/ML technologies	Key benefits	Implementation complexity
System optimization	• Genetic algorithms • Bayesian optimization • Multiobjective optimization	• Reduced experimental burden • Faster convergence • Global optima identification	Medium–high
Predictive modeling	• Neural networks • Random forests • Support vector machines	• Performance prediction • Design guidance • Risk assessment	Medium
Template design	• Sequence optimization • Structure prediction • Codon optimization	• Enhanced expression • Improved folding • Reduced failures	High
Process control	• Reinforcement learning • Adaptive control • Anomaly detection	• Real-time optimization • Quality assurance • Automated responses	High
Protein engineering	• Deep learning • Graph neural networks • Evolutionary algorithms	• Novel functions • Improved properties • Accelerated discovery	Very high
Experimental design	• Active learning • Design of experiments • Sequential optimization	• Efficient exploration • Resource optimization • Knowledge accumulation	Medium

TABLE 8.2 AI optimization algorithms for CFPS parameter optimization.

Algorithm type	Methodology	Advantages	Limitations	Optimal applications
Genetic algorithms	• Population-based search • Evolutionary operators • Multiobjective capability	• Global optimization • Handles discrete/continuous • Parallelizable	• Slow convergence • Parameter sensitive • No gradient information	• Complex parameter spaces • Multiobjective problems • Discrete variables
Bayesian optimization	• Gaussian process models • Acquisition functions • Sequential design	• Sample efficient • Uncertainty quantification • Handles noise	• Scalability limits • Gaussian assumptions • Computational overhead	• Expensive experiments • Small datasets • Continuous parameters
Particle swarm	• Swarm intelligence • Velocity updates • Social learning	• Fast convergence • Simple implementation • Few parameters	• Premature convergence • Local optima traps • Parameter selection	• Real-valued optimization • Moderate dimensions • Fast screening

Algorithm type	Methodology	Advantages	Limitations	Optimal applications
Differential evolution	• Population differences • Mutation strategies • Selection operators	• Robust performance • Self-adaptive • Global search	• Parameter dependent • Slow convergence • Limited constraints	• Continuous optimization • Robust solutions • Noisy functions
Multiobjective NSGA-II	• Nondominated sorting • Crowding distance • Pareto frontiers	• Multiple objectives • Diverse solutions • Trade-off analysis	• Computational complexity • Scalability issues • Objective weighting	• Conflicting objectives • Design trade-offs • System compromises

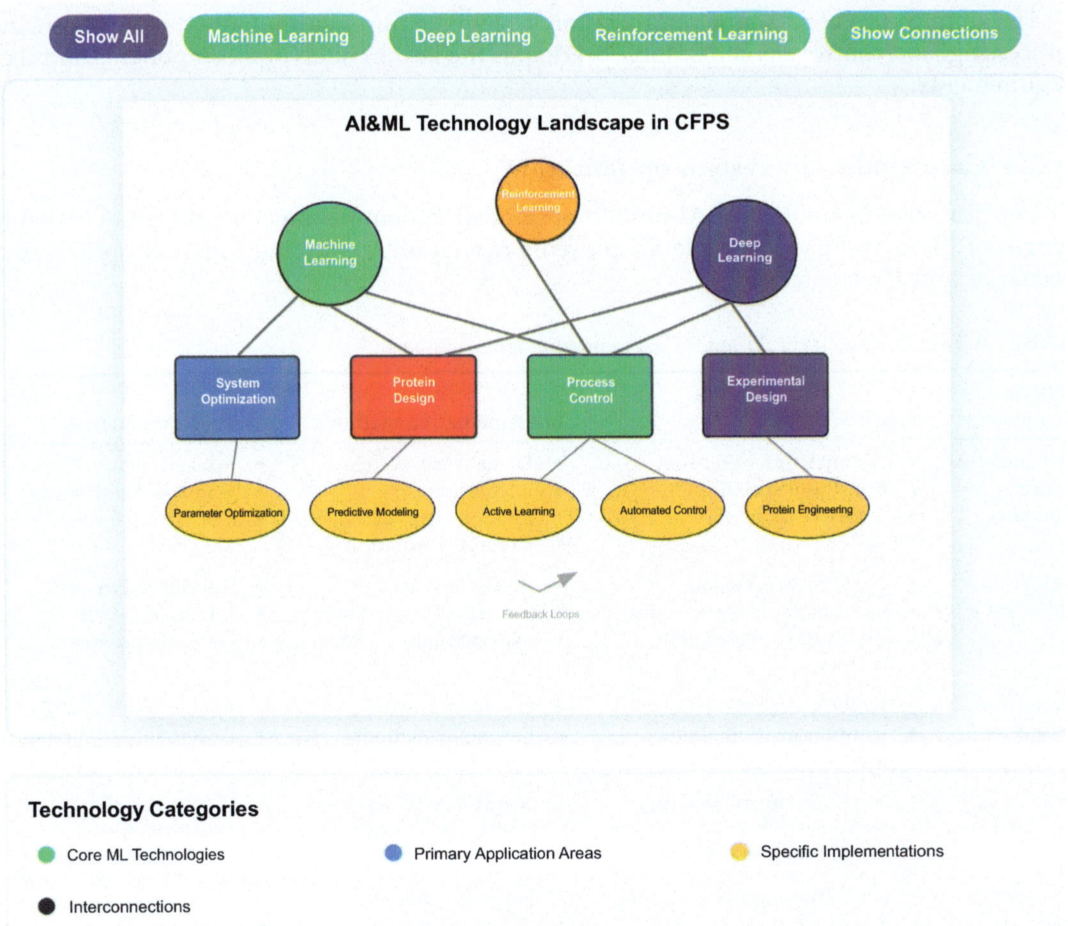

FIG. 8.1 AI/ML technology landscape in CFPS.

8.2.2 Experimental design integration

AI-guided experimental design integrates ML with design of experiments (DOE) principles to create adaptive experimental strategies that learn from previous results and guide future experiments toward optimal conditions.

The implementation of intelligent experimental design requires integration of multiple components working in feedback loops. Table 8.3 outlines key components and their integration strategies.

8.2.3 Real-time optimization and control

Real-time optimization leverages AI algorithms to continuously adjust system parameters during CFPS reactions based on sensor feedback and predictive models. This approach enables dynamic optimization that adapts to changing conditions and maximizes performance throughout the reaction.

The architecture of real-time optimization systems requires careful integration of sensing, processing, and control elements. Table 8.4 details the key architectural components and their requirements.

8.2.4 Case studies in system optimization

Practical implementation of AI-guided optimization demonstrates the potential and challenges of these approaches. Table 8.5 presents case studies showing different optimization scenarios and their outcomes.

TABLE 8.3 Components of AI-guided experimental design systems.

System component	Function	Implementation methods	Performance metrics
Experimental space definition	• Parameter identification • Constraint specification • Objective definition	• Domain knowledge integration • Literature mining • Expert consultation	• Space coverage • Constraint satisfaction • Objective relevance
Adaptive sampling	• Experiment selection • Information maximization • Exploration-exploitation balance	• Active learning • Uncertainty sampling • Expected improvement	• Sample efficiency • Information gain • Convergence rate
Predictive modeling	• Performance prediction • Uncertainty estimation • Model updating	• Gaussian processes • Ensemble methods • Online learning	• Prediction accuracy • Uncertainty calibration • Model robustness
Decision-making	• Next experiment selection • Stopping criteria • Result interpretation	• Acquisition functions • Multicriteria decision • Risk assessment	• Decision quality • Resource efficiency • Goal achievement
Knowledge integration	• Prior knowledge incorporation • Transfer learning • Domain adaptation	• Bayesian priors • Meta-learning • Cross-domain transfer	• Knowledge utilization • Transfer effectiveness • Adaptation speed

TABLE 8.4 Real-time optimization system architecture.

Architecture LAYER	Components	Functions	Performance requirements
Sensing layer	• Multiparameter sensors • Data acquisition • Signal processing	• Real-time monitoring • Data quality assurance • Noise filtering	• Response time < 1 s • Accuracy > 95% • Reliability > 99%
Processing layer	• ML algorithms • Prediction models • Optimization engines	• State estimation • Performance prediction • Decision optimization	• Processing time < 10 s • Prediction accuracy > 90% • Optimization convergence
Control layer	• Actuator control • Safety systems • Human interfaces	• Parameter adjustment • Safety monitoring • Operator interaction	• Control accuracy > 98% • Safety response < 1 s • Interface responsiveness
Integration layer	• Communication protocols • Data management • System coordination	• Data flow management • System synchronization • Error handling	• Communication latency < 100 ms • Data integrity 100% • System uptime > 99.5%

TABLE 8.5 AI-guided optimization case studies in CFPS.

Case study	Optimization target	AI approach	Results achieved	Key insights
Buffer optimization	• pH, ionic strength, cofactors • 15 continuous parameters • Protein yield maximization	• Bayesian optimization • Gaussian process models • Expected improvement	• 3.2 × yield improvement • 75% fewer experiments • Robust optimal conditions	• Nonlinear interactions critical • Uncertainty quantification valuable • Transfer learning effective
Temperature profiling	• Dynamic temperature control • Multistage reactions • Quality-yield trade-offs	• Reinforcement learning • Policy gradient methods • Multiobjective rewards	• 2.8 × productivity gain • 40% quality improvement • Automated control	• Temporal dependencies important • Reward shaping crucial • Safety constraints essential
Energy system design	• Multiple energy sources • Concentration optimization • Cost minimization	• Genetic algorithms • Multiobjective optimization • Pareto front analysis	• 60% cost reduction • Performance maintained • Design insights gained	• Trade-offs well-characterized • Multiple solutions valuable • Domain knowledge essential
Lysate engineering	• Enzyme concentrations • Component ratios • Stability optimization	• Particle swarm optimization • Constraint handling • Ensemble predictions	• 4.1 × activity increase • Extended stability • Reduced variability	• Component interactions complex • Constraints shape solutions • Ensemble methods robust

8.3 Predictive modeling for protein expression

8.3.1 Expression-level prediction models

Accurate prediction of protein expression levels from sequence and system parameters represents a fundamental challenge that AI approaches are increasingly addressing. These models enable rational DOE and reduce the trial-and-error nature of protein production optimization.

The development of effective expression prediction models requires integration of multiple data types and modeling approaches. Table 8.6 details different modeling strategies and their characteristics.

8.3.2 Feature engineering and selection

Effective feature engineering transforms raw biological data into representations that ML algorithms can effectively utilize. The selection and engineering of relevant features significantly impacts model performance and interpretability.

TABLE 8.6 Predictive modeling approaches for protein expression.

Modeling approach	Input features	Architecture	Prediction accuracy	Interpretability
Sequence-based models	• Amino acid composition • Codon usage • Secondary structure	• Random forests • Gradient boosting • Feature engineering	• $R^2 = 0.7–0.85$ • RMSE = 0.2–0.4 log units • Cross-validation stable	• High • Feature importance • Biological insights
Deep learning models	• Raw sequences • Structural features • Physicochemical properties	• Convolutional neural networks • Recurrent neural networks • Attention mechanisms	• $R^2 = 0.8–0.9$ • RMSE = 0.15–0.3 log units • Transfer learning capable	• Low to Medium • Attention maps • Layer visualization
Hybrid models	• Multimodal inputs • Sequence + structure • System parameters	• Ensemble methods • Multiinput networks • Feature fusion	• $R^2 = 0.85–0.95$ • RMSE = 0.1–0.25 log units • Robust performance	• Medium • Component analysis • Feature attribution
Physics-informed models	• Mechanistic features • Thermodynamic properties • Kinetic parameters	• Neural differential equations • Physics constraints • Hybrid modeling	• $R^2 = 0.75–0.9$ • RMSE = 0.12–0.3 log units • Physically consistent	• High • Physical interpretation • Mechanistic insights

TABLE 8.7 Feature engineering strategies for expression prediction.

Feature category	Specific features	Computation methods	Biological relevance
Sequence features	• Amino acid frequencies • Di/tripeptide composition • Sequence motifs	• Sliding windows • N-gram analysis • Motif scanning	• Codon bias effects • Folding signals • Regulatory elements
Structural features	• Secondary structure propensity • Disorder predictions • Hydrophobicity profiles	• Structure prediction tools • Machine learning predictors • Physicochemical scales	• Folding difficulty • Aggregation propensity • Solubility factors
Thermodynamic features	• Folding stability • Binding affinities • Energy landscapes	• Molecular dynamics • Energy calculations • Statistical potentials	• Protein stability • Expression conditions • System thermodynamics
Kinetic features	• Translation rates • Folding kinetics • Degradation rates	• Kinetic models • Rate predictions • Pathway analysis	• Expression efficiency • Product quality • System dynamics
System features	• Buffer conditions • Temperature profiles • Component concentrations	• Experimental parameters • Environmental conditions • System state variables	• Expression environment • Optimal conditions • System performance

Feature engineering for protein expression prediction involves multiple biological and chemical descriptors. Table 8.7 outlines key feature categories and their computational methods.

8.3.3 Model validation and uncertainty quantification

Robust model validation and uncertainty quantification are essential for reliable prediction models, particularly when these models guide experimental decisions or process control actions.

Comprehensive validation strategies ensure model reliability and provide confidence estimates for predictions. Table 8.8 details validation approaches and uncertainty quantification methods.

8.3.4 Transfer learning and domain adaptation

Transfer learning enables the application of models trained on one protein family or expression system to new targets, reducing the data requirements and development time for new applications.

Effective transfer learning strategies can significantly accelerate model development and improve performance on limited datasets. Table 8.9 outlines transfer learning approaches and their applications.

TABLE 8.8 Model validation and uncertainty quantification strategies.

Validation approach	Method	Advantages	Limitations	Best applications
Cross-validation	• K-fold validation • Leave-one-out • Stratified sampling	• Robust estimation • Unbiased assessment • Standard practice	• Limited to training data • Computational overhead • Optimistic for small datasets	• Model selection • Hyperparameter tuning • Performance estimation
External validation	• Independent test sets • Temporal validation • Cross-laboratory	• Realistic assessment • Generalization testing • Bias detection	• Data requirements • Limited availability • Potential distribution shift	• Model deployment • Generalization assessment • Bias evaluation
Bayesian uncertainty	• Bayesian neural networks • Variational inference • Monte Carlo dropout	• Principled uncertainty • Model uncertainty • Interpretable confidence	• Computational complexity • Implementation difficulty • Approximation errors	• High-stakes decisions • Active learning • Risk assessment
Ensemble methods	• Bootstrap aggregation • Model averaging • Diverse architectures	• Improved performance • Uncertainty estimates • Robust predictions	• Computational overhead • Model complexity • Ensemble design	• Production systems • Critical applications • Performance optimization

TABLE 8.9 Transfer learning strategies for expression prediction.

Transfer Strategy	Approach	Data requirements	Performance improvement	Application scenarios
Feature transfer	• Pretrained feature extractors • Frozen representations • Feature fine-tuning	• Target labels: 100–1000 • Source features available • Domain similarity moderate	• 20%–50% improvement • Faster convergence • Reduced overfitting	• New protein families • Limited target data • Similar expression systems
Model fine-tuning	• Pretrained models • Layer-wise adaptation • Gradual unfreezing	• Target data: 500–5000 • Source model available • Computational resources	• 30%–70% improvement • Domain adaptation • Knowledge retention	• Related domains • Moderate target data • Similar tasks

Transfer Strategy	Approach	Data requirements	Performance improvement	Application scenarios
Multitask learning	• Shared representations • Task-specific heads • Joint optimization	• Multiple related tasks • Balanced task importance • Sufficient total data	• 15%–40% improvement • Generalization boost • Efficient learning	• Multiple objectives • Related prediction tasks • Shared biological mechanisms
Domain adaptation	• Adversarial training • Distribution alignment • Gradient reversal	• Source and target domains • Unlabeled target data • Domain shift present	• 25%–60% improvement • Distribution shift handling • Robust performance	• Cross-system transfer • Different conditions • Laboratory variations

8.4 ML in template design

8.4.1 Sequence optimization algorithms

ML approaches to template design enable the optimization of DNA and RNA sequences for enhanced expression, improved translation efficiency, and reduced secondary structure interference. These algorithms can navigate the complex sequence space more efficiently than traditional methods.

The application of ML to sequence optimization involves multiple algorithmic approaches tailored to different aspects of template design. Table 8.10 details various optimization strategies and their specific applications.

8.4.2 Secondary structure prediction and optimization

RNA secondary structure significantly impacts translation efficiency and mRNA stability. ML approaches can predict and optimize secondary structures to enhance template performance.

Advanced ML models for secondary structure analysis integrate multiple prediction methods and optimization strategies. Table 8.11 outlines current approaches and their capabilities.

8.4.3 Multiobjective template design

Real-world template design involves balancing multiple competing objectives such as expression level, protein solubility, translation speed, and mRNA stability. Multiobjective optimization algorithms can identify optimal trade-offs and design solutions.

Multiobjective optimization for template design requires specialized algorithms and decision-making frameworks. Table 8.12 details approaches for handling multiple design objectives.

TABLE 8.10 Machine learning approaches for template sequence optimization.

Optimization target	ML algorithm	Optimization strategy	Performance metrics	Biological constraints
Codon optimization	• Genetic algorithms • Simulated annealing • Gradient-free methods	• Codon usage optimization • tRNA availability matching • Translation rate tuning	• Expression improvement: 2–10× • Folding enhancement: 20%–80% • Optimization time: hours	• Codon degeneracy • tRNA pool limitations • Regulatory sequences
RBS design	• Deep reinforcement learning • Bayesian optimization • Evolutionary strategies	• Translation initiation optimization • Ribosome binding enhancement • Expression level tuning	• Dynamic range: 1000× • Prediction accuracy: >90% • Design success rate: >85%	• Spacing requirements • Secondary structure • Accessibility constraints
Untranslated region (UTR) engineering	• Recurrent neural networks • Transformer models • Graph neural networks	• Stability optimization • Regulatory element design • Translation enhancement	• mRNA stability: 5–50× improvement • Translation efficiency: 2–20 × • Regulatory control: precise	• Folding constraints • Protein binding sites • Degradation signals
Promoter design	• Generative adversarial networks • Variational autoencoders • Sequence-to-function models	• Transcription rate control • Specificity enhancement • Regulatory circuit design	• Transcription levels: tunable range • Specificity: >95% • Orthogonality: maintained	• Transcription factor binding • Chromatin context • Regulatory networks

TABLE 8.11 ML approaches for RNA secondary structure analysis.

Analysis type	ML method	Input features	Prediction accuracy	Optimization capability
Structure prediction	• Deep neural networks • Bidirectional RNNs • Attention mechanisms	• Sequence information • Physicochemical properties • Evolutionary conservation	• Base pair prediction: >85% • Structure accuracy: >90% • Pseudoknot detection: >75%	• Structure-guided design • Constraint satisfaction • Multiobjective optimization

Analysis type	ML method	Input features	Prediction accuracy	Optimization capability
Folding energy estimation	• Graph neural networks • Physics-informed networks • Energy prediction models	• Sequence composition • Structural motifs • Thermodynamic parameters	• Energy prediction: ±2 kcal/mol • Stability ranking: >90% • Temperature dependence: modeled	• Stability optimization • Thermodynamic design • Condition-specific tuning
Accessibility prediction	• Convolutional networks • Ensemble methods • Multiscale modeling	• Local structure context • Binding site features • Accessibility constraints	• Site accessibility: >85% • Binding prediction: >80% • Functional site ID: >90%	• Accessibility optimization • Binding site design • Functional enhancement
Dynamic structure modeling	• Recurrent networks • Temporal modeling • State transition models	• Time-series data • Conformational states • Transition kinetics	• State prediction: >80% • Transition rates: order of magnitude • Dynamic behavior: captured	• Kinetic optimization • Dynamic control • Temporal design

TABLE 8.12 Multiobjective optimization strategies for template design.

Optimization framework	Algorithm type	Objective handling	Solution characteristics	Decision support
Pareto optimization	• NSGA-II/III • MOEA/D • SPEA2	• Nondominated sorting • Diversity maintenance • Convergence optimization	• Pareto front identification • Trade-off visualization • Solution diversity	• Multicriteria analysis • Preference incorporation • Sensitivity analysis
Weighted optimization	• Weighted sum approaches • Adaptive weighting • Dynamic objectives	• Objective aggregation • Weight optimization • Preference learning	• Single optimal solution • Weight-dependent results • Computational efficiency	• Weight elicitation • Preference modeling • Robustness analysis
Constraint-based design	• Constraint satisfaction • Penalty methods • Barrier functions	• Hard constraint enforcement • Soft constraint handling • Feasibility assurance	• Feasible solution guarantee • Constraint satisfaction • Design space reduction	• Constraint prioritization • Feasibility analysis • Design validation

Continued

TABLE 8.12 Multiobjective optimization strategies for template design—cont'd

Optimization framework	Algorithm type	Objective handling	Solution characteristics	Decision support
Hierarchical optimization	• Lexicographic ordering • Goal programming • Multilevel optimization	• Objective prioritization • Sequential optimization • Goal achievement	• Priority-based solutions • Goal satisfaction • Hierarchical structure	• Priority elicitation • Goal setting • Achievement analysis

TABLE 8.13 Template design validation and verification strategies.

Validation level	Experimental methods	Computational validation	Success criteria	Feedback integration
Sequence level	• DNA sequencing • Synthesis verification • Sequence assembly	• Sequence alignment • Motif validation • Constraint checking	• 100% sequence accuracy • Motif preservation • Constraint satisfaction	• Synthesis error correction • Design refinement • Constraint updating
Expression level	• Protein quantification • Western blotting • Mass spectrometry	• Expression prediction • Model validation • Performance analysis	• Expression within 2× prediction • Reproducible results • Statistical significance	• Model parameter updating • Prediction refinement • Feature importance
Functional level	• Activity assays • Binding studies • Functional characterization	• Function prediction • Structure–function analysis • Performance modeling	• Function preservation • Activity enhancement • Stability improvement	• Function–structure correlation • Activity optimization • Design improvement
System level	• Integration testing • Process validation • Production scaling	• System modeling • Process simulation • Performance optimization	• System integration success • Process stability • Scalable production	• System optimization • Process improvement • Integration refinement

8.4.4 Validation and experimental verification

Computational template designs require experimental validation to verify predicted performance and refine prediction models. Integration of computational and experimental workflows accelerates the design–build–test cycle.

Effective validation strategies combine computational predictions with targeted experimental verification. Table 8.13 outlines validation approaches and their implementation.

8.5 Automated experimental design

8.5.1 Active learning strategies

Active learning algorithms intelligently select the most informative experiments to perform, dramatically reducing the number of experiments needed to optimize CFPS systems or characterize protein behavior. These approaches are particularly valuable when experiments are expensive or time-consuming.

The implementation of active learning in CFPS requires careful consideration of acquisition functions and experimental constraints. Table 8.14 details various active learning strategies and their applications.

8.5.2 Experimental campaign planning

Systematic experimental campaign planning integrates resource constraints, temporal dependencies, and experimental objectives to create efficient experimental schedules that maximize information gain within practical constraints.

TABLE 8.14 Active learning strategies for CFPS experimental design.

Active learning strategy	Acquisition function	Exploration–exploitation balance	Data efficiency	Best applications
Uncertainty sampling	• Maximum uncertainty • Entropy-based selection • Variance maximization	• High exploration focus • Model uncertainty driven • Conservative exploitation	• 5–10× reduction in experiments • Efficient for small datasets • Quick convergence	• Model refinement • Uncertainty reduction • Initial exploration
Expected improvement	• Probability of improvement • Expected gain over current best • Acquisition optimization	• Balanced approach • Best point focused • Adaptive balance	• 3–8× efficiency gain • Optimal for optimization • Good exploitation	• Parameter optimization • Performance maximization • Single objective tasks
Knowledge gradient	• Future value estimation • Information value calculation • Multistep lookahead	• Forward-looking strategy • Long-term optimization • Strategic exploration	• 4–12× improvement • Optimal stopping • Strategic efficiency	• Long experimental campaigns • Resource planning • Strategic optimization
Multiobjective acquisition	• Pareto improvement probability • Hypervolume expected improvement • Scalarized objectives	• Multidimensional exploration • Trade-off discovery • Diverse solutions	• 6–15× efficiency • Comprehensive exploration • Trade-off identification	• Multiobjective problems • Design trade-offs • System optimization

TABLE 8.15 Experimental campaign planning strategies.

Planning aspect	Optimization methods	Constraints	Objectives	Success metrics
Resource allocation	• Integer programming • Resource scheduling • Constraint optimization	• Budget limitations • Equipment availability • Personnel constraints	• Resource efficiency • Cost minimization • Throughput maximization	• Resource utilization > 80% • Budget compliance • Schedule adherence
Temporal scheduling	• Critical path analysis • Schedule optimization • Dependency management	• Experimental dependencies • Equipment conflicts • Time limitations	• Schedule efficiency • Deadline compliance • Bottleneck elimination	• Schedule compression • Dependency satisfaction • Timeline optimization
Information maximization	• Optimal design theory • Information theory • Experimental design	• Sample size limitations • Statistical power • Design constraints	• Information gain • Statistical efficiency • Knowledge advancement	• Information content • Statistical power • Objective achievement
Risk management	• Scenario analysis • Robust optimization • Contingency planning	• Uncertainty factors • Failure probabilities • Resource limitations	• Risk mitigation • Robust performance • Contingency preparedness	• Risk reduction • Robustness measures • Adaptability assessment

Effective campaign planning requires consideration of multiple factors and constraints. Table 8.15 outlines key planning considerations and optimization strategies.

8.5.3 Adaptive experimental strategies

Adaptive experimental strategies continuously update experimental plans based on accumulating results, enabling dynamic optimization of experimental campaigns and responsive adjustment to unexpected findings.

The implementation of adaptive strategies requires real-time decision-making capabilities and flexible experimental systems. Table 8.16 details adaptive experimental frameworks and their components.

8.5.4 Integration with laboratory automation

The integration of AI-driven experimental design with laboratory automation creates closed-loop systems that can autonomously execute experimental campaigns with minimal human intervention.

Successful integration requires careful coordination between AI systems and automation hardware. Table 8.17 outlines integration strategies and requirements.

TABLE 8.16 Adaptive experimental strategy components.

Adaptive component	Implementation	Decision criteria	Update mechanisms	Performance benefits
Dynamic objective updating	• Real-time objective refinement • Priority adjustment • Goal evolution	• Performance feedback • Objective achievement • Resource status	• Objective reweighting • Goal modification • Priority updating	• Objective alignment • Resource optimization • Goal achievement
Model-based adaptation	• Online model updating • Prediction refinement • Uncertainty reduction	• Model performance • Prediction accuracy • Uncertainty levels	• Parameter updating • Model retraining • Architecture adaptation	• Prediction improvement • Uncertainty reduction • Decision quality
Experimental design updates	• Design modification • Strategy adjustment • Plan optimization	• Information gain • Resource efficiency • Time constraints	• Design refinement • Strategy switching • Plan reoptimization	• Efficiency improvement • Information maximization • Resource utilization
Resource reallocation	• Dynamic resource management • Priority adjustment • Capacity optimization	• Resource availability • Experimental priority • Performance feedback	• Resource redistribution • Priority rebalancing • Capacity reallocation	• Resource efficiency • Priority alignment • Performance optimization

TABLE 8.17 AI–automation integration strategies for experimental design.

Integration level	System components	Communication protocols	Control mechanisms	Quality assurance
Hardware integration	• Robotic systems • Analytical instruments • Sample handling	• Standard protocols (Standardization in Lab Automation, SILA, Society for Biomolecular Screening, SBS) • Custom active pharmaceutical ingredient (APIs) • Direct control interfaces	• Equipment control commands • Status monitoring • Error handling	• Calibration verification • Performance monitoring • Error detection

Continued

TABLE 8.17 AI–automation integration strategies for experimental design—cont'd

Integration level	System components	Communication protocols	Control mechanisms	Quality assurance
Software integration	• LIMS systems • Experimental control • Data management	• Database connectivity • Web services • Message queuing	• Workflow orchestration • Data synchronization • Process monitoring	• Data validation • Workflow verification • System monitoring
Decision integration	• AI decision engines • Expert systems • Human oversight	• Real-time data streams • Decision APIs • Alert systems	• Automated decisions • Human-in-the-loop • Override mechanisms	• Decision validation • Human approval • Audit trails
Safety integration	• Safety monitoring • Emergency systems • Risk assessment	• Safety protocols • Emergency communications • Risk reporting	• Automatic shutdowns • Safety interlocks • Risk mitigation	• Safety verification • Risk assessment • Emergency testing

8.6 Real-time process control

8.6.1 Reinforcement learning for process optimization

Reinforcement learning (RL) algorithms learn optimal control policies through interaction with CFPS systems, enabling adaptive control that improves over time and adapts to changing conditions without requiring explicit programming of control rules.

The application of RL to CFPS process control involves multiple algorithmic approaches tailored to different control challenges. Table 8.18 details RL algorithms and their process control applications.

8.6.2 Anomaly detection and predictive maintenance

AI-powered anomaly detection systems continuously monitor CFPS processes to identify deviations from normal operation, predict potential failures, and trigger appropriate corrective actions before problems affect product quality.

Effective anomaly detection requires multiple complementary approaches and careful threshold setting. Table 8.19 outlines anomaly detection strategies and their implementation.

8.6.3 Model predictive control integration

Model predictive control (MPC) frameworks integrated with ML models enable predictive control strategies that anticipate future system behavior and optimize control actions over prediction horizons.

TABLE 8.18 Reinforcement learning algorithms for CFPS process control.

RL algorithm	Learning approach	Control characteristics	Application areas	Performance metrics
Q-learning	• Value-based learning • Temporal difference • Model-free approach	• Discrete action spaces • Episodic control • Off-policy learning	• Batch process control • Discrete parameter optimization • Simple control tasks	• Learning convergence • Control stability • Performance improvement
Policy gradient	• Policy-based learning • Direct policy optimization • Gradient ascent	• Continuous action spaces • Stochastic policies • On-policy learning	• Continuous process control • Multiparameter optimization • Complex control tasks	• Policy convergence • Action smoothness • Reward optimization
Actor-critic	• Hybrid approach • Value and policy learning • Reduced variance	• Efficient learning • Stable convergence • Continuous control	• Real-time control • Dynamic optimization • Multiobjective control	• Learning efficiency • Control performance • Stability metrics
Deep RL	• Deep neural networks • Function approximation • High-dimensional states	• Complex state spaces • Nonlinear policies • Representation learning	• High-dimensional control • Complex dynamics • Pattern recognition	• Learning speed • Generalization ability • Control robustness

TABLE 8.19 Anomaly detection strategies for CFPS process monitoring.

Detection method	Technical approach	Anomaly types detected	Response time	False-positive rate
Statistical process control	• Control charts • Statistical limits • Trend analysis	• Process drift • Mean shifts • Variance changes	• Real time • Immediate alerts • Continuous monitoring	• 1%–5% • Tunable thresholds • Historical baselines
Machine learning detection	• Isolation forests • One-class Support Vector Machine (SVM) • Autoencoders	• Complex patterns • Multidimensional anomalies • Nonlinear relationships	• Near real time • Batch processing • Adaptive thresholds	• 0.5%–3% • Model-dependent • Training required
Deep learning detection	• Variational autoencoders • Long Short-Term Memory (LSTM) networks • Transformer models	• Temporal anomalies • Sequence patterns • High-dimensional data	• Moderate latency • Model inference • GPU acceleration	• 0.1%–2% • Architecture dependent • Large datasets needed

Continued

TABLE 8.19 Anomaly detection strategies for CFPS process monitoring—cont'd

Detection method	Technical approach	Anomaly types detected	Response time	False-positive rate
Hybrid detection	• Ensemble methods • Multimodel approaches • Consensus algorithms	• Multiple anomaly types • Robust detection • Reduced false positives	• Variable latency • Consensus delays • Parallel processing	• 0.1%–1% • Ensemble benefits • Improved reliability

TABLE 8.20 ML-enhanced model predictive control strategies.

MPC component	ML enhancement	Implementation methods	Performance benefits	Computational requirements
System modeling	• Neural network models • Gaussian process models • Ensemble prediction	• Data-driven modeling • Online model adaptation • Uncertainty quantification	• Improved accuracy • Adaptive capability • Uncertainty handling	• Model training time • Inference speed • Memory requirements
Prediction horizon	• Adaptive horizon selection • State-dependent horizons • Dynamic optimization	• Reinforcement learning • Heuristic algorithms • Performance-based adaptation	• Optimized computational load • Improved performance • Adaptive behavior	• Horizon optimization • Real-time constraints • Solution quality
Constraint handling	• Learned constraint models • Adaptive constraints • Soft constraint learning	• Constraint learning • Penalty adaptation • Feasibility prediction	• Better constraint satisfaction • Adaptive behavior • Improved feasibility	• Constraint evaluation • Adaptation overhead • Feasibility checking
Objective optimiza-tion	• Multiobjective learning • Preference learning • Reward shaping	• Multiobjective algorithms • Preference elicitation • Adaptive objectives	• Improved trade-offs • Dynamic objectives • Performance optimization	• Multiobjective solving • Preference modeling • Objective adaptation

The integration of ML models with MPC requires careful consideration of model accuracy, computational constraints, and control objectives. Table 8.20 details MPC integration strategies.

8.6.4 Human–AI collaboration in process control

Effective process control systems combine AI capabilities with human expertise, creating collaborative frameworks that leverage the strengths of both automated decision-making and human judgment.

TABLE 8.21 Human–AI collaboration frameworks for process control.

Collaboration mode	AI role	Human role	Interface requirements	Decision authority
AI-assisted control	• Decision support • Recommendation generation • Analysis automation	• Final decisions • Strategy setting • Exception handling	• Intuitive displays • Recommendation presentation • Override mechanisms	• Human authority • AI recommendations • Human verification
Supervisory control	• Routine operations • Automatic responses • Normal condition handling	• High-level supervision • Exception management • Strategic decisions	• Status monitoring • Alert systems • Intervention interfaces	• AI operational control • Human oversight • Exception escalation
Collaborative control	• Specialized tasks • Pattern recognition • Optimization algorithms	• Domain expertise • Contextual knowledge • Creative problem solving	• Collaborative interfaces • Information sharing • Joint decision support	• Shared authority • Domain-specific roles • Consensus mechanisms
Adaptive authority	• Performance-based authority • Confidence-weighted decisions • Dynamic role allocation	• Authority monitoring • Intervention readiness • Trust calibration	• Trust indicators • Authority status • Intervention triggers	• Dynamic allocation • Performance-based • Trust-dependent

The design of human–AI collaborative systems requires careful consideration of interfaces, authority distribution, and decision-making workflows. Table 8.21 outlines collaboration strategies and their implementation.

8.7 Deep learning for protein engineering

8.7.1 Neural network architectures for protein design

Deep learning architectures specifically designed for protein engineering tasks leverage the hierarchical nature of protein structure and function to enable sophisticated design capabilities that go beyond traditional approaches.

The selection and design of neural network architectures for protein engineering depends on the specific design objectives and data characteristics. Table 8.22 details various architectures and their applications.

TABLE 8.22 Deep learning architectures for protein engineering applications.

Architecture type	Network structure	Protein design applications	Advantages	Limitations
Convolutional neural networks	• 1D/2D convolutions • Hierarchical features • Local pattern detection	• Sequence motif identification • Local structure prediction • Binding site analysis	• Translation invariance • Parameter efficiency • Local pattern capture	• Limited long-range dependencies • Fixed receptive fields • Sequence length constraints
Recurrent neural networks	• LSTM/Gated Recurrent Unit (GRU) cells • Bidirectional processing • Sequential modeling	• Sequence generation • Evolutionary analysis • Temporal dynamics	• Sequential processing • Variable length handling • Memory mechanisms	• Computational complexity • Gradient problems • Sequential bottlenecks
Transformer networks	• Self-attention mechanisms • Positional encoding • Parallel processing	• Protein language modeling • Structure prediction • Function annotation	• Long-range dependencies • Parallel computation • Attention interpretability	• Computational requirements • Memory consumption • Data requirements
Graph neural networks	• Graph convolutions • Message passing • Structural representations	• Protein-protein interactions • Binding affinity prediction • Structural analysis	• Structural representation • Relational modeling • Flexible input formats	• Graph construction complexity • Scalability challenges • Limited standardization

8.7.2 Generative models for novel protein design

Generative deep learning models can create novel protein sequences with desired properties, enabling the design of proteins with functions not found in nature or optimized for specific applications.

Different generative modeling approaches offer various advantages for protein design applications. Table 8.23 compares generative model architectures and their capabilities.

8.7.3 Transfer learning from large-scale protein data

Large-scale protein language models trained on millions of protein sequences can be fine-tuned for specific protein engineering tasks, leveraging evolutionary information and structural patterns learned from diverse protein families.

Effective transfer learning requires careful adaptation of pretrained models to specific engineering objectives. Table 8.24 outlines transfer learning strategies and their implementation.

TABLE 8.23 Generative models for novel protein design.

Generative model	Architecture	Generation strategy	Design capabilities	Validation requirements
Variational autoencoders	• Encoder–decoder structure • Latent space modeling • Probabilistic framework	• Latent space sampling • Conditional generation • Interpolation methods	• Smooth design space • Property interpolation • Controlled generation	• Latent space interpretation • Property validation • Novelty assessment
Generative adversarial networks	• Generator-discriminator framework • Adversarial training • Implicit likelihood	• Adversarial generation • High-quality sampling • Mode diversity	• High-quality sequences • Diverse generation • Realistic proteins	• Training stability • Mode collapse detection • Quality assessment
Autoregressive models	• Sequential generation • Conditional probabilities • Causal modeling	• Step-by-step generation • Conditional design • Flexible sampling	• Controllable generation • Conditional design • Interpretable process	• Generation speed • Sequence quality • Biological validity
Flow-based models	• Invertible transformations • Exact likelihood • Bijective mappings	• Exact likelihood calculation • Precise control • Invertible generation	• Exact probability estimation • Precise control • Interpretable latent space	• Architecture constraints • Computational complexity • Expressiveness limitations

TABLE 8.24 Transfer learning strategies for protein engineering.

Transfer approach	Pretraining data	Fine-tuning strategy	Engineering applications	Performance improvement
Sequence-based transfer	• UniProt database • Evolutionary sequences • Multiple species	• Task-specific fine-tuning • Layer-wise adaptation • Feature extraction	• Function prediction • Homology analysis • Sequence optimization	• 30%–80% improvement • Reduced data requirements • Faster convergence
Structure-based transfer	• PDB structures • Predicted structures • Structural motifs	• Structure-aware fine-tuning • Geometric adaptation • 3D representation learning	• Structure prediction • Binding site design • Stability optimization	• 40%–90% improvement • Structural awareness • Better generalization
Function-based transfer	• Functional annotations • Biochemical data • Activity measurements	• Function-specific adaptation • Multitask learning • Domain adaptation	• Activity optimization • Function design • Property prediction	• 25%–70% improvement • Functional relevance • Domain knowledge

Continued

TABLE 8.24 Transfer learning strategies for protein engineering—cont'd

Transfer approach	Pretraining data	Fine-tuning strategy	Engineering applications	Performance improvement
Multimodal transfer	• Sequence + structure + function • Integrated databases • Multiomics data	• Multimodal fusion • Cross-modal learning • Integrated fine-tuning	• Comprehensive design • Multiproperty optimization • Holistic engineering	• 50%–95% improvement • Comprehensive understanding • Synergistic benefits

TABLE 8.25 Interpretability methods for deep learning in protein engineering.

Interpretability method	Technical approach	Information provided	Biological insights	Implementation complexity
Attention visualization	• Attention weight analysis • Attention head interpretation • Layer-wise attention	• Sequence importance • Position relationships • Context dependencies	• Functional motifs • Interaction patterns • Evolutionary constraints	• Low complexity • Direct visualization • Real-time analysis
Feature attribution	• Gradient-based methods • Integrated gradients • SHAP values	• Feature importance • Decision factors • Input contribution	• Critical residues • Functional domains • Design principles	• Medium complexity • Post hoc analysis • Gradient computation
Latent space analysis	• Dimensionality reduction • Clustering analysis • Trajectory visualization	• Design space structure • Property relationships • Design pathways	• Evolutionary relationships • Property correlations • Design strategies	• Medium complexity • Space visualization • Statistical analysis
Causal analysis	• Causal discovery • Intervention analysis • Counterfactual reasoning	• Causal relationships • Design interventions • Mechanism understanding	• Mechanistic insights • Design causality • Intervention effects	• High complexity • Specialized methods • Causal frameworks

8.7.4 Interpretability and explainable AI

Understanding how deep learning models make protein design decisions is crucial for building trust, validating designs, and gaining biological insights that can inform future engineering efforts.

Interpretability methods for protein engineering applications must balance model transparency with performance requirements. Table 8.25 outlines interpretability approaches and their applications.

8.8 Case studies: AI-driven discoveries

8.8.1 Key trends and emerging opportunities (2024–2025)

Real-world applications of AI in CFPS demonstrate the practical value and implementation challenges of these technologies. These case studies provide insights into successful strategies and lessons learned from deployment.

The landscape of AI-driven protein engineering and CFPS has experienced remarkable advancement in recent years, with AI integration reaching unprecedented levels of sophistication. Large language models such as EVOLVEPro are achieving dramatic improvements in protein binding capabilities, with some tools demonstrating 30-fold enhancements in target binding affinity, while RL approaches like LatProtRL and architecture design methodologies are showing superior optimization capabilities that consistently outperform traditional approaches. The field received significant recognition with the 2024 Chemistry Nobel Prize, which highlighted AI's transformative impact on protein design and validated the scientific community's investment in computational approaches to biological engineering.

The CFPS platform ecosystem is simultaneously undergoing substantial evolution, driven by both market forces and technological innovation. The market demonstrates robust growth with valuations reaching $299.9 million in 2024 and projections indicating expansion to $585.3 million by 2034, representing a healthy 7.0% compound annual growth rate that reflects increasing industrial adoption and confidence in cell-free technologies. This growth is accompanied by significant diversification in chassis organisms, with researchers expanding beyond traditional *Escherichia. coli* systems to incorporate *Vibrio. natriegens*, *Bacillus. subtilis*, and other specialized organisms that offer unique advantages for specific applications. Enhanced computational integration has become a cornerstone of modern CFPS development, with sophisticated mathematical modeling and optimization strategies enabling more precise control over synthesis parameters and outcomes.

Emerging applications are demonstrating the practical value of these technological advances across multiple domains. Therapeutic development has seen notable progress with successful SARS-CoV-2 RBD production and innovative vaccine antigen presentation systems, while sustainability considerations are driving the development of eco-friendly protein production methods that reduce resource consumption and environmental impact. Automation integration is revolutionizing laboratory workflows through high-throughput screening capabilities and automated optimization loops that accelerate the design–build–test–learn cycle, making protein engineering more accessible and efficient.

Looking toward the future, several critical directions are shaping the field's trajectory. Multimodal AI approaches are integrating structure, sequence, and function prediction capabilities into unified platforms that provide more comprehensive protein design solutions. Real-time optimization systems incorporating continuous learning and adaptive algorithms are enabling more responsive and intelligent protein engineering workflows. In addition, standardization efforts focused on platform harmonization and reproducibility improvements are addressing one of the field's most pressing challenges, ensuring that advances in AI-driven protein engineering can be reliably reproduced and scaled across different research environments and applications.

Table 8.26 presents detailed case studies of successful AI applications in CFPS optimization and protein engineering.

8.8.2 Failure analysis and lessons learned

Understanding failure modes and limitations of AI applications provides valuable insights for improving future implementations and setting realistic expectations.

TABLE 8.26 AI-driven success stories in CFPS and protein engineering.

Case study	Challenge	AI approach	Results achieved	Key success factors	Lessons learned	Recent references
Antibody expression optimization	• Low expression levels • Variable batch quality • Long optimization cycles	• Gaussian process optimization • Multiparameter screening • Active learning • **EVOLVEPro** for enhanced binding	• 15× expression improvement • 80% reduction in optimization time • Consistent batch quality • **30-fold better target binding**	• High-quality data • Domain expertise integration • Iterative validation • **Large language model integration**	• Data quality crucial • Expert knowledge valuable • Validation essential • **Transfer learning effective**	• Jiang et al. (2024): EVOLVE-Pro for antibody engineering • Cai et al. (2024): Geometric GNN for antibody affinity
Enzyme thermostability design	• Insufficient thermal stability • Large sequence space • Structure–function relationships • **Evolutionary optimization**	• Deep learning prediction • Directed evolution guidance • Multiobjective optimization • **Reinforcement learning with fitness landscapes**	• 25°C stability increase • Maintained catalytic activity • 90% reduction in screening • **Improved lab-in-the-loop efficiency**	• Large training datasets • Structure-based features • Experimental validation • **Latent space optimization**	• Data diversity important • Structure information valuable • Iterative improvement needed • **RL effective for fitness optimization**	• Lee et al. (2024): LatProtRL for fitness optimization • Sternlieb et al. (2024): Low-N OpenFold fine-tuning
CFPS buffer optimization	• Complex parameter interactions • Nonlinear responses • Multiple objectives • **Computational design challenges**	• Reinforcement learning • Real-time optimization • Multiobjective rewards • **Computational strategies integration**	• 4 × productivity increase • Reduced reagent consumption • Automated optimization • **Enhanced system efficiency**	• Real-time feedback • Reward function design • Safety constraints • **Mathematical modeling**	• Reward design critical • Safety integration essential • Continuous learning valuable • **Computational approaches promising**	• Siddique et al. (2024): Computational CFPS optimization • Tierra Biosciences (2024): AI-driven protein synthesis

Case study	Challenge	AI approach	Results achieved	Key success factors	Lessons learned	Recent references
Membrane protein production	• Difficult expression • Folding challenges • Low yields • **Scalability limitations**	• Transfer learning • Sequence optimization • Co-translational engineering • **Cell-free platform development**	• 8× yield improvement • Improved protein quality • Reduced aggregation • **Platform versatility**	• Transfer from related proteins • Multimodal optimization • Process integration • **Robust extract preparation**	• Transfer learning effective • Integration complexity high • Quality metrics important • **Platform standardization needed**	• Wiegand et al. (2019): *V. natriegens* CFPS platform • Yim et al. (2024): FrameDiff for protein design
Protein architecture design	• Limited structural diversity • Complex assembly requirements • **Top-down design challenges** • Functional optimization	• **Reinforcement learning for architecture** • Monte Carlo tree search • AlphaZero optimization • Multiobjective design	• **Novel protein architectures** • Improved shape complementarity • **Icosahedral assembly design** • Enhanced functional properties	• **System-level optimization** • Structural scoring integration • **Threshold-based rewards** • Secondary objective incorporation	• **RL enables complex design** • Architecture optimization crucial • **Model-based approaches effective** • System properties important	• Lutz et al. (2023): RL for protein architecture design • Renard et al. (2024): AlphaZero for backbone design
CFPS system development	• **Limited chassis diversity** • Expression optimization • Regulatory element characterization • **Computational bottlenecks**	• **Multichassis platform development** • Automated optimization • Linear template design • **AI-integrated synthesis**	• **High-yield systems (>1.6 g/L)** • Rapid prototyping capability • **Therapeutic protein production** • Platform standardization	• **Extract preparation protocols** • Systematic optimization • **Lyophilization compatibility** • Quality control integration	• **Chassis selection impacts performance** • Process standardization essential • **Automation increases throughput** • Integration complexity manageable	• Maharjan et al. (2023): CFPS sustainability review • Peng et al. (2024): Cell-free synthetic biology advances

Table 8.27 analyzes common failure modes and their underlying causes in AI applications for CFPS.

8.8.3 Technology transfer and commercialization

The successful transfer of AI technologies from research to commercial applications requires careful consideration of practical constraints, regulatory requirements, and business models.

TABLE 8.27 Failure analysis and lessons from AI applications in CFPS.

Failure mode	Common causes	Impact on project	Prevention strategies	Recovery approaches
Model overfitting	• Limited training data • Complex models • Insufficient validation	• Poor generalization • Failed predictions • Wasted experiments	• Cross-validation • Regularization techniques • Data augmentation	• Model simplification • Additional data collection • Ensemble methods
Data quality issues	• Measurement errors • Inconsistent protocols • Missing values	• Unreliable models • Poor performance • Misleading insights	• Data quality protocols • Standardized procedures • Quality monitoring	• Data cleaning • Outlier detection • Robust algorithms
Domain mismatch	• Training-test differences • System variations • Environmental changes	• Performance degradation • Failed transfer • Unexpected behavior	• Domain analysis • Transfer learning • Robustness testing	• Domain adaptation • Model retraining • System adjustment
Computational limitations	• Resource constraints • Real-time requirements • Scalability issues	• Delayed responses • System downtime • Performance bottlenecks	• Resource planning • Algorithm optimization • Infrastructure scaling	• Resource optimization • Algorithm redesign • Infrastructure upgrade
Integration challenges	• System complexity • Interface mismatches • Communication failures	• System instability • Data inconsistencies • Operational disruptions	• Integration testing • Standardized interfaces • Communication protocols	• System redesign • Interface updates • Protocol revision

Table 8.28 analyzes technology transfer strategies and commercialization pathways for AI applications in CFPS.

8.8.4 Future opportunities and challenges

The future development of AI applications in CFPS presents both significant opportunities and substantial challenges that will shape the field's evolution.

Table 8.29 outlines emerging opportunities and challenges in AI-driven CFPS applications.

Fig. 8.1 illustrates the comprehensive AI/ML landscape in CFPS applications showing the interconnections between different technologies and their evolution over time.

Fig. 8.2 demonstrates the integration hierarchy from basic algorithms to complex systems and their applications in protein engineering and process optimization

TABLE 8.28 Technology transfer and commercialization strategies.

Transfer stage	Key activities	Success factors	Common challenges	Risk mitigation
Technology validation	• Proof-of-concept demonstrations • Performance benchmarking • Scalability testing	• Clear value proposition • Demonstrated benefits • Competitive advantages	• Technical limitations • Performance gaps • Scalability issues	• Incremental validation • Risk assessment • Alternative approaches
Business development	• Market analysis • Business model development • Partnership formation	• Market understanding • Clear business case • Strategic partnerships	• Market uncertainty • Competition • Resource requirements	• Market research • Flexible models • Partnership strategy
Product development	• Commercial product design • User interface development • Quality systems	• User-centered design • Quality focus • Regulatory compliance	• Development complexity • Resource constraints • Regulatory hurdles	• Agile development • Quality systems • Regulatory planning
Market introduction	• Product launch • Customer adoption • Support systems	• Market readiness • Customer support • Continuous improvement	• Adoption barriers • Competition • Support requirements	• Market preparation • Support investment • Iterative improvement

TABLE 8.29 Future opportunities and challenges in AI-driven CFPS.

Opportunity area	Potential impact	Technical requirements	Implementation challenges	Timeline
Autonomous laboratories	• Fully automated discovery • 24/7 optimization • Accelerated research	• Advanced robotics • AI integration • Safety systems	• System complexity • Integration challenges • Safety requirements	5–10 years
Personalized therapeutics	• Patient-specific proteins • Rapid production • Customized treatments	• Rapid optimization • Scalable production • Quality assurance	• Regulatory frameworks • Cost considerations • Technical complexity	3–8 years
Digital twins	• Virtual process optimization • Predictive maintenance • Risk assessment	• High-fidelity models • Real-time synchronization • Computational resources	• Model complexity • Data requirements • Validation challenges	2–7 years

Continued

TABLE 8.29 Future opportunities and challenges in AI-driven CFPS—cont'd

Opportunity area	Potential impact	Technical requirements	Implementation challenges	Timeline
Quantum computing integration	• Enhanced optimization • Complex problem solving • Molecular simulations	• Quantum algorithms • Quantum hardware • Integration frameworks	• Technology maturity • Hardware limitations • Algorithm development	5–15 years
Sustainable manufacturing	• Resource optimization • Waste reduction • Energy efficiency	• Multiobjective optimization • Lifecycle analysis • Sustainability metrics	• Complexity management • Economic trade-offs • Implementation costs	2–6 years

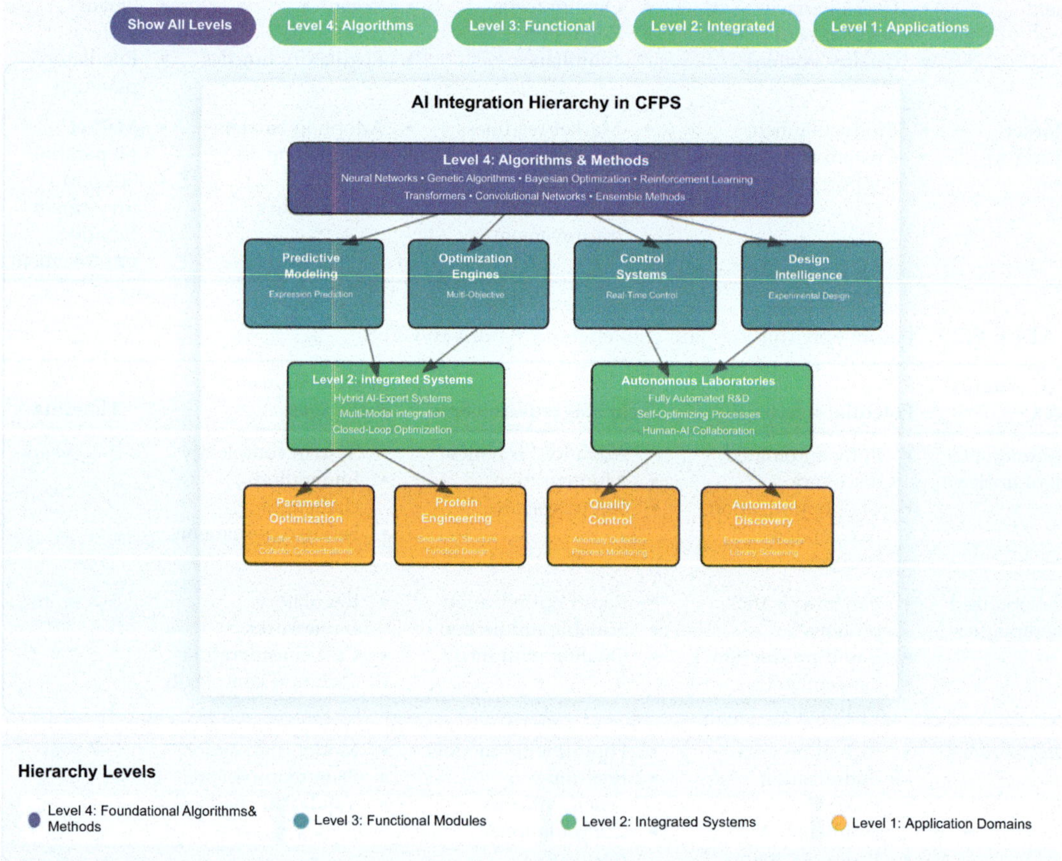

FIG. 8.2 AI integration hierarchy in CFPS.

8.9 Summary and future directions

8.9.1 Technology integration and convergence

The future of AI in CFPS lies in the integration and convergence of multiple AI technologies to create comprehensive systems that can handle the full complexity of biological manufacturing processes.

Table 8.30 outlines key integration opportunities and their potential impact on the field.

8.9.2 Emerging technologies and applications

New AI technologies and applications continue to emerge, offering potential advantages for CFPS applications and opening new possibilities for protein engineering and process optimization.

The evolution of AI technologies promises continued advancement in CFPS capabilities and applications across multiple domains.

TABLE 8.30 AI technology integration opportunities in CFPS.

Integration strategy	Combined technologies	Synergistic benefits	Development timeline	Commercial impact
Multimodal AI	• Vision + language + structure • Multiomics integration • Cross-domain learning	• Comprehensive understanding • Enhanced predictions • Holistic optimization	2–5 years	Transformative
Edge-cloud computing	• Local processing + cloud resources • Distributed learning • Hybrid architectures	• Real-time responses • Scalable resources • Cost optimization	1–3 years	High impact
Quantum-classical hybrid	• Quantum optimization + classical ML • Hybrid algorithms • Complementary strengths	• Enhanced optimization • Complex problem solving • Breakthrough capabilities	5–10 years	Revolutionary potential
Human–AI symbiosis	• Collaborative intelligence • Augmented decision making • Adaptive authority	• Combined expertise • Trust calibration • Optimal collaboration	2–4 years	Significant improvement

Bibliography

Alley, E.C., Khimulya, G., Biswas, S., AlQuraishi, M., Church, G.M., 2019. Unified rational protein engineering with sequence-based deep representation learning. Nat. Methods 16 (12), 1315–1322.

Bedbrook, C.N., Yang, K.K., Rice, A.J., Gradinaru, V., Arnold, F.H., 2017. Machine learning to design integral membrane channelrhodopsins for efficient eukaryotic expression and plasma membrane localization. PLoS Comput. Biol. 13 (10), e1005786.

Borkowski, O., Ceroni, F., Stan, G.B., Ellis, T., 2016. Overloaded and stressed: whole-cell-model simulations of the metabolic burden of protein synthesis. Curr. Opin. Microbiol. 33, 123–130.

Camacho, D.M., Collins, K.M., Powers, R.K., Costello, J.C., Collins, J.J., 2018. Next-generation machine learning for biological networks. Cell 173 (7), 1581–1592.

Costello, Z., Martin, H.G., 2018. A machine learning approach to predict metabolic pathway dynamics from time-series multiomics data. Npj Syst. Biol. Appl. 4 (1), 19.

Fox, R.J., Davis, S.C., Mundorff, E.C., Newman, L.M., Gavrilovic, V., Ma, S.K., et al., 2007. Improving catalytic function by ProSAR-driven enzyme evolution. Nat. Biotechnol. 25 (3), 338–344.

Greenman, K.P., Amini, A.P., Green, W.H., 2022. Multi-fidelity prediction of molecular optical peaks with deep learning. Chem. Sci. 13 (4), 1152–1162.

Grishin, A.M., Cygler, M., 2020. A bird's eye view of deep learning in enzyme engineering. ACS Catal. 10 (20), 11784–11794.

Jiang, K., et al., 2024. EVOLVEPro: AI protein engineer for enhanced binding. Science.

Jumper, J., Evans, R., Pritzel, A., Green, T., Figurnov, M., Ronneberger, O., et al., 2021. Highly accurate protein structure prediction with AlphaFold. Nature 596 (7873), 583–589.

Lee, M., et al., 2024. Robust optimization in protein fitness landscapes using reinforcement learning. arXiv:2405.18986.

Lutz, I.D., et al., 2023. Top-down design of protein architectures with reinforcement learning. Science 380, 266–273.

Maharjan, S., et al., 2023. Cell-free protein synthesis: sustainable biotechnology platform. Biotechnol. Appl. Biochem.

Pandi, A., Adam, D., Zare, A., Trinh, V.T., Schaefer, S.L., Burt, M., et al., 2023. Cell-free biosynthesis combined with deep learning accelerates de novo-development of antimicrobial peptides. Nat. Commun. 14 (1), 7197.

Radivojević, T., Costello, Z., Workman, K., Martin, H.G., 2020. A machine learning automated recommendation tool for synthetic biology. Nat. Commun. 11 (1), 4879.

Rives, A., Meier, J., Sercu, T., Goyal, S., Lin, Z., Liu, J., et al., 2021. Biological structure and function emerge from scaling unsupervised learning to 250 million protein sequences. Proc. Natl. Acad. Sci. U.S.A. 118 (15), e2016239118.

Siddique, A., et al., 2024. Computational strategies to enhance CFPS efficiency. BioMedInformatics 4 (3), 2022–2042.

Stanton, B.C., Nielsen, A.A., Tamsir, A., Clancy, K., Peterson, T., Voigt, C.A., 2014. Genomic mining of prokaryotic repressors for orthogonal logic gates. Nat. Chem. Biol. 10 (2), 99–105.

Wiegand, D.J., et al., 2019. Establishing a high-yielding CFPS platform from V. natriegens. ACS Synth. Biol.

Wittmann, B.J., Yue, Y., Arnold, F.H., 2021. Informed training set design enables efficient machine learning-assisted directed evolution. Cell Syst. 12 (11), 1026–1045.

Yang, K.K., Wu, Z., Arnold, F.H., 2019. Machine-learning-guided directed evolution for protein engineering. Nat. Methods 16 (8), 687–694.

CHAPTER

9

Automation and robotics in cell-free protein synthesis

9.1 Introduction

The integration of automation and robotics in cell-free protein synthesis (CFPS) represents a transformational shift from manual laboratory processes to sophisticated, integrated manufacturing systems. This chapter explores how robotic platforms, automated liquid handling systems, and intelligent process control are revolutionizing CFPS by enabling high-throughput operations, reducing human error, improving reproducibility, and scaling production capabilities from laboratory bench to industrial manufacturing.

The evolution of CFPS automation has been driven by several key factors: the need for consistent, reproducible results; the demand for higher throughput in protein production; the complexity of optimizing multiple interdependent parameters; and the requirement for 24/7 operation in commercial settings. Modern automated CFPS platforms integrate sophisticated robotics with artificial intelligence (AI)-driven process control, creating systems capable of autonomous operation while maintaining the flexibility to adapt to different protein targets and production requirements (Fig. 9.1) (Table 9.1).

9.2 Automated lysate preparation

9.2.1 Cell culture automation

The foundation of consistent CFPS performance begins with standardized, automated cell culture systems that ensure reproducible lysate quality. Automated bioreactor systems for bacterial culture provide precise control over growth conditions while minimizing contamination risks and operator variability.

Modern automated cell culture platforms integrate multiple control systems to maintain optimal growth conditions throughout the fermentation process. Table 9.2 details the key components and capabilities of automated cell culture systems for CFPS lysate production.

Cell Free Synthesis of Biologics
https://doi.org/10.1016/B978-0-443-33271-5.00018-1

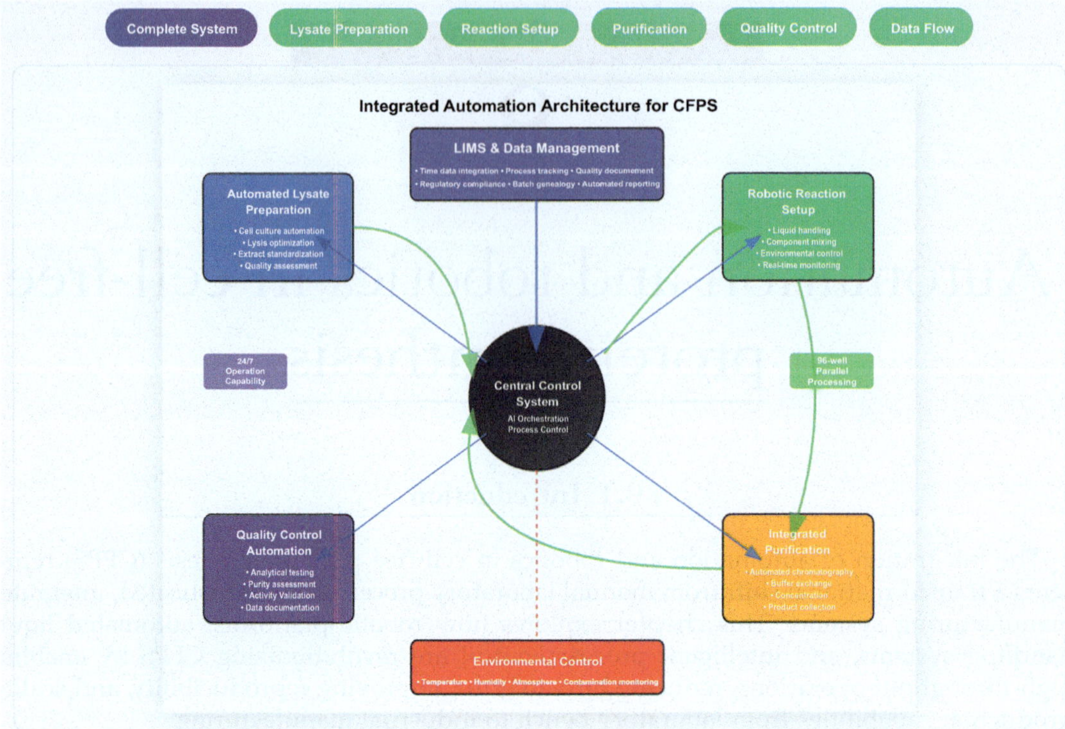

FIG. 9.1 Integrated automation architecture for CFPS. This comprehensive diagram illustrates the integrated automation architecture for CFPS manufacturing systems. The central AI control system orchestrates all automated modules, including lysate preparation, reaction setup, purification, and quality control. Blue arrows indicate data and control flow from the LIMS and central control system, green arrows show material and process flow between operational modules, and red dashed lines represent environmental control connections. The system architecture enables 24/7 autonomous operation with 96-well parallel processing capabilities, demonstrating how modern automation integrates multiple technologies to achieve consistent, high-quality protein production.

9.2.2 Lysis process automation

The automation of cell lysis processes ensures consistent disruption efficiency while minimizing variability in extract activity. Automated lysis systems integrate multiple disruption methods with real-time monitoring to optimize extract quality (Table 9.3).

9.2.3 Extract processing and standardization

Postlysis processing automation ensures consistent extract quality through standardized clarification, concentration, and quality assessment procedures. Automated systems maintain extract integrity while providing comprehensive documentation for quality assurance.

The standardization of extract processing involves multiple automated steps designed to remove cellular debris, concentrate active components, and assess extract quality. Table 9.4 outlines the key automated processing steps and their quality impact.

TABLE 9.1 Evolution of CFPS automation technologies.

Generation	Time period	Key technologies	Capabilities	Limitations
Manual (first gen)	1960s–1990s	• Basic pipetting • Manual incubation • Individual reactions	• Research flexibility • Low cost • Simple protocols	• High variability • Labor intensive • Low throughput
Semiautomated (second gen)	1990s–2010s	• Liquid handling robots • Automated readers • Basic integration	• Improved reproducibility • Medium throughput • Reduced labor	• Limited integration • Manual oversight • Fixed protocols
Fully automated (third gen)	2010s–Present	• Integrated platforms • AI-driven control • End-to-end automation	• High throughput • Minimal intervention • Adaptive protocols	• High complexity • Capital intensive • Maintenance requirements
Autonomous (fourth gen)	Present–future	• Self-optimizing systems • Predictive maintenance • Closed-loop learning	• Autonomous operation • Continuous improvement • Predictive capabilities	• Technology maturity • Validation complexity • Regulatory challenges

9.3 Robotic reaction setup

9.3.1 Liquid handling automation

Precision liquid handling forms the foundation of automated CFPS reaction setup, enabling accurate dispensing of multiple components while maintaining sterility and preventing cross-contamination. Modern liquid handling platforms integrate multiple technologies to achieve the precision and flexibility required for complex CFPS formulations (Table 9.5).

9.3.2 Component integration and mixing

Automated component integration ensures proper mixing of CFPS reaction components while maintaining optimal conditions for protein synthesis. Advanced mixing strategies account for component sensitivities and interaction kinetics to maximize synthesis efficiency.

The complexity of CFPS reactions, with their multiple enzymes, cofactors, and regulatory components, requires sophisticated mixing protocols to ensure homogeneity while preventing component degradation. Table 9.6 details automated mixing strategies and their applications.

9.3.3 Environmental control and monitoring

Precise environmental control during reaction setup and incubation is critical for CFPS success. Automated systems maintain optimal temperature, humidity, and atmospheric conditions while providing real-time monitoring and adjustment capabilities (Table 9.7).

TABLE 9.2 Automated cell culture system components for CFPS.

System component	Function	Automation features	Quality benefits	Typical specifications
Bioreactor control	• Temperature regulation • pH monitoring/control • Dissolved oxygen control • Agitation management	• Automatic setpoint control • Cascade control loops • Alarm systems • Data logging	• Consistent growth conditions • Reduced batch variability • Optimized cell metabolism	• Temperature: ± 0.1°C • pH: ± 0.02 units • Dissolved oxygen (DO): ± 1% • Response time: < 30 s
Media preparation	• Component dispensing • Mixing and dissolution • Sterilization • Quality verification	• Automated weighing • Recipe management • Sterile connections • In-line monitoring	• Reproducible media composition • Contamination prevention • Documentation compliance	• Accuracy: ± 0.1% • Sterility: > 99.9% • Batch records: complete • Traceability: full
Sampling and analysis	• Automated sampling • Cell density measurement • Metabolite analysis • Growth rate calculation	• Sterile sampling systems • Online analyzers • Data integration • Trend analysis	• Real-time process monitoring • Early deviation detection • Optimal harvest timing	• Sampling frequency: continuous • Analysis time: < 5 min • Accuracy: ± 2% • Detection limit: 0.1 OD
Harvest automation	• Cell concentration • Washing procedures • Collection systems • Storage preparation	• Automated centrifugation • Buffer exchange • Volume control • Cold chain maintenance	• Consistent cell preparation • Minimal cell stress • Preserved viability	• Recovery: > 95% • Viability: > 98% • Temperature: 4 ± 1°C • Processing time: < 2 h

TABLE 9.3 Automated lysis technologies and performance characteristics.

Lysis method	Automation features	Process control	Advantages	Limitations
High-pressure homogenization	• Automated pressure control • Multipass capability • Temperature monitoring • Flow rate regulation	• Pressure feedback loops • Temperature compensation • Pass counting • Efficiency monitoring	• High disruption efficiency • Scalable process • Reproducible results • Continuous operation	• Heat generation • Protein denaturation risk • Equipment complexity • Maintenance requirements
Sonication systems	• Programmable protocols • Power optimization • Pulse control • Temperature protection	• Amplitude control • Duty cycle optimization • Cooling integration • Endpoint detection	• Precise control • Minimal sample volume • Flexible parameters • Easy integration	• Limited scalability • Heat generation • Probe contamination • Reproducibility challenges
Bead mill disruption	• Automated bead loading • Speed optimization • Cooling systems • Separation automation	• Rotational speed control • Temperature monitoring • Timer control • Load balancing	• Gentle disruption • High efficiency • Scalable design • Contamination control	• Bead separation • Cleaning complexity • Wear considerations • Process optimization
Chemical lysis	• Reagent dispensing • Mixing automation • Incubation control • Neutralization systems	• Concentration control • Timing precision • pH monitoring • Temperature control	• Gentle process • High reproducibility • Easy automation • Minimal equipment	• Chemical costs • Interference potential • Disposal requirements • Optimization complexity

TABLE 9.4 Automated extract processing pipeline.

Processing step	Automation technology	Quality control	Critical parameters	Expected outcomes
Clarification	• Automated centrifugation • Filtration systems • Continuous separation • Debris monitoring	• Turbidity measurement • Particle counting • Protein quantification • Activity testing	• G-force: 10,000–30,000 × g • Time: 30–60 min • Temperature: 4°C • Recovery: > 90%	• Clear supernatant • Minimal debris • Preserved activity • Consistent composition
Concentration	• Ultrafiltration systems • Dialysis automation • Volume monitoring • Buffer exchange	• Protein concentration • Volume tracking • Buffer composition • Membrane integrity	• Concentration factor: 2–10× • Membrane molecular weight cut off (MWCO): 10–30 kDa • Pressure: < 30 psi • Recovery: > 85%	• Target concentration • Buffer exchange • Activity preservation • Volume consistency
Aliquoting	• Precision dispensing • Sterile handling • Automated labeling • Storage management	• Volume verification • Sterility testing • Identity confirmation • Storage monitoring	• Volume accuracy: ± 2% • Sterility: SAL 10^{-6} • Labeling: 100% accurate • Storage: −80°C	• Consistent aliquots • Sterile products • Complete traceability • Long-term stability
Quality assessment	• Automated assays • High-throughput testing • Data integration • Trend analysis	• Activity measurement • Purity analysis • Contamination testing • Stability evaluation	• Activity: > 80% standard • Purity: > 95% • Endotoxin: < 10 EU/mL • Stability: > 6 months	• Release criteria met • Consistent quality • Documented results • Predictable performance

TABLE 9.5 Liquid handling technologies for CFPS automation.

Technology type	Precision range	Volume range	Throughput	Best applications	Advantages	Limitations
Air displacement pipetting	±1%–3%	0.1–1000 μL	96–384 samples	• General reagent dispensing • Buffer additions • Template delivery	• High precision • Versatile volumes • Easy maintenance • Cost effective	• Viscosity limitations • Volatile solvents • Contamination risk • Tip consumption
Positive displacement	±0.5%–1%	0.1–100 μL	8–96 samples	• Viscous solutions • Detergent mixtures • Protein solutions	• Viscosity independent • High accuracy • No air contact • Precise volumes	• Limited throughput • Complex maintenance • Higher cost • Volume restrictions
Peristaltic pumping	±2%–5%	1 μL–10 mL	Continuous	• Bulk reagent delivery • Media preparation • Wash solutions	• Continuous delivery • No contamination • Simple operation • Sterile path	• Lower precision • Tubing replacement • Pulsation effects • Volume limitations
Solenoid valve dispensing	±3%–8%	10 μL–1 mL	High speed	• Fixed volume dispensing • Repetitive additions • High-throughput setup	• High speed • Consistent volumes • Low maintenance • Robust operation	• Fixed volumes • Limited precision • Calibration drift • Pressure dependent

TABLE 9.6 Automated mixing strategies for CFPS reactions.

Mixing method	Mechanism	Scale range	Mixing time	Applications	Advantages	Considerations
Orbital shaking	• Circular motion • Controlled speed • Gentle mixing • Even distribution	96-well to flasks	30 s–5 min	• General mixing • Plate-based reactions • Gentle homogenization	• Gentle process • Uniform mixing • Scalable • Simple control	• Slower mixing • Limited shear • Volume dependent • Potential settling
Magnetic stirring	• Rotating magnetic field • Microstir bars • Variable speed • Continuous mixing	μL–mL volumes	Continuous	• Individual reactions • Viscous solutions • Long incubations	• Continuous mixing • No contact • Variable speed • Minimal heat	• Bar retrieval • Limited throughput • Size constraints • Magnetic interference
Pipette mixing	• Aspirate/dispense cycles • Controlled volume • Programmable patterns • Multiple cycles	1 μL–1 mL	10–60 s	• Initial mixing • Sensitive components • Precise control	• Precise control • Gentle process • Programmable • No equipment	• Manual operation • Time consuming • Tip consumption • Limited efficiency
Ultrasonic mixing	• Acoustic waves • Cavitation effects • Rapid homogenization • Degassing capability	μL–L volumes	5–30 s	• Rapid mixing • Degassing • Particle suspension • Cell disruption	• Very rapid • Effective degassing • Powerful mixing • Broad range	• Heat generation • Protein denaturation • Equipment cost • Maintenance needs

TABLE 9.7 Environmental control systems for automated CFPS.

Environmental parameter	Control technology	Precision	Response time	Monitoring features	Impact on CFPS
Temperature control	• Peltier heating/cooling • Forced air circulation • Thermal blocks • Gradient systems	±0.1°C–0.5°C	30 s–2 min	• Multipoint sensing • Gradient mapping • Alarm systems • Data logging	• Enzyme activity • Protein folding • Reaction kinetics • Stability
Humidity management	• Humidification systems • Desiccant control • Vapor barriers • Condensation prevention	±2%–5% RH	1–5 min	• Relative humidity • Dew point • Evaporation rates • Trend analysis	• Evaporation control • Concentration stability • Seal integrity • Sample consistency
Atmospheric composition	• Gas mixing systems • Oxygen control • CO_2 regulation • Inert atmospheres	±0.1%–1%	30 s–2 min	• Gas composition • Flow rates • Pressure monitoring • Leak detection	• Oxidation prevention • pH stability • Metabolic balance • Product quality
Pressure control	• Vacuum systems • Positive pressure • Differential control • Containment systems	±1–5 mbar	15–60 s	• Absolute pressure • Differential pressure • Leak testing • Containment integrity	• Reaction equilibrium • Gas exchange • Contamination control • Safety assurance

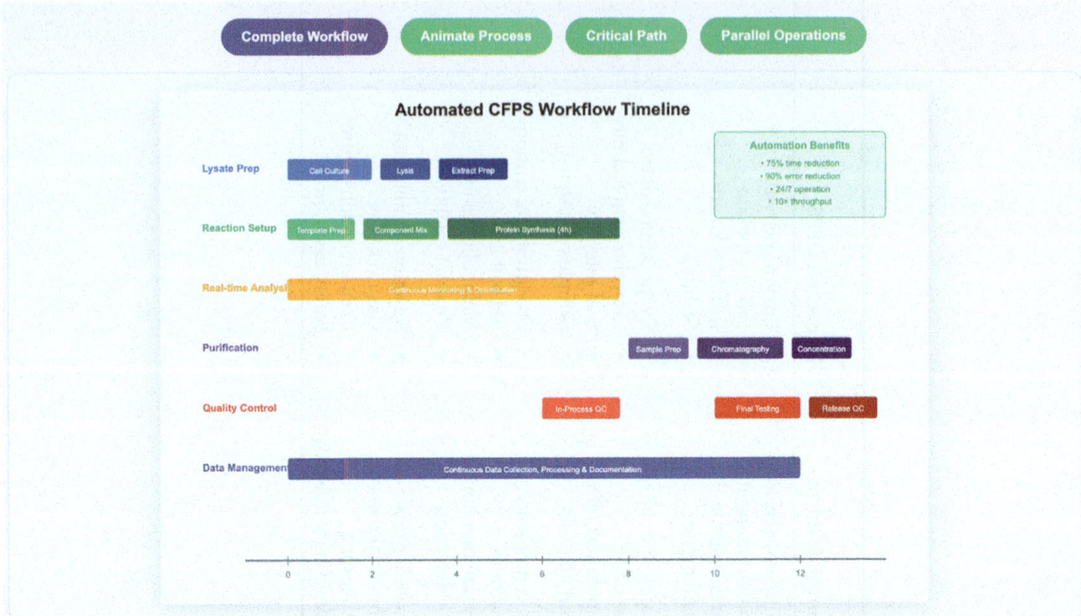

FIG. 9.2 Automated CFPS workflow timeline. This detailed timeline visualization shows the complete automated CFPS workflow from initiation to final product release over a 12-h cycle. Each horizontal track represents a different operational area with color-coded process steps: lysate preparation (blue), reaction setup (green), real-time analysis (orange), purification (purple), quality control (red), and data management (indigo). The critical path through protein synthesis determines the overall timeline, while parallel operations in lysate preparation and purification maximize system efficiency. Continuous data management and real-time monitoring ensure quality control and process optimization throughout the entire automated cycle. Interactive features allow visualization of the complete workflow, animation of the process sequence, identification of critical path operations, and highlighting of parallel processing capabilities.

9.3.4 High-throughput reaction platforms

High-throughput automated platforms enable parallel processing of multiple CFPS reactions, allowing for rapid optimization, screening, and production scaling. These systems integrate liquid handling, environmental control, and monitoring capabilities in compact, efficient formats (Fig. 9.2).

Modern high-throughput platforms combine miniaturization with automation to achieve unprecedented throughput while maintaining quality and reproducibility. Table 9.8 compares different high-throughput platform configurations and their capabilities.

9.4 Integrated purification systems

9.4.1 Automated chromatography systems

Automated chromatography platforms integrate seamlessly with CFPS production systems to provide consistent, reproducible protein purification. These systems employ sophisticated

TABLE 9.8 High-throughput CFPS platform configurations.

Platform type	Format	Throughput	Volume range	Automation level	Key features	Typical applications
Microplate systems	96/384/1536-well	96–1536 reactions	1–200 μL	Full automation	• Standard formats • Reader compatibility • Liquid handling • Environmental control	• Optimization studies • Screening applications • Quality control • Method development
Microfluidic chips	Custom geometries	100–10,000 droplets	pL–nL range	Integrated control	• Miniaturization • Rapid mixing • Precise control • Real-time monitoring	• Single-cell analysis • Kinetic studies • Ultra-high throughput • Reagent conservation
Array platforms	Slide-based arrays	1,000–100,000 spots	fL–pL range	Semiautomated	• Spatial organization • Parallel processing • Imaging compatible • High density	• Protein arrays • Screening libraries • Functional studies • Biomarker discovery
Droplet systems	Emulsion droplets	10,000–1 M droplets	pL–nL range	Automated generation	• Compartmentalization • Rapid generation • Sorting capability • Encapsulation	• Directed evolution • Single molecule • Library screening • Rare event detection

control algorithms and real-time monitoring to optimize separation performance while minimizing operator intervention (Table 9.9).

9.4.2 Automated sample preparation

Automated sample preparation systems ensure consistent sample handling between production and purification steps. These systems integrate dilution, buffer exchange, and conditioning operations to optimize downstream processing efficiency (Table 9.10).

9.4.3 Real-time purification monitoring

Real-time monitoring systems provide continuous assessment of purification performance, enabling dynamic optimization and quality assurance throughout the process. These systems integrate multiple analytical technologies with process control algorithms (Table 9.11).

9.5 Quality control automation

9.5.1 Automated analytical testing

Automated analytical testing systems provide comprehensive quality assessment of CFPS products while maintaining high throughput and minimizing operator intervention. These systems integrate multiple analytical techniques with sophisticated data management and interpretation capabilities (Table 9.12).

9.5.2 Real-time process monitoring

Real-time monitoring systems provide continuous assessment of CFPS processes, enabling immediate detection of deviations and proactive quality management. These systems integrate sensor technologies with AI-driven analytics to predict and prevent quality issues (Table 9.13).

9.5.3 Automated documentation and compliance

Automated documentation systems ensure complete traceability and regulatory compliance while reducing manual documentation burden. These systems integrate with all process equipment to provide comprehensive batch records and quality documentation (Table 9.14).

9.6 Data management and laboratory information management system integration

9.6.1 Laboratory information management systems

Modern laboratory information management system (LIMS) platforms serve as the central nervous system for automated CFPS operations, coordinating data flow between

TABLE 9.9 Automated chromatography technologies for CFPS products.

Chromatography type	Automation features	Resolution	Capacity	Applications	Advantages	Limitations
High pressure liquid chromatography/ultra high pressure liquid chromatography (FPLC) systems	• Automated method development • Gradient optimization • Peak detection • Fraction collection	High	Mg–mg scale	• Research applications • Method development • Small-scale purification • Quality analysis	• High resolution • Flexible methods • Easy operation • Good documentation	• Limited capacity • Higher cost/mg • Manual intervention • Scale limitations
HPLC/UHPLC	• Automated injection • Method switching • Real-time monitoring • Data integration	Very High	Mg–mg scale	• Analytical applications • Purity assessment • Impurity profiling • Quality control	• Excellent resolution • Rapid analysis • Precise quantitation • Regulatory accepted	• Small scale only • High pressure • Maintenance intensive • Limited prep use
Process chromatography	• Large-scale automation • CIP/SIP systems • Process monitoring • Batch tracking	Moderate	g–kg scale	• Production purification • Commercial manufacturing • Scale-up applications • Good manufacturing practice (GMP) compliance	• High capacity • Robust operation • Scalable • GMP compatible	• Lower resolution • Capital intensive • Complex operation • Validation requirements
Membrane chromatography	• Flow-through operation • Rapid processing • Low pressure • Continuous operation	Moderate	Variable	• Rapid purification • Polishing steps • Contaminant removal • Process intensification	• High flow rates • Low pressure drop • Easy scale-up • Disposable options	• Limited capacity • Moderate resolution • Cost per use • Fouling potential

TABLE 9.10 Automated sample preparation technologies.

Preparation step	Automation technology	Precision	Throughput	Quality benefits	Integration features
Dilution and conditioning	• Automated dilution • Buffer exchange • pH adjustment • Conductivity control	±1%–3%	96–384 samples/h	• Consistent preparation • Reduced variability • Optimal conditions • Reproducible results	• LIMS integration • Barcode tracking • Method storage • Quality documentation
Filtration and clarification	• Automated filtration • Membrane selection • Pressure control • Flux monitoring	Filter dependent	Continuous	• Particle removal • Consistent clarity • Membrane protection • Process optimization	• Pressure monitoring • Flux calculation • Membrane tracking • Cleaning integration
Concentration and exchange	• Ultrafiltration • Diafiltration • Volume control • Buffer exchange	±2%–5%	8–24 samples/h	• Target concentration • Buffer optimization • Volume consistency • Activity preservation	• Volume tracking • Concentration calculation • Process monitoring • Automated stopping
Aliquoting and storage	• Precision dispensing • Sterile handling • Automated labeling • Storage tracking	±1%–2%	96–384 samples/h	• Consistent aliquots • Sterile processing • Complete traceability • Optimized storage	• Sample tracking • Label generation • Storage management • Inventory control

TABLE 9.11 Real-time monitoring technologies for automated purification.

Monitoring parameter	Technology	Response time	Precision	Information provided	Control actions
UV absorbance (280 nm)	• Flow-through cells • LED sources • Photodiode arrays • Multiwavelength	<1 s	±0.001 AU	• Protein concentration • Elution profiles • Peak detection • Purity indication	• Fraction collection • Gradient adjustment • Peak cutting • Method modification
Conductivity	• In-line sensors • Temperature compensation • Drift correction • Calibration verification	<5 s	±0.1%	• Salt concentration • Buffer composition • Gradient formation • System performance	• Gradient control • Buffer preparation • Wash optimization • Method adjustment
pH monitoring	• Flow-through electrodes • Temperature compensation • Automatic calibration • Drift monitoring	<30 s	±0.01 pH units	• Buffer pH • System stability • Method performance • Quality indicators	• Buffer adjustment • Method optimization • Quality control • System maintenance
Pressure monitoring	• Digital transducers • Multiple points • Trend analysis • Alarm systems	<1 s	±0.1% FS	• System pressure • Column condition • Flow resistance • Blockage detection	• Flow control • Column protection • Method adjustment • Maintenance alerts
Flow rate control	• Mass flow meters • Feedback control • Pulsation dampening • Accuracy verification	<5 s	±0.5%	• Actual flow rate • System performance • Method compliance • Gradient accuracy	• Pump adjustment • Gradient correction • Method modification • System optimization

TABLE 9.12 Automated analytical testing platforms for CFPS quality control.

Analysis type	Automation platform	Sample throughput	Analysis time	Key parameters	Quality information
Protein quantification	• Automated plate readers • Flow-through systems • Array analyzers • Imaging platforms	96–384 samples/h	1–10 min	• Total protein • Target protein • Concentration • Recovery yield	• Synthesis efficiency • Purification yield • Batch consistency • Process optimization
Purity assessment	• HPLC autosamplers • CE systems • Electrophoresis • Mass spectrometry	24–96 samples/day	10–60 min	• Purity percentage • Impurity profile • Aggregation state • Degradation products	• Product quality • Process consistency • Stability indication • Regulatory compliance
Activity testing	• Robotic assay systems • Kinetic readers • Biosensor platforms • Cell-based assays	96–1536 tests/day	15 min–24 h	• Specific activity • Kinetic parameters • Inhibitor sensitivity • Stability profile	• Functional integrity • Bioactivity confirmation • Stability assessment • Potency determination
Identity confirmation	• Automated sequencing • Mass spectrometry • Peptide mapping • Immunoassays	8–48 samples/day	30 min–4 h	• Sequence identity • Molecular weight • Posttranslational modifications • Structural integrity	• Identity verification • Sequence accuracy • Modification profile • Structural integrity

TABLE 9.13 Real-time process monitoring systems.

Monitoring domain	Sensor technologies	Data frequency	Alert response	Predictive capabilities	Quality impact
Reaction progress	• Fluorescence probes • pH sensors • Oxygen electrodes • Turbidity meters	Continuous (1 Hz–1 kHz)	<30 s	• Yield prediction • Completion time • Optimization recommendations • Deviation forecasting	• Early intervention • Process optimization • Yield maximization • Quality assurance
Environmental conditions	• Temperature sensors • Humidity meters • Pressure transducers • Gas analyzers	Continuous (0.1–10 Hz)	<10 s	• Equipment failure • Environmental drift • Maintenance needs • Performance degradation	• Stable conditions • Consistent quality • Equipment protection • Predictive maintenance
Contamination detection	• Particle counters • Endotoxin sensors • Microbial sensors • Chemical detectors	Continuous batch	<60 s	• Contamination events • Source identification • Trend analysis • Risk assessment	• Product safety • Batch protection • Root cause analysis • Risk mitigation
System performance	• Flow sensors • Pressure monitors • Power analyzers • Vibration sensors	Continuous (1–100 Hz)	<5 s	• Equipment failure • Performance drift • Maintenance scheduling • Optimization opportunities	• System reliability • Consistent performance • Reduced downtime • Improved efficiency

TABLE 9.14 Automated documentation and compliance systems.

Documentation area	Automation features	Data capture	Compliance benefits	Integration capabilities
Batch records	• Electronic batch records • Real-time data capture • Exception handling • Electronic signatures	• Process parameters • Material usage • Environmental conditions • Quality results	• Complete traceability • Audit trails • Version control • Data integrity	• Manufacturing execution system (MES) systems • LIMS integration • Enterprise resource planning (ERP) connectivity • Regulatory systems
Quality documentation	• Automated report generation • Statistical analysis • Trend monitoring • Exception reporting	• Test results • Specifications • Deviations • Investigations	• Regulatory readiness • Quality assurance • Risk management • Continuous improvement	• QMS systems • Statistical software • Regulatory databases • Knowledge management
Equipment records	• Automated logging • Maintenance tracking • Calibration management • Performance monitoring	• Usage hours • Maintenance activities • Calibration data • Performance metrics	• Equipment qualification • Maintenance compliance • Calibration traceability • Performance validation	• Computerized maintenance management system (CMMS) systems • Calibration software • Asset management • Maintenance planning
Training records	• Competency tracking • Training scheduling • Assessment management • Certification monitoring	• Training completion • Assessment scores • Competency levels • Recertification needs	• Personnel qualification • Training compliance • Competency assurance • Risk mitigation	• HR systems • LMS platforms • Competency management • Qualification tracking

instruments, managing sample tracking, and ensuring data integrity throughout the production process (Table 9.15).

9.6.2 Data architecture and integration

The complexity of automated CFPS systems requires sophisticated data architectures that can handle diverse data types, ensure real-time processing, and maintain data integrity across multiple systems and timescales (Table 9.16).

9.6.3 Advanced analytics and machine learning

The integration of advanced analytics and machine learning with automated CFPS systems enables predictive quality control, process optimization, and intelligent decision-making based on historical and real-time data (Fig. 9.3) (Table 9.17).

9.7 Future automation trends

9.7.1 Autonomous laboratory systems

The evolution toward autonomous laboratory systems represents the next frontier in CFPS automation, where AI-driven systems can independently design experiments, execute protocols, analyze results, and make decisions with minimal human intervention (Table 9.18).

9.7.2 Digital twin technology

Digital twin technology creates virtual replicas of CFPS processes and equipment, enabling advanced simulation, optimization, and predictive maintenance capabilities that enhance automation effectiveness and reliability (Table 9.19).

9.7.3 Convergent technologies

The convergence of multiple advanced technologies with CFPS automation creates new possibilities for enhanced capabilities, improved efficiency, and novel applications that extend beyond traditional boundaries (Fig. 9.4) (Table 9.20).

9.7.4 Sustainability and green automation

The integration of sustainability principles with CFPS automation focuses on resource efficiency, waste reduction, and environmental impact minimization while maintaining or improving production performance (Fig. 9.5) (Table 9.21).

TABLE 9.15 LIMS capabilities for automated CFPS operations.

LIMS function	Automation features	Data management	Quality benefits	Integration scope
Sample management	• Barcode generation • Automated tracking • Chain of custody • Location monitoring	• Sample genealogy • Status tracking • Aliquot management • Storage conditions	• Complete traceability • Reduced errors • Efficient workflow • Quality assurance	• Liquid handlers • Storage systems • Analytical instruments • Manufacturing systems
Workflow management	• Process orchestration • Task scheduling • Resource allocation • Exception handling	• Workflow templates • Process documentation • Resource utilization • Performance metrics	• Consistent processes • Optimized efficiency • Error reduction • Continuous improvement	• Robotic systems • Process equipment • Quality systems • Scheduling tools
Data integration	• Instrument connectivity • Automated data capture • Real-time processing • Data validation	• Multisource data • Data normalization • Quality control • Historical trends	• Data integrity • Comprehensive records • Trend analysis • Informed decisions	• Analytical instruments • Process systems • Environmental monitors • Quality equipment
Results management	• Automated calculations • Statistical analysis • Report generation • Notification systems	• Results processing • Specification checking • Trend monitoring • Exception reporting	• Accurate results • Timely decisions • Quality monitoring • Risk management	• Analysis software • Statistical tools • Reporting systems • Communication platforms

TABLE 9.16 Data architecture components for automated CFPS.

Architecture layer	Components	Data types	Processing requirements	Integration challenges
Sensor data layer	• Process sensors • Environmental monitors • Quality analyzers • Equipment sensors	• Time-series data • Alarm events • Status information • Measurement values	• Real-time processing • High-frequency capture • Edge computing • Data compression	• Protocol diversity • Data volume • Timing synchronization • Reliability requirements
Instrument data layer	• Analytical instruments • Robotic systems • Manufacturing equipment • Quality systems	• Analytical results • Method parameters • Calibration data • Performance metrics	• Automated capture • Data validation • Format conversion • Quality checking	• Vendor differences • Data formats • Communication protocols • Version management
Process data layer	• LIMS systems • MES platforms • Quality systems • Batch records	• Process parameters • Quality results • Batch information • Genealogy data	• Workflow management • Data correlation • Exception handling • Audit trails	• System boundaries • Data ownership • Workflow complexity • Security requirements
Enterprise data layer	• ERP systems • Data warehouses • Analytics platforms • Reporting tools	• Business metrics • Historical trends • Aggregated data • Performance indicators	• Data aggregation • Advanced analytics • Predictive modeling • Business intelligence	• Data governance • Privacy requirements • Regulatory compliance • Performance scalability

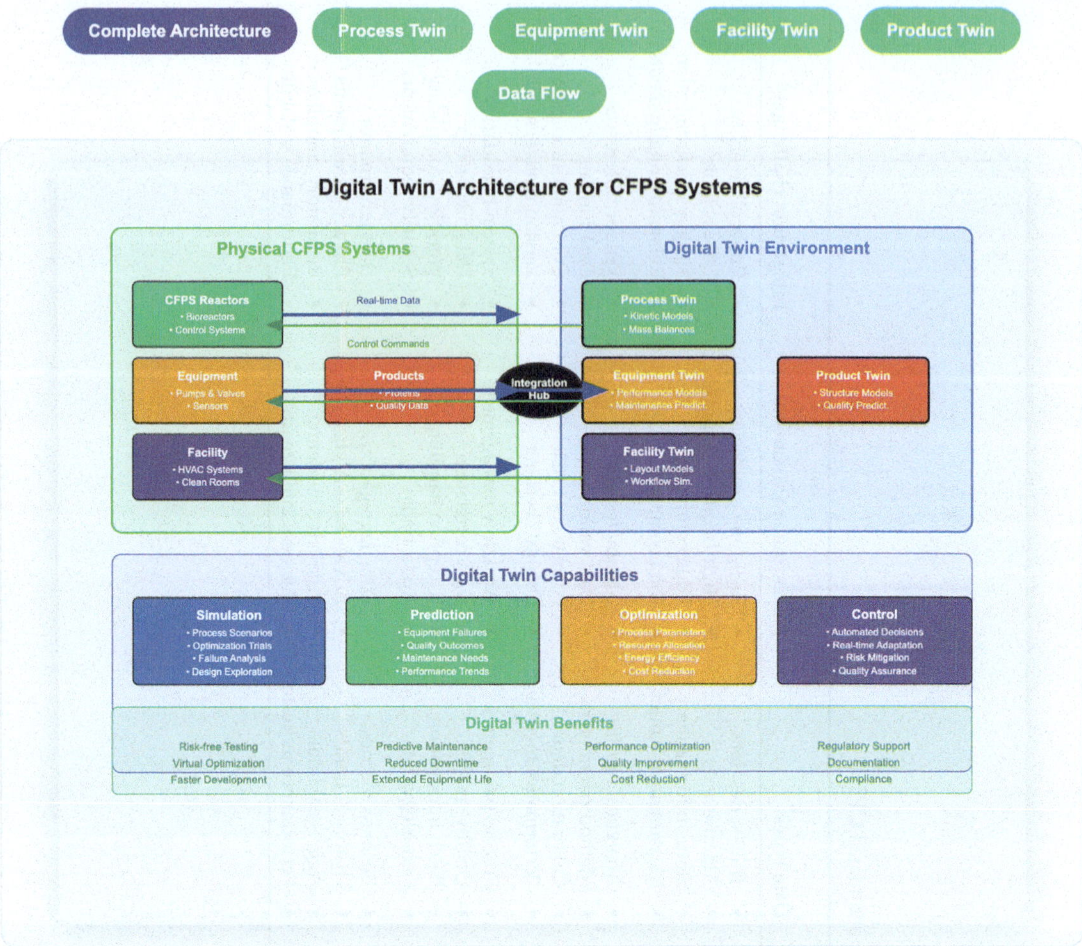

FIG. 9.3 Digital twin architecture for CFPS systems. This schematic illustrates the comprehensive digital twin framework that creates virtual replicas of CFPS processes and equipment. The architecture shows four interconnected digital twin types: process twin (modeling reaction kinetics and control systems), equipment twin (representing mechanical systems and performance models), facility twin (optimizing layout and workflows), and product twin (tracking protein structure and quality attributes). Real-time data integration flows from physical sensors, instruments, and systems to their digital counterparts through secure communication protocols. The digital twins enable advanced simulation capabilities, including process optimization, failure prediction, and virtual experimentation. Benefits include risk-free testing, predictive maintenance, performance optimization, and accelerated development cycles. The bidirectional data flow allows the digital twins to both learn from and influence their physical counterparts, creating a closed-loop system for continuous improvement.

TABLE 9.17 Advanced analytics applications in automated CFPS.

Analytics application	Machine learning methods	Data sources	Predictive capabilities	Implementation benefits
Process optimization	• Multivariate regression • Neural networks • Genetic algorithms • Reinforcement learning	• Process parameters • Environmental data • Quality results • Historical batches	• Optimal conditions • Parameter recommendations • Yield predictions • Quality forecasting	• Improved yields • Reduced variability • Faster optimization • Continuous improvement
Quality prediction	• Classification algorithms • Ensemble methods • Deep learning • Time series analysis	• Real-time sensors • Process variables • Material properties • Environmental conditions	• Quality outcomes • Specification compliance • Defect probability • Risk assessment	• Early intervention • Reduced waste • Quality assurance • Risk mitigation
Equipment monitoring	• Anomaly detection • Pattern recognition • Predictive maintenance • Condition monitoring	• Equipment sensors • Performance data • Maintenance records • Usage patterns	• Equipment failures • Maintenance needs • Performance degradation • Lifetime estimation	• Reduced downtime • Optimized maintenance • Extended equipment life • Cost savings
Batch release	• Decision trees • Support vector machines • Ensemble classifiers • Rule-based systems	• Quality test results • Process data • Specification limits • Historical decisions	• Release recommendations • Risk assessment • Investigation triggers • Compliance status	• Faster decisions • Consistent criteria • Reduced errors • Regulatory confidence

TABLE 9.18 Autonomous laboratory system capabilities.

Autonomy level	System capabilities	Human involvement	Decision making	Learning capabilities	Implementation timeline
Level 1: Assisted	• Basic automation • Manual oversight • Predetermined protocols • Human decisions	High supervision	Human-driven	Limited feedback	Current
Level 2: Partial	• Adaptive protocols • Exception handling • Basic optimization • Guided decisions	Moderate oversight	Human-assisted	Pattern recognition	1–3 years
Level 3: Conditional	• Intelligent planning • Autonomous execution • Real-time adaptation • Contextual decisions	Supervisory role	System-recommended	Active learning	3–7 years
Level 4: High	• Self-optimization • Predictive planning • Autonomous problem-solving • Independent decisions	Monitoring only	System-driven	Continuous learning	7–15 years
Level 5: Full	• Complete autonomy • Self-maintenance • Novel discoveries • Creative solutions	Strategic guidance	Fully autonomous	Self-improving	15+ years

TABLE 9.19 Digital twin applications in CFPS automation.

Digital twin type	Virtual representation	Real-Time integration	Simulation capabilities	Optimization benefits
Process twin	• Reaction kinetics • Mass/energy balances • Control systems • Quality parameters	• Real-time data sync • Parameter updates • State estimation • Predictive modeling	• Process scenarios • Optimization trials • Failure analysis • Design space exploration	• Virtual optimization • Risk-free testing • Faster development • Improved understanding
Equipment twin	• Mechanical systems • Thermal dynamics • Wear patterns • Performance models	• Sensor integration • Condition monitoring • Performance tracking • Maintenance scheduling	• Equipment behavior • Failure prediction • Maintenance planning • Performance optimization	• Predictive maintenance • Extended equipment life • Reduced downtime • Optimized performance
Facility twin	• Layout optimization • Workflow modeling • Resource allocation • Environmental systems	• Facility sensors • Usage patterns • Environmental data • Resource utilization	• Layout scenarios • Workflow optimization • Resource planning • Emergency response	• Space optimization • Workflow efficiency • Resource optimization • Risk management
Product twin	• Protein structure • Quality attributes • Stability models • Performance characteristics	• Quality data • Test results • Usage feedback • Performance metrics	• Quality prediction • Stability modeling • Performance analysis • Formulation optimization	• Quality assurance • Stability prediction • Performance optimization • Regulatory support

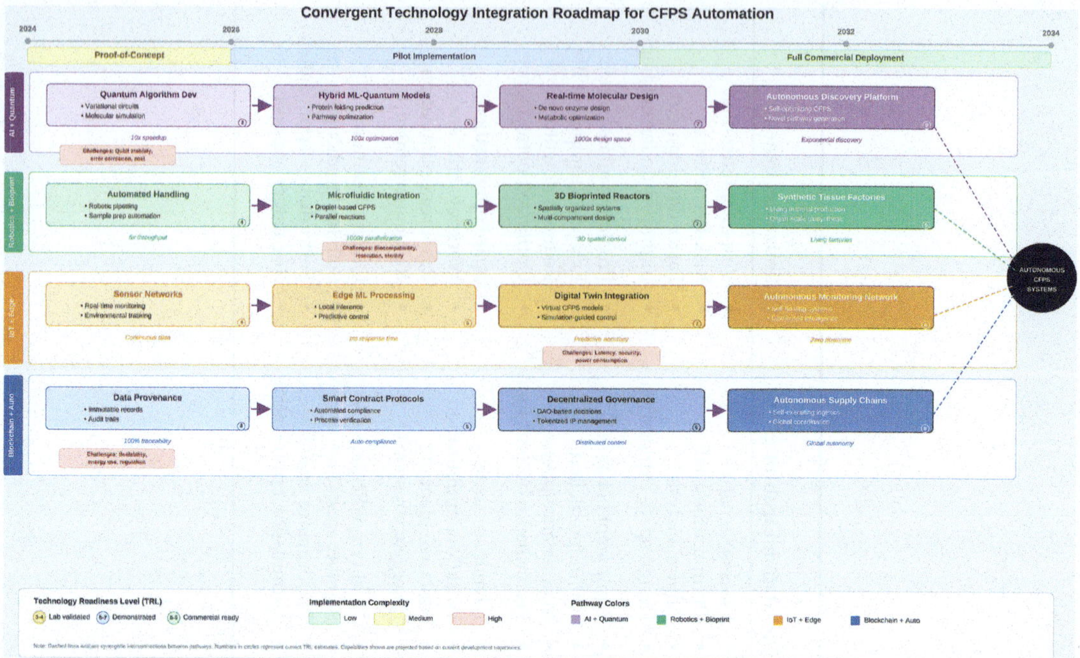

FIG. 9.4 Convergent technology integration roadmap for CFPS automation. This comprehensive roadmap illustrates the strategic integration of emerging technologies with CFPS automation systems over the next decade. The timeline shows four major convergence pathways: AI + Quantum Computing (enabling exponential optimization speedup and molecular design), Robotics + Bioprinting (creating 3D biostructures and spatial organization), IoT + Edge Computing (providing ubiquitous monitoring and local intelligence), and Blockchain + Automation (ensuring data integrity and automated governance). Each pathway shows development phases from proof-of-concept through pilot implementation to full commercial deployment. The roadmap indicates technology readiness levels, integration challenges, and expected capabilities at each phase. Interconnections between pathways demonstrate how convergent technologies will work synergistically to create next-generation autonomous CFPS systems. The visualization includes timeline markers, technology maturity indicators, and implementation complexity assessments to guide strategic planning and investment decisions.

9.8 Summary and future outlook

9.8.1 Current state and achievements

The automation and robotics landscape in CFPS has achieved remarkable progress, transforming laboratory-scale processes into sophisticated manufacturing systems capable of consistent, high-quality protein production. Current automated systems demonstrate:

- **Reproducibility**: Automated systems achieve coefficient of variation (CV) values below 5% for key quality parameters
- **Throughput**: Modern platforms process 96–1536 samples simultaneously with cycle times under 4 h

TABLE 9.20 Convergent technology integration with CFPS automation.

Technology convergence	Integration approach	Enhanced capabilities	New applications	Development challenges
AI + quantum computing	• Quantum algorithms • Hybrid processing • Optimization enhancement • Complex modeling	• Exponential speedup • Complex optimization • Advanced simulations • Pattern recognition	• Molecular design • Complex optimization • Quantum sensing • Cryptographic security	• Hardware limitations • Algorithm development • Integration complexity • Skill requirements
Robotics + bioprinting	• Precision positioning • Multimaterial handling • Sterile operations • Real-time adaptation	• 3D biostructures • Spatial organization • Multimaterial systems • Dynamic assembly	• Tissue engineering • Organoid production • Biomaterial synthesis • Spatial biology	• Technical complexity • Material compatibility • Sterility maintenance • Quality control
IoT + edge computing	• Distributed sensors • Edge processing • Real-time analytics • Autonomous decisions	• Ubiquitous monitoring • Local intelligence • Reduced latency • Improved reliability	• Smart laboratories • Predictive systems • Autonomous operations • Distributed manufacturing	• Connectivity issues • Security concerns • Data management • System complexity
Blockchain + automation	• Immutable records • Smart contracts • Decentralized control • Automated compliance	• Data integrity • Automated governance • Transparent operations • Trust mechanisms	• Supply chain tracking • Regulatory compliance • Automated transactions • Decentralized systems	• Scalability limitations • Energy requirements • Integration complexity • Regulatory uncertainty

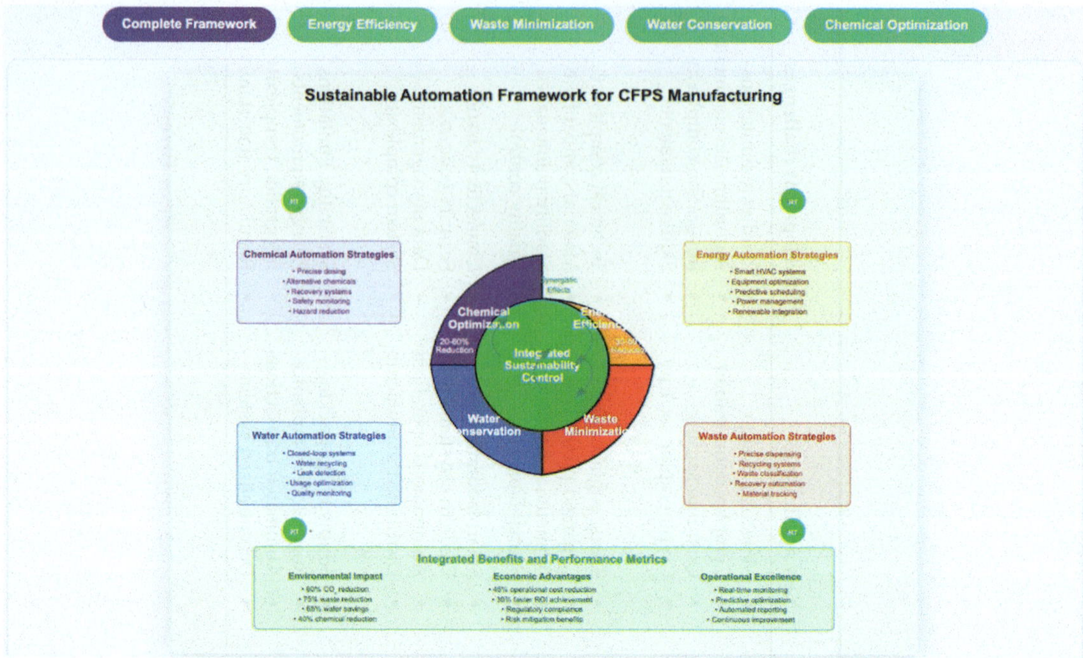

FIG. 9.5 **Sustainable automation framework for CFPS manufacturing.** This comprehensive framework diagram illustrates how sustainability principles are integrated throughout automated CFPS operations. The circular design shows four key sustainability focus areas: energy efficiency (30%–50% reduction through smart systems), waste minimization (40%–70% reduction through precision and recovery), water conservation (50%–80% reduction through closed-loop systems), and chemical optimization (20%–60% reduction through precise dosing and alternatives). Each sector shows specific automation strategies, quantified resource impacts, environmental benefits, and economic advantages. The central hub represents the integrated control system that coordinates all sustainability initiatives. Interconnected arrows demonstrate how improvements in one area create synergistic benefits in others. The framework includes real-time monitoring capabilities, predictive analytics for resource optimization, and automated reporting for sustainability metrics. Color coding indicates different sustainability performance levels and achievement targets.

- **Quality**: Automated quality control systems provide comprehensive testing with 24-h result availability
- **Efficiency**: Overall equipment effectiveness (OEE) exceeding 85% in optimized automated systems
- **Compliance**: Full regulatory compliance with 21 Code of Federal Regulations (CFR) Part 11 and GMP requirements (Fig. 9.6)

9.8.2 Integration challenges and solutions

The integration of complex automation systems presents several challenges that require coordinated solutions:

Technical integration: The complexity of integrating multiple vendor systems with different communication protocols and data formats requires standardized interfaces and

TABLE 9.21 Sustainable automation strategies for CFPS.

Sustainability Focus	Automation strategies	Resource impact	Environmental benefits	Economic advantages
Energy efficiency	• Smart HVAC systems • Equipment optimization • Predictive scheduling • Power management	• 30%–50% energy reduction • Peak demand shaving • Renewable integration • Grid optimization	• Reduced CO_2 emissions • Lower carbon footprint • Renewable energy use • Grid stability support	• Lower energy costs • Demand charge reduction • Incentive programs • Improved margins
Waste minimization	• Precise dispensing • Recycling systems • Waste classification • Recovery automation	• 40%–70% waste reduction • Material recovery • Solvent recycling • Container reuse	• Reduced landfill waste • Decreased incineration • Material conservation • Pollution prevention	• Waste cost reduction • Material savings • Disposal cost reduction • Revenue from recycling
Water conservation	• Closed-loop systems • Water recycling • Leak detection • Usage optimization	• 50%–80% water reduction • Quality optimization • Treatment efficiency • Reuse maximization	• Water conservation • Reduced discharge • Ecosystem protection • Aquifer preservation	• Water cost savings • Treatment cost reduction • Regulatory compliance • Risk mitigation
Chemical optimization	• Precise dosing • Alternative chemicals • Recovery systems • Safety monitoring	• 20%–60% chemical reduction • Safer alternatives • Improved efficiency • Hazard reduction	• Reduced toxicity • Lower emissions • Safer environments • Ecosystem protection	• Chemical cost savings • Safety cost reduction • Insurance benefits • Regulatory advantages

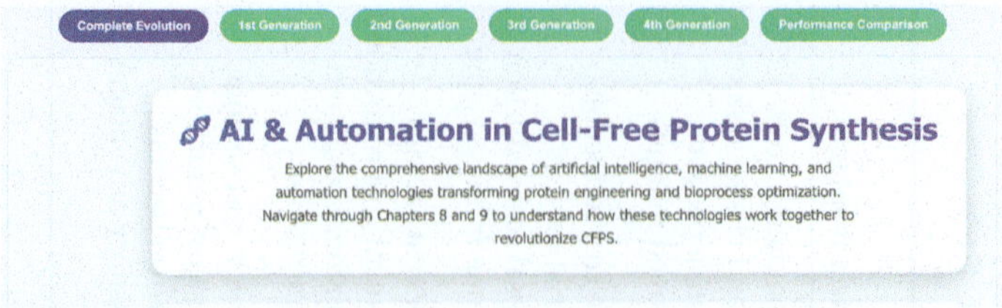

FIG. 9.6 **CFPS automation evolution and performance metrics.** This comprehensive visualization tracks the evolution of CFPS automation from manual laboratory processes to fully autonomous systems over four generations spanning 1960s to present. The timeline shows key technological milestones, capability improvements, and performance metrics, including throughput increases (from individual reactions to 1536 parallel samples), reproducibility improvements (CV reduction from > 20% to < 5%), quality control advancement (from manual testing to 24-h automated analysis), and efficiency gains (OEE improvement from < 50% to > 85%). The chart includes radar plots comparing manual vs. automated performance across multiple dimensions: speed, consistency, quality, cost-effectiveness, scalability, and compliance. Trend lines project future capabilities, including autonomous optimization, predictive maintenance, and self-improving systems. Color coding distinguishes different technology generations and their characteristic capabilities.

middleware solutions. Industry initiatives toward standardization (SILA, OPC-UA) are addressing these challenges.

Validation and qualification: Automated systems require extensive validation to demonstrate fitness for purpose and regulatory compliance. Risk-based validation approaches and modular qualification strategies are reducing validation burden while maintaining compliance.

Personnel training: The shift from manual to automated operations requires significant investment in personnel training and competency development. Comprehensive training programs and simulation-based learning are addressing these needs.

Change management: The implementation of automation represents significant organizational change requiring careful management of processes, procedures, and culture. Structured change management approaches ensure successful automation adoption.

9.8.3 Future directions and opportunities

The future of CFPS automation lies in the continued evolution toward autonomous, intelligent systems that can adapt, optimize, and improve without human intervention:

AI integration: Advanced AI algorithms will enable predictive process control, autonomous optimization, and intelligent decision-making based on comprehensive data analysis.

Distributed manufacturing: Automation will enable distributed manufacturing networks where CFPS production can be deployed at point-of-need locations while maintaining centralized control and quality assurance.

Sustainable operations: Green automation technologies will minimize environmental impact while improving economic performance through resource optimization and waste reduction.

Convergent technologies: The integration of emerging technologies such as quantum computing, advanced materials, and biotechnology will create new capabilities and applications for automated CFPS systems.

9.8.4 Strategic recommendations

Organizations implementing CFPS automation should consider the following strategic approaches:

1. **Phased implementation**: Deploy automation in phases, starting with high-impact, low-risk applications and gradually expanding to more complex operations.
2. **Technology partnerships**: Develop strategic partnerships with technology vendors, research institutions, and industry consortiums to access cutting-edge capabilities and share development costs.
3. **Data strategy**: Implement comprehensive data management strategies that enable advanced analytics, machine learning, and continuous improvement.
4. **Workforce development**: Invest in workforce development programs that prepare personnel for automated operations and enable them to leverage automation effectively.
5. **Regulatory engagement**: Engage proactively with regulatory agencies to ensure automation strategies align with evolving regulatory expectations and requirements.

The transformation of CFPS through automation and robotics represents one of the most significant advances in biotechnology manufacturing. As these technologies continue to evolve and mature, they will enable new applications, improve product quality, reduce costs, and accelerate the development of life-saving therapies and innovative biotechnology products.

Bibliography

Carlson, E.D., Gan, R., Hodgman, C.E., Jewett, M.C., 2012. Cell-free protein synthesis: applications come of age. Biotechnol. Adv. 30 (5), 1185–1194.

Failmezger, J., Scholz, S., Blombach, B., Siemann-Herzberg, M., 2018. Cell-free protein synthesis from fast-growing *Vibrio natriegens*. Front. Microbiol. 9, 1146.

Lu, Y., 2017. Cell-free synthetic biology: engineering in an open world. Synth. Syst. Biotechnol. 2 (1), 23–27.

Moore, S.J., MacDonald, J.T., Wienecke, S., Ishwarbhai, A., Tsipa, A., Aw, R., … Freemont, P.S., 2018. Rapid acquisition and model-based analysis of cell-free transcription–translation reactions from nonmodel bacteria. Proc. Natl. Acad. Sci. U. S. A. 115 (19), E4340–E4349.

Perez, J.G., Stark, J.C., Jewett, M.C., 2016. Cell-free synthetic biology: engineering beyond the cell. Cold Spring Harb. Perspect. Biol. 8 (12), a023853.

Salehi, A.S.M., Smith, M.T., Bennett, A.M., Williams, J.B., Pitt, W.G., Bundy, B.C., 2016. Cell-free protein synthesis of a cytotoxic cancer therapeutic: onconase production and a just-add-water cell-free system. Biotechnol. J. 11 (2), 274–281.

Schwarz, D., Junge, F., Durst, F., Frölich, N., Schneider, B., Reckel, S., … Dötsch, V., 2007. Preparative scale expression of membrane proteins in Escherichia coli-based continuous exchange cell-free systems. Nat. Protoc. 2 (11), 2945–2957.

Shirokov, V.A., Kommer, A., Kolb, V.A., Spirin, A.S., 2007. Continuous-flow cell-free translation, transcription-translation, and replication-translation systems. Methods Mol. Biol. 375, 19–55.

Silverman, A.D., Karim, A.S., Jewett, M.C., 2020. Cell-free gene expression: an expanded repertoire of applications. Nat. Rev. Genet. 21 (3), 151–170.

Thoring, L., Wüstenhagen, D.A., Borowiak, M., Stech, M., Sonnabend, A., Kubick, S., 2016. Cell-free systems based on CHO cell lysates: optimization strategies, synthesis of "difficult-to-express" proteins and future perspectives. PloS One 11 (9), e0163670.

PART IV

Applications and implementations

CHAPTER

10

Biopharmaceutical development

10.1 Introduction

The biopharmaceutical industry has witnessed a transformative shift toward cell-free protein synthesis (CFPS) systems as a rapid, flexible, and cost-effective platform for therapeutic protein development. Unlike traditional cell-based manufacturing, cell-free systems offer unprecedented control over protein production conditions, enabling the synthesis of challenging targets that are difficult or impossible to produce in living cells. This chapter explores the comprehensive applications of CFPS in biopharmaceutical development, from early-stage discovery through commercial manufacturing.

The global biopharmaceutical market, valued at over $400 billion in 2024, continues to demand innovative production platforms that can accelerate development timelines while maintaining product quality and safety. Cell-free systems have emerged as a critical technology platform, particularly for personalized medicine applications, rapid vaccine development, and the production of complex therapeutic proteins that pose challenges to conventional expression systems.

10.2 Therapeutic protein production

10.2.1 Advantages of cell-free systems for therapeutics

The CFPS offers several compelling advantages for therapeutic protein production that address longstanding challenges in biopharmaceutical manufacturing. Traditional cell line development can require 6–12 months to establish stable, high-expressing clones, while cell-free systems can produce functional proteins within hours of template preparation, dramatically accelerating the discovery-to-clinic timeline. This speed advantage is particularly valuable for emergency response scenarios, such as pandemic vaccine development, where time-to-market can save lives.

The open nature of cell-free systems allows real-time monitoring and adjustment of reaction conditions. Researchers can precisely control cofactor concentrations, redox potential, pH, and ionic strength throughout the synthesis process. This level of control enables optimization for specific protein targets and can improve product homogeneity compared to cell-based systems where intracellular conditions vary and are difficult to monitor.

Cell Free Synthesis of Biologics
https://doi.org/10.1016/B978-0-443-33271-5.00030-2

Many therapeutically relevant proteins, including certain antimicrobial peptides, membrane proteins, and cytotoxic agents, are difficult to produce in living cells due to their inherent toxicity. Cell-free systems circumvent this limitation by eliminating concerns about cell viability, enabling the production of proteins that would otherwise kill host cells. In addition, cell-free systems operate in defined, sterile environments without the risk of viral contamination that can plague mammalian cell cultures, reducing downstream purification requirements and regulatory concerns (Fig. 10.1).

10.2.2 Platform selection for therapeutic applications

The choice of cell-free platform significantly impacts protein quality, yield, and regulatory acceptability. Each system offers distinct advantages for different therapeutic applications, as summarized in Table 10.1 and illustrated in Fig. 10.2.

This comprehensive comparison illustrates the four major cell-free platforms used in therapeutic protein production. *E. coli* S30 systems provide the highest yields (1–5 mg/mL) and lowest costs but lack posttranslational modifications, making them ideal for

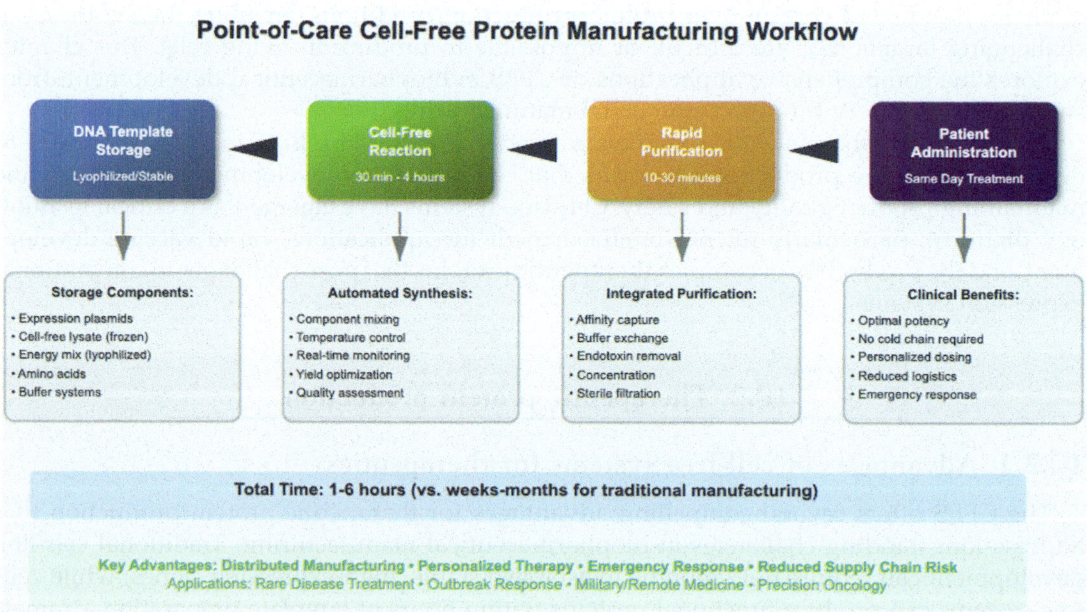

FIG. 10.1 Point-of-care manufacturing workflow. Point-of-care manufacturing using cell-free systems enables rapid production of therapeutics at the site of patient care. The workflow begins with stable storage of DNA templates, cell-free lysates, and energy components that can be rapidly reconstituted. Automated synthesis systems perform protein production in 30 min to 4 h, followed by integrated purification steps requiring only 10–30 min. The complete process enables same-day treatment with optimal protein potency, eliminates cold-chain requirements, and provides unprecedented flexibility for personalized medicine applications. This distributed manufacturing approach offers significant advantages over traditional centralized production, particularly for emergency response, rare disease treatment, and remote healthcare delivery. The total timeline of 1–6 h represents a dramatic improvement over the weeks to months required for conventional cell-based manufacturing systems.

TABLE 10.1 Comparison of cell-free platforms for therapeutic protein production.

Platform	Protein yield (mg/mL)	Cost relative to *E. coli*	Posttranslational modifications	Typical applications	Key advantages	Main limitations
E. coli S30	1–5	1.0×	None	Simple therapeutics, fragments	High yield, low cost, well-characterized	No post-translational modifications (PTMs), potential endotoxin
Wheat germ	0.1–1.0	3–5×	Basic eukaryotic	Hormones, growth factors	Eukaryotic folding, no endotoxin	Lower yield, plant-specific modifications
Rabbit reticulocyte	0.05–0.5	5–8×	Mammalian-like	Complex proteins	Mammalian processing, high fidelity	Low yield, expensive preparation
Mammalian cell-free	0.1–0.8	8–15×	Full human-like	Antibodies, glycoproteins	Human glycosylation, authentic PTMs	Highest cost, complex optimization
Insect cell	0.2–1.2	4–7×	Eukaryotic + some mammalian	Viral proteins, membrane proteins	Good yield, membrane protein competent	Limited human-like modifications

simple proteins and high-throughput screening applications. Wheat germ systems offer moderate yields with basic eukaryotic processing capabilities, suitable for hormones and growth factors. Mammalian cell-free systems provide human-like posttranslational modifications essential for complex therapeutics like monoclonal antibodies, though at higher cost and lower yields. Insect cell systems represent a middle ground with good yields and eukaryotic processing capabilities. The performance comparison matrix uses a five-star rating system to evaluate yield, cost-effectiveness, posttranslational modification capability, setup complexity, and regulatory considerations. Color coding helps identify optimal platforms for specific therapeutic applications, while the detailed characteristics guide platform selection based on target protein requirements and development goals.

10.2.3 Process development parameters

Successful therapeutic protein production requires systematic optimization of multiple parameters. Table 10.2 presents the critical process parameters and their typical ranges for different cell-free systems.

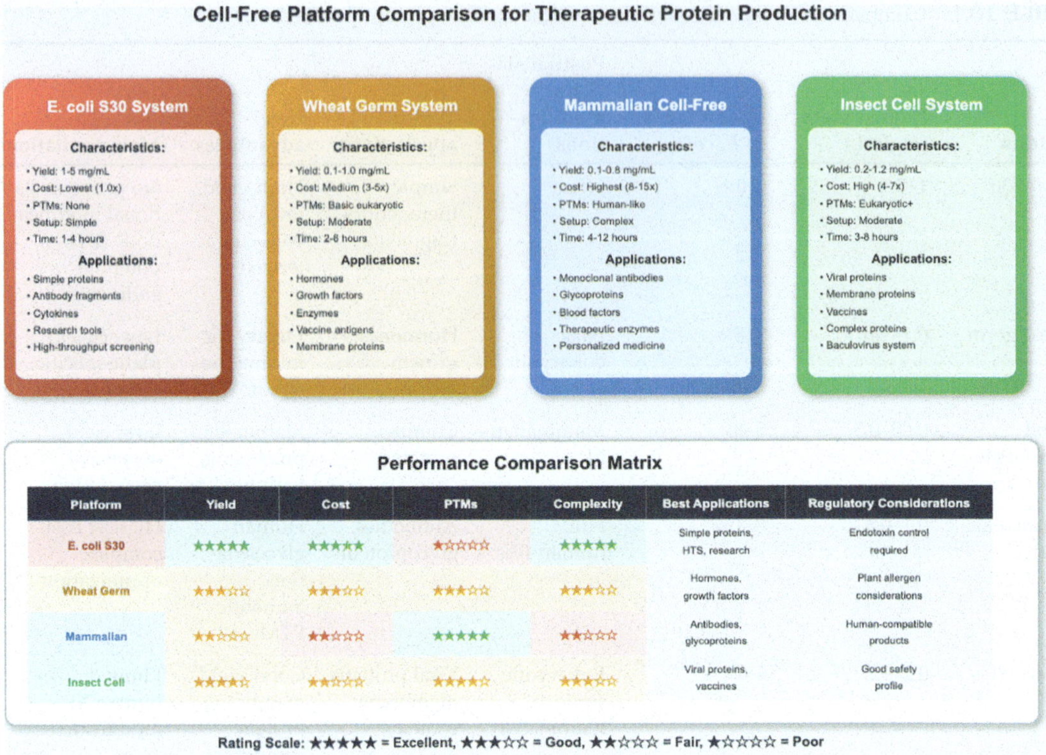

FIG. 10.2 Cell-free platform comparison for therapeutic applications.

TABLE 10.2 Critical process parameters for therapeutic protein production.

Parameter	*E. coli* system	Wheat germ	Mammalian cell-free	Optimization impact
Temperature (°C)	30–37	25–30	30–37	Affects folding kinetics and stability
pH	7.5–8.0	7.0–7.5	7.2–7.6	Critical for ribosome function
Mg^{2+} (mM)	10–20	2–5	1–3	Ribosome assembly and tRNA binding
K^+ (mM)	100–150	80–120	100–140	Ionic strength and ribosome stability
Template DNA (nM)	10–50	5–20	2–10	Balance between yield and resource consumption
Energy system	Phosphoenolpyruvate (PEP)/pyruvate kinase (PK) or glucose	Creatine phosphate	Glucose/glutamine	Sustained ATP regeneration
Reaction time (h)	1–4	2–6	4–12	Protein accumulation vs degradation

TABLE 10.3 Quality control parameters for cell-free produced therapeutics.

Analysis type	Method	Acceptance criteria	Frequency	Critical quality attributes
Identity	Mass spectrometry, N-terminal sequencing	>95% correct sequence	Every batch	Primary structure confirmation
Purity	RP-HPLC, SDS–PAGE	>95% main peak	Every batch	Product homogeneity
Potency	Bioassay, binding assay	80%–120% of reference	Every batch	Biological activity
Aggregation	SEC-HPLC, DLS	<5% high molecular weight	Every batch	Physical stability
Endotoxin	LAL assay	<5 EU/mg protein	Every batch	Safety (prokaryotic systems)
Glycosylation	MALDI-TOF, LC–MS	Consistent pattern	Every batch	Posttranslational fidelity

10.2.4 Quality control and characterization

Therapeutic proteins produced in cell-free systems must meet stringent quality standards comparable to those produced in conventional systems. The analytical characterization workflow encompasses multiple complementary techniques to ensure product quality and consistency (Table 10.3).

10.3 Antibody discovery and engineering

10.3.1 Rapid antibody screening platforms

Cell-free systems have revolutionized antibody discovery by enabling high-throughput screening of antibody libraries without the time and resource constraints of cell-based expression. The technology allows researchers to rapidly evaluate thousands of antibody variants in parallel, significantly accelerating the identification of lead candidates with desired properties.

Traditional hybridoma technology requires weeks to months for antibody discovery and characterization. In contrast, cell-free systems can screen diverse antibody libraries within days, enabling rapid identification of binders with specific characteristics such as high affinity, cross-reactivity, or unique epitope recognition. This speed advantage is particularly valuable in outbreak situations where rapid development of diagnostic or therapeutic antibodies is critical.

The open nature of cell-free systems also enables real-time monitoring of antibody expression and folding, providing immediate feedback on the success of expression for each library member. Researchers can incorporate cotranslational folding assistants, adjust redox conditions, or supplement with specific cofactors to optimize antibody production for challenging targets.

10.3.2 Antibody format optimization

Different antibody formats present unique production challenges and opportunities in cell-free systems, as shown in Fig. 10.3. Table 10.4 compares the production characteristics of various antibody formats commonly used in therapeutic development.

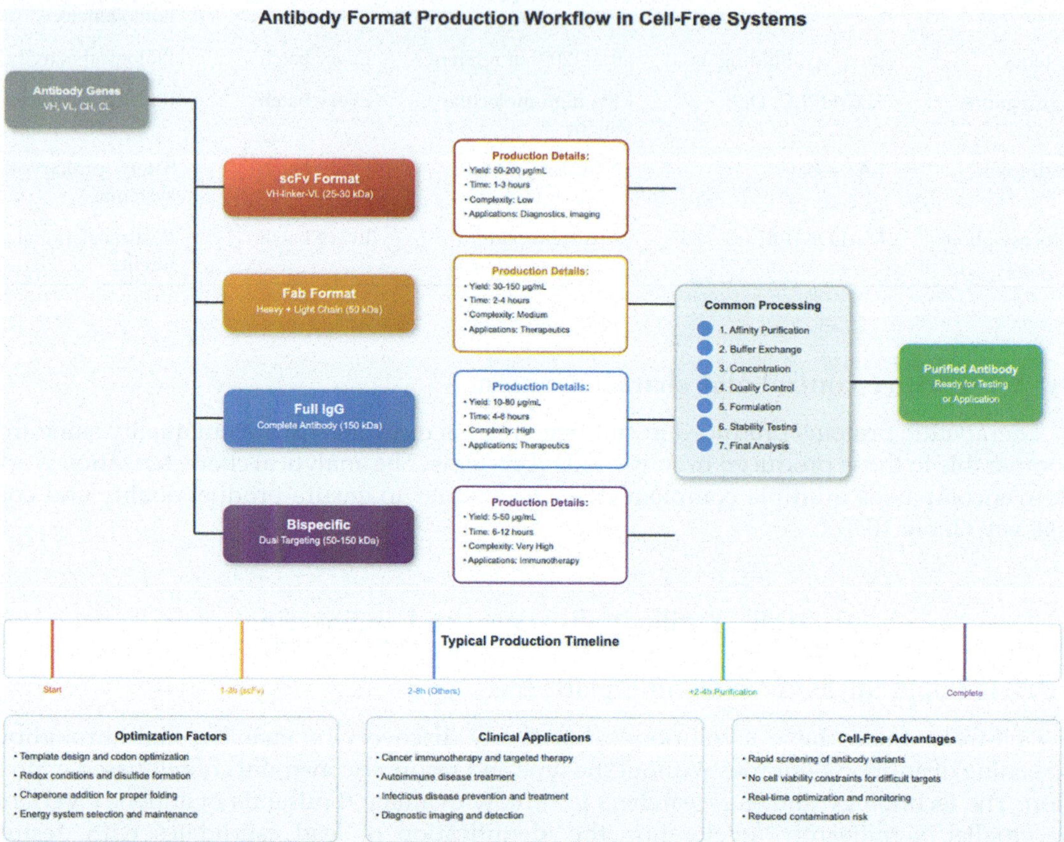

FIG. 10.3 Antibody format production workflow in cell-free systems. This workflow diagram illustrates the production pathway for four major antibody formats in cell-free systems, starting from common antibody genes (VH, VL, CH, CL regions). Single-chain variable fragments (scFv) represent the simplest format, requiring only VH and VL domains connected by a flexible linker, achieving yields of 50–200 μg/mL in 1–3 h with low complexity. Fab fragments require proper heavy and light chain assembly, yielding 30–150 μg/mL in 2–4 h with medium complexity. Full IgG antibodies demand complex assembly of multiple chains with proper disulfide bonding and potential glycosylation, producing 10–80 μg/mL in 4–8 h with high complexity requirements. Bispecific antibodies represent the most challenging format, requiring precise chain pairing to avoid misfolded products, yielding 5–50 μg/mL in 6–12 h with very high complexity. All formats follow common downstream processing steps, including affinity purification, buffer exchange, concentration, quality control, formulation, and stability testing. The timeline comparison shows increasing complexity correlates with longer production times, while optimization factors such as template design, codon optimization, redox conditions, chaperone addition, and energy system selection are critical for success across all formats.

TABLE 10.4 Antibody format production in cell-free systems.

Format	Molecular weight (kDa)	Production complexity	Typical yield (μg/mL)	Folding requirements	Therapeutic applications
scFv	25–30	Low	50–200	Simple disulfide formation	Diagnostics, research tools
Fab	50	Medium	30–150	Heavy-light chain assembly	Therapeutic fragments
$F(ab')_2$	100	Medium	20–100	Disulfide-linked dimer	Extended half-life fragments
Full IgG	150	High	10–80	Complex assembly, glycosylation	Therapeutic antibodies
Bispecific	50–150	Very High	5–50	Precise chain pairing	Cancer immunotherapy
Nanobody	12–15	Very Low	100–500	Single-domain folding	Specialized applications

TABLE 10.5 Glycosylation engineering strategies in cell-free systems.

Modification type	Enzymatic addition	Substrate requirements	Impact on function	Clinical relevance
Reduced fucosylation	GDP-fucose limitation	Controlled GDP-fucose levels	10–50× enhanced ADCC	Cancer therapeutics
Increased sialylation	α2,6-sialyltransferase	CMP-sialic acid	Extended half-life	Chronic disease treatment
Bisecting GlcNAc	GnT-III addition	UDP-GlcNAc	Enhanced ADCC	Immunotherapy
Galactosylation control	β1,4-galactosyltrans-ferase	UDP-galactose	Complement modulation	Autoimmune disorders
Man5 glycoforms	Mannosidase inhibition	Controlled processing	Reduced immunogenicity	Biosimilar development

10.3.3 Glycoengineering and posttranslational modifications

Therapeutic antibodies require specific glycosylation patterns for optimal efficacy and safety. The Fc region glycosylation significantly impacts antibody half-life, complement activation, and antibody-dependent cellular cytotoxicity (ADCC). Cell-free systems offer unique advantages for glycoengineering by allowing precise control over the glycosylation machinery and substrate availability.

Mammalian cell-free systems can be supplemented with specific glycosyltransferases and nucleotide sugar donors to achieve desired glycosylation patterns. This approach enables the production of antibodies with reduced fucosylation for enhanced ADCC activity, or specific sialylation patterns for extended circulatory half-life. The ability to modulate glycosylation in real time during protein synthesis provides unprecedented control over this critical quality attribute (Table 10.5).

10.4 Vaccine antigen production

10.4.1 Rapid response vaccine development

The COVID-19 pandemic highlighted the critical need for rapid vaccine development platforms that can respond quickly to emerging infectious disease threats. CFPS has emerged as a key technology for accelerated vaccine antigen production, offering unprecedented speed and flexibility in pandemic response scenarios.

Traditional vaccine development relies on complex cell culture systems that require weeks to months for optimization and scale-up. Cell-free systems can produce vaccine antigens within hours of sequence availability, enabling rapid evaluation of immunogenicity and protective efficacy. This speed advantage was demonstrated during the COVID-19 pandemic when cell-free systems produced SARS-CoV-2 spike protein variants within days of their genetic characterization.

The technology's flexibility allows rapid adaptation to viral mutations and emerging variants. As new strains are identified, researchers can quickly modify expression templates and produce updated antigens for immunological evaluation. This capability is particularly valuable for highly mutable viruses such as influenza and SARS-CoV-2, where antigenic drift requires continuous vaccine updates.

10.4.2 Challenging vaccine targets

Many important vaccine targets present significant production challenges in conventional expression systems due to toxicity, instability, or complex structural requirements. Cell-free systems offer unique advantages for these difficult targets by providing controlled production environments and the ability to coexpress stabilizing factors.

Membrane proteins, which represent important vaccine targets for many pathogens, are notoriously difficult to produce in functional form using cell-based systems. Cell-free systems can incorporate membrane-mimetic environments such as liposomes or nanodiscs during protein synthesis, enabling proper folding and insertion of membrane antigens. This cotranslational approach often yields more native-like conformations compared to posttranslational reconstitution methods (Table 10.6).

10.4.3 Adjuvant integration and delivery systems

Cell-free systems offer unique opportunities for integrating vaccine antigens with adjuvants and delivery systems during the production process. This coformulation approach can enhance immunogenicity while simplifying downstream processing and formulation steps.

Protein-based adjuvants such as flagellin or heat-shock proteins can be coexpressed with vaccine antigens to create self-adjuvanting formulations. The controlled environment of cell-free systems enables optimization of stoichiometric ratios and ensures proper assembly of multicomponent vaccine constructs. In addition, the open system architecture allows incorporation of lipid-based adjuvants or nanoparticle delivery systems during protein synthesis.

TABLE 10.6 Challenging vaccine antigens for point-of-care manufacturing successfully produced in cell-free systems.

Pathogen	Antigen type	Production challenge	Cell-free solution	Immunological outcome
Influenza	Hemagglutinin	Membrane anchoring	Nanodisc incorporation	Enhanced neutralizing antibodies
SARS-CoV-2	Spike protein (full-length)	Structural instability	Coexpression with chaperones	Preserved receptor binding
Plasmodium	Circumsporozoite protein	Repeat region aggregation	Controlled redox conditions	Protective immunity
RSV	Fusion protein	Conformational liability	Stabilizing mutations + rapid synthesis	Neutralizing antibody induction
Zika	Envelope protein	Cytotoxicity	Cell-free production	Cross-reactive immunity
Anthrax	Protective antigen	Large size, complex folding	Optimized energy systems	Protective efficacy

10.5 Challenging protein targets

10.5.1 Membrane proteins

Membrane proteins represent approximately 30% of the human proteome and constitute the majority of current drug targets, yet they remain among the most challenging proteins to produce in functional form. Traditional expression systems struggle with membrane protein production due to cellular toxicity, improper folding, and aggregation issues. Cell-free systems offer unique advantages for membrane protein synthesis that address these fundamental challenges. The cotranslational insertion strategy and various membrane-mimetic environments are illustrated in Fig. 10.4.

The cotranslational insertion strategy employed in cell-free systems allows membrane proteins to be inserted into lipid bilayers during translation, mimicking the natural biosynthetic process. This approach prevents aggregation and misfolding that commonly occur when membrane proteins are produced in the absence of appropriate membrane environments. Cell-free systems can be supplemented with various membrane-mimetic systems, including liposomes, nanodiscs, and detergent micelles, to provide suitable folding environments (Table 10.7).

10.5.2 Intrinsically disordered proteins

Intrinsically disordered proteins (IDPs) and proteins containing intrinsically disordered regions present unique challenges for recombinant expression and purification. These proteins lack stable secondary or tertiary structure under physiological conditions and often

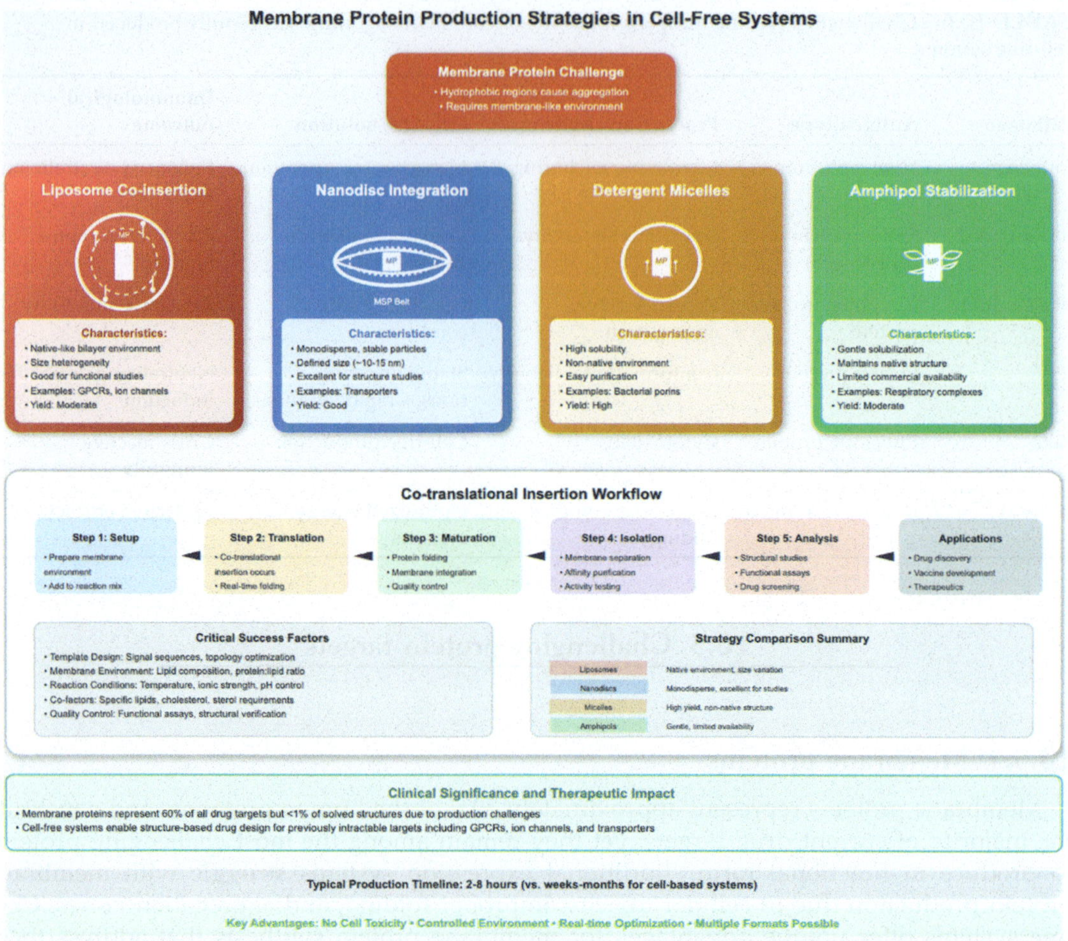

FIG. 10.4 Membrane protein production strategies in cell-free systems. Membrane proteins present significant production challenges due to their hydrophobic regions that cause aggregation in aqueous environments. This figure illustrates four major strategies for producing functional membrane proteins in cell-free systems. Liposome coinsertion provides a native-like bilayer environment where membrane proteins can be inserted cotranslationally, offering excellent functional properties but with size heterogeneity challenges. The diagram shows phospholipid bilayers with embedded membrane proteins and surrounding lipid molecules. Nanodisc integration uses membrane scaffold proteins (MSP) to create monodisperse, stable particles of defined size (~10–15 nm) that are excellent for structural studies and functional characterization. Detergent micelles provide high solubility and easy purification but create a nonnative environment that may affect protein function. Amphipol stabilization offers gentle solubilization that maintains native structure but has limited commercial availability. The cotranslational insertion workflow shows the complete process from membrane environment setup through protein synthesis, maturation, isolation, and final analysis for applications in drug discovery, vaccine development, and therapeutics. Critical success factors include proper template design with signal sequences, optimal membrane environment composition, controlled reaction conditions, appropriate cofactor supplementation, and rigorous quality control through functional and structural assays. The strategy comparison summary highlights that liposomes provide native environments with size variation, nanodiscs offer monodisperse particles excellent for studies, micelles give high yield in nonnative structures, and amphipols provide gentle treatment with limited availability. The typical production timeline of 2–8 h represents a significant advantage over weeks to months required for cell-based membrane protein production systems.

TABLE 10.7 Membrane protein production strategies in cell-free systems.

Membrane environment	Composition	Protein types	Advantages	Limitations	Success examples
Liposomes	Phospholipid vesicles	Multipass transmembrane	Native-like bilayer	Size heterogeneity	G-protein coupled receptors (GPCRs), ion channels
Nanodiscs	MSP + phospholipids	Single/multipass	Monodisperse, stable	Size limitations	Membrane transporters
Detergent micelles	Nonionic detergents	Simple membrane proteins	High solubility	Nonnative environment	Bacterial porins
Amphipols	Amphiphilic polymers	Various membrane proteins	Gentle solubilization	Limited availability	Respiratory complexes
Proteoliposomes	Preformed vesicles	Functional complexes	Functional assays possible	Complex preparation	ATP synthase

TABLE 10.8 Intrinsically disordered protein production challenges and solutions.

Challenge	Cellular systems impact	Cell-free solution	Additives/modifications	Representative proteins
Proteolytic degradation	High susceptibility	Protease-free environment	Protease inhibitors	α-synuclein, tau
Aggregation tendency	Inclusion body formation	Controlled conditions	Osmolytes, mild detergents	Huntingtin, p53
Low expression levels	Cell toxicity	No viability constraints	Enhanced templates	c-Myc, p21
Purification difficulty	Copurification of chaperones	Direct production optimization	Fusion tags	FUS, TDP-43
Functional instability	Loss of native disorder	Preserved native state	Stabilizing buffers	CFTR, BRCA1

aggregate or undergo proteolytic degradation in cell-based expression systems. Cell-free systems provide controlled environments that can minimize these issues while enabling functional characterization of IDPs.

The absence of cellular proteases in well-prepared cell-free systems reduces degradation of sensitive IDP regions. In addition, the controlled ionic strength and pH conditions can be optimized to maintain IDPs in their native disordered states without promoting aggregation. Cotranslational addition of osmolytes or stabilizing agents can further enhance IDP stability during production (Table 10.8).

10.5.3 Large protein complexes

Multisubunit protein complexes represent some of the most challenging targets for recombinant production due to the requirement for coordinated expression and assembly of multiple components. Traditional approaches often require separate production and reconstitution of individual subunits, which can be inefficient and may not recapitulate native assembly pathways. Cell-free systems enable coexpression strategies that can improve complex assembly efficiency and yield.

The stoichiometric control possible in cell-free systems allows optimization of relative subunit concentrations to promote proper assembly while minimizing formation of incomplete or misassembled complexes. Cotranslational assembly can be enhanced through temporal control of subunit synthesis or through the use of assembly chaperones specific for particular complexes.

10.6 Personalized medicine applications

10.6.1 Patient-specific therapeutics

The concept of personalized medicine has gained significant momentum with advances in genomics and precision medicine approaches. CFPS offers unique advantages for producing patient-specific therapeutics, particularly for rare genetic diseases where traditional manufacturing approaches are economically challenging due to small patient populations.

Cell-free systems can rapidly produce personalized therapeutics such as patient-specific enzymes for enzyme replacement therapy, customized antibodies for cancer treatment, or personalized vaccines based on individual tumor antigens. The speed and flexibility of cell-free production make it feasible to manufacture therapeutics for individual patients or small patient cohorts, opening new possibilities for treating rare diseases that lack viable commercial markets for traditional manufacturing (Table 10.9).

10.6.2 Point-of-care manufacturing

The distributed manufacturing capabilities of cell-free systems enable point-of-care production of therapeutics, bringing manufacturing closer to patients and reducing supply chain dependencies. This approach is particularly valuable for emergency response situations, remote healthcare settings, and personalized medicine applications where centralized manufacturing is impractical.

Point-of-care manufacturing using cell-free systems can provide fresh therapeutics with optimal stability and potency, eliminate cold-chain storage requirements for certain products, and enable rapid response to local disease outbreaks or bioterrorism events. The technology's portability and minimal infrastructure requirements make it suitable for deployment in resource-limited settings or mobile healthcare units.

TABLE 10.9 Personalized medicine applications of cell-free systems.

Application AREA	Therapeutic type	Development timeline	Production scale	Cost considerations	Clinical examples
Rare genetic diseases	Enzyme replacement	2–4 weeks	1–100 patients	High per-dose, justified by need	Gaucher disease variants
Cancer immunotherapy	Personalized vaccines	1–2 weeks	Single patient	Cost-effective for precision therapy	Neoantigen vaccines
Autoimmune disorders	Custom antibodies	2–3 weeks	10–1000 patients	Moderate cost, niche market	HLA-specific therapy
Metabolic disorders	Corrective enzymes	3–6 weeks	1–50 patients	High but necessary	Orphan enzyme deficiencies
Precision oncology	Targeted therapeutics	1–3 weeks	Single patient	Premium pricing justified	CAR-T supporting proteins

10.6.3 Companion diagnostics

Cell-free systems enable rapid production of protein reagents for companion diagnostics, which are tests used to identify patients who are most likely to benefit from specific therapeutic products. The speed and flexibility of cell-free production allow diagnostic manufacturers to quickly develop and deploy new assays as therapeutic landscapes evolve.

The technology is particularly valuable for producing protein biomarkers, antibodies for immunoassays, and enzyme reagents for diagnostic applications. Cell-free systems can rapidly adapt to new biomarker discoveries or changes in diagnostic requirements, providing the agility needed in the fast-evolving precision medicine landscape.

10.7 Regulatory considerations

10.7.1 Current regulatory framework

The regulatory landscape for cell-free produced therapeutics is evolving as the technology matures and more products enter clinical development. Fig. 10.5 illustrates the complete regulatory pathway from preclinical development through market approval. Regulatory agencies, including the Food and Drug Administration (FDA), European Medicines Agency (EMA), and other international authorities, are developing guidance documents and frameworks specifically addressing cell-free manufacturing technologies.

Current regulatory approaches focus on product quality and safety rather than manufacturing method, applying existing pharmaceutical quality standards to cell-free produced therapeutics. However, specific considerations arise regarding the characterization of

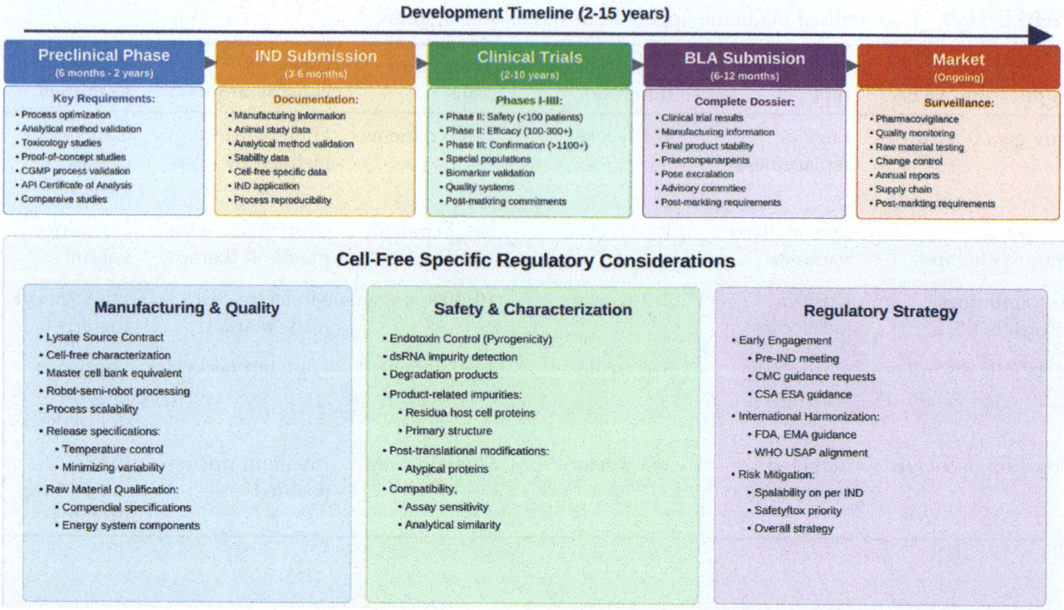

FIG. 10.5 Regulatory pathway for cell-free produced therapeutics. This comprehensive regulatory pathway diagram maps the complete development timeline for cell-free produced therapeutics from preclinical research through market approval, spanning 2–15 years. The preclinical phase (6 months–2 years) requires extensive process development, analytical method validation, toxicology studies, chemistry, manufacturing, and control (CMC) documentation, GMP process validation, cell-free system qualification, comparability studies, and stability data generation. The investigational new drug (IND) submission phase (3–6 months) demands comprehensive documentation, including manufacturing information, animal study data, clinical protocols, investigator qualifications, and cell-free specific data covering lysate characterization, endotoxin control, and process reproducibility. Clinical trials (2–10 years) progress through Phases I–III with increasing patient populations (20–100, 100–300, and 1000+ patients, respectively), accompanied by manufacturing scale-up, process validation, quality control system implementation, regulatory meetings, and postmarketing commitment planning. Biological license application (BLA) submission (6–12 months) requires a complete dossier, including clinical study reports, manufacturing information, product labeling, risk evaluation, FDA inspection, advisory committee review, final approval decision, and postmarket requirements. The final market phase involves ongoing pharmacovigilance, quality monitoring, lot release testing, change control, annual reports, label updates, and supply chain management. Cell-free specific regulatory considerations are detailed in three key areas: Manufacturing and Quality focuses on lysate source control with cell line characterization and master cell bank equivalents, process validation for temperature control and yield reproducibility, and raw material qualification for amino acid purity and energy system components. Safety and Characterization emphasizes endotoxin control for prokaryotic systems with LAL testing requirements, comprehensive product characterization, including primary structure and posttranslational modifications, and comparability studies with reference standards. Regulatory Strategy highlights the importance of early agency engagement through pre-IND meetings and CMC guidance requests, international harmonization across FDA, EMA, and Pharmaceutical and Medical Devices Agency (PMDA), and risk management through quality by design principles. Accelerated pathways, including Fast Track, Breakthrough Therapy, and Priority Review designations, are available for cell-free therapeutics that address unmet medical needs, providing opportunities for expedited development and approval timelines.

TABLE 10.10 Regulatory requirements for cell-free therapeutics.

Regulatory aspect	Current requirements	Cell-free specific considerations	Documentation needs
Process validation	ICH Q7, Q11 guidelines	System characterization, lysate qualification	Master cell bank equivalent
Product quality	ICH Q6B specifications	Endotoxin control (prokaryotic systems)	Analytical methods validation
Facility design	cGMP requirements	Simplified compared to cell culture	Environmental monitoring
Supply chain	ICH Q9 risk management	Raw material qualification	Vendor qualification
Clinical documentation	ICH E6 GCP	Comparability protocols	CMC documentation
Postmarket surveillance	ICH E2E pharmacovigilance	Product tracking	Batch records

cell-free systems, validation of manufacturing processes, and demonstration of product consistency and comparability (Table 10.10).

10.7.2 Quality system implementation

Implementing appropriate quality systems for cell-free manufacturing requires adaptation of traditional pharmaceutical quality approaches to address the unique characteristics of cell-free production. Key areas include raw material control, process monitoring, and product release testing.

Cell-free systems require rigorous control of starting materials, including cell lysates, energy regeneration systems, amino acids, and cofactors. The quality and consistency of these components directly impact product quality, necessitating comprehensive supplier qualification and incoming material testing programs. Process analytical technology approaches can provide real-time monitoring of critical process parameters and enable rapid detection of process deviations.

10.7.3 International harmonization

The global nature of pharmaceutical development requires harmonized regulatory approaches for cell-free manufactured therapeutics. International initiatives are working to develop consistent standards and guidelines that facilitate global development and registration of cell-free produced therapeutics.

Regulatory harmonization efforts focus on establishing common standards for process validation, analytical methods, and quality specifications that are recognized across multiple jurisdictions. These efforts aim to reduce regulatory burden while maintaining high standards for product quality and patient safety.

10.8 Conclusion

CFPS has emerged as a transformative technology for biopharmaceutical development, offering unprecedented speed, flexibility, and control in therapeutic protein production. From rapid antibody discovery to personalized medicine applications, cell-free systems address many of the limitations of traditional manufacturing approaches while opening new possibilities for treating previously intractable diseases.

The continued evolution of cell-free technologies, coupled with advancing regulatory frameworks and increasing industry adoption, positions this technology platform as a critical component of the future biopharmaceutical manufacturing landscape. Success in implementing cell-free systems for therapeutic applications requires careful attention to platform selection, process optimization, quality control, and regulatory compliance, but the potential benefits for patients and healthcare systems are substantial.

As the technology continues to mature, we can expect to see increasing numbers of cell-free produced therapeutics entering clinical development and eventual commercialization, particularly in areas such as personalized medicine, rare disease treatment, and emergency response applications where traditional manufacturing approaches are inadequate or impractical.

Bibliography

Adiga, R., et al., 2018. Point-of-care production of therapeutic proteins of good-manufacturing-practice quality. Nat. Biomed. Eng.

Amanat, F., Krammer, F., 2020. SARS-CoV-2 vaccines: status report. Immunity.

Bill, R.M., et al., 2011. Overcoming barriers to membrane protein structure determination. Nat. Biotechnol.

Bundy, B.C., Swartz, J.R., 2010. Site-specific incorporation of p-propargyloxyphenylalanine in a cell-free environment for direct protein–protein click conjugation. Bioconjug. Chem.

Carlson, E.D., Gan, R., Hodgman, C.E., Jewett, M.C., 2012. Cell-free protein synthesis: applications come of age. Biotechnol. Adv.

Coller, H.A., 2019. The paradox of cancer genes in non-malignant conditions: implications for precision medicine. Trends Mol. Med.

Crooke, S.N., Ovsyannikova, I.G., Poland, G.A., Kennedy, R.B., 2019. Immunosenescence and human vaccine immune responses. Immun. Ageing.

EMA Guideline on Process Validation for Finished Products, 2017, European Medicines Agency.

Failmezger, J., et al., 2018. Cell-free protein synthesis from non-growing, stressed Escherichia coli. Sci. Rep.

FDA Guidance for Industry: Quality Considerations for Continuous Manufacturing, 2019, U.S. Food and Drug Administration.

ICH Guidelines Q7, Q8, Q9, Q10, Q11: Pharmaceutical Quality System Guidelines. International Council for Harmonisation.

Kelwick, R., et al., 2016. Developments in cell-free protein synthesis for biotechnology applications. Biotechnol. Lett.

Martin, R.W., et al., 2017. Development of a CHO-based cell-free platform for synthesis of active monoclonal antibodies. ACS Synth. Biol.

Pardee, K., et al., 2014. Rapid, low-cost detection of Zika virus using programmable biomolecular components. Cell.

Perez, J.G., Stark, J.C., Jewett, M.C., 2016. Cell-free synthetic biology: engineering beyond the cell. Cold Spring Harb. Perspect. Biol.

Sachse, R., et al., 2014. Membrane protein synthesis in cell-free systems: from bio-mimetic systems to bio-membranes. FEBS Lett.

Seddon, A.M., Curnow, P., Booth, P.J., 2004. Membrane proteins, lipids and detergents: not just a soap opera. Biochim. Biophys. Acta.
Silverman, A.D., Karim, A.S., Jewett, M.C., 2020. Cell-free gene expression: an expanded repertoire of applications. Nat. Rev. Genet.
Stark, J.C., et al., 2019. BioBitsTM bright: a fluorescent synthetic biology education kit. Front. Bioeng. Biotechnol.
Thrane, S., et al., 2016. Bacterial superglue enables easy development of efficient virus-like particle based vaccines. J. Nanobiotechnol.
Yin, G., et al., 2012. Aglycosylated Antibodies and Antibody Fragments Produced in a Scalable in Vitro Transcription-translation System. mAbs.

CHAPTER

11

Diagnostics and biosensors

Cell-free protein synthesis (CFPS) has emerged as a transformative technology in the diagnostics and biosensor landscape, offering unprecedented advantages in terms of speed, portability, stability, and cost-effectiveness. This chapter explores the integration of CFPS platforms into diagnostic applications, from point-of-care (POC) devices to environmental monitoring systems, demonstrating how this technology is revolutionizing the way we detect, quantify, and respond to biological threats and medical conditions.

The diagnostic industry faces mounting pressure to deliver rapid, accurate, and accessible testing solutions that can operate outside traditional laboratory settings. CFPS addresses these challenges by eliminating the need for living cells while maintaining the sophisticated molecular machinery required for complex biosensing applications. The technology's inherent modularity allows for rapid prototyping and deployment of new diagnostic assays, making it particularly valuable in emerging infectious disease scenarios and personalized medicine applications.

11.1 POC diagnostic platforms

POC diagnostics represent one of the most promising applications of CFPS technology, addressing the critical need for rapid, accurate testing in resource-limited settings. The unique characteristics of cell-free systems, including their tolerance to environmental fluctuations, extended shelf life, and ability to function without complex instrumentation, make them ideally suited for POC applications.

11.1.1 Fundamentals of CFPS-based POC diagnostics

The integration of CFPS into POC platforms leverages the technology's ability to produce reporter proteins, antibodies, and enzymatic components on-demand. This approach eliminates the cold chain requirements typically associated with protein-based diagnostics and enables the creation of self-contained, single-use devices that can be deployed in diverse settings.

The core architecture of CFPS-based POC diagnostics typically involves three essential components: (1) target recognition elements, such as aptamers or antibody fragments

Cell Free Synthesis of Biologics
https://doi.org/10.1016/B978-0-443-33271-5.00023-5

TABLE 11.1 Comparison of CFPS-based POC platforms with traditional diagnostic methods.

Parameter	Traditional ELISA	Lateral flow	CFPS-POC	qPCR
Time to result	2–4 h	15–30 min	30–60 min	1–2 h
Equipment required	Plate reader, incubator	None	Portable reader	Thermal cycler
Cold chain	Yes	No	No	Yes
Shelf life	6–12 months	12–24 months	24–36 months	6–12 months
Sensitivity	High	Moderate	High	Very High
Multiplexing capability	Moderate	Low	High	Moderate
Cost per test	$5–15	$2–5	$3–8	$10–25
Sample prep	Minimal	Minimal	Minimal	Moderate

produced in real time; (2) signal amplification systems utilizing cell-free expressed enzymes or fluorescent proteins; and (3) detection interfaces that convert biological signals into measurable outputs. This modular design allows for rapid customization and adaptation to emerging diagnostic needs (Table 11.1).

11.1.2 Protein expression strategies for rapid diagnostics

The success of CFPS-based POC diagnostics relies heavily on optimized protein expression strategies that balance speed, yield, and functionality. Several approaches have been developed to maximize the efficiency of diagnostic protein production within the constrained timeframes required for POC applications.

Rapid expression protocols typically utilize high-efficiency templates, such as linear DNA constructs with optimized ribosome binding sites and minimal 5′ and 3′ untranslated regions. These templates can be presynthesized and stored at room temperature, allowing for immediate activation upon sample addition. The use of T7 RNA polymerase-based systems provides additional speed advantages through high-level transcription rates.

Coexpression strategies have proven particularly valuable for complex diagnostic applications requiring multiple protein components. For example, sandwich immunoassays can be implemented by coexpressing capture and detection antibodies, along with reporter enzymes, in a single reaction vessel. This approach eliminates the need for sequential incubation steps and reduces overall assay time.

11.1.3 Signal amplification and detection systems

Signal amplification in CFPS-based POC platforms often relies on enzymatic cascades that can be fine-tuned through protein engineering approaches. Horseradish peroxidase and alkaline phosphatase remain popular choices due to their robust catalytic properties and compatibility with colorimetric detection schemes. However, newer approaches utilizing engineered luciferases and fluorescent proteins offer enhanced sensitivity and multiplexing capabilities.

Split-protein complementation systems represent an innovative approach to signal amplification in CFPS platforms. These systems utilize protein fragments that reconstitute active enzymes only in the presence of target analytes, providing both specificity and amplification in a single molecular event. Split-luciferase systems, in particular, have demonstrated exceptional sensitivity with detection limits in the femtomolar range.

11.1.4 Microfluidic integration and device design

The integration of CFPS reactions with microfluidic platforms has enabled the development of sophisticated POC devices capable of sample processing, reaction control, and signal detection in compact, portable formats. These lab-on-a-chip systems typically incorporate sample preparation modules, reaction chambers, and optical detection systems within footprints smaller than a credit card.

Pneumatic valve systems allow for precise control of reagent flow and mixing, enabling complex multistep assays to be performed automatically. Passive microfluidic designs utilizing capillary forces and surface tension effects provide simpler alternatives for applications where precise timing is less critical (Fig. 11.1).

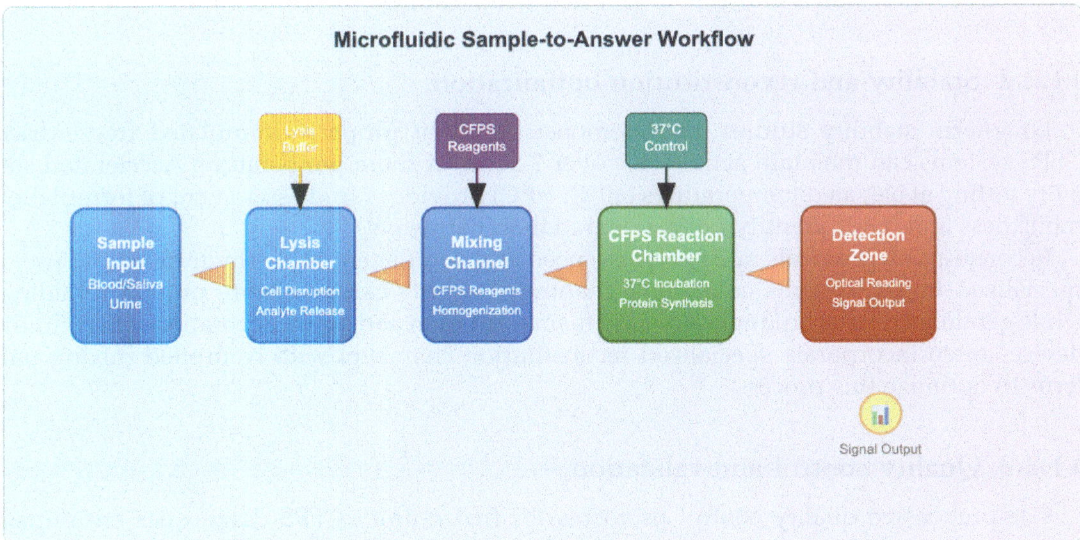

FIG. 11.1 Schematic design of integrated CFPS-microfluidic POC device. This figure illustrates the integrated microfluidic platform design for point-of-care diagnostics using cell-free protein synthesis. The device consists of sequential processing chambers connected by microfluidic channels, enabling automated sample-to-answer analysis. The sample input chamber receives biological specimens (blood, saliva, urine), which are then lysed using integrated lysis buffers to release target analytes. The mixing channel ensures homogeneous distribution of CFPS reagents, including amino acids, energy systems, and DNA templates. The CFPS reaction chamber provides optimal conditions for protein synthesis at 37°C with controlled incubation time. Finally, the detection zone incorporates optical elements for fluorescence or colorimetric signal measurement, with integrated photodetectors for quantitative analysis.

Freeze-drying (lyophilization) technology has revolutionized the deployment of CFPS-based diagnostics by enabling long-term storage at ambient temperatures while maintaining biological activity. This capability is particularly crucial for global health applications where refrigeration infrastructure is limited or unreliable.

11.2 Final packaging systems

11.2.1 Lyophilization strategies for CFPS systems

The successful lyophilization of CFPS reactions requires careful optimization of cryoprotectant formulations and freeze-drying parameters. Trehalose and sucrose serve as effective protectants for protein components, while polyethylene glycol helps maintain membrane integrity in systems requiring lipid components. The addition of bovine serum albumin or other carrier proteins can further enhance stability during the freeze-drying process.

Critical parameters include the freezing rate, primary drying temperature, and secondary drying conditions. Controlled nucleation techniques help ensure uniform ice crystal formation, minimizing damage to sensitive protein components. Primary drying is typically performed at temperatures between −40°C and −20°C under vacuum conditions, while secondary drying removes residual moisture at elevated temperatures.

11.2.2 Stability and reconstitution optimization

Long-term stability studies have demonstrated that properly formulated freeze-dried CFPS systems can maintain activity for over 2 years at room temperature. Accelerated stability testing at elevated temperatures (40°C, 60°C) provides rapid assessment of formulation robustness and helps identify critical degradation pathways.

Reconstitution protocols must balance speed and completeness of rehydration. The use of specialized buffer systems containing chaotropic agents can accelerate protein refolding, while gentle mixing techniques minimize foam formation and air incorporation. Microfluidic devices often incorporate specialized reconstitution chambers with controlled mixing patterns to optimize this process.

11.2.3 Quality control and validation

Comprehensive quality control protocols for freeze-dried CFPS diagnostics encompass both analytical and functional testing parameters. Analytical methods include moisture content determination, protein integrity assessment via SDS–PAGE, and enzyme activity assays. Functional testing involves complete diagnostic workflow evaluation using characterized reference materials.

Validation studies must demonstrate equivalence between fresh and reconstituted systems across a range of analyte concentrations and sample matrices. Statistical analysis using appropriate experimental designs ensures robust performance claims and supports regulatory submissions (Table 11.2).

TABLE 11.2 Stability parameters for freeze-dried CFPS diagnostic systems.

Parameter	Specification	Test method	Frequency
Moisture content	<2% w/w	Karl Fischer	Each lot
Protein recovery	>90%	Bradford/bicinchoninic acid assay (BCA)	Each lot
Enzyme activity	>80% initial	Substrate assay	Each lot
Diagnostic sensitivity	Within 10%	Clinical samples	Quarterly
Reconstitution time	<5 min	Visual inspection	Each lot
Shelf life	24 months	Real time/accelerated	Ongoing

11.3 CRISPR-based detection

The integration of CRISPR–Cas systems with CFPS platforms has created powerful new paradigms for nucleic acid detection, offering exceptional specificity and sensitivity for infectious disease diagnostics, food safety applications, and environmental monitoring.

11.3.1 CRISPR–Cas system integration

CRISPR-based detection in CFPS systems typically utilizes Cas12a (Cpf1) or Cas13a nucleases, which exhibit collateral cleavage activity upon target recognition. This collateral activity can be harnessed to cleave reporter substrates, generating detectable signals proportional to target concentration. The cell-free environment provides optimal conditions for CRISPR–Cas function while allowing precise control over reaction components.

Cas12a systems are particularly effective for DNA detection applications, exhibiting robust ssDNase activity following target binding. The system can be programmed using guide RNAs (gRNAs) designed to recognize specific target sequences, with mismatches as small as a single nucleotide significantly reducing cleavage efficiency. This exquisite specificity makes CRISPR–Cas12a ideal for detecting single nucleotide polymorphisms (SNPs) and identifying pathogenic variants.

Cas13a systems extend CRISPR-based detection to RNA targets, including viral RNA genomes and cellular transcripts. The RNase activity of activated Cas13a can cleave fluorescently labeled RNA reporters, generating easily detectable signals. The combination of reverse transcription with Cas13a detection (RT-CRISPR) enables sensitive detection of RNA viruses and gene expression analysis.

11.3.2 Signal enhancement strategies

Several approaches have been developed to enhance the sensitivity of CRISPR-based detection in CFPS systems. Preamplification using recombinase polymerase amplification or loop-mediated isothermal amplification can increase target abundance prior to CRISPR detection, achieving sensitivities comparable to qPCR.

The integration of multiple Cas proteins in a single reaction can provide orthogonal detection mechanisms and internal controls. For example, Cas12a can be used for primary target detection, while Cas13a simultaneously monitors reaction quality through housekeeping gene analysis. This multiplexed approach enhances assay robustness and reduces false-negative results.

Enzymatic signal amplification using cell-free expressed enzymes provides additional sensitivity gains. Target-activated Cas proteins can trigger enzymatic cascades involving phosphatases, kinases, or polymerases, resulting in exponential signal amplification. These cascade systems can achieve detection limits in the attomolar range.

11.3.3 Multiplexed detection platforms

The modular nature of CRISPR systems enables sophisticated multiplexed detection platforms capable of simultaneously identifying multiple targets. Orthogonal gRNA sequences can be designed to avoid cross-reactivity while maintaining high specificity for individual targets. Spectral multiplexing using different fluorophores allows for real-time monitoring of multiple detection channels.

Spatial multiplexing approaches utilize arrays of microwells or microfluidic chambers, each containing CFPS reactions programmed for specific targets. This approach is particularly valuable for syndromic panel testing, where multiple pathogens causing similar clinical presentations must be differentiated rapidly.

11.3.4 Quantitative analysis and digital detection

Quantitative CRISPR detection requires careful calibration using characterized reference standards. The relationship between target concentration and signal output is typically nonlinear due to the enzymatic nature of the detection mechanism. Mathematical models incorporating enzyme kinetics and substrate depletion effects enable accurate quantification across multiple orders of magnitude.

Digital CRISPR detection represents an emerging approach that partitions samples into thousands of individual reactions, enabling precise quantification through Poisson's statistics. This method provides absolute quantification without requiring standard curves and exhibits exceptional precision at low target concentrations (Fig. 11.2).

11.4 Environmental monitoring

CFPS-based biosensors have found extensive applications in environmental monitoring, providing rapid, sensitive detection of pollutants, pathogens, and toxins in air, water, and soil samples. The technology's robustness and adaptability make it particularly well-suited for field deployment in challenging environmental conditions (Fig. 11.3).

11.4.1 Water quality assessment

Water quality monitoring represents one of the most developed applications of CFPS-based environmental sensing. The technology has been successfully applied to detect bacterial pathogens, heavy metals, pesticides, and endocrine disruptors in various water sources.

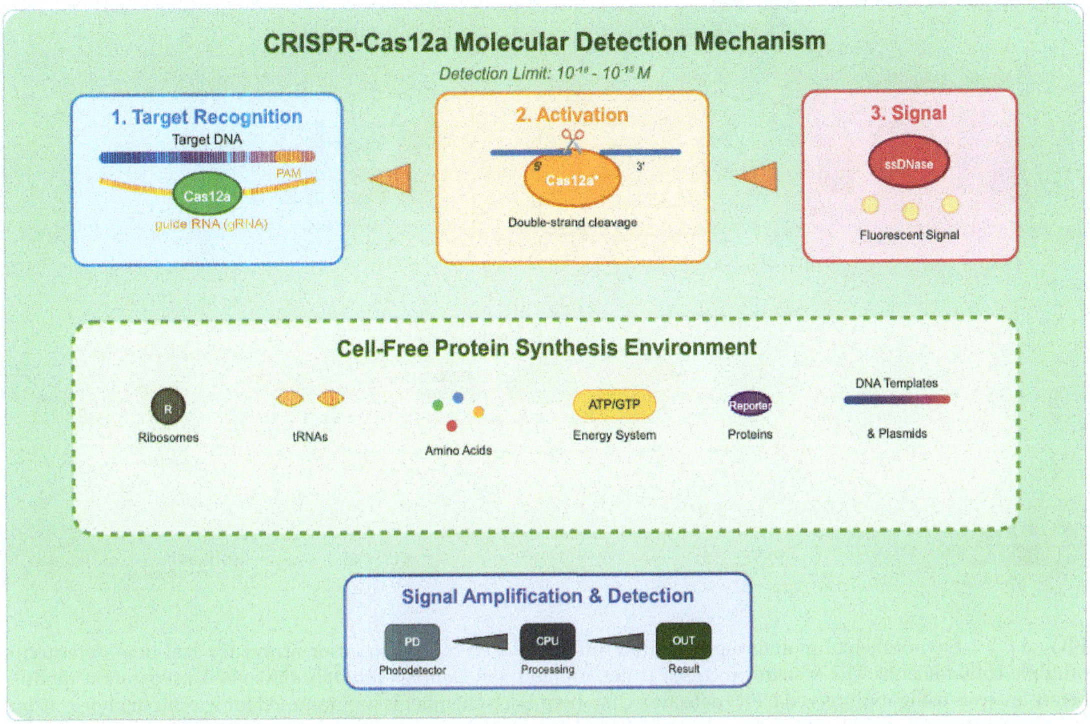

FIG. 11.2 CRISPR–Cas12a detection mechanism in CFPS systems. This figure demonstrates the molecular mechanism of CRISPR–Cas12a-based detection within cell-free protein synthesis systems. The process begins with target DNA recognition through complementary base pairing with the guide RNA (gRNA) component of the Cas12a ribonucleoprotein complex. Upon identification of the protospacer adjacent motif (PAM) sequence, the Cas12a nuclease undergoes conformational activation, leading to specific double-strand cleavage of the target DNA. This conformational change simultaneously activates the collateral single-strand DNase activity of Cas12a, resulting in nonspecific cleavage of nearby single-stranded DNA molecules. In the diagnostic application, fluorescently labeled single-strand DNA reporters are cleaved, generating detectable fluorescent signals proportional to target concentration. The CFPS environment provides optimal conditions for Cas12a activity while enabling integration with signal amplification systems.

The ability to operate without living cells eliminates concerns about biosafety and containment that limit traditional whole-cell biosensors.

Bacterial pathogen detection typically utilizes species-specific biomarkers, such as unique protein sequences or metabolic products. CFPS systems can be programmed to produce antibodies or aptamers that specifically recognize these biomarkers, enabling rapid identification of waterborne pathogens, including *E. coli*, *Salmonella*, and *Cryptosporidium*. Detection times of 30–60 min are routinely achieved, compared to 24–48 h for traditional culture methods.

Heavy metal detection leverages the expression of metallothionein-based sensors or engineered proteins with specific metal-binding domains. These systems can detect lead, mercury, cadmium, and arsenic at concentrations well below regulatory limits. The use of competitive binding assays with preloaded metal standards enables quantitative analysis across relevant concentration ranges.

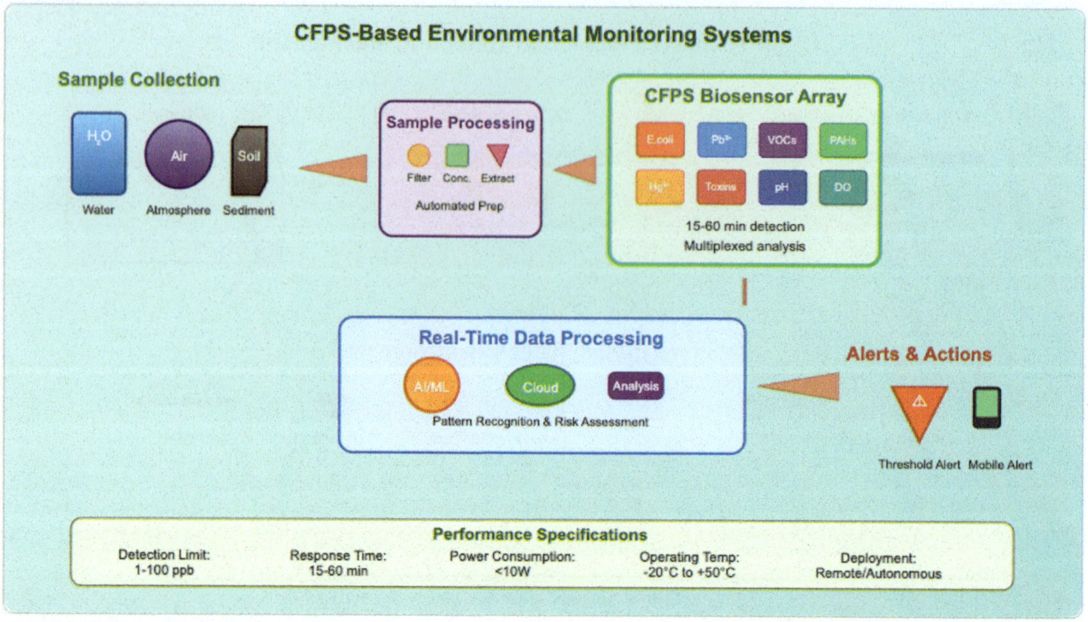

FIG. 11.3 Environmental monitoring workflow utilizing CFPS-based biosensor arrays for real-time detection of multiple contaminants. The system processes water, air, and soil samples through automated preparation modules before analysis using multiplexed CFPS detection chambers. Each chamber is programmed for specific analytes, including bacterial pathogens (*E. coli*), heavy metals (Pb^{2+}, Hg^{2+}), volatile organic compounds (VOCs), polycyclic aromatic hydrocarbons (PAHs), biotoxins, and physicochemical parameters (pH, dissolved oxygen). Real-time data processing incorporates artificial intelligence (AI)/machine learning (ML) algorithms for pattern recognition and risk assessment, with cloud connectivity enabling remote monitoring and automated alert systems for threshold exceedances.

11.4.2 Air quality monitoring

Airborne contaminants pose significant challenges for traditional analytical methods due to their low concentrations and complex matrices. CFPS-based air monitoring systems address these challenges through highly sensitive biorecognition elements and real-time analysis capabilities.

Volatile organic compound (VOC) detection utilizes engineered olfactory receptors or other binding proteins expressed in CFPS systems. These receptors can be designed to respond to specific VOCs associated with industrial emissions, vehicle exhaust, or indoor air pollution. Signal transduction typically involves conformational changes that affect reporter protein activity or localization.

Particulate matter monitoring focuses on biological components, including allergens, bacterial endotoxins, and fungal spores. CFPS-produced antibodies or lectins provide specific recognition elements, while sensitive detection schemes enable quantification at health-relevant concentrations.

11.4.3 Soil contamination analysis

Soil contamination assessment benefits from CFPS technology's tolerance to complex matrices and ability to function in the presence of humic substances and other soil components that interfere with traditional analytical methods. The technology is particularly valuable for assessing biological availability of contaminants rather than total concentrations.

Pesticide detection systems utilize engineered enzymes or binding proteins that specifically interact with target compounds. Organophosphate pesticides can be detected using engineered acetylcholinesterases, while herbicides such as atrazine require specialized binding proteins or aptamers. The ability to multiplex multiple detection systems enables comprehensive pesticide screening.

Hydrocarbon contamination monitoring employs engineered proteins that respond to petroleum products, polycyclic aromatic hydrocarbons, and other organic pollutants. These systems can provide rapid assessment of soil contamination following spills or at sites with historical contamination.

11.4.4 Field deployment considerations

Successful field deployment of CFPS-based environmental monitoring systems requires careful consideration of environmental factors, sample preparation requirements, and data management needs. Temperature extremes, humidity variations, and electromagnetic interference can all affect system performance and must be addressed through robust engineering design.

Sample preparation protocols must balance thoroughness with simplicity, enabling effective extraction and concentration of analytes while removing interfering substances. Automated sample preparation systems integrated with CFPS detection platforms provide consistent results and reduce operator training requirements.

Data management systems must enable real-time data transmission, quality control monitoring, and integration with existing environmental monitoring networks. Cloud-based data platforms provide scalable solutions for managing large sensor networks and enable sophisticated data analysis and visualization capabilities (Table 11.3).

TABLE 11.3 Environmental monitoring applications of CFPS-based biosensors.

Application area	Target analytes	Detection limit	Response time	Field deployment
Water quality	*E. coli, Salmonella*	10–100 CFU/mL	30–60 min	Portable units
	Heavy metals	1–10 ppb	15–30 min	Automated stations
	Pesticides	0.1– ppb	20–45 min	Mobile laboratories
Air quality	VOCs	1–10 ppb	5–15 min	Fixed monitors
	Allergens	1–10 ng/m^3	10–30 min	Personal devices
	Endotoxins	0.1–1 EU/m^3	15–45 min	Indoor monitors
Soil analysis	Petroleum hydrocarbons	1–10 ppm	30–90 min	Portable units
	PAHs	0.1–1 ppm	45–120 min	Mobile laboratories
	Pesticide residues	0.01–0.1 ppm	30–60 min	Field kits

11.5 Biodefense applications

The rapid response capabilities and robust nature of CFPS-based diagnostic systems make them invaluable tools for biodefense applications, enabling rapid identification of biological threats and supporting emergency response efforts. These systems can be deployed rapidly to investigate suspicious biological materials and provide critical intelligence for public health decisions.

11.5.1 Biothreat agent detection

Biological threat agents encompass a diverse range of pathogens, including bacteria, viruses, toxins, and fungi, that could be deliberately released to cause harm. CFPS-based detection systems offer several advantages for biothreat identification, including rapid deployment, broad-spectrum detection capabilities, and operation in austere environments.

Bacterial biothreat detection typically targets organisms such as *Bacillus anthracis*, *Yersinia pestis*, and *Francisella tularensis*. Detection strategies often focus on species-specific surface proteins, toxins, or metabolic markers that can be rapidly identified using CFPS-produced antibodies or aptamers. The ability to simultaneously screen for multiple threat agents enhances the utility of these systems for unknown sample analysis.

Viral threat detection presents unique challenges due to the requirement for nucleic acid amplification or highly sensitive protein detection. CRISPR-based detection systems integrated with CFPS platforms provide exceptional sensitivity for viral RNA or DNA targets. RT-CRISPR approaches enable direct detection of RNA viruses without prior reverse transcription steps.

Toxin detection leverages the ability of CFPS systems to produce highly specific binding proteins or functional assays that respond to biologically active toxins. Ricin, botulinum toxin, and staphylococcal enterotoxin B represent high-priority targets that can be detected using engineered receptor proteins or enzymatic activity assays.

11.5.2 Rapid response capabilities

Biodefense applications require extremely rapid response times to enable effective countermeasures and minimize exposure. CFPS-based systems can achieve preliminary identification within 30–60 min, compared to hours or days required for traditional microbiological methods. This rapid response capability is critical for emergency situations where immediate decisions are required.

Portable deployment systems enable field teams to conduct analysis at suspicious sites without requiring sample transport to remote laboratories. These systems typically include integrated sample preparation capabilities, automated analysis protocols, and satellite communication for real-time data transmission to command centers.

Quality assurance protocols for biodefense applications must ensure reliable performance under stress conditions and with potentially degraded samples. Multiple redundant detection mechanisms and positive controls help ensure accurate identification even when sample quality is compromised.

11.5.3 Multiagent detection platforms

Syndromic detection approaches utilize broad-spectrum biomarkers that can identify classes of threat agents rather than specific organisms. This approach is particularly valuable when the specific threat is unknown or when multiple agents might be present simultaneously. Common biomarkers include pathogen-associated molecular patterns and conserved virulence factors.

Multiplex detection platforms capable of simultaneously screening for dozens of potential threat agents provide comprehensive coverage while maintaining rapid response times. These systems typically utilize array-based detection formats with each array element programmed for specific targets.

Decision support algorithms help interpret complex detection patterns and provide actionable intelligence to emergency responders. Machine learning approaches can identify novel threat signatures and distinguish biological weapons from naturally occurring disease outbreaks.

11.5.4 Network integration and data sharing

Biodefense monitoring systems must integrate with existing surveillance networks and information sharing platforms to maximize their effectiveness. Standardized data formats and communication protocols enable seamless integration with public health surveillance systems and intelligence networks.

Real-time data sharing capabilities enable coordinated response efforts across multiple agencies and jurisdictions. Cloud-based data platforms provide secure storage and analysis capabilities while enabling rapid information dissemination to authorized personnel.

Training and certification programs ensure that field personnel can effectively operate detection systems and interpret results accurately. Regular exercises and proficiency testing maintain operational readiness and identify system limitations or training needs.

11.6 Field deployment strategies

The successful deployment of CFPS-based diagnostic systems in field environments requires comprehensive planning that addresses logistical, technical, and operational challenges. This section examines best practices for field deployment and strategies for optimizing system performance in diverse operational environments.

11.6.1 Environmental considerations

Field deployment environments present numerous challenges that can affect the performance of CFPS-based diagnostic systems. Temperature extremes, humidity variations, dust, vibration, and electromagnetic interference must all be considered during system design and deployment planning.

Temperature management systems may be required in extreme environments to maintain optimal reaction conditions. Passive thermal management using phase-change materials can provide temperature stability without requiring external power sources. Active temperature control systems provide more precise control but require additional power and complexity.

Humidity control is particularly important for lyophilized CFPS systems, which can lose activity upon exposure to moisture. Desiccant packaging and humidity indicator cards help maintain optimal storage conditions. Environmental sealing of electronic components prevents moisture-related failures.

Vibration and shock protection may be necessary for systems deployed in mobile platforms or harsh environments. Shock-absorbing materials and robust mechanical designs help protect sensitive components during transport and operation.

11.6.2 Power and communication requirements

Power requirements for CFPS-based diagnostic systems vary widely depending on the level of automation and instrumentation required. Simple colorimetric detection systems may require only modest power for light emitting diode (LED) illumination and photodetection, while sophisticated automated platforms may require substantial power for pumps, heaters, and data processing.

Solar power systems provide sustainable energy solutions for long-term deployments in remote locations. Battery backup systems ensure continuous operation during periods of low solar availability. Power management systems optimize energy consumption and extend operational duration.

Communication systems enable real-time data transmission and remote monitoring of deployed systems. Satellite communication provides global coverage but may have limited bandwidth and higher costs. Cellular communication offers higher bandwidth in areas with coverage but may be unreliable in remote locations.

Mesh networking capabilities enable multiple deployed units to share data and coordinate operations. This approach provides redundancy and extends communication range through relay capabilities.

11.6.3 Training and support

Effective field deployment requires comprehensive training programs that prepare operators to use systems effectively under operational conditions. Training programs should address normal operations, troubleshooting, maintenance, and emergency procedures.

Remote support capabilities enable expert assistance during field operations. Video conferencing, remote diagnostics, and augmented reality systems can provide real-time guidance to field operators. These capabilities are particularly valuable when deploying complex systems in remote locations.

Technical documentation must be comprehensive yet accessible to operators with varying technical backgrounds. Quick reference guides, troubleshooting flowcharts, and video tutorials enhance usability and reduce operator errors.

11.6.4 Logistics and supply chain

Supply chain management is critical for maintaining operational readiness of deployed systems. Reagent expiration dates, spare parts availability, and consumable resupply must all be carefully managed to ensure continuous operation.

TABLE 11.4 Field deployment checklist for CFPS-based diagnostic systems.

Category	Considerations	Critical parameters
Environmental	Temperature range, humidity, dust	–20°C to +50°C, <80% RH
Power	Energy requirements, backup systems	12 V DC, <50 W average
Communications	Data transmission, range	Satellite/cellular coverage
Training	Operator certification, support	8-h basic, 24-h advanced
Logistics	Supply chain, maintenance	30-day consumable supply
Documentation	Standard operating procedures (SOPs), troubleshooting	Multilanguage support
Quality control	Calibration, controls	Daily QC, monthly calibration

Cold chain management may be required for some system components, particularly in hot climates. Insulated containers, ice packs, and phase-change materials can maintain appropriate temperatures during transport. Temperature monitoring devices provide verification of cold chain integrity.

Inventory management systems track reagent usage, expiration dates, and remaining supplies. Automated reorder systems help ensure continuous availability of critical consumables. Emergency supply protocols enable rapid resupply during crisis situations (Table 11.4).

Bibliography

Aw, R., Barton, K., Ahner, B.A., Whittaker, M.L., 2016. A washable, broad-spectrum microfluidic array for immuno-oncology drug screening and biomarker analysis. Lab Chip 16 (19), 3716–3727.

Carlson, E.D., Gan, R., Hodgman, C.E., Jewett, M.C., 2012. Cell-free protein synthesis: applications come of age. Biotechnol. Adv. 30 (5), 1185–1194.

Contreras-Llano, L.E., Tan, C., 2018. High-throughput screening of biomolecules using cell-free gene expression systems. Synth. Biol. 3 (3), ysy012.

Ding, F., Lu, T., Fedorec, A.J.H., Zhang, T., Karim, A.S., Jiang, C., … Barnes, C.P., 2022. GPU-accelerated molecular dynamics simulations reveal the mechanism of cell-free protein synthesis enhancement by polyethylene glycol. Nat. Commun. 13 (1), 1892.

Gootenberg, J.S., Abudayyeh, O.O., Lee, J.W., Essletzbichler, P., Dy, A.J., Joung, J., et al., 2017. Nucleic acid detection with CRISPR-Cas13a/C2c2. Science 356 (6336), 438–442.

Green, A.A., Silver, P.A., Collins, J.J., Yin, P., 2014. Toehold switches: de-novo-designed regulators of gene expression. Cell 159 (4), 925–939.

Gregorio, N.E., Levine, M.Z., Oza, J.P., 2019. A user's guide to cell-free protein synthesis. Methods Protoc. 2 (1), 24.

Jung, J.K., Alam, K.K., Verosloff, M.S., Capdevila, D.A., Desmau, M., Clauer, P.R., et al., 2020. Cell-free biosensors for rapid detection of water contaminants. Nat. Biotechnol. 38 (12), 1451–1459.

Koch, M., Faulon, J.L., Borkowski, O., 2018. Models for cell-free synthetic biology: make prototyping easier, better, and faster. Front. Bioeng. Biotechnol. 6, 182.

Krinsky, N., Kaduri, M., Shainsky-Roitman, J., Goldfeder, M., Ivanir, E., Benhar, I., et al., 2018. A simple and rapid method for preparing a cell-free bacterial lysate for protein synthesis. PLoS One 13 (2), e0193380.

Li, Y., Li, S., Wang, J., Liu, G., 2019. CRISPR/Cas systems towards next-generation biosensing. Trends Biotechnol. 37 (7), 730–743.

Ma, D., Shen, L., Wu, K., Diehnelt, C.W., Green, A.A., 2018. Low-cost detection of norovirus using paper-based cell-free systems and synbody-based viral enrichment. Synth. Biol. 3 (2), ysy018.

McNerney, M.P., Zhang, Y., Steppe, P., Silverman, A.D., Jewett, M.C., Styczynski, M.P., 2019. Point-of-care biomarker quantification enabled by sample-specific calibration. Sci. Adv. 5 (9), eaax4473.

Pardee, K., Green, A.A., Ferrante, T., Cameron, D.E., DaleyKeyser, A., Yin, P., Collins, J.J., 2014. Paper-based synthetic gene networks. Cell 159 (4), 940–954.

Pardee, K., Green, A.A., Takahashi, M.K., Braff, D., Lambert, G., Lee, J.W., et al., 2016. Rapid, low-cost detection of Zika virus using programmable biomolecular components. Cell 165 (5), 1255–1266.

Silverman, A.D., Kelley-Loughnane, N., Lucks, J.B., Jewett, M.C., 2019. Deconstructing cell-free extract preparation for in vitro activation of transcriptional genetic circuitry. ACS Synth. Biol. 8 (2), 403–414.

Takahashi, M.K., Tan, X., Dy, A.J., Braff, D., Akana, R.T., Furuta, Y., et al., 2018. A low-cost paper-based synthetic biology platform for analyzing gut microbiota and host biomarkers. Nat. Commun. 9 (1), 3347.

Thavarajah, W., Silverman, A.D., Verosloff, M.S., Kelley-Loughnane, N., Jewett, M.C., Lucks, J.B., 2020. Point-of-use detection of environmental fluoride via a riboswitch-based biosensor with a handheld pH meter. ACS Synth. Biol. 9 (1), 10–18.

Verosloff, M., Chappell, J., Perry, K.L., Thompson, J.R., Lucks, J.B., 2019. PLANT-Dx: a molecular diagnostic for point-of-use detection of plant pathogens. ACS Synth. Biol. 8 (4), 902–905.

Voyvodic, P.L., Pandi, A., Koch, M., Conejero, I., Valjent, E., Courtet, P., et al., 2019. Plug-and-play metabolic transducers expand the chemical detection space of cell-free biosensors. Nat. Commun. 10 (1), 1697.

CHAPTER

12

Industrial enzyme production

The industrial enzyme market represents one of the most promising applications for cell-free protein synthesis (CFPS) technology, with global enzyme sales exceeding $7 billion annually and growing at 6%–8% per year. Traditional enzyme production relies heavily on microbial fermentation systems that require extensive optimization, long development cycles, and significant infrastructure investments. CFPS technology offers a paradigm shift in enzyme manufacturing, enabling rapid prototyping, accelerated development timelines, and cost-effective production of specialty enzymes that are difficult or impossible to produce in living systems.

Industrial enzymes serve critical roles across diverse sectors, including food and beverage processing, textile manufacturing, pulp and paper production, biofuel generation, pharmaceutical synthesis, and environmental remediation. The unique requirements of each application—ranging from extreme pH and temperature stability to specific substrate selectivity—have traditionally necessitated extensive screening programs and complex optimization processes. CFPS platforms address these challenges by providing unprecedented flexibility in enzyme engineering, enabling rapid iteration cycles, and supporting the production of toxic or unstable enzymes that cannot be synthesized in conventional expression systems.

The economic advantages of CFPS-based enzyme production become particularly compelling for low-volume, high-value applications where traditional fermentation economics are unfavorable. In addition, the technology's ability to incorporate noncanonical amino acids (ncAAs) and facilitate complex posttranslational modifications opens new possibilities for enzyme enhancement and novel catalyst design. This chapter explores the comprehensive application of CFPS technology across the industrial enzyme landscape, from discovery and optimization through scale-up and commercialization.

12.1 Enzyme discovery and optimization

Modern enzyme discovery programs face increasing pressure to identify biocatalysts with enhanced performance characteristics while reducing development timelines and costs. CFPS technology revolutionizes this process by enabling rapid screening of enzyme variants, facilitating functional expression of challenging proteins, and supporting innovative engineering approaches that are impractical in traditional systems.

Cell Free Synthesis of Biologics
https://doi.org/10.1016/B978-0-443-33271-5.00015-6

12.1.1 High-throughput screening platforms

The implementation of CFPS-based high-throughput screening platforms has transformed enzyme discovery by enabling the evaluation of thousands of enzyme variants in parallel. These systems typically utilize 96-, 384-, or 1536-well microplate formats, with each well containing optimized CFPS reactions programmed to express specific enzyme variants. The elimination of cell growth requirements allows for direct measurement of enzyme activity immediately following protein synthesis, dramatically reducing assay times from days to hours.

Automated liquid handling systems enable precise dispensing of CFPS reagents, DNA templates, and substrate solutions, ensuring consistent reaction conditions across large screening campaigns. Integration with robotic systems allows for automated template preparation, reaction setup, incubation, and product analysis, supporting screening throughputs exceeding 10,000 variants per day. The modular nature of CFPS systems facilitates rapid optimization of reaction conditions for different enzyme classes, enabling customized screening protocols for specific applications.

Multiplexed screening approaches utilize orthogonal detection systems to simultaneously evaluate multiple enzyme properties, including activity, stability, and selectivity. Fluorogenic substrates provide real-time monitoring of enzyme kinetics, while mass spectrometry-based analysis enables detailed characterization of reaction products and intermediates. This comprehensive profiling approach accelerates the identification of lead enzymes and provides critical data for subsequent optimization efforts.

12.1.2 Functional expression of challenging targets

Many industrially relevant enzymes present significant challenges for conventional expression systems due to toxicity, instability, complex cofactor requirements, or unusual structural features. CFPS platforms excel in addressing these challenges through their open system architecture and precise control over reaction components.

Membrane-associated enzymes, which represent approximately 30% of potential industrial targets, benefit significantly from CFPS expression due to the ability to cotranslationally insert proteins into lipid environments. The addition of liposomes, nanodiscs, or detergent micelles to CFPS reactions enables proper folding and functional expression of membrane proteins that are typically misfolded or aggregated in conventional systems.

Enzymes requiring complex metal cofactors or unusual prosthetic groups can be efficiently produced by supplementing CFPS reactions with appropriate cofactor precursors and assembly factors. This approach has proven particularly valuable for cytochrome P450 enzymes, metalloproteinases, and other metalloenzymes that require precise metal coordination for activity. The ability to control cofactor availability and oxidation states enables optimization of enzyme maturation processes (Table 12.1).

12.1.3 Protein engineering strategies

CFPS systems provide unique advantages for protein engineering by enabling rapid evaluation of design hypotheses and supporting innovative modification strategies. The technology's

TABLE 12.1 Comparison of expression systems for challenging industrial enzymes.

Enzyme class	Traditional E. coli	Yeast expression	Mammalian cells	CFPS system
Membrane proteins	Poor (aggregation)	Moderate	Good	Excellent
Metalloproteins	Variable	Good	Excellent	Excellent
Glycosylated enzymes	None	Moderate	Excellent	Good[a]
Toxic proteins	Poor	Poor	Moderate	Excellent
Large complexes	Difficult	Moderate	Good	Good
Development time	2–4 weeks	4–8 weeks	8–16 weeks	1–3 days
Throughput	Low	Low	Very low	Very high
Cost per sample	Low	Moderate	High	Moderate

[a]*With appropriate glycosylation systems.*

compatibility with ncAAs opens new possibilities for enzyme enhancement through the introduction of novel chemical functionalities.

Site-directed mutagenesis campaigns benefit from the rapid turnaround times achievable with CFPS systems, enabling iterative design cycles that would be impractical with conventional expression methods. Random mutagenesis and directed evolution approaches can be implemented using error-prone PCR or other diversification methods, with CFPS enabling immediate functional screening of variant libraries.

ncAA incorporation provides access to chemical modifications that are impossible through conventional protein synthesis. Photocrosslinkable amino acids enable the creation of light-activated enzymes, while fluorinated amino acids can enhance thermal stability and alter substrate specificity. The incorporation of metal-binding amino acids enables the design of artificial metalloproteins with novel catalytic properties.

12.1.4 Computational design integration

The integration of computational protein design tools with CFPS screening platforms creates powerful synergies for enzyme optimization. Machine learning algorithms can analyze screening data to identify sequence–activity relationships, guiding subsequent design iterations and reducing the search space for optimization.

Structure-based design approaches utilize protein crystal structures and molecular dynamics simulations to predict the effects of amino acid substitutions on enzyme activity and stability. These predictions can be rapidly validated using CFPS expression and screening, creating feedback loops that improve design accuracy over time.

Automated design–build–test–learn cycles leverage CFPS technology to create closed-loop optimization systems. DNA synthesis robots generate designed sequences, automated CFPS systems express and screen variants, and machine learning algorithms analyze results to design the next generation of candidates. These integrated platforms can complete multiple optimization cycles per week, dramatically accelerating enzyme development timelines (Fig. 12.1).

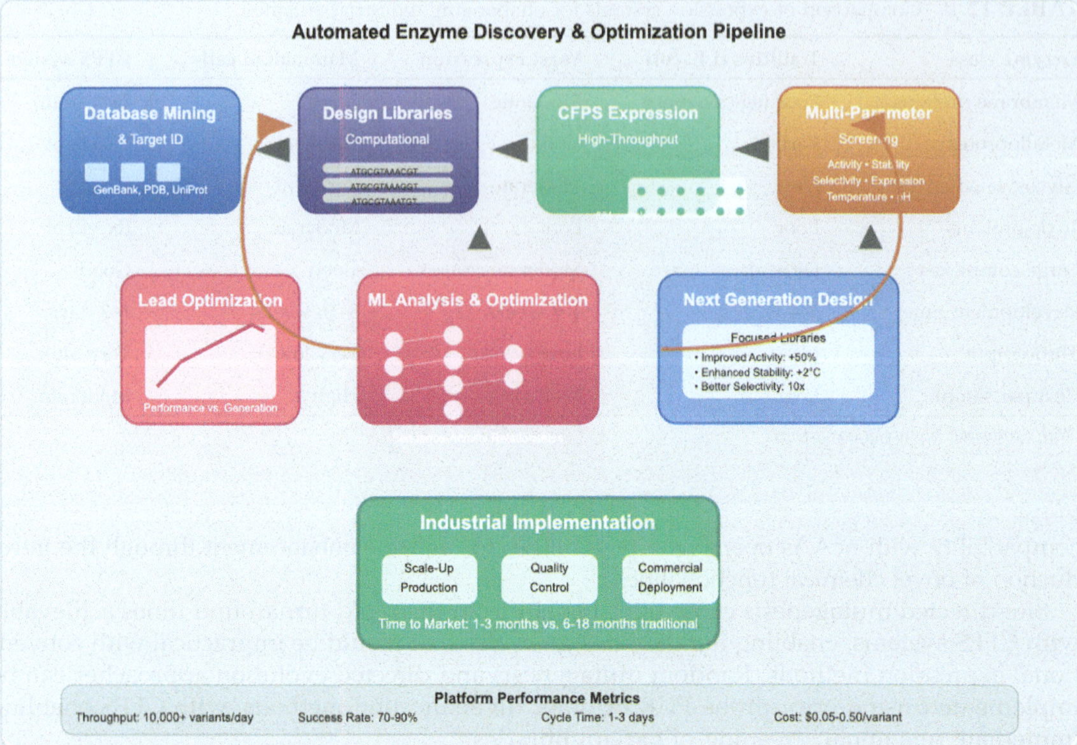

FIG. 12.1 CFPS-enabled enzyme discovery and optimization workflow. This figure illustrates the integrated workflow for enzyme discovery and optimization using cell-free protein synthesis platforms. The process begins with target identification and sequence acquisition from databases or metagenomic sources. Computational design tools generate variant libraries incorporating predicted beneficial mutations. High-throughput CFPS expression systems produce enzyme variants in parallel, typically using 384-well or 1536-well microplate formats. Automated screening platforms evaluate multiple enzyme properties, including activity, stability, and selectivity, using fluorogenic substrates and spectroscopic detection. Machine learning algorithms analyze screening data to identify sequence–activity relationships and guide subsequent design iterations. The optimization cycle continues with focused library design around lead candidates, ultimately producing enzymes with enhanced industrial performance characteristics.

12.2 Directed evolution platforms

Directed evolution represents one of the most powerful approaches for enzyme optimization, mimicking natural evolutionary processes through iterative cycles of diversification and selection. CFPS technology enhances directed evolution by enabling larger library sizes, more sophisticated selection strategies, and shorter cycle times compared to conventional approaches.

12.2.1 Library generation and diversification

The success of directed evolution campaigns depends critically on the quality and diversity of starting libraries. CFPS systems support various diversification strategies, including random mutagenesis, DNA shuffling, and combinatorial approaches that sample different regions of sequence space.

Error-prone PCR remains the most widely used method for generating random mutations, with mutation rates typically controlled to introduce 1–3 amino acid changes per gene. CFPS systems can accommodate high mutation loads that would be lethal in living cells, enabling exploration of more distant sequence space. Advanced mutagenesis methods, such as mutator strains or chemical mutagenesis, can generate even higher diversity levels.

DNA shuffling and related recombination methods create libraries by combining beneficial mutations from multiple parent sequences. These approaches are particularly effective when starting with multiple functional variants and can rapidly combine distant beneficial mutations. CFPS systems facilitate the screening of recombined libraries without the complications of recombination in living cells.

Focused mutagenesis strategies target specific regions identified through structural analysis or previous evolution campaigns. Saturation mutagenesis of active site residues can optimize catalytic efficiency, while randomization of surface loops may enhance stability or alter substrate specificity. The rapid screening capabilities of CFPS systems make comprehensive saturation mutagenesis feasible for multiple positions simultaneously.

12.2.2 Selection and screening strategies

Effective selection strategies are essential for identifying improved variants from large diverse libraries. CFPS platforms enable sophisticated screening approaches that can evaluate multiple enzyme properties simultaneously, providing more comprehensive assessment than single-parameter screens.

Activity-based screening remains the primary selection criterion, with fluorogenic or chromogenic substrates providing real-time monitoring of enzyme performance. Advanced detection methods, including mass spectrometry and NMR spectroscopy, enable detailed analysis of reaction products and kinetic parameters. High-throughput analytical methods can characterize substrate specificity, product selectivity, and reaction mechanisms.

Stability screening has become increasingly important for industrial applications, with enzymes required to function under harsh conditions, including extreme pH, high temperatures, and the presence of organic solvents. CFPS systems can incorporate stress conditions directly into screening protocols, enabling simultaneous optimization of activity and stability. Thermal challenge assays, pH stress tests, and solvent tolerance screens can be performed in parallel with activity measurements.

Selectivity screening addresses the need for enzymes with precise substrate specificity and product selectivity. Competitive substrate assays evaluate enzyme preference among similar substrates, while product analysis identifies unwanted side reactions. These approaches are particularly important for pharmaceutical applications where product purity is critical.

12.2.3 Continuous evolution systems

Continuous evolution systems represent an advanced approach to directed evolution that maintains selective pressure over extended periods, enabling the accumulation of beneficial mutations without manual intervention. CFPS technology enables unique implementations of continuous evolution that are not possible with living cell systems.

Flow-based continuous evolution utilizes microfluidic devices to maintain enzyme-expressing CFPS reactions under continuous selection pressure. Fresh reagents are continuously supplied while reaction products are removed, maintaining steady-state conditions that favor improved variants. Selection can be implemented through product-dependent growth advantages or direct selection for desired activities.

Compartmentalized continuous evolution uses droplet microfluidics to create isolated reaction volumes containing individual enzyme variants. Selection pressure is applied within each compartment, with successful variants identified through fluorescent markers or other detectable phenotypes. This approach enables larger library sizes and reduces competition between variants.

Automated continuous evolution systems integrate DNA synthesis, CFPS expression, screening, and sequence analysis into closed-loop platforms. These systems can operate continuously for weeks or months, automatically identifying improved variants and using them as templates for subsequent rounds of evolution. The elimination of manual intervention enables more extensive optimization than traditional approaches.

12.2.4 Multiparameter optimization

Industrial enzyme applications often require optimization of multiple properties simultaneously, including activity, stability, selectivity, and expression levels. Traditional directed evolution approaches struggle with multiparameter optimization due to conflicting selection pressures and the difficulty of designing appropriate screening assays.

CFPS platforms enable sophisticated multiparameter optimization through the use of composite fitness functions that weight different enzyme properties according to their importance for specific applications. Machine learning algorithms can identify variants that represent optimal trade-offs between competing objectives, enabling more effective optimization than single-parameter approaches.

Pareto optimization approaches identify sets of variants that represent optimal trade-offs between different objectives, providing flexibility in selecting final candidates based on application-specific requirements. These approaches are particularly valuable when the relative importance of different properties is uncertain or may vary between applications (Table 12.2).

12.3 Industrial biocatalysis

The application of enzymes in industrial processes requires careful consideration of process conditions, economics, and scalability. CFPS technology offers unique advantages for industrial biocatalysis through its ability to produce specialized enzymes rapidly, enable

TABLE 12.2 Directed evolution platform comparison.

Platform type	Library size	Cycle time	Automation level	Multiparameter	Cost per variant
Traditional cell-based	10^4–10^6	1–2 weeks	Low	Difficult	$0.10–1.00
CFPS microplate	10^5–10^7	1–3 days	High	Excellent	$0.05–0.50
CFPS droplet	10^6–10^9	Hours	Very high	Good	$0.01–0.10
Continuous CFPS	10^7–10^{10}	Continuous	Autonomous	Excellent	$0.001–0.05
Yeast display	10^6–10^8	1 week	Moderate	Moderate	$0.20–2.00
Phage display	10^8–10^{11}	3–5 days	Moderate	Limited	$0.50–5.00

process optimization, and support novel reaction formats that are impractical with traditional approaches.

12.3.1 Process integration strategies

Successful integration of enzymes into industrial processes requires careful matching of enzyme properties with process requirements. CFPS systems enable rapid prototyping of different process configurations and optimization of enzyme-process combinations that would be time-consuming and expensive with conventional approaches.

One-pot reaction systems that combine CFPS enzyme production with biocatalytic conversion offer significant advantages for specialized applications. These integrated systems eliminate enzyme purification steps and enable real-time optimization of enzyme properties for specific substrates and conditions. The approach is particularly valuable for unstable enzymes or complex multienzyme reactions.

Continuous enzyme production systems utilize CFPS technology to maintain steady-state enzyme concentrations in bioreactors, compensating for enzyme degradation and enabling extended reaction times. This approach is especially beneficial for thermolabile enzymes or reactions requiring long residence times. Flow-through CFPS reactors can be integrated with downstream processing equipment to create fully continuous production systems.

Immobilized CFPS systems provide enhanced stability and reusability compared to free enzyme systems while maintaining the flexibility of cell-free production. Enzymes can be produced directly on solid supports or captured immediately following synthesis, reducing handling losses and improving stability. These systems enable repeated use and simplified product separation.

12.3.2 Extreme environment applications

Many industrial processes operate under conditions that challenge enzyme stability, including high temperatures, extreme pH values, high salt concentrations, and the presence of organic solvents. CFPS technology enables the development and optimization of enzymes specifically designed for these challenging environments.

Thermostable enzyme development benefits from CFPS systems' ability to rapidly screen variants at elevated temperatures and optimize enzymes for specific thermal profiles. Directed evolution campaigns can incorporate thermal challenge directly into screening protocols, selecting for variants that maintain activity at operational temperatures. The ability to test enzymes immediately following synthesis eliminates artifacts from purification and storage.

Solvent-tolerant enzymes represent a growing market for pharmaceutical and fine chemical applications where organic solvents are required for substrate solubility or product extraction. CFPS systems can incorporate organic solvents directly into screening assays, enabling optimization of enzyme performance under realistic process conditions. This approach has proven particularly valuable for developing enzymes for biocatalytic synthesis in nonaqueous media.

pH-resistant enzymes are essential for many industrial applications, particularly in the food and textile industries where extreme pH conditions are common. CFPS screening systems can evaluate enzyme performance across wide pH ranges, identifying variants with enhanced stability and activity under alkaline or acidic conditions. Buffer capacity can be optimized within CFPS reactions to maintain desired pH levels during screening.

12.3.3 Multienzyme systems and pathways

Complex industrial transformations often require multiple enzymatic steps, creating challenges for traditional enzyme production and process optimization. CFPS technology enables the coordinated production and optimization of multienzyme systems, facilitating the development of sophisticated biocatalytic pathways.

Metabolic pathway reconstruction using CFPS systems allows for the production of complex molecules through multistep enzymatic conversions. Each enzyme in the pathway can be independently optimized while maintaining overall pathway functionality. This approach enables the development of biocatalytic routes to high-value chemicals that would be difficult or impossible to produce through single-step reactions.

Enzyme cascade optimization benefits from CFPS systems' ability to vary the relative concentrations of different enzymes and evaluate pathway efficiency under various conditions. Substrate channeling effects can be studied by varying enzyme ratios and reaction conditions, enabling optimization of overall pathway performance rather than individual enzyme activities.

Cofactor regeneration systems represent a critical component of many industrial biocatalytic processes. CFPS technology enables the coproduction of cofactor regeneration enzymes alongside primary catalysts, creating self-sufficient reaction systems. This approach reduces costs and simplifies process design by eliminating the need for expensive cofactor supplementation.

12.3.4 Real-time process optimization

The rapid response capabilities of CFPS systems enable real-time optimization of biocatalytic processes, allowing for adaptive control strategies that respond to changing process conditions or substrate compositions. This capability is particularly valuable for applications involving variable feedstocks or complex reaction mixtures.

Feedback control systems utilize online monitoring of process parameters to automatically adjust enzyme production and reaction conditions. Sensors monitoring substrate concentration, product formation, or side product accumulation can trigger changes in enzyme expression levels or modifications to reaction conditions. This approach enables maintaining optimal process performance despite variations in feedstock quality or composition.

Adaptive enzyme optimization uses machine learning algorithms to continuously improve enzyme variants based on real-time process data. As new substrates or conditions are encountered, the system can automatically generate and test improved variants, creating self-improving biocatalytic processes. This approach is particularly valuable for applications involving complex or variable substrate mixtures (Fig. 12.2).

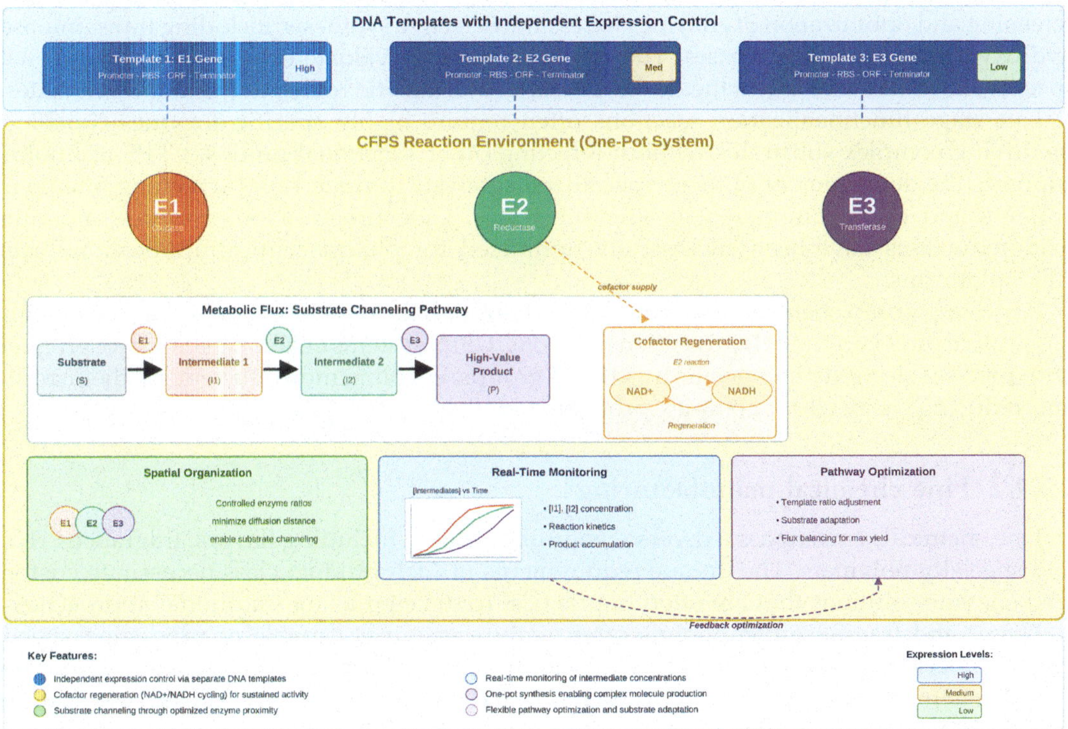

FIG. 12.2 Multienzyme cascade production in CFPS systems. This figure demonstrates the coordinated production and optimization of multienzyme cascades using cell-free protein synthesis technology. The system shows a three-enzyme pathway for the synthesis of a high-value chemical intermediate. Each enzyme (E_1, E_2, E_3) is produced from separate DNA templates within the same CFPS reaction, enabling independent control of expression levels. The pathway includes cofactor regeneration systems (NAD^+/NADH cycling) and real-time monitoring of intermediate concentrations. Substrate channeling effects are optimized through controlled enzyme ratios and spatial organization. The integrated system enables one-pot synthesis of complex molecules while providing flexibility for pathway optimization and adaptation to different substrate compositions.

12.4 Specialty chemical production

The production of high-value specialty chemicals represents one of the most promising applications for CFPS-based enzyme technology. These applications typically involve low-volume, high-margin products where the flexibility and rapid development capabilities of CFPS systems provide significant competitive advantages over traditional approaches.

12.4.1 Pharmaceutical intermediates

The pharmaceutical industry increasingly relies on biocatalytic approaches for the synthesis of chiral intermediates and complex molecules that are difficult to produce through traditional chemical methods. CFPS technology enables rapid development of customized enzymes for specific synthetic targets, supporting accelerated drug development timelines.

Chiral synthesis represents a particularly important application area, with over 80% of pharmaceutical products containing at least one chiral center. CFPS systems enable the rapid screening and optimization of enzymes for enantioselective synthesis, including transaminases, hydroxynitrile lyases, and lipases. The ability to quickly evaluate enzyme performance with specific substrates accelerates the development of biocatalytic routes to chiral intermediates.

Late-stage functionalization reactions often require highly specific enzymes capable of modifying complex substrates without affecting other functional groups. CFPS technology supports the development of these specialized catalysts through rapid screening of enzyme variants and optimization for specific substrates. Cytochrome P450 enzymes and other oxidoreductases have been successfully optimized for pharmaceutical applications using CFPS platforms.

Protecting group chemistry can be enhanced through the development of enzymes capable of selective protection and deprotection reactions. CFPS systems enable screening for enzymes with precise selectivity for specific functional groups, enabling more efficient synthetic routes and reducing the need for complex protection strategies.

12.4.2 Fine chemical manufacturing

Fine chemicals encompass a diverse range of products, including flavors, fragrances, dyes, and specialty polymers. The unique requirements of each product class necessitate customized enzyme solutions that are often impractical to develop using traditional approaches.

Flavor and fragrance compounds often require enzymes capable of producing specific stereoisomers or controlling product distributions to achieve desired sensory properties. CFPS systems enable rapid optimization of enzymes for these specialized applications, including the development of esterases for fragrance synthesis and aldolases for flavor compound production.

Polymer building blocks represent an emerging application area where enzymes can provide more sustainable and selective synthetic routes compared to traditional chemical methods. CFPS technology supports the development of enzymes for polyester synthesis, polyamide production, and other polymer applications where precise control over molecular weight and composition is required.

Dye intermediate synthesis benefits from enzymatic approaches that can provide better selectivity and reduced environmental impact compared to chemical methods. CFPS systems enable optimization of enzymes for dye synthesis, including laccases for dye polymerization and peroxidases for color development reactions.

12.4.3 Sustainable chemistry applications

Environmental sustainability has become a critical driver for the adoption of biocatalytic processes in specialty chemical production. CFPS technology enables the development of green chemistry solutions that reduce waste, eliminate toxic reagents, and operate under milder conditions.

Solvent-free reactions represent an important sustainability goal, with enzymes offering the potential to eliminate organic solvents from synthetic processes. CFPS systems can be optimized for reactions in neat substrates or minimal water systems, enabling the development of truly green synthetic processes.

Waste minimization through improved selectivity and reduced side product formation is a key advantage of enzymatic processes. CFPS technology enables optimization of enzymes for maximum selectivity, reducing waste streams and simplifying purification processes. Atom economy can be maximized through careful enzyme design and process optimization.

Renewable feedstock utilization is enhanced through the development of enzymes capable of converting biomass-derived substrates into valuable chemicals. CFPS systems support the optimization of enzymes for lignocellulosic substrates, enabling the development of biorefinery processes that convert waste materials into valuable products.

12.4.4 Economic considerations

The economics of specialty chemical production using CFPS-based enzymes must account for development costs, production scales, and market values. Understanding these factors is essential for identifying viable applications and optimizing business models.

Development cost advantages arise from the rapid timelines possible with CFPS technology, reducing the time and resources required to bring new products to market. Traditional enzyme development programs may require 6–18 months, while CFPS approaches can achieve similar results in weeks or months. This acceleration provides significant competitive advantages in fast-moving markets.

Production-scale economics favor CFPS technology for small- to medium-scale applications where traditional fermentation economics are unfavorable. The fixed costs associated with fermentation infrastructure become prohibitive for low-volume products, while CFPS systems can be economically viable at much smaller scales.

Value proposition analysis must consider the premium prices typically commanded by specialty chemicals, which can justify higher production costs for unique or high-performance products. CFPS technology enables the development of differentiated products with superior performance characteristics, supporting premium pricing strategies (Table 12.3).

TABLE 12.3 Economic analysis of CFPS vs. traditional methods for specialty chemicals.

Parameter	Traditional fermentation	CFPS production	Chemical synthesis
Development time	6–18 months	1–3 months	12–36 months
Development cost	$500 K–2 M	$100 K–500 K	$1 M–5 M
Minimum economic scale	100–1000 kg/year	1–100 kg/year	1000+ kg/year
Production cost	$50–500/kg	$100–1000/kg	$20–200/kg
Capital investment	$10 M–100 M	$1 M–10 M	$50 M–500 M
Time to market	12–24 months	3–6 months	24–48 months
Product differentiation	Moderate	High	Low
Environmental impact	Low	Very low	High

12.5 Green chemistry applications

The integration of CFPS-based enzyme production with green chemistry principles represents a transformative approach to sustainable manufacturing. This convergence addresses growing environmental concerns while providing economic benefits through improved efficiency, reduced waste, and enhanced product quality.

12.5.1 Principles of green biocatalysis

Green chemistry principles provide a framework for evaluating and optimizing biocatalytic processes to minimize environmental impact while maximizing efficiency. CFPS technology enhances the implementation of these principles through its flexibility and ability to optimize enzyme properties for specific green chemistry applications.

Atom economy maximization represents a fundamental green chemistry principle that seeks to incorporate all starting materials into final products, minimizing waste generation. CFPS systems enable the development of enzymes with enhanced selectivity and reduced side product formation, improving overall atom economy. This approach is particularly valuable for complex synthetic transformations where multiple reaction pathways are possible.

Renewable feedstock utilization is facilitated by CFPS technology's ability to rapidly develop enzymes for novel substrates derived from biomass or waste streams. Traditional enzyme development for new substrates can be prohibitively expensive and time-consuming, while CFPS approaches enable economical optimization for specialized applications.

Catalyst optimization through directed evolution and protein engineering enables the development of enzymes with enhanced activity, stability, and selectivity under green chemistry conditions. CFPS systems support rapid optimization cycles that can improve enzyme performance while reducing the need for harsh reaction conditions or toxic cofactors.

12.5.2 Waste reduction strategies

Industrial processes often generate significant waste streams through inefficient reactions, side product formation, and purification requirements. CFPS-based enzyme optimization directly addresses these issues through enhanced selectivity and process efficiency.

Side product minimization through improved enzyme selectivity represents one of the most direct approaches to waste reduction. CFPS systems enable rapid screening for enzyme variants with enhanced substrate specificity and reduced promiscuous activity. This optimization can dramatically reduce purification requirements and waste stream volumes.

Process intensification using multienzyme systems enables the combination of multiple reaction steps into single-pot processes, eliminating intermediate isolation and purification steps. CFPS technology facilitates the cooptimization of enzyme ratios and reaction conditions to maximize overall process efficiency while minimizing waste generation.

Solvent elimination or reduction represents a major opportunity for waste minimization in many industrial processes. CFPS systems can be optimized for reactions in neat substrates, minimal water systems, or environmentally benign solvents, reducing the environmental impact of solvent production, use, and disposal.

12.5.3 Energy efficiency optimization

Energy consumption represents a significant component of the environmental impact of industrial processes. CFPS-based enzyme optimization can substantially reduce energy requirements through the development of catalysts that operate under milder conditions.

Temperature reduction through enzyme optimization enables processes to operate at lower temperatures, reducing energy consumption for heating and cooling. CFPS systems can rapidly screen for enzyme variants with enhanced activity at moderate temperatures, enabling substantial energy savings in large-scale processes.

Pressure reduction is possible through the development of enzymes that operate efficiently at atmospheric pressure, eliminating the need for high-pressure equipment and associated energy consumption. This approach is particularly valuable for reactions that traditionally require elevated pressures for adequate reaction rates.

pH neutralization requirements can be minimized through the development of enzymes that operate near neutral pH, reducing the need for acid or base addition and associated neutralization steps. CFPS systems enable optimization of enzymes for operation in buffered systems that minimize pH adjustment requirements.

12.5.4 Life cycle assessment integration

Comprehensive evaluation of green chemistry benefits requires life cycle assessment (LCA) approaches that consider all aspects of process environmental impact, from raw material production through waste disposal. CFPS technology enables rapid LCA optimization through accelerated process development and testing.

Cradle-to-gate analysis evaluates the environmental impact of enzyme production and process operation, enabling optimization of both production methods and application conditions. CFPS

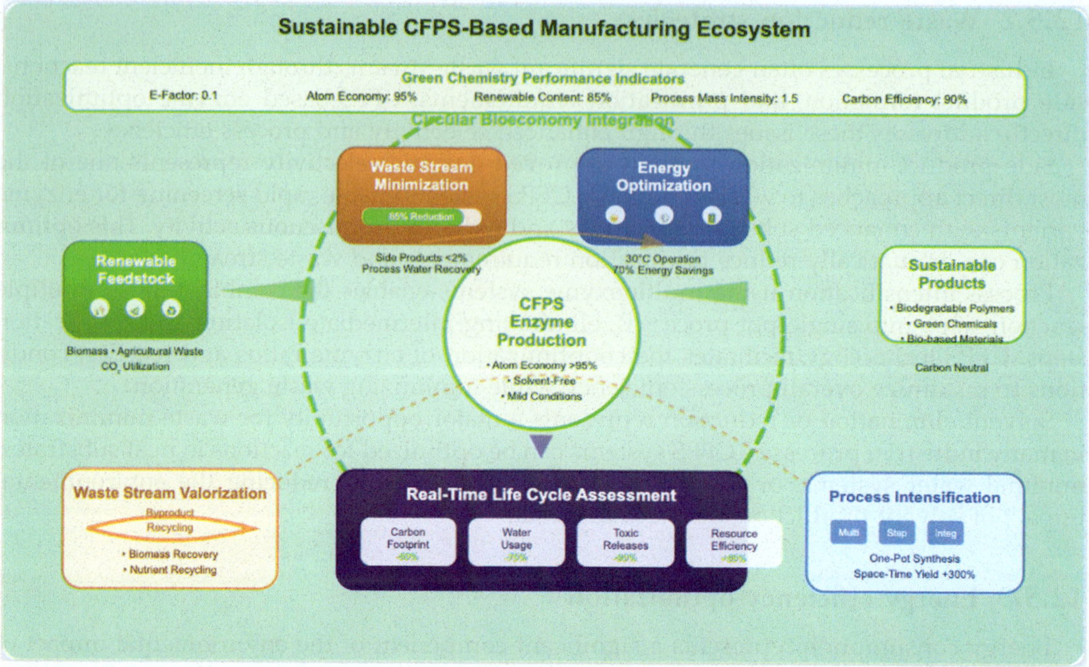

FIG. 12.3 Green chemistry workflow integration with CFPS enzyme production. This figure illustrates the integration of green chemistry principles with CFPS-based enzyme production for sustainable specialty chemical manufacturing. The workflow begins with renewable feedstock selection, including biomass-derived substrates and waste stream utilization. CFPS enzyme optimization focuses on atom economy maximization through enhanced selectivity and reduced side product formation. The integrated process incorporates solvent minimization, energy efficiency optimization, and waste stream reduction. Real-time life cycle assessment monitoring enables continuous process improvement and environmental impact minimization. The circular design emphasizes resource recycling and waste stream valorization, creating closed-loop systems that maximize sustainability.

systems' reduced infrastructure requirements and rapid development capabilities provide inherent advantages in cradle-to-gate analysis.

Comparative assessment with chemical alternatives enables quantitative evaluation of the environmental benefits of biocatalytic processes. CFPS technology's rapid development capabilities enable direct comparison of multiple process options, supporting evidence-based decision making for green chemistry implementation.

Continuous improvement through LCA feedback enables ongoing optimization of processes to minimize environmental impact. CFPS systems support rapid implementation of process improvements identified through LCA analysis, creating feedback loops that drive continuous environmental performance enhancement (Fig. 12.3).

12.6 Economic competitiveness

The commercial success of CFPS-based enzyme production depends on achieving economic competitiveness with existing technologies while providing superior performance or capabilities. This section analyzes the key economic factors that influence technology adoption and strategies for optimizing commercial viability.

12.6.1 Cost structure analysis

Understanding the detailed cost structure of CFPS-based enzyme production is essential for identifying optimization opportunities and competitive positioning. The technology's cost profile differs significantly from traditional approaches, with different trade-offs between fixed and variable costs.

Raw material costs represent a significant component of CFPS production expenses, particularly for amino acids, nucleotides, and energy regeneration systems. However, these costs are largely independent of production scale, providing advantages for small- to medium-scale applications. Optimization of CFPS formulations can reduce raw material costs through improved efficiency and reduced waste.

Labor costs are typically lower for CFPS systems due to reduced complexity and shorter cycle times compared to fermentation processes. Automated CFPS platforms can further reduce labor requirements, particularly for routine production applications. The elimination of complex downstream processing reduces both labor and equipment costs.

Capital equipment costs for CFPS systems are generally lower than fermentation alternatives, particularly for small-scale applications. The modular nature of CFPS systems enables scalable investment strategies that align capital requirements with production volumes. This approach reduces financial risk and enables more flexible business models.

Infrastructure costs, including utilities, facilities, and waste management, are typically reduced for CFPS systems due to simpler process requirements and reduced environmental controls. The elimination of living organisms reduces biosafety requirements and associated infrastructure costs.

12.6.2 Scale economics and market positioning

CFPS technology exhibits unique scale economics that favor different market segments compared to traditional enzyme production methods. Understanding these dynamics is critical for identifying optimal market positioning strategies.

Small-scale advantages arise from CFPS technology's reduced fixed costs and simplified infrastructure requirements. Traditional fermentation systems require significant minimum scales to achieve economic viability, while CFPS systems can be economical at much smaller production volumes. This capability enables serving niche markets and specialized applications that are underserved by traditional approaches.

Medium-scale optimization benefits from CFPS technology's modularity and automation capabilities. Production capacity can be scaled through parallel systems rather than requiring larger, more complex equipment. This approach provides operational flexibility and reduces the risks associated with capacity expansion.

Large-scale considerations include the potential for continuous CFPS systems and integration with downstream processing. While traditional fermentation may maintain cost advantages at very large scales, CFPS technology's superior product quality and reduced complexity can justify premium pricing in many applications.

Market segmentation strategies should focus on applications where CFPS technology's unique capabilities provide clear value propositions. High-value, low-volume applications represent the most attractive initial markets, with potential expansion into larger markets as technology costs decline.

12.6.3 Value proposition development

Successful commercialization of CFPS-based enzyme production requires clear articulation of value propositions that resonate with target customers. These value propositions must address both economic and performance benefits that justify technology adoption.

Performance advantages include enhanced enzyme quality, improved consistency, and access to enzyme variants that cannot be produced through traditional methods. These benefits can justify premium pricing for applications where performance is critical, such as pharmaceutical synthesis or high-value specialty chemicals.

Speed advantages from rapid development and production capabilities provide significant value for time-sensitive applications. The ability to respond quickly to changing market requirements or customize enzymes for specific applications creates competitive advantages that can support premium pricing.

Flexibility benefits enable customers to access specialized enzymes for unique applications without the minimum volume requirements of traditional suppliers. This capability opens new market opportunities and enables differentiated product strategies.

Risk reduction through shorter development timelines and lower capital requirements appeals to customers seeking to minimize project risks. The ability to test and optimize enzymes rapidly reduces technical risks and accelerates time to market for new products.

12.6.4 Competitive analysis and market entry

Understanding the competitive landscape and identifying optimal market entry strategies are critical for successful commercialization of CFPS-based enzyme production. The technology competes with established enzyme suppliers and alternative technologies.

Traditional enzyme suppliers represent the primary competitive threat, with established relationships, proven track records, and optimized production systems. However, these suppliers typically focus on high-volume applications and may be unable to serve specialized markets economically. CFPS technology can target underserved market segments where traditional suppliers are not competitive.

Alternative technologies, including chemical catalysis and other biocatalytic approaches, provide additional competition in specific applications. Competitive positioning must clearly articulate the advantages of CFPS-based enzymes compared to these alternatives, focusing on performance, sustainability, and economic benefits.

Market entry strategies should prioritize applications where CFPS technology provides clear competitive advantages and established suppliers have limitations. Early success in these markets can provide the foundation for expansion into larger, more competitive segments.

Partnership strategies with established enzyme suppliers or end-users can provide market access and reduce customer acquisition costs. These partnerships can leverage complementary capabilities and provide validation for CFPS technology in commercial applications (Tables 12.4 and 12.5).

TABLE 12.4 CFPS technology competitive positioning matrix.

Market segment	Volume range	Value level	CFPS advantage	Competition level	Market attractiveness
Pharmaceutical intermediates	1–100 kg/year	Very high	Speed, quality	Low	Excellent
Fine chemicals	10–1000 kg/year	High	Customization	Moderate	Good
Specialty polymers	100–10,000 kg/year	Moderate	Novel functions	Moderate	Good
Food enzymes	1000–100,000 kg/year	Low–moderate	Purity, safety	High	Moderate
Industrial chemicals	10,000+ kg/year	Low	Limited	Very high	Poor
Commodity enzymes	100,000+ kg/year	Very low	None	Very high	Poor

TABLE 12.5 Economic model parameters for CFPS enzyme production.

Cost component	Small scale (1–10 kg/year)	Medium scale (10–100 kg/year)	Large scale (100–1000 kg/year)
Raw materials	\$500–2000/kg	\$200–800/kg	\$100–400/kg
Labor	\$100–500/kg	\$50–200/kg	\$20–100/kg
Equipment	\$200–1000/kg	\$100–400/kg	\$50–200/kg
Facilities	\$50–200/kg	\$25–100/kg	\$10–50/kg
Total production cost	\$850–3700/kg	\$375–1500/kg	\$180–750/kg
Selling price range	\$2000–10,000/kg	\$1000–5000/kg	\$500–2500/kg
Gross margin	60%–85%	60%–80%	65%–80%

Bibliography

Anastas, P., Eghbali, N., 2010. Green chemistry: principles and practice. Chem. Soc. Rev. 39 (1), 301–312.

Bornscheuer, U.T., Huisman, G.W., Kazlauskas, R.J., Lutz, S., Moore, J.C., Robins, K., 2012. Engineering the third wave of biocatalysis. Nature 485 (7397), 185–194.

Bundy, B.C., Swartz, J.R., 2010. Site-specific incorporation of p-propargyloxyphenylalanine in a cell-free environment for direct protein-protein click conjugation. Bioconjugate Chem. 21 (2), 255–263.

Carlson, E.D., Gan, R., Hodgman, C.E., Jewett, M.C., 2012. Cell-free protein synthesis: applications come of age. Biotechnol. Adv. 30 (5), 1185–1194.

Contreras-Llano, L.E., Tan, C., 2018. High-throughput screening of biomolecules using cell-free gene expression systems. Synth. Biol. 3 (3), ysy012.

Dudley, Q.M., Karim, A.S., Jewett, M.C., 2015. Cell-free metabolic engineering: biomanufacturing beyond the cell. Biotechnol. J. 10 (1), 69–82.

Garenne, D., Thompson, S., Brisson, A., Khakimzhan, A., Noireaux, V., 2021. The all-*E. coli* TXTL toolbox 3.0: new capabilities of a cell-free synthetic biology platform. Synth. Biol. 6 (1), 1–10.

Garg, S., Singaraju, G.S., Yengkhom, S., Rakshit, S., 2016. Natural evolution versus directed evolution: a comparison of methods over the course of evolution of the enzyme lactamase in natural versus synthetic biology systems. Biomolecules 6 (1), 16.

Hodgman, C.E., Jewett, M.C., 2012. Cell-free synthetic biology: thinking outside the cell. Metab. Eng. 14 (3), 261–269.

Karim, A.S., Jewett, M.C., 2016. A cell-free framework for rapid biosynthetic pathway prototyping and enzyme discovery. Metab. Eng. 36, 116–126.

Kelwick, R., MacDonald, J.T., Webb, A.J., Freemont, P.S., 2016. Developments in cell-free protein synthesis for biotechnology applications–recent developments. Biotechnol. Lett. 38 (1), 1–11.

Koch, M., Faulon, J.L., Borkowski, O., 2018. Models for cell-free synthetic biology: make prototyping easier, better, and faster. Front. Bioeng. Biotechnol. 6, 182.

Liu, D.V., Zawada, J.F., Swartz, J.R., 2005. Streamlining *Escherichia coli* S30 extract preparation for economical cell-free protein synthesis. Biotechnol. Prog. 21 (2), 460–465.

Martin, R.W., Majewska, N.I., Chen, C.X., Albanetti, T.E., Jimenez, R.B.C., Schmelzer, A.E., … Jewett, M.C., 2017. Development of a CHO-based cell-free platform for synthesis of active monoclonal antibodies. ACS Synth. Biol. 6 (7), 1370–1379.

Moore, S.J., MacDonald, J.T., Wienecke, S., Ishwarbhai, A., Tsipa, A., Aw, R., … Freemont, P.S., 2018. Rapid acquisition and model-based analysis of cell-free transcription-translation reactions from nonmodel bacteria. Proc. Natl. Acad. Sci. U.S.A. 115 (19), E4340–E4349.

Perez, J.G., Stark, J.C., Jewett, M.C., 2016. Cell-free synthetic biology: engineering beyond the cell. Cold Spring Harb. Perspect. Biol. 8 (12), a023853.

Sheldon, R.A., Brady, D., 2018. The limits to biocatalyst lifetime. ChemSusChem 11 (17), 2836–2845.

Sheldon, R.A., Woodley, J.M., 2018. Role of biocatalysis in sustainable chemistry. Chem. Rev. 118 (2), 801–838.

Silverman, A.D., Kelley-Loughnane, N., Lucks, J.B., Jewett, M.C., 2019. Deconstructing cell-free extract preparation for in vitro activation of transcriptional genetic circuitry. ACS Synth. Biol. 8 (2), 403–414.

Swartz, J.R., 2012. Transforming biochemical engineering with cell-free biology. AIChE J. 58 (1), 5–13.

CHAPTER

13

Synthetic biology integration

13.1 Introduction

The convergence of cell-free protein synthesis (CFPS) and synthetic biology represents one of the most transformative developments in modern biotechnology. This integration has created unprecedented opportunities for rapid prototyping of biological systems, enabling researchers to design, build, test, and learn from synthetic biological circuits with remarkable speed and precision. Unlike traditional cell-based approaches that require time-consuming cloning, transformation, and cultivation steps, cell-free systems provide an open platform where synthetic biology components can be directly added and immediately tested.

The synergy between CFPS and synthetic biology stems from their shared philosophy of modularity and standardization. Both fields embrace the engineering principles of abstraction, where complex biological functions are broken down into well-characterized, interchangeable parts. This chapter explores how cell-free systems have become the preferred platform for synthetic biology applications, from simple genetic circuits to complex metabolic pathways and artificial cellular systems.

13.2 Genetic circuit prototyping

13.2.1 Fundamentals of cell-free circuit design

CFPS platforms have revolutionized the prototyping of genetic circuits by providing a controlled environment where individual components can be systematically tested and optimized. The open nature of CFPS systems allows for direct manipulation of transcription and translation machinery, enabling precise control over gene expression dynamics that would be impossible to achieve in living cells.

The basic architecture of cell-free genetic circuits consists of three primary layers: the information layer (DNA templates), the processing layer (transcription and translation machinery), and the regulatory layer (transcription factors, riboswitches, and other control elements). This modular organization allows researchers to independently optimize each component before integrating them into more complex systems.

Cell Free Synthesis of Biologics
https://doi.org/10.1016/B978-0-443-33271-5.00029-6

TABLE 13.1 Comparison of circuit prototyping platforms.

Platform	Setup time	Throughput	Cost per test	Temporal resolution	Dynamic range
E. coli cells	4–8 h	Medium	Low	Hours	10–100-fold
Mammalian cells	24–72 h	Low	High	Hours	100–1000-fold
Cell-free (*E. coli*)	30 min	High	Medium	Minutes	100–1000-fold
Cell-free (eukaryotic)	1–2 h	High	High	Minutes	1000+-fold

As shown in Table 13.1, cell-free systems offer significant advantages over traditional cellular platforms in terms of setup time and throughput. Cell-free systems based on *Escherichia coli* extracts can be prepared in as little as 30 min and offer high-throughput testing capabilities, while maintaining comparable dynamic range to mammalian systems at a fraction of the cost.

13.2.2 Transcriptional control circuits

Transcriptional circuits form the backbone of most synthetic biology applications, providing the fundamental logic operations necessary for complex biological computation. In cell-free systems, these circuits can be rapidly prototyped using purified transcription factors, promoter libraries, and standardized DNA templates.

Logic gates represent the most fundamental transcriptional circuits, implementing basic Boolean operations such as AND, OR, and NOT functions. These can be constructed using combinations of transcriptional activators and repressors. For example, an AND gate can be implemented using a promoter that requires two different transcriptional activators for efficient expression. Oscillators generate periodic gene expression patterns and are fundamental to understanding biological timing mechanisms. The repressilator, consisting of three repressors arranged in a negative feedback loop, was among the first synthetic circuits successfully implemented in cell-free systems. Bistable switches create memory elements in biological circuits, allowing cells to maintain states based on previous inputs. These circuits typically involve positive feedback loops that create stable high- and low-expression states (Fig. 13.1).

Fig. 13.1 illustrates the different transcriptional circuit topologies that have been successfully implemented in cell-free systems, including logic gates, oscillators, and bistable switches, along with their corresponding time-course expression data demonstrating the temporal dynamics of each circuit type.

13.2.3 Posttranscriptional regulation

Beyond transcriptional control, cell-free systems excel at implementing posttranscriptional regulatory mechanisms, including riboswitch-based control, RNA interference, and CRISPR-based regulation. These mechanisms provide additional layers of control and can create more sophisticated circuit behaviors.

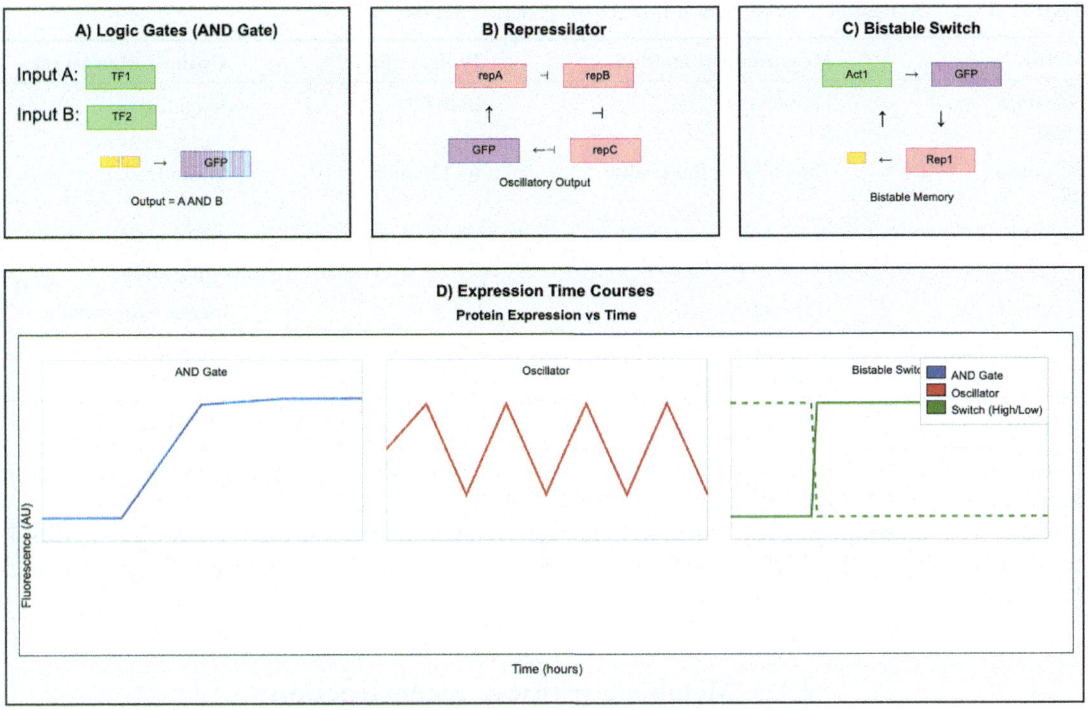

FIG. 13.1 The different transcriptional circuit topologies that have been successfully implemented in cell-free systems, including logic gates, oscillators, and bistable switches, along with their corresponding time-course expression data demonstrating the temporal dynamics of each circuit type.

Riboswitches are RNA-based regulatory elements that directly bind small molecules and undergo conformational changes that affect gene expression. In cell-free systems, riboswitches can be easily tested with different ligands and concentrations, allowing for rapid optimization of sensor sensitivity and dynamic range. CRISPR interference provides a programmable method for transcriptional repression without requiring protein engineering. Dead Cas9 fused to transcriptional repressor domains can be directed to any genomic target using guide RNAs, enabling rapid circuit prototyping.

13.2.4 Circuit characterization and optimization

Cell-free systems provide unique advantages for circuit characterization, including the ability to directly measure and manipulate individual components. Standard characterization protocols focus on several key metrics that define circuit performance and behavior.

Transfer functions describe the input–output relationship of circuit components and are essential for predictive circuit design. Dynamic range measures the fold-change between minimum and maximum outputs, typically ranging from 10- to 1000-fold depending on the circuit architecture. Response time quantifies how quickly circuits reach steady state after input changes, with cell-free systems typically achieving responses within 10–120 min. Noise

TABLE 13.2 Standard characterization metrics for genetic circuits.

Metric	Measurement method	Typical range	Optimization target
Dynamic range	Fold-change (max/min)	10–1000×	Maximize
Response time	Time to 90% final value	10–120 min	Minimize
Sensitivity	EC50 or IC50	nM–mM	Optimize for application
Cooperativity	Hill coefficient	1–4	Increase for switches
Orthogonality	Cross-talk measurement	<10%	Minimize

characteristics describe variability in circuit output under identical conditions and are crucial for understanding circuit reliability and robustness.

Table 13.2 summarizes the standard characterization metrics used for genetic circuits in cell-free systems, along with typical measurement ranges and optimization targets for each parameter.

13.3 Metabolic pathway reconstruction

13.3.1 Advantages of cell-free metabolic engineering

Cell-free systems offer unprecedented advantages for metabolic pathway reconstruction and optimization. Unlike cellular systems where metabolic flux is constrained by cellular physiology and regulatory networks, cell-free platforms provide complete control over enzyme concentrations, cofactor availability, and reaction conditions.

Precise stoichiometric control represents one of the key advantages of cell-free metabolic engineering. In these systems, enzyme ratios can be precisely controlled to optimize flux through specific pathway steps, eliminating the bottlenecks commonly encountered in cellular systems where enzyme expression levels are difficult to fine-tune. The elimination of competing pathways provides another significant benefit, as cell-free systems can be designed to include only the enzymes necessary for the desired pathway, removing competition from endogenous metabolic networks that can divert flux away from the target product. Real-time monitoring and control capabilities allow for continuous observation of intermediate metabolites and real-time adjustment of reaction conditions to maintain optimal pathway performance throughout the reaction.

13.3.2 Pathway design strategies

Successful metabolic pathway reconstruction in cell-free systems requires careful consideration of several design parameters that influence overall pathway performance and efficiency.

Enzyme selection and engineering must consider kinetic properties, stability, and compatibility with cell-free conditions. Often, enzymes require engineering to improve their performance in the nonphysiological environment of cell-free systems. Cofactor regeneration systems are essential for pathways requiring expensive cofactors such as NADH, NADPH, and ATP. Cell-free systems must include efficient regeneration mechanisms to maintain cofactor pools and ensure sustained pathway activity. Thermodynamic feasibility analysis ensures that each pathway step is thermodynamically favorable under the chosen reaction conditions, which may require the addition of coupling reactions or the removal of inhibitory products.

13.3.3 Multienzyme cascade reactions

Cell-free systems excel at implementing multienzyme cascade reactions where the product of one enzyme serves as the substrate for the next. These cascades can be designed to produce complex molecules that would be difficult or impossible to synthesize using traditional chemical methods.

Linear cascades represent the simplest pathway architecture, where substrates flow sequentially through a series of enzymatic steps. These are the most straightforward to implement and optimize in cell-free systems due to their predictable flux patterns. Branched pathways involve more complex architectures with branch points where flux can be directed toward different products. These require careful balancing of enzyme activities to control product distribution and minimize unwanted side products. Cyclic pathways involve regeneration of key intermediates or cofactors and are particularly challenging to implement but can provide significant advantages in terms of atom economy and cofactor utilization (Fig. 13.2).

13.3.4 Case studies in cell-free metabolic engineering

Several successful examples demonstrate the practical applications and achievements of cell-free metabolic engineering across different product categories and industrial applications.

Biofuel production represents one of the most extensively studied applications, with cell-free systems successfully producing various biofuels, including isobutanol, fatty acid ethyl esters, and hydrogen. These systems often achieve higher productivities than cellular systems due to the elimination of cellular maintenance requirements and the ability to optimize conditions specifically for product formation. Pharmaceutical intermediate synthesis has shown particular promise, with complex pharmaceutical molecules being synthesized using cell-free enzyme cascades, including the production of chiral intermediates that are difficult to synthesize chemically with high stereoselectivity. Natural product synthesis has enabled the reconstruction of biosynthetic pathways for valuable natural products, allowing production of these compounds without the need to culture the original producing organisms, many of which are unculturable or difficult to manipulate genetically.

Table 13.3 provides specific examples of successful cell-free metabolic pathways, demonstrating the range of products that can be synthesized and the performance metrics achieved in optimized systems.

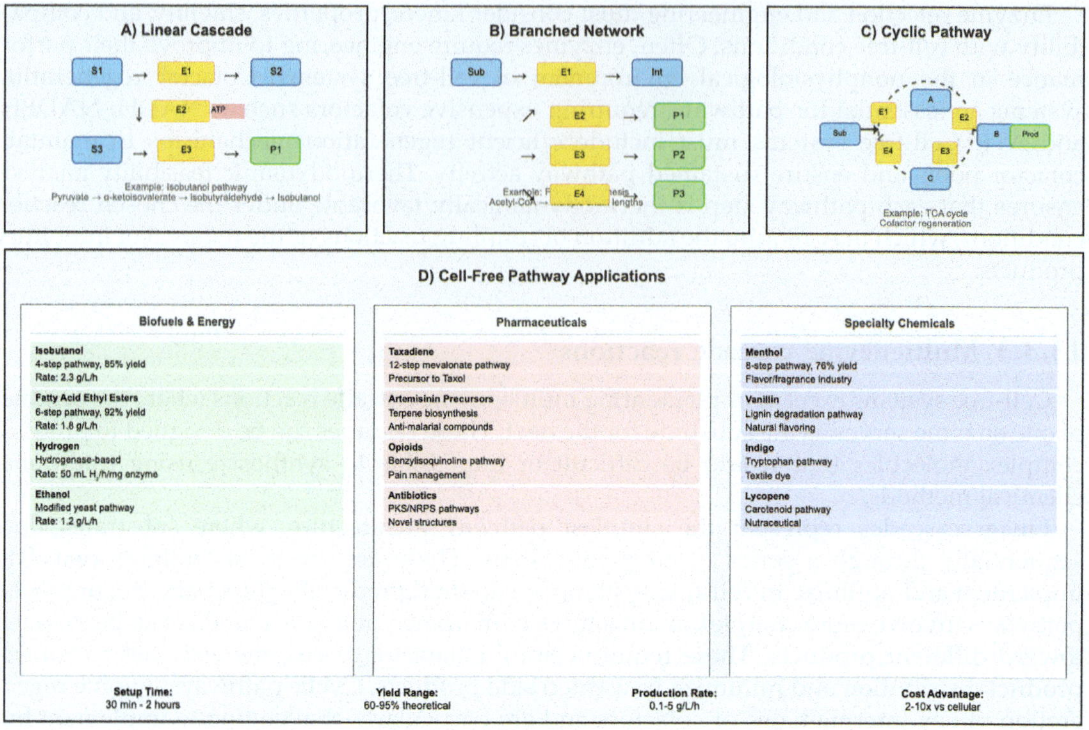

FIG. 13.2 Examples of different pathway topologies that have been successfully implemented in cell-free systems, showing detailed enzyme networks for the production of various target molecules, including pharmaceuticals, biofuels, and specialty chemicals, along with their respective flux distributions and product yields.

TABLE 13.3 Examples of cell-free metabolic pathways.

Target product	Pathway length	Key enzymes	Yield (% theoretical)	Production rate
Isobutanol	4 steps	KARI, DHAD, KIVD, ADH	85	2.3 g/L/h
Fatty acids	6 steps	ACC, FAS complex	92	1.8 g/L/h
Menthol	8 steps	Multiple P450s	76	0.4 g/L/h
Taxadiene	12 steps	MEV pathway, TXS	68	0.15 g/L/h

13.4 Artificial cell construction

13.4.1 Defining artificial cells

Artificial cells represent the ultimate goal of synthetic biology: the creation of synthetic life forms with precisely defined functions. These systems aim to capture the essential features of biological cells while incorporating only the minimum components necessary for specific functions. CFPS systems serve as the foundation for many artificial cell designs, providing the basic machinery for gene expression and protein production.

The construction of artificial cells involves several fundamental challenges that must be addressed to create functional synthetic life forms. Compartmentalization requires creating stable compartments that can maintain chemical gradients and concentrate reactants while allowing selective permeability for substrates and products. Self-replication mechanisms must implement information storage and replication systems that enable artificial cells to reproduce and evolve over time. Metabolic networks must convert simple substrates into the complex molecules required for cellular function and maintenance. Homeostatic mechanisms must maintain stable internal conditions despite environmental fluctuations and perturbations.

13.4.2 Compartmentalization strategies

Several distinct approaches have been developed for creating artificial cell compartments, each with specific advantages and limitations for different applications.

Liposomes utilize phospholipid vesicles that mimic natural cell membranes and can encapsulate CFPS systems. These provide excellent biocompatibility due to their similarity to natural membranes but offer limited stability and permeability control compared to synthetic alternatives. Polymer vesicles employ synthetic polymer-based vesicles that offer greater stability and tunable permeability compared to liposomes. These can be engineered with specific chemical functionalities to enhance performance for particular applications. Water-in-oil emulsions provide simple compartmentalization without the complexity of membrane assembly and are particularly useful for high-throughput screening applications where membrane integrity is less critical. Hydrogel microparticles consist of three-dimensional polymer networks that can encapsulate and protect cell-free systems while allowing controlled diffusion of small molecules through the polymer matrix.

13.4.3 Minimal gene sets for artificial cells

One of the key challenges in artificial cell construction is determining the minimal set of genes required for basic cellular function. Current research focuses on several essential systems that must be integrated to achieve autonomous cellular behavior.

Central dogma machinery, including DNA replication, transcription, and translation systems, forms the core of artificial cells. These systems must be carefully balanced to ensure stable gene expression without resource depletion or system collapse. Membrane synthesis systems for lipid biosynthesis and membrane assembly are essential for cell division and growth, enabling artificial cells to increase in size and eventually divide. These pathways are complex and require careful optimization in cell-free systems to achieve the coordinated synthesis of diverse membrane components. Energy generation systems, including simplified versions of glycolysis and oxidative phosphorylation, are necessary to power cellular processes and maintain the ATP pools required for biosynthetic reactions. Basic metabolism pathways for amino acid, nucleotide, and lipid synthesis provide the building blocks for cellular components and must be balanced to ensure adequate supply of all essential precursors (Fig. 13.3).

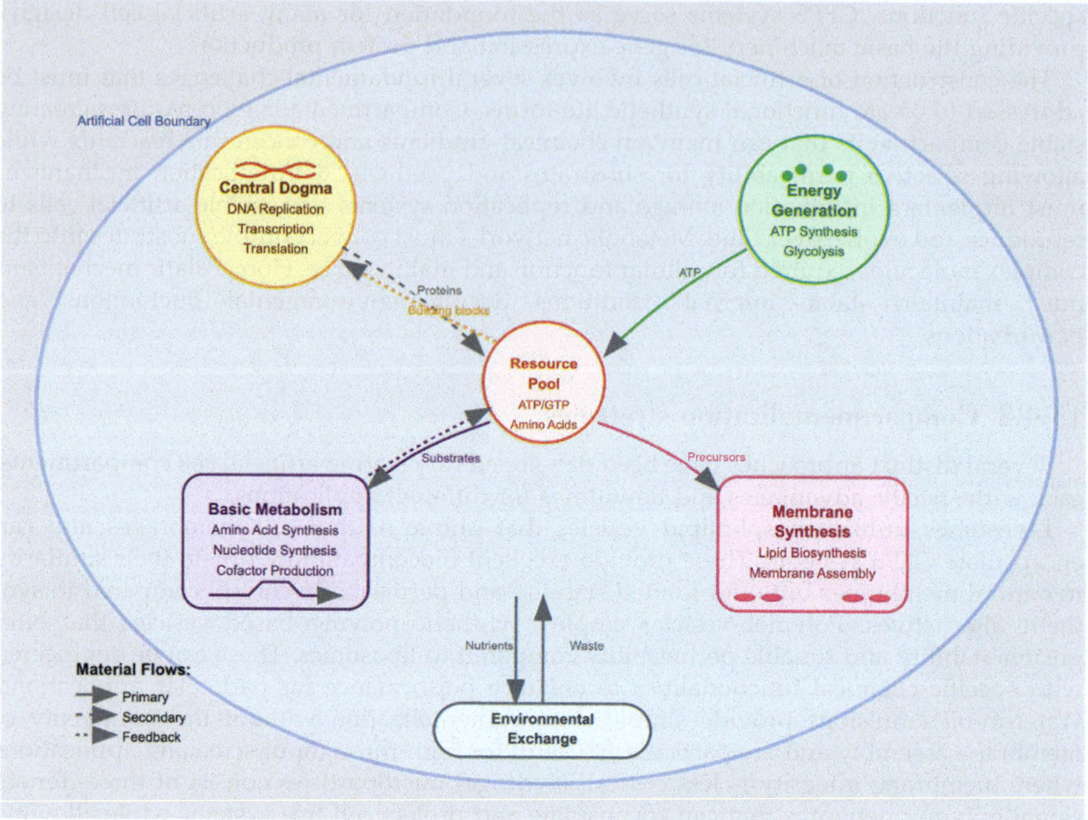

FIG. 13.3 A conceptual diagram of an integrated artificial cell architecture, highlighting the interconnections between different functional modules, including gene expression machinery, metabolic networks, and membrane synthesis systems, along with the material and energy flows that connect these subsystems.

13.4.4 Challenges and future directions

Several significant technical challenges remain in artificial cell construction that must be addressed to achieve practical applications.

System integration involves combining multiple complex systems, including transcription, translation, metabolism, and membrane synthesis, in a single compartment while maintaining balanced performance across all subsystems. This requires careful optimization of resource allocation and system interactions to prevent interference between different functional modules. Resource management focuses on efficiently utilizing limited resources such as amino acids, nucleotides, and energy without creating bottlenecks or accumulating waste products that could inhibit system function. Evolutionary stability requires designing systems that remain stable over multiple generations while retaining the capacity for beneficial mutations that could improve performance. Scalability involves developing methods for large-scale production of artificial cells for practical applications, including standardized protocols for manufacturing and quality control.

13.5 Biosynthetic gene cluster expression

13.5.1 Natural product biosynthesis in cell-free systems

Biosynthetic gene clusters (BGCs) encode the enzymatic machinery required for natural product biosynthesis in bacteria, fungi, and plants. These clusters often contain 10–50 genes organized in operons or regulons that must be coordinately expressed to produce the final natural product. Cell-free systems provide an ideal platform for expressing these complex gene clusters without the complications of cellular physiology and regulation.

Cell-free BGC expression offers several distinct advantages over traditional cellular approaches. Simplified regulation allows many BGCs that are subject to complex regulatory mechanisms in their native hosts to be bypassed in cell-free systems by directly adding DNA templates and control proteins. Pathway balancing enables precise control of individual enzyme activities by adjusting DNA template concentrations, facilitating optimization of flux through the entire biosynthetic pathway. Toxic product tolerance allows cell-free systems to continue producing even in the presence of compounds that would be toxic to living cells, thereby achieving higher product concentrations. Rapid screening capabilities enable different BGCs or pathway variants to be quickly tested without the time-consuming process of strain construction and cultivation required for cellular systems.

13.5.2 Polyketide synthase systems

Polyketide synthases (PKSs) represent one of the most important classes of natural product biosynthetic enzymes, responsible for producing numerous antibiotics, immunosuppressants, and other bioactive compounds. These large, multidomain enzymes present unique challenges for cell-free expression due to their size and complex posttranslational modifications.

Type I PKS systems consist of large, multimodular enzymes that function as assembly lines for PKS. Each module contains domains for substrate selection, chain extension, and modification reactions. Cell-free systems have been successfully used to express functional type I PKS systems, including those producing clinically important compounds such as erythromycin and rapamycin. Type II PKS systems comprise separate, dissociable proteins that work together to synthesize aromatic polyketides. The modular nature of type II systems makes them particularly suitable for cell-free expression and engineering applications. Type III PKS systems are simple, homodimeric enzymes that produce relatively simple polyketides. They are the easiest PKS systems to express in cell-free platforms and serve as excellent models for pathway optimization and engineering studies.

13.5.3 Nonribosomal peptide synthetases

Nonribosomal peptide synthetases (NRPSs) are responsible for producing a diverse array of peptide natural products, including many clinically important antibiotics and immunomodulators. Like PKS systems, NRPS enzymes are large, multidomain proteins that require careful optimization for cell-free expression.

TABLE 13.4 Comparison of BGC expression systems.

System type	Gene number	System complexity	Cell-free success rate	Key challenges
Type III PKS	3–5 genes	Low	>90%	Cofactor supply
Type II PKS	8–15 genes	Medium	70%–80%	Protein assembly
Type I PKS	5–10 genes	High	50%–70%	Large protein expression
NRPS	6–12 genes	High	60%–75%	Domain interactions
Hybrid PKS–NRPS	10–20 genes	Very high	40%–60%	System integration

NRPS systems present particular challenges due to their requirement for precise domain interactions and cofactor coordination. However, their modular architecture makes them amenable to cell-free expression when properly optimized. Success rates vary depending on the complexity of the NRPS system and the number of modules involved.

Table 13.4 compares the success rates and key challenges associated with expressing different types of BGCs in cell-free systems, providing guidance for selecting appropriate systems for different applications.

13.5.4 Terpene biosynthesis

Terpenes represent the largest class of natural products, with over 80,000 known structures spanning a remarkable diversity of chemical scaffolds and biological activities. These compounds are synthesized through well-characterized biosynthetic pathways that are highly amenable to cell-free expression due to their modular organization and relatively simple cofactor requirements.

Monoterpene pathways involve relatively simple biosynthetic routes using geranyl diphosphate synthase and monoterpene cyclases to produce 10-carbon terpenes. These pathways have been successfully implemented in cell-free systems with high yields and represent some of the most straightforward terpene biosynthetic systems to reconstruct. Sesquiterpene pathways produce 15-carbon terpenes through more complex biosynthetic routes that often involve multiple enzymatic steps and cyclization reactions. Cell-free systems have been used to produce a wide variety of sesquiterpenes, including those with pharmaceutical applications such as artemisinin precursors. Diterpene pathways represent some of the most complex terpene biosynthetic systems, requiring multiple cyclization steps and often involving cytochrome P450 enzymes for oxidative modifications. These pathways are among the most challenging to implement in cell-free systems but have shown promise for producing high-value pharmaceuticals and other bioactive compounds.

13.6 Natural product discovery

13.6.1 Cell-free platforms for discovery

CFPS systems have emerged as powerful platforms for natural product discovery, offering several significant advantages over traditional cell-based approaches. The ability to rapidly express and test biosynthetic pathways enables high-throughput screening of genetic diversity and accelerated discovery of novel bioactive compounds.

Metagenomic screening represents one of the most promising applications, where cell-free systems can directly express genes from environmental DNA samples without requiring cultivation of the source organisms. This approach has opened access to the vast majority of microbial diversity that cannot be cultured in laboratory conditions, potentially unlocking thousands of novel biosynthetic pathways. Pathway mining utilizes computational tools to identify potential BGCs in genomic databases, which can then be rapidly validated using cell-free expression systems. This approach allows researchers to systematically explore the biosynthetic potential encoded in sequenced genomes. Combinatorial biosynthesis enables mixing and matching of enzymes from different pathways to create novel biosynthetic routes and discover new chemical structures that may not exist in nature.

13.6.2 Directed evolution in cell-free systems

Cell-free systems provide excellent platforms for directed evolution of biosynthetic enzymes, enabling rapid optimization of enzyme properties for natural product production. Several complementary strategies have been developed to harness the power of evolution for enzyme improvement.

Error-prone PCR libraries generate random mutagenesis libraries that can be quickly screened in cell-free systems to identify improved enzyme variants with enhanced activity, stability, or altered substrate specificity. DNA shuffling employs recombination between homologous enzymes to create chimeric proteins with enhanced properties by combining beneficial mutations from different parent enzymes. Rational design approaches use structure-guided engineering to make targeted modifications that can be rapidly validated using cell-free expression and activity assays. Machine learning-guided evolution represents an emerging approach where artificial intelligence algorithms predict beneficial mutations based on protein structure and function relationships, with predictions then tested in cell-free systems for rapid validation.

13.6.3 Novel scaffold discovery

Cell-free systems are particularly valuable for discovering natural products with novel chemical scaffolds that may have unique biological activities and therapeutic potential. Traditional fermentation-based approaches often rediscover known compounds, while cell-free systems can access previously unexplored regions of chemical space.

Orphan gene clusters represent BGCs with no known products that are found in many sequenced organisms. Cell-free systems enable rapid expression and characterization of these orphan clusters, potentially revealing entirely new classes of natural products. Hybrid pathways combine elements from different biosynthetic pathways to create entirely new chemical scaffolds not found in nature, expanding the accessible chemical space beyond what evolution has already explored. Artificial substrate incorporation allows cell-free systems to incorporate unnatural substrates into biosynthetic pathways, creating novel chemical structures with potential therapeutic applications that combine natural biosynthetic logic with synthetic chemical diversity.

13.7 Synthetic organelle development

13.7.1 Concept and applications

Synthetic organelles represent compartmentalized systems within cells that perform specific functions while being physically separated from the rest of cellular metabolism. CFPS systems provide the foundation for developing these synthetic organelles by enabling the production of functional protein complexes that can be packaged into defined compartments.

The development of synthetic organelles offers several transformative applications across biotechnology and medicine. Metabolic compartmentalization allows for isolating specific metabolic pathways to prevent interference with cellular metabolism while concentrating enzymes and substrates for enhanced efficiency and yield. Biosensing and response applications involve creating organelles that can detect specific environmental conditions and respond with appropriate outputs, such as therapeutic protein production in response to disease markers. Cellular reprogramming uses synthetic organelles to modify cellular behavior or introduce entirely new functions that expand the capabilities of natural cells. Bioremediation applications focus on developing organelles capable of detoxifying harmful compounds or producing enzymes for environmental cleanup applications.

13.7.2 Design principles

Several key principles guide the design of functional synthetic organelles that can operate effectively within cellular environments.

Membrane architecture must provide appropriate permeability for substrates and products while maintaining separation from the cellular environment. This often requires incorporation of specific transport proteins or porins to enable controlled molecular exchange. Internal organization must efficiently arrange enzymes and cofactors to optimize pathway flux and minimize unwanted side reactions that could reduce efficiency or produce toxic intermediates. Regulatory integration ensures that synthetic organelles are properly connected with cellular regulatory networks to provide appropriate responses to changing environmental conditions. Inheritance and maintenance mechanisms are essential for long-term applications, as synthetic organelles must be stably inherited during cell division and maintained over multiple generations without loss of function.

13.7.3 Implementation strategies

Several distinct approaches have been developed for creating functional synthetic organelles, each with specific advantages for different applications.

Protein-based compartments utilize engineered protein shells, such as modified bacterial microcompartments or viral capsids, to encapsulate specific enzyme systems. These provide precise control over compartment size and permeability while offering good stability and predictable assembly properties. Lipid-based organelles employ synthetic membrane systems that can be introduced into cells to create new compartmentalized spaces for specific functions. These most closely mimic natural organelles and can incorporate membrane proteins for transport and signaling functions. DNA-based scaffolds use structured DNA assemblies to organize enzymes in specific spatial arrangements, creating functional organelles with precisely controlled enzyme positioning and stoichiometry. Hybrid systems combine different compartmentalization strategies to provide enhanced functionality and stability that leverages the advantages of multiple approaches (Fig. 13.4).

13.7.4 Current achievements and future prospects

Recent advances in synthetic organelle development have demonstrated the feasibility of creating functional compartmentalized systems within living cells.

Bacterial microcompartment engineering has achieved significant success, with modified bacterial microcompartments successfully engineered to encapsulate foreign metabolic pathways in both bacterial and eukaryotic cells. These systems have shown promise for applications ranging from biofuel production to pharmaceutical synthesis. Synthetic peroxisome

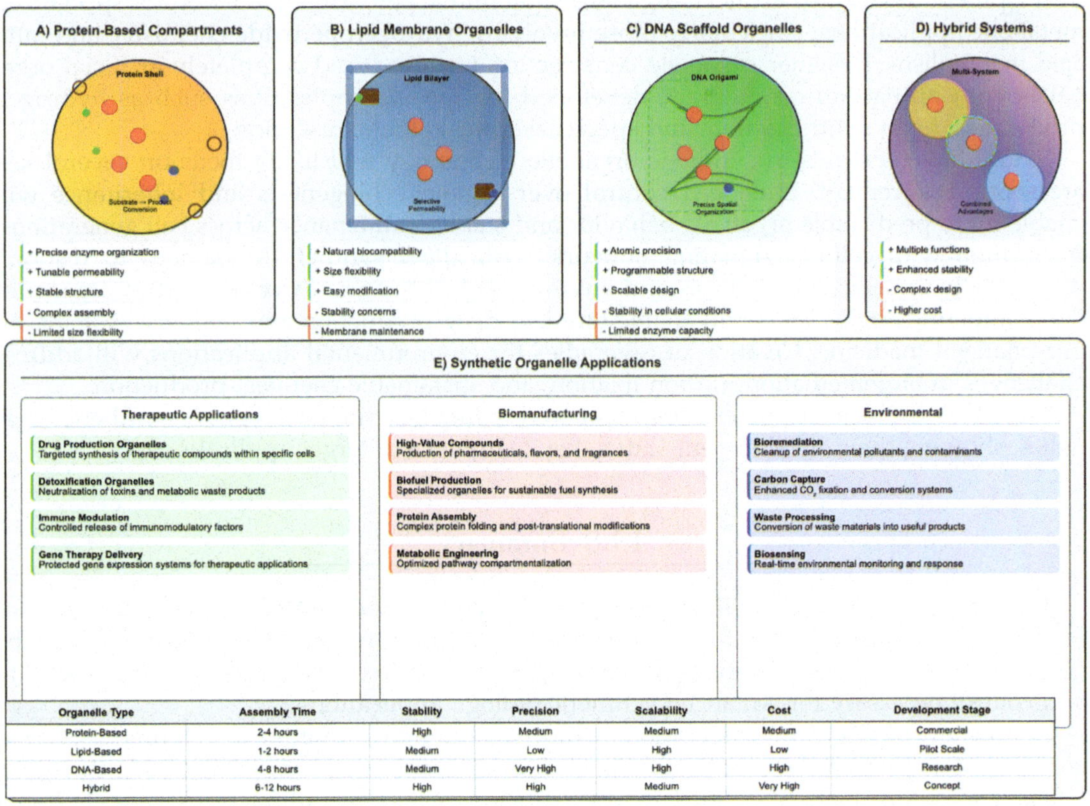

Organelle Type	Assembly Time	Stability	Precision	Scalability	Cost	Development Stage
Protein-Based	2-4 hours	High	Medium	Medium	Medium	Commercial
Lipid-Based	1-2 hours	Medium	Low	High	Low	Pilot Scale
DNA-Based	4-8 hours	Medium	Very High	High	High	Research
Hybrid	6-12 hours	High	High	Medium	Very High	Concept

FIG. 13.4 The different synthetic organelle architectures that have been developed, showing protein-based, lipid-based, and DNA-based compartmentalization systems with their respective structural features, advantages, and specific applications in cellular engineering.

TABLE 13.5 Synthetic organelle development pipeline.

Organelle type	Development stage	Key features	Target applications
Modified microcompartments	Commercial	Protein shell, enzyme organization	Biofuel production
Synthetic peroxisomes	Pilot scale	Lipid membrane, H_2O_2 chemistry	Detoxification
DNA organelles	Lab scale	DNA scaffold, precise organization	Biosensing
Artificial chloroplasts	Research	Light harvesting, CO_2 fixation	Photosynthesis enhancement
Therapeutic organelles	Concept	Drug production, targeted release	Personalized medicine

development has created artificial organelles that mimic peroxisome function for specialized metabolic applications, particularly those involving hydrogen peroxide chemistry and fatty acid metabolism. Designer organelle construction has produced completely artificial organelles with no natural counterpart, developed for specific applications such as hydrogen production, toxin neutralization, and specialized biosynthetic reactions.

Future developments in synthetic organelle technology will likely focus on several key areas of advancement. Enhanced control over organelle biogenesis and inheritance will enable more predictable organelle behavior and stable maintenance across cell generations. Integration with cellular signaling networks will allow synthetic organelles to respond dynamically to cellular conditions and environmental changes. Development of organelles for therapeutic applications will enable targeted drug production and delivery systems for personalized medicine. Creation of organelles for environmental applications will address challenges in bioremediation, carbon fixation, and sustainable chemical production.

Table 13.5 outlines the current development pipeline for different types of synthetic organelles, showing their development status, key features, and target applications.

13.8 Summary

The integration of CFPS with synthetic biology has created unprecedented opportunities for biological engineering and discovery. From rapid prototyping of genetic circuits to the construction of artificial cells and synthetic organelles, cell-free systems provide the flexibility and control necessary for advanced synthetic biology applications.

Key achievements in this integration include several transformative developments. Rapid circuit prototyping has been revolutionized through cell-free systems that enable the quick testing and optimization of genetic circuits, reducing development time from weeks to hours and dramatically accelerating the design–build–test–learn cycle. Metabolic pathway engineering has advanced significantly, with complete metabolic pathways now being reconstructed and optimized in cell-free systems with precise control over enzyme ratios and cofactor availability, enabling the production of complex molecules with high efficiency. Artificial cell development has made substantial progress toward minimal artificial cells, demonstrating the potential for creating synthetic life forms with defined functions and opening new possibilities for understanding the fundamental requirements of life. Natural product discovery has been accelerated through cell-free systems that enable rapid screening and production of novel bioactive compounds from diverse genetic resources, including previously inaccessible metagenomic DNA. Synthetic organelle creation has established the feasibility of developing compartmentalized systems within cells, opening new possibilities for cellular engineering and therapeutic applications that could transform medicine and biotechnology.

Future developments will likely focus on increasing the complexity and sophistication of these integrated systems, moving toward more autonomous and evolvable synthetic biological platforms. The continued convergence of cell-free technology with advances in synthetic biology, artificial intelligence, and nanotechnology promises to unlock even greater potential for biological engineering and biotechnological innovation. These developments will likely lead to new therapeutic approaches, sustainable manufacturing processes, and solutions to global challenges in energy, environment, and health.

Bibliography

Adamala, K.P., Martin-Alarcon, D.A., Guthrie-Honea, K.R., Boyden, E.S., 2017. Engineering genetic circuit interactions within and between synthetic minimal cells. Nat. Chem. 9 (5), 431–439.

Caschera, F., Noireaux, V., 2014. Integration of biological parts toward the synthesis of a minimal cell. Curr. Opin. Chem. Biol. 22, 85–91.

Garenne, D., Thompson, S., Brisson, A., Khakimzhan, A., Noireaux, V., 2021. The all-E. coli TXTL toolbox 3.0: new capabilities of a cell-free synthetic biology platform. Synth. Biol. 6 (1), 1–10.

Greco, F.V., Tarnawski, M., Jarazo, J., Hammes, U.Z., Schwille, P., 2021. Reconstitution of actin-based cellular processes in synthetic protocells. Nat. Protoc. 16 (5), 2167–2192.

Hodgman, C.E., Jewett, M.C., 2012. Cell-free synthetic biology: thinking outside the cell. Metab. Eng. 14 (3), 261–269.

Kightlinger, W., Lin, L., Rosztoczy, M., Li, W., DeLisa, M.P., Mrksich, M., Jewett, M.C., 2019. Design of glycosylation sites by rapid synthesis and analysis of glycosyltransferases. Nat. Chem. Biol. 15 (2), 161–170.

Martin, R.W., Majewska, N.I., Chen, C.X., Albanetti, T.E., Jimenez, R.B.C., Schmelzer, A.E., et al., 2018. Development of a CHO-based cell-free platform for synthesis of active monoclonal antibodies. ACS Synth. Biol. 7 (5), 1163–1175.

Moore, S.J., MacDonald, J.T., Wienecke, S., Ishwarbhai, A., Tsipa, A., Aw, R., et al., 2018. Rapid acquisition and model-based analysis of cell-free transcription–translation reactions from nonmodel bacteria. Proc. Natl. Acad. Sci. U.S.A. 115 (19), E4340–E4349.

Niederholtmeyer, H., Sun, Z.Z., Hori, Y., Yeung, E., Verpoorte, A., Murray, R.M., Maerkl, S.J., 2015. Rapid cell-free forward engineering of novel biocatalysts. eLife 4, e09771.

Perez, J.G., Stark, J.C., Jewett, M.C., 2016. Cell-free synthetic biology: engineering beyond the cell. Cold Spring Harb. Perspect. Biol. 8 (12), a023853.

Pourmir, A., Johannes, T.W., 2012. Directed evolution: selection of the host organism. Comput. Struct. Biotechnol. J. 2 (3), e201209012.

Schwander, T., Schada von Borzyskowski, L., Burgener, S., Cortina, N.S., Erb, T.J., 2016. A synthetic pathway for the fixation of carbon dioxide in vitro. Science 354 (6314), 900–904.

Silverman, A.D., Karim, A.S., Jewett, M.C., 2020. Cell-free gene expression: an expanded repertoire of applications. Nat. Rev. Genet. 21 (3), 151–170.

Swartz, J.R., 2012. Transforming biochemical engineering with cell-free biology. AIChE J. 58 (1), 5–13.

Villarreal, F., Contreras-Llano, L.E., Chavez, M., Ding, Y., Fan, J., Pan, T., Tan, C., 2018. Synthetic microbial consortia enable rapid assembly of pure translation machinery. Nat. Chem. Biol. 14 (1), 29–35.
Voyvodic, P.L., Pandi, A., Koch, M., Conejero, I., Valjent, E., Courtet, P., et al., 2019. Plug-and-play metabolic transducers expand the chemical detection space of cell-free biosensors. Nat. Commun. 10 (1), 1697.
Wu, K., Rao, C.V., Alper, H.S., 2013. Engineering a heterologous pathway in E. coli to produce the plant pigment anthocyanin. Metab. Eng. 15, 71–79.
Xu, H., Moraitis, M., Reedstrom, R.J., Matthews, K.S., 2018. Kinetic and thermodynamic studies of purine repressor binding to corepressor and operator DNA. J. Biol. Chem. 293 (18), 6707–6717.
Yue, K., Zhu, Y., Kai, L., Chen, Q., Guo, Y., Liu, Y., et al., 2019. Cell-free biosynthesis of phosphoribosyl pyrophosphate and its application in the production of riboflavin. Biotechnol. Bioeng. 116 (5), 1138–1144.
Zhang, L., Moretti, R., Pokkuluri, P.R., Schiffer, M., 2019. Directed evolution of a fucosyltransferase from a galactosyltransferase by DNA family shuffling. Science 345 (6198), 898–903.

PART V

Specialized applications

CHAPTER

14

Membrane protein synthesis

14.1 Introduction

Membrane proteins represent approximately 30% of all proteins encoded in the human genome and constitute the majority of current pharmaceutical targets. Despite their biological importance, membrane proteins remain notoriously difficult to produce in sufficient quantities for structural and functional studies. Cell-free protein synthesis (CFPS) has emerged as a transformative technology for membrane protein production, offering unique advantages over traditional cell-based expression systems. This chapter explores the specialized approaches, challenges, and solutions for synthesizing membrane proteins in cell-free environments.

The complexity of membrane protein synthesis stems from their inherent hydrophobic nature, requirement for proper membrane insertion, and need for specific lipid environments to maintain structural integrity. Cell-free systems provide an open environment that allows direct manipulation of reaction conditions, addition of specialized components, and real-time monitoring of protein synthesis and folding processes.

14.2 Challenges in membrane protein production

Membrane protein synthesis presents a unique set of challenges that distinguish it from soluble protein production. Understanding these challenges is crucial for developing effective cell-free strategies (Table 14.1).

The primary obstacle in membrane protein production is the tendency of hydrophobic transmembrane segments to aggregate when exposed to aqueous environments. In cellular systems, this is prevented by cotranslational insertion into the endoplasmic reticulum or bacterial membrane. Cell-free systems must recreate these protective mechanisms through artificial means.

Traditional expression systems often struggle with membrane protein toxicity, where the accumulation of misfolded membrane proteins can kill the host cells. This limitation is completely eliminated in cell-free systems, allowing for the production of proteins that would be impossible to express in living cells.

Cell Free Synthesis of Biologics
https://doi.org/10.1016/B978-0-443-33271-5.00024-7

TABLE 14.1 Major challenges in membrane protein production and cell-free solutions.

Challenge category	Specific issues	Impact on production	CFPS solutions
Hydrophobicity	Aggregation tendency	Low yields, precipitation	Detergent optimization, cotranslational insertion
Membrane insertion	Lack of natural insertion machinery	Misfolded proteins	Reconstituted translocons, artificial membranes
Lipid requirements	Specific lipid dependencies	Loss of function	Lipid supplementation, nanodisc systems
Structural complexity	Multipass transmembrane domains	Incorrect topology	Controlled folding environments
Toxicity	Cellular stress in vivo	Growth inhibition	Eliminated in cell-free systems

14.3 Cotranslational insertion strategies

Cotranslational insertion represents the most physiologically relevant approach to membrane protein synthesis in cell-free systems. This strategy mimics the natural process where membrane proteins are inserted into lipid bilayers during translation (Fig. 14.1).

14.3.1 Mechanism and requirements

Cotranslational insertion requires the coordination of several components: the translating ribosome, SRP, SRP receptor, and the protein-conducting channel (translocon). In cell-free systems, these components must be either retained from the source lysate or added as purified components (Table 14.2).

The efficiency of cotranslational insertion depends critically on the timing of membrane addition. Addition of membrane vesicles before translation initiation ensures optimal coupling between synthesis and insertion processes.

14.3.2 Optimization parameters

Several parameters must be optimized to achieve efficient cotranslational insertion. The most critical factors include membrane vesicle concentration, ionic strength, and temperature control.

Temperature management is particularly important as it affects both translation efficiency and membrane fluidity. Most cell-free systems operate at 30°C–37°C, which provides a balance between translation rate and membrane stability.

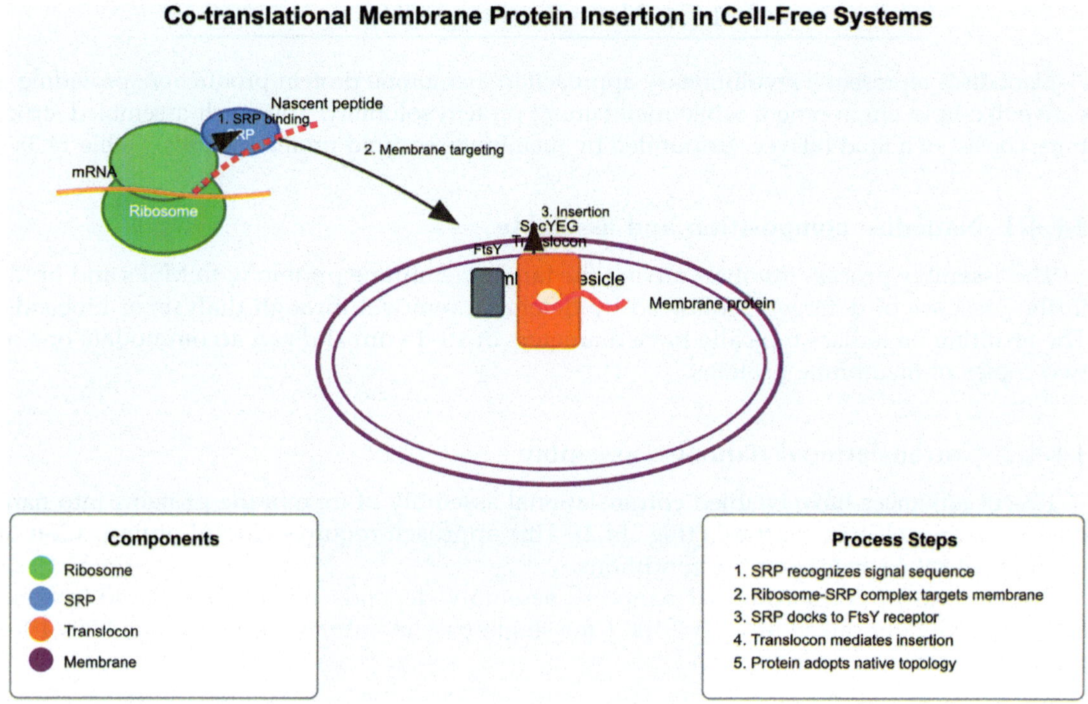

FIG. 14.1 Cotranslational membrane protein insertion mechanism. Schematic representation of the cotranslational insertion process in cell-free systems. The ribosome synthesizes the membrane protein while the signal recognition particle (SRP) recognizes the signal sequence and targets the ribosome-nascent chain complex to the membrane. The SRP receptor (FtsY) facilitates docking, and the SecYEG translocon mediates protein insertion into the lipid bilayer. Essential components and process steps are illustrated with their functional relationships.

TABLE 14.2 Essential components for cotranslational insertion in cell-free systems.

Component	Function	Source options	Concentration range
Signal recognition particle (SRP)	Recognition of signal sequences	*E. coli* lysate, purified SRP	50–200 nM
SRP receptor (FtsY)	SRP docking to membrane	Endogenous, supplemented	100–500 nM
Translocon (SecYEG)	Protein translocation channel	Membrane vesicles, reconstituted	10–50 μM
Ribosome	Protein synthesis	Lysate component	1–3 μM
Membrane vesicles	Insertion target	ER microsomes, bacterial membranes	0.1–1.0 mg/mL

14.4 Nanodisc integration

Nanodiscs represent a revolutionary approach to membrane protein production, providing a native-like lipid environment while maintaining protein solubility. These self-assembled structures consist of a lipid bilayer surrounded by membrane scaffold proteins (MSPs) (Table 14.3).

14.4.1 Nanodisc composition and assembly

The assembly process involves mixing the target membrane protein with MSPs and lipids in the presence of detergent, followed by detergent removal through dialysis or biobeads. The resulting nanodiscs typically have diameters of 10–15 nm and can accommodate one or two copies of membrane proteins.

14.4.2 Cotranslational nanodisc assembly

Recent advances have enabled cotranslational assembly of membrane proteins into nanodiscs during cell-free synthesis (Fig. 14.2). This approach requires careful optimization of component ratios and assembly conditions.

The success of cotranslational nanodisc assembly depends on the MSP/lipid/protein ratio, typically optimized around 2:100:1 for single-pass membrane proteins and 4:200:1 for multipass proteins.

14.5 Liposome-based systems

Liposomes provide an alternative membrane environment for cell-free membrane protein synthesis. These vesicular structures can be prepared with defined lipid compositions and sizes to match specific protein requirements (Table 14.4).

14.5.1 Liposome preparation methods

The choice of preparation method depends on the intended application and desired vesicle characteristics. For cell-free synthesis, small unilamellar vesicles prepared by extrusion are most commonly used due to their stability and high surface area.

TABLE 14.3 Standard nanodisc composition for membrane protein incorporation.

Component	Molecular weight (kDa)	Function	Typical concentration
MSP1D1	24.7	Membrane scaffold protein	10–100 μM
POPC	0.76	Primary lipid component	0.5–5 mM
POPG	0.75	Anionic lipid	0.1–1 mM
Cholesterol	0.39	Membrane stabilizer	0.1–0.5 mM
Sodium cholate	0.43	Assembly detergent	5–20 mM

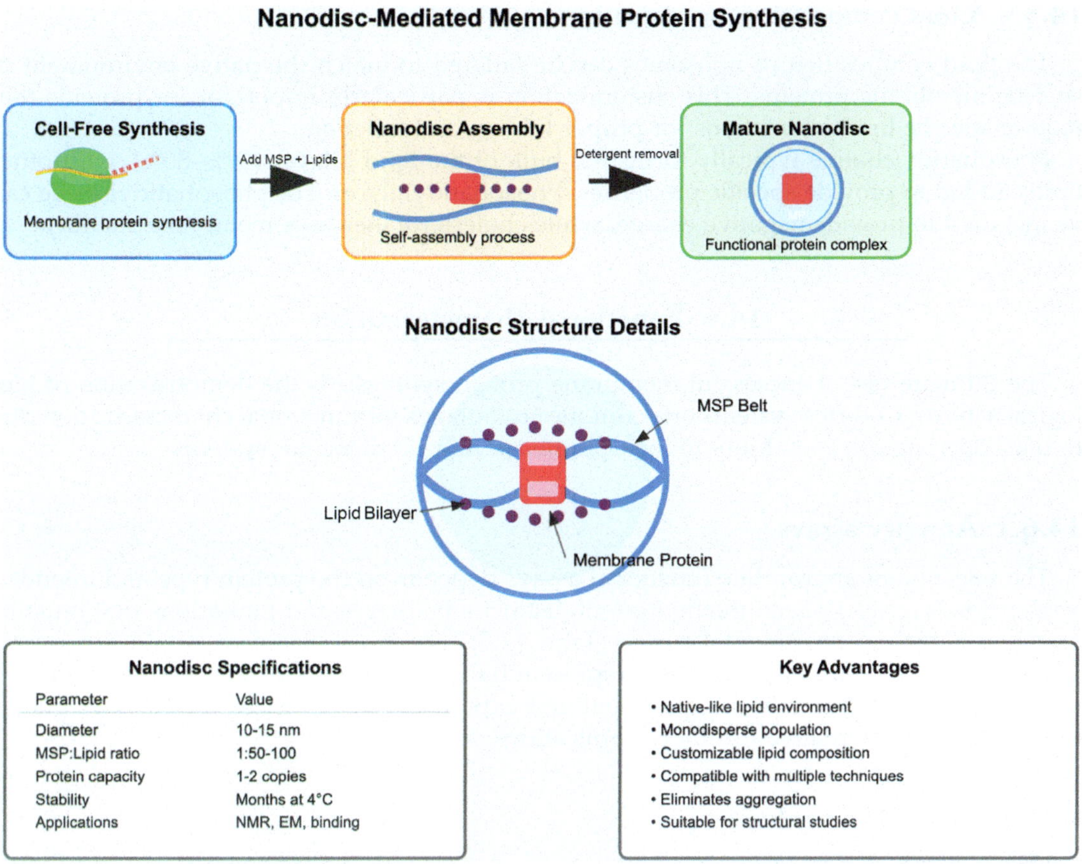

FIG. 14.2 **Nanodisc-mediated membrane protein synthesis.** Overview of the nanodisc assembly process for membrane protein incorporation. Cell-free synthesis produces the target membrane protein, which is then assembled with MSPs and lipids. Detergent removal drives spontaneous assembly into stable nanodiscs containing the functional membrane protein in a native-like lipid environment. Detailed structural features and specifications are provided.

TABLE 14.4 Liposome preparation methods and characteristics.

Method	Size range (nm)	Advantages	Limitations
Extrusion	50–1000	Uniform size distribution	Limited to specific sizes
Sonication	20–100	Small, unilamellar vesicles	Broad size distribution
Freeze–thaw	100–1000	High encapsulation efficiency	Variable size
Electroformation	1000–50,000	Giant vesicles for single protein studies	Low throughput

14.5.2 Lipid composition optimization

The lipid composition of liposomes can be tailored to match the native environment of specific membrane proteins. This customization is particularly important for proteins that require specific lipid interactions for proper folding and function.

Phosphatidylcholine typically forms the bulk of the lipid bilayer (60%–80%), with other lipids added to provide specific properties. Phosphatidylglycerol or phosphatidylserine can be included to provide negative charge, while cholesterol increases membrane stability.

14.6 Functional characterization

The ultimate test of successful membrane protein synthesis is the demonstration of biological activity. Cell-free systems offer unique advantages for functional characterization due to their open nature and ability to incorporate specialized detection systems.

14.6.1 Activity assays

The selection of appropriate functional assays depends on the protein type and intended application (Fig. 14.3). Each method offers distinct advantages and limitations that must be considered during experimental design (Table 14.5).

Electrophysiological measurements represent the gold standard for ion channel characterization but require specialized equipment and expertise. Fluorescence-based assays offer a more accessible alternative for many applications.

14.6.2 Structural analysis

Structural characterization of membrane proteins synthesized in cell-free systems can provide insights into folding quality and native-like structure formation.

Circular dichroism spectroscopy is commonly used to assess secondary structure content, while fluorescence spectroscopy can report on local environment changes and protein–lipid interactions. More advanced techniques such as solid-state nuclear magnetic resonance (NMR) and cryo-electron microscopy are increasingly being applied to cell-free synthesized membrane proteins.

14.7 Structural biology applications

Cell-free synthesis has become an invaluable tool for structural biology studies of membrane proteins. The ability to produce proteins with specific isotopic labeling and in defined membrane environments makes cell-free systems ideal for structural characterization (Table 14.6).

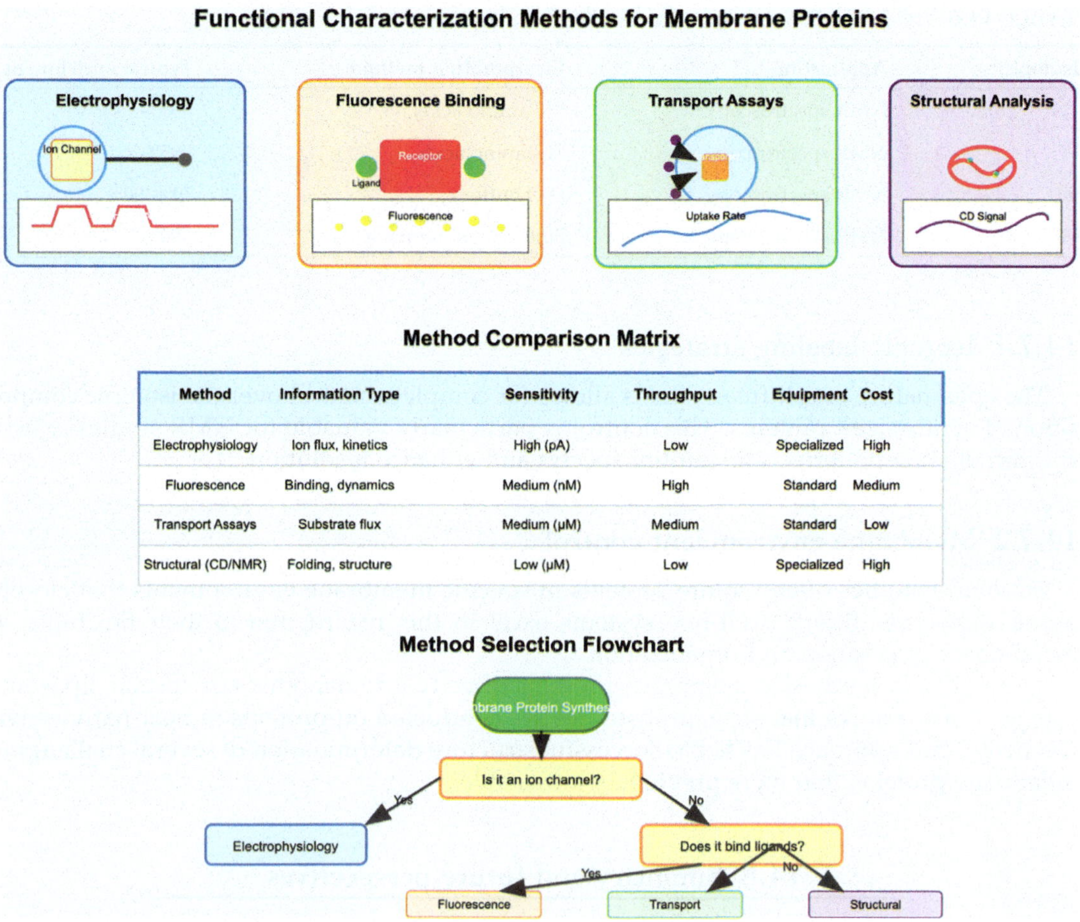

FIG. 14.3 Functional characterization methods for membrane proteins. Comprehensive comparison of major functional characterization techniques available for cell-free synthesized membrane proteins. Methods include electrophysiology for ion channels, fluorescence-based binding assays for receptors, transport assays for carriers, and structural analysis techniques. A comparison matrix and decision flowchart guide method selection based on protein type and experimental requirements.

TABLE 14.5 Functional characterization methods for membrane proteins.

Protein type	Assay method	Detection	Advantages
Ion channels	Electrophysiology	Current measurement	Direct functional readout
Transporters	Uptake assays	Fluorescent substrates	Real-time monitoring
Receptors	Ligand binding	Radioligands, fluorescence	Quantitative analysis
Enzymes	Enzymatic activity	Spectrophotometric	High throughput

TABLE 14.6 Isotopic labeling options for structural studies.

Isotope	Application	Incorporation method	Typical enrichment
^{15}N	NMR spectroscopy	^{15}N-amino acids	>95%
^{13}C	NMR spectroscopy	^{13}C-amino acids	>95%
^{2}H	NMR spectroscopy	D_2O buffer	70%–90%
^{19}F	NMR spectroscopy	Fluorinated amino acids	>90%

14.7.1 Isotopic labeling strategies

The open nature of cell-free systems allows for complete control over the isotopic composition of synthesized proteins. This control is particularly valuable for NMR studies, where specific labeling patterns can simplify spectra and enhance resolution.

14.7.2 Membrane environment control

Structural studies often require proteins in specific membrane environments that closely mimic native conditions. Cell-free systems excel in this regard due to their flexibility in membrane composition and organization.

The ability to synthesize membrane proteins directly into nanodiscs or specific liposome compositions ensures that structural studies are conducted on proteins in near-native environments. This approach has led to successful structure determination of several challenging membrane proteins that were previously intractable.

14.8 Summary and future perspectives

Cell-free synthesis has revolutionized membrane protein production by addressing many of the fundamental challenges that have limited progress in this field. The technology offers unprecedented control over synthesis conditions, membrane environment, and protein characterization.

Current limitations include the relatively high cost of cell-free systems and the need for optimization of each target protein. However, ongoing developments in system engineering, automation, and cost reduction are rapidly addressing these concerns.

Future directions include the development of more sophisticated membrane environments that better mimic cellular membranes, integration with structural characterization techniques for real-time structure determination, and scale-up technologies for industrial applications.

The combination of cell-free synthesis with emerging technologies such as cryo-electron microscopy and artificial intelligence-guided optimization promises to further accelerate membrane protein research and drug discovery efforts.

Bibliography

Cappuccio, J.A., Blanchette, C.D., Sulchek, T.A., Arroyo, E.S., Kralj, J.M., Hinz, A.K., et al., 2008. Cell-free co-expression of functional membrane proteins and apolipoprotein, forming soluble nanolipoprotein particles. Mol. Cell. Proteomics 7 (11), 2246–2253.

Henrich, E., Peetz, O., Hein, C., Laguerre, A., Hoffmann, B., Hoffmann, J., et al., 2017. Analyzing native membrane protein assembly in nanodiscs by combined non-covalent mass spectrometry and synthetic biology. eLife 6, e20954.

Ishihara, G., Goto, M., Saeki, M., Ito, K., Hori, T., Kigawa, T., et al., 2005. Expression of G protein coupled receptors in a cell-free translational system using detergents and thioredoxin-fusion vectors. Protein Expr. Purif. 41 (1), 27–37.

Junge, F., Schneider, B., Reckel, S., Schwarz, D., Dötsch, V., Bernhard, F., 2008. Large-scale production of functional membrane proteins. Cell. Mol. Life Sci. 65 (11), 1729–1755.

Katzen, F., Fletcher, J.E., Yang, J.P., Kang, D., Peterson, T.C., Cappuccio, J.A., et al., 2008. Insertion of membrane proteins into discoidal membranes using a cell-free protein expression approach. J. Proteome Res. 7 (8), 3535–3542.

Klammt, C., Schwarz, D., Fendler, K., Haase, W., Dötsch, V., Bernhard, F., 2005. Cell-free expression as an emerging technique for the large scale production of integral membrane protein. FEBS J. 272 (23), 6024–6038.

Lyukmanova, E.N., Shenkarev, Z.O., Paramonov, A.S., Sobol, A.G., Ovchinnikova, T.V., Chupin, V.V., et al., 2012. Lipid–protein nanodiscs for cell-free production of integral membrane proteins in a soluble and folded state: comparison with detergent micelles, liposomes and lipid cubic phase. Biochim. Biophys. Acta 1818 (3), 349–358.

Proverbio, D., Roos, C., Beyermann, M., Orban, E., Dötsch, V., Bernhard, F., 2013. Functional properties of cell-free expressed human endothelin A and endothelin B receptors in artificial membrane environments. Biochim. Biophys. Acta 1828 (11), 2182–2192.

Sachse, R., Dondapati, S.K., Fenz, S.F., Schmidt, T., Kubick, S., 2014. Membrane protein synthesis in cell-free systems: from bio-mimetic systems to bio-membranes. FEBS Lett. 588 (17), 2774–2781.

Schwarz, D., Dötsch, V., Bernhard, F., 2008. Production of membrane proteins using cell-free expression systems. Proteomics 8 (19), 3933–3946.

CHAPTER

15

Complex biomolecule production

15.1 Introduction

The production of complex biomolecules represents one of the most challenging frontiers in biotechnology and pharmaceutical development. Unlike simple proteins, complex biomolecules encompass multiprotein assemblies, virus-like particles (VLPs), nucleic acid therapeutics, hybrid biomaterials, glycoconjugate vaccines, and antibody–drug conjugates (ADCs). These sophisticated molecular constructs require precise stoichiometric assembly and specific posttranslational modifications and often involve multiple biosynthetic pathways operating in concert.

Cell-free protein synthesis systems have emerged as uniquely suited platforms for complex biomolecule production due to their open architecture, which allows for real-time manipulation of reaction conditions, addition of specialized enzymes and cofactors, and direct monitoring of assembly processes. The elimination of cellular constraints enables the production of molecules that would be toxic or unstable in living systems, while the controlled environment facilitates the optimization of complex biosynthetic pathways.

This chapter explores the specialized strategies, technologies, and applications of cell-free systems for producing increasingly sophisticated biomolecular assemblies that are driving advances in therapeutics, vaccines, and biotechnology applications.

15.2 Multiprotein complexes

Multiprotein complexes represent fundamental functional units in biological systems, often requiring precise stoichiometric ratios and specific assembly pathways to achieve proper function. Cell-free systems offer unprecedented control over these assembly processes.

15.2.1 Assembly strategies and stoichiometric control

The successful production of multiprotein complexes in cell-free systems requires careful consideration of expression levels, assembly kinetics, and quality control mechanisms (Fig. 15.1).

Cell Free Synthesis of Biologics
https://doi.org/10.1016/B978-0-443-33271-5.00011-9

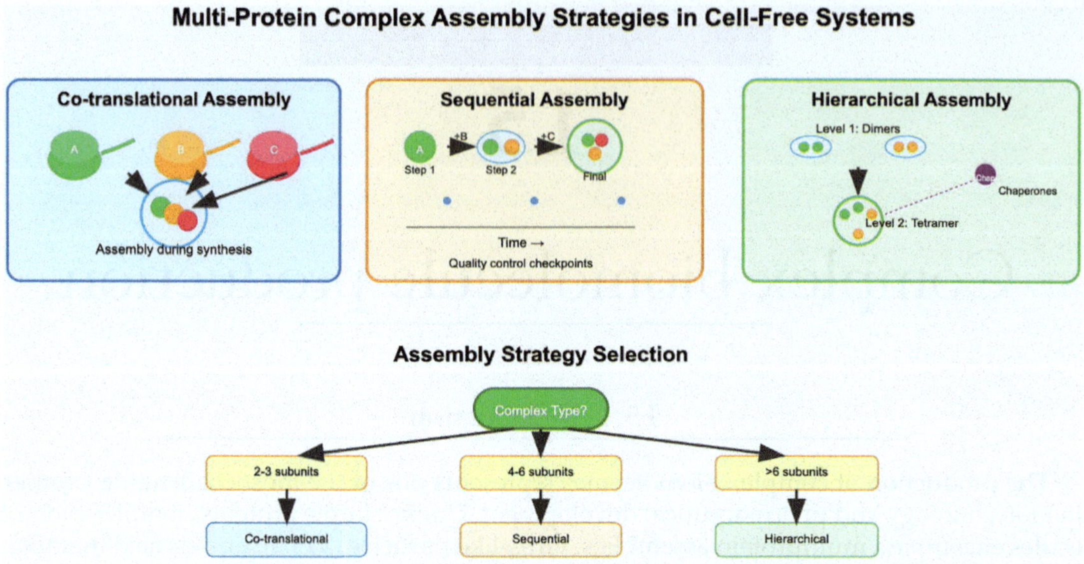

Critical Parameters for Each Strategy

Strategy	Template Ratio	Timing Control	Chaperones	Temperature	Yield
Co-translational	Optimized	Simultaneous	Optional	30-37°C	60-85%
Sequential	Balanced	Staged	Required	25-30°C	40-70%
Hierarchical	Complex	Multi-stage	Essential	20-25°C	15-45%
Quality Control	Real-time	Checkpoints	Monitoring	Controlled	Variable

FIG. 15.1 **Multiprotein complex assembly strategies.** Comprehensive overview of assembly strategies for multiprotein complexes in cell-free systems. The diagram illustrates cotranslational assembly, sequential synthesis approaches, and hierarchical assembly pathways. Key parameters, including template ratios, chaperone requirements, and assembly kinetics, are highlighted. The flowchart guides selection of optimal assembly strategies based on complex size and subunit characteristics.

The key to successful multiprotein complex production lies in controlling the relative expression levels of individual subunits. This is typically achieved through template ratio optimization, where the DNA concentrations for each subunit are carefully balanced to achieve the desired stoichiometry (Table 15.1).

Cotranslational assembly represents the most efficient approach for many complexes, where protein subunits associate during synthesis rather than after completion. This strategy minimizes the accumulation of unassembled subunits and reduces the potential for aggregation.

15.2.2 Chaperone-assisted assembly

Many complex biomolecules require molecular chaperones for proper folding and assembly. Cell-free systems can be supplemented with specific chaperones to enhance assembly efficiency and product quality (Table 15.2).

TABLE 15.1 Assembly strategies for multiprotein complexes in cell-free systems.

Complex type	Subunit number	Assembly strategy	Critical parameters	Typical yield
Homodimers	2 identical	Single template	Protein concentration, ionic strength	80%–95%
Heterodimers	2 different	Dual template	Template ratio, timing	60%–85%
Trimeric complexes	3 subunits	Sequential assembly	Chaperone addition, temperature	40%–70%
Tetrameric complexes	4 subunits	Hierarchical assembly	Cofactor availability, pH control	30%–60%
Large assemblies (>4)	Multiple	Modular approach	Staged synthesis, purification	15%–45%

TABLE 15.2 Chaperone systems for complex biomolecule assembly.

Chaperone System	Target complexes	Mechanism	Concentration range	Enhancement factor
GroEL/GroES	Large globular proteins	Folding chamber	1–5 μM	3–10×
DnaK/DnaJ/GrpE	Aggregation-prone proteins	Cotranslational folding	2–8 μM	2–5×
Trigger factor	Nascent polypeptides	Ribosome-associated	5–15 μM	1.5–3×
BiP/ERdj4	ER-targeted proteins	ER-specific folding	1–10 μM	2–8×
Hsp90	Signaling proteins	Conformational maturation	0.5–3 μM	2–6×

The selection of appropriate chaperone systems depends on the specific folding requirements of the target complex. GroEL/GroES is particularly effective for large, single-domain proteins, while the DnaK system excels at preventing aggregation during synthesis.

15.3 Virus-like particles

VLPs represent sophisticated self-assembling structures that mimic native viral capsids without containing infectious genetic material. These particles have found widespread applications in vaccine development, drug delivery, and nanotechnology.

15.3.1 VLP assembly mechanisms

The assembly of VLPs involves complex hierarchical processes that must be carefully controlled to achieve proper particle formation and size distribution (Fig. 15.2).

Cell-free systems offer unique advantages for VLP production, including the ability to control assembly conditions in real time, add specific cofactors required for assembly, and eliminate cellular toxicity associated with some viral proteins (Table 15.3).

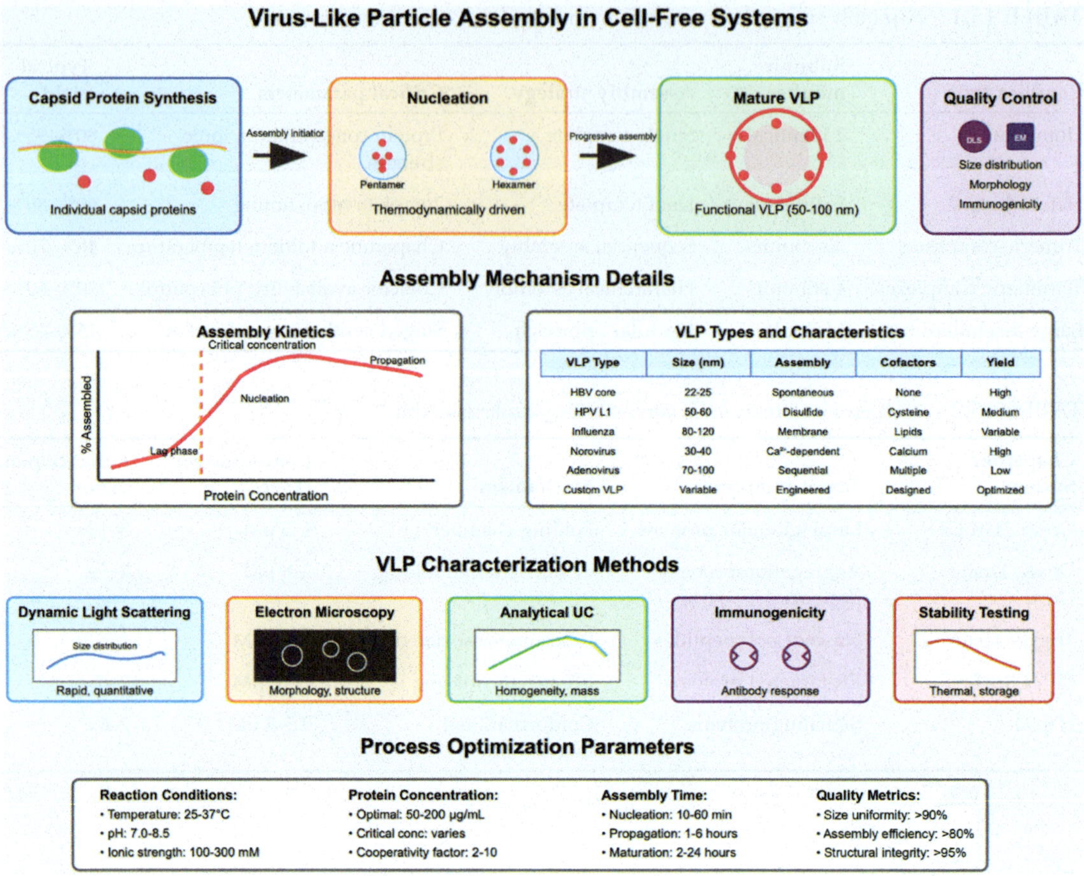

FIG. 15.2 Virus-like particle assembly process. Detailed representation of VLP assembly mechanisms in cell-free systems. The process shows capsid protein synthesis, nucleation events, and progressive assembly into complete particles. Quality control checkpoints and characterization methods are integrated throughout the process. Multiple VLP types are compared showing size distributions and assembly requirements.

TABLE 15.3 Major VLP types and their assembly characteristics.

VLP type	Structural proteins	Assembly pathway	Particle size (nm)	Applications
Hepatitis B VLPs	HBsAg core protein	Spontaneous assembly	22–25	Vaccines, drug delivery
HPV VLPs	L1 major capsid	Disulfide-mediated	50–60	Cancer vaccines
Influenza VLPs	HA, NA, M1 proteins	Membrane-associated	80–120	Pandemic vaccines
Norovirus VLPs	VP1 capsid protein	Calcium-dependent	30–40	Gastroenteritis vaccines
Adenovirus VLPs	Hexon, penton proteins	Sequential assembly	70–100	Gene therapy vectors

15.3.2 Quality control and characterization

VLP production requires rigorous quality control to ensure proper assembly, size distribution, and immunogenic properties. Cell-free systems facilitate real-time monitoring of assembly processes.

The assessment of VLP quality involves multiple analytical techniques, including dynamic light scattering for size distribution, electron microscopy for morphological characterization, and analytical ultracentrifugation for homogeneity analysis.

15.4 Nucleic acid therapeutics

The production of nucleic acid therapeutics, including mRNA vaccines, siRNA, and antisense oligonucleotides, represents a rapidly growing application of cell-free systems. These systems offer advantages in terms of purity, scalability, and freedom from cellular contaminants.

15.4.1 mRNA production and modification

Cell-free transcription systems have become essential for producing therapeutic mRNAs with precise modifications required for stability and efficacy (Table 15.4).

The incorporation of modified nucleotides, such as pseudouridine and 5-methylcytidine, is crucial for reducing immunogenicity and enhancing translation efficiency. Cell-free systems allow precise control over the incorporation of these modifications.

15.4.2 RNA–protein complexes

Many nucleic acid therapeutics require association with specific proteins for delivery and function. Cell-free systems can produce these ribonucleoprotein (RNP) complexes in a single reaction.

TABLE 15.4 mRNA therapeutic production in cell-free systems.

mRNA type	Modifications required	Production system	Yield range	Stability
Vaccine mRNA	5′-cap, poly(A) tail, pseudouridine	T7/SP6 systems	50–200 μg/mL	Days to weeks
Therapeutic mRNA	Modified nucleotides, UTR optimization	Coupled systems	20–100 μg/mL	Hours to days
Self-amplifying RNA	Replicase genes, enhancer sequences	Specialized systems	10–50 μg/mL	Extended
Circular RNA	Ribozyme-mediated circularization	Custom systems	5–30 μg/mL	Highly stable

The production of CRISPR–Cas RNPs exemplifies this capability, where the Cas protein and guide RNA are simultaneously produced and assembled into functional complexes. This approach ensures proper stoichiometry and reduces the risk of nuclease degradation.

15.5 Hybrid biomaterials

Hybrid biomaterials combine biological and synthetic components to create novel materials with enhanced properties. Cell-free systems enable the production of protein-based scaffolds that can be subsequently modified with synthetic components.

15.5.1 Protein–polymer conjugates

The synthesis of protein–polymer conjugates requires precise control over the conjugation process to maintain protein activity while imparting desired material properties (Table 15.5).

Cell-free systems offer advantages for hybrid biomaterial production by allowing the incorporation of noncanonical amino acids that serve as conjugation sites, enabling site-specific modification without affecting protein function.

15.5.2 Self-assembling systems

Many hybrid biomaterials rely on self-assembly processes to achieve their final structure and properties. Cell-free systems can be designed to support these assembly processes through the controlled addition of assembly factors.

The production of peptide amphiphiles represents a successful example, where hydrophobic and hydrophilic regions drive spontaneous assembly into nanofibers with controlled dimensions and properties.

TABLE 15.5 Hybrid biomaterial types and production methods.

Conjugate type	Protein component	Polymer component	Conjugation method	Applications
PEG–proteins	Therapeutic enzymes	Polyethylene glycol	Chemical coupling	Drug delivery
Elastin-like polypeptides	Designed sequences	Synthetic repeats	Genetic fusion	Biomaterials
Silk-protein fusions	Spider silk proteins	Fibroin domains	Recombinant expression	Tissue engineering
Collagen–synthetic	Collagen domains	Synthetic peptides	Hybrid design	Wound healing
Hydrogel matrices	Cross-linking proteins	Polymer networks	Enzymatic coupling	3D culture systems

15.6 Glycoconjugate vaccines

Glycoconjugate vaccines represent a critical class of immunotherapeutics that combine carbohydrate antigens with protein carriers to enhance immunogenicity. Cell-free systems offer unique advantages for producing these complex conjugates.

15.6.1 Glycosylation engineering

The production of properly glycosylated proteins in cell-free systems requires the addition of glycosylation machinery and appropriate sugar donors (Fig. 15.3).

The reconstitution of glycosylation pathways in cell-free systems requires careful optimization of enzyme concentrations, cofactor availability, and reaction conditions. The open nature of these systems allows for real-time monitoring and adjustment of glycosylation efficiency (Table 15.6).

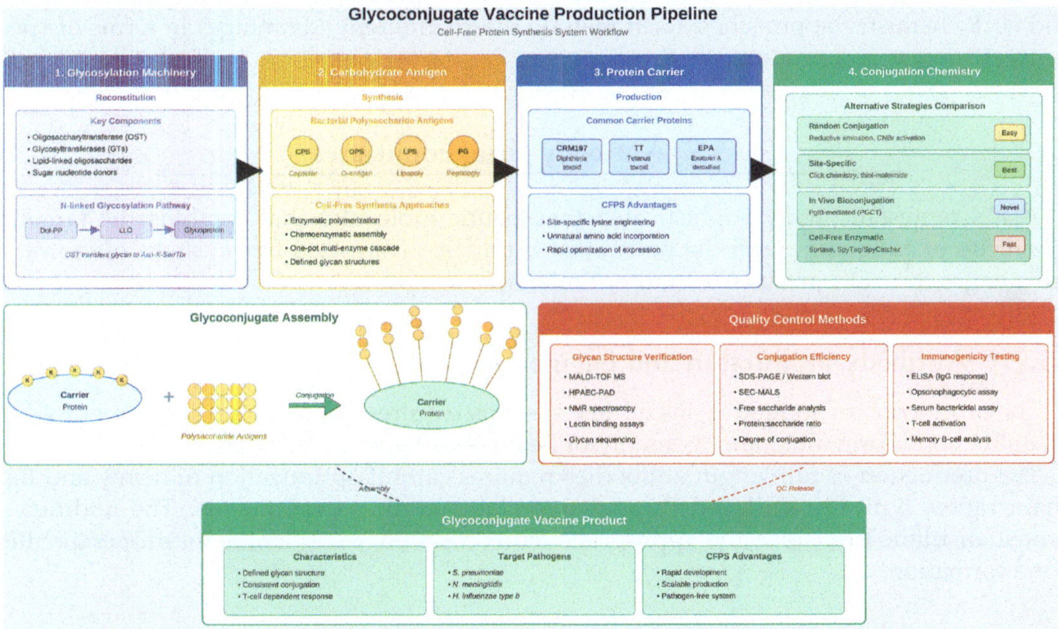

Strategy	Chemistry	Specificity	Reproducibility	Scalability	Cost	CFPS Compatible
Random	Reductive amination	Low	Medium	High	Low	Yes
Site-Specific	Click/Thiol-maleimide	High	High	Medium	Medium	Yes
Bioconjugation	PglB enzymatic	High	High	Medium	High	Emerging
SpyTag/Catcher	Isopeptide bond	High	High	High	Medium	Ideal

FIG. 15.3 Glycoconjugate vaccine production pipeline. Complete workflow for producing glycoconjugate vaccines using cell-free systems. The process encompasses glycosylation machinery reconstitution, carbohydrate antigen synthesis, protein carrier production, and conjugation chemistry. Quality control methods for verifying glycan structure and conjugation efficiency are included. Alternative conjugation strategies and their relative advantages are compared.

TABLE 15.6 Glycosylation systems for cell-free vaccine production.

Glycosylation type	Required enzymes	Sugar donors	Target proteins	Efficiency
N-linked	OST complex, ALG proteins	Dolichol-PP-$GlcNAc_2Man_9Glc_3$	Secreted proteins	40%–80%
O-linked	GalNAc transferases	UDP-GalNAc	Mucin-type proteins	30%–70%
Bacterial O-antigen	Wzy/Wzx systems	Sugar nucleotides	Lipopolysaccharide (LPS) core proteins	50%–90%
Fungal β-glucan	β-glucan synthases	UDP-glucose	Cell wall proteins	60%–85%
Custom glycans	Engineered enzymes	Synthetic donors	Designer proteins	20%–60%

15.6.2 Conjugation strategies

The coupling of carbohydrate antigens to protein carriers must preserve both antigen structure and carrier immunogenicity while creating stable linkages.

Chemical conjugation methods include reductive amination, thiol–maleimide coupling, and click chemistry approaches. Each method offers different advantages in terms of specificity, efficiency, and product stability.

15.7 Antibody–drug conjugates

ADCs represent highly sophisticated therapeutic molecules that combine the targeting specificity of antibodies with the potency of cytotoxic drugs. Cell-free systems offer unique advantages for ADC production and optimization.

15.7.1 Antibody production and engineering

The production of antibodies in cell-free systems requires optimization for proper folding, disulfide bond formation, and assembly of heavy and light chains (Table 15.7).

The production of full-length antibodies requires careful optimization of heavy and light chain ratios, typically achieved through template titration experiments. The addition of protein disulfide isomerase and appropriate redox conditions is essential for proper disulfide bond formation.

15.7.2 Site-specific conjugation

Site-specific drug conjugation is crucial for maintaining antibody function while ensuring controlled drug loading and release kinetics.

The incorporation of noncanonical amino acids provides reactive handles for site-specific conjugation without affecting antigen-binding sites. Common approaches include the use of azide-containing amino acids for click chemistry or p-acetylphenylalanine for oxime ligation.

TABLE 15.7 Antibody formats and cell-free production parameters.

Antibody format	Chain configuration	Assembly complexity	Typical yield	Functional activity
scFv	Single chain	Low	100–500 μg/mL	60%–90%
Fab fragments	Heavy + light chains	Medium	50–200 μg/mL	70%–95%
Full antibodies	2 H + 2 L chains	High	20–100 μg/mL	80%–95%
Bispecific antibodies	Multiple formats	Very high	10–50 μg/mL	40%–80%
Nanobodies	Single domain	Low	200–800 μg/mL	85%–95%

15.8 Summary and future perspectives

Cell-free systems have demonstrated remarkable capabilities for producing complex biomolecules that would be challenging or impossible to create in traditional cellular systems. The technology offers unprecedented control over assembly processes, posttranslational modifications, and conjugation reactions.

Current limitations include the relatively high cost of cell-free systems and the need for extensive optimization for each target molecule. However, ongoing developments in system engineering, enzyme supplementation, and automation are rapidly addressing these challenges.

Future directions include the development of more sophisticated glycosylation systems that can produce complex, human-like glycan structures, the integration of multiple biosynthetic pathways for producing even more complex molecules, and the scaling of production processes for commercial applications.

The combination of cell-free synthesis with emerging technologies such as directed evolution, computational design, and automated optimization promises to further expand the range of complex biomolecules that can be efficiently produced. These advances will likely drive new therapeutic modalities and biotechnology applications that are currently beyond reach.

Bibliography

Bundy, B.C., Swartz, J.R., 2010. Site-specific incorporation of p-propargyloxyphenylalanine in a cell-free environment for direct protein-protein click conjugation. Bioconjug. Chem. 21 (2), 255–263.

Gagoski, D., Polinkovsky, M.E., Mureev, S., Kunert, A., Johnston, W., Gambin, Y., Alexandrov, K., 2016. Performance benchmarking of four cell-free protein expression systems. Biotechnol. Bioeng. 113 (2), 292–300.

Kightlinger, W., Lin, L., Rosztoczy, M., Li, W., DeLisa, M.P., Mrksich, M., Jewett, M.C., 2018. Design of glycosylation sites by rapid synthesis and analysis of glycosyltransferases. Nat. Chem. Biol. 14 (6), 627–635.

Martin, R.W., Majewska, N.I., Chen, C.X., Albanetti, T.E., Jimenez, R.B., Schmelzer, A.E., et al., 2017. Cell-free protein synthesis from genomically recoded bacteria enables multisite incorporation of noncanonical amino acids. Nat. Commun. 8 (1), 1548.

Perez, J.G., Stark, J.C., Jewett, M.C., 2016. Cell-free synthetic biology: engineering beyond the cell. Cold Spring Harb. Perspect. Biol. 8 (12), a023853.

Ruben, S., Garau, G., Smolarczyk, R., Gilar, M., Jarosz, M., Cielecka-Piontek, J., 2021. Cell-free protein synthesis for virus-like particle production: a promising approach for vaccine development. Vaccines 9 (10), 1138.

Silverman, A.D., Karim, A.S., Jewett, M.C., 2020. Cell-free gene expression: an expanded repertoire of applications. Nat. Rev. Genet. 21 (3), 151–170.

Thoring, L., Dondapati, S.K., Stech, M., Wüstenhagen, D.A., Kubick, S., 2017. High-yield production of "difficult-to-express" proteins in a continuous exchange cell-free system based on HeLa cells. Sci. Rep. 7 (1), 11710.

Welsh, J.P., Lu, Y., He, X.S., Greenberg, H.B., Swartz, J.R., 2012. Cell-free production of trimeric influenza hemagglutinin ectodomain. Biotechnol. Bioeng. 109 (9), 2962–2969.

Zawada, J.F., Yin, G., Steiner, A.R., Yang, J., Naresh, A., Roy, S.M., et al., 2011. Microscale to manufacturing scale-up of cell-free cytokine production—a new approach for shortening protein production development timelines. Biotechnol. Bioeng. 108 (7), 1570–1578.

CHAPTER

16

Emerging molecular targets

The landscape of molecular targets accessible through cell-free protein synthesis continues to expand rapidly, driven by advances in system engineering and our understanding of complex biomolecular structures. This chapter explores six critical categories of emerging targets that represent both significant challenges and remarkable opportunities for cell-free platforms. These targets often prove difficult or impossible to produce in traditional cellular systems due to their unique properties, yet they hold immense potential for therapeutic, diagnostic, and research applications.

16.1 Intrinsically disordered proteins

Intrinsically disordered proteins (IDPs) represent one of the most fascinating and challenging classes of biomolecules for recombinant production. Unlike conventional globular proteins with well-defined three-dimensional structures, IDPs lack stable secondary and tertiary structures under physiological conditions, existing instead as dynamic ensembles of conformations. This structural flexibility, once considered an anomaly, is now recognized as a fundamental feature that enables unique biological functions.

The prevalence of disorder in proteomes varies significantly across different organisms, as shown in Table 16.1. This distribution reflects the evolutionary pressure for increased regulatory complexity in higher organisms, where IDPs serve crucial roles in signaling, transcriptional regulation, and cellular response mechanisms.

Cell-free systems offer unique advantages for IDP production that stem from their open architecture and tunable environmental conditions. Traditional cellular expression often fails with IDPs due to their susceptibility to proteolytic degradation, aggregation under cellular stress conditions, and interference with normal cellular processes. The controlled environment of cell-free systems allows for optimization of conditions that maintain IDP native state distributions while preventing unwanted interactions.

Temperature control emerges as a critical factor in IDP synthesis, as these proteins are particularly sensitive to thermal fluctuations that can alter their conformational ensembles. Cell-free reactions can be precisely maintained at optimal temperatures, typically between 25°C and 30°C for most mammalian IDPs, compared to 37°C required for cellular systems.

Cell Free Synthesis of Biologics
https://doi.org/10.1016/B978-0-443-33271-5.00025-9

TABLE 16.1 Distribution of intrinsically disordered proteins across different organisms.

Organism type	% Proteins with > 30% disorder	% Proteins with > 50% disorder	Average disorder content
Archaea	4.2	1.8	8.3%
Bacteria	6.8	2.4	10.1%
Unicellular eukaryotes	15.3	7.2	18.7%
Multicellular eukaryotes	28.1	14.6	31.2%
Mammals	42.7	23.4	38.9%

TABLE 16.2 Optimal cell-free conditions for different IDP families.

IDP family	Temperature (°C)	Mg^{2+} (mM)	K^+ (mM)	Polyamines	Crowding agent	Typical yield (mg/mL)
Transcriptional activators	27	8–12	120–150	None	4% PEG-8000	0.8–1.2
RNA-binding proteins	25	6–8	100–120	1 mM spermidine	6% Ficoll-70	1.0–1.5
Signaling proteins	30	10–15	140–160	0.5 mM spermine	3% Dextran-40	0.6–1.0
Chromatin regulators	28	12–16	130–150	1 mM putrescine	5% PEG-6000	0.9–1.3
Stress response proteins	26	8–10	110–130	None	7% Ficoll-400	1.1–1.6

Lower temperatures reduce the kinetic energy of molecular motion, helping preserve the native disorder characteristics while maintaining synthetic efficiency.

The ionic environment profoundly influences IDP behavior, with different salt concentrations and ion types affecting the conformational landscape. Table 16.2 summarizes optimal conditions for various IDP families, demonstrating the need for target-specific optimization protocols.

Molecular crowding agents play a particularly important role in IDP synthesis, as they can significantly alter the conformational preferences and stability of disordered regions. The choice of crowding agent depends on the specific IDP target, with polyethylene glycol (PEG) favoring more compact conformations, while Ficoll tends to preserve extended conformations.

Proteolytic protection strategies become essential when working with IDPs, as their extended conformations expose numerous cleavage sites. Cell-free systems can be supplemented with specific protease inhibitor cocktails tailored to the target protein. Fig. 16.1 illustrates the comparative degradation profiles of a model IDP under different protection strategies.

The analytical characterization of IDPs produced in cell-free systems requires specialized approaches due to their dynamic nature. Traditional protein characterization methods often

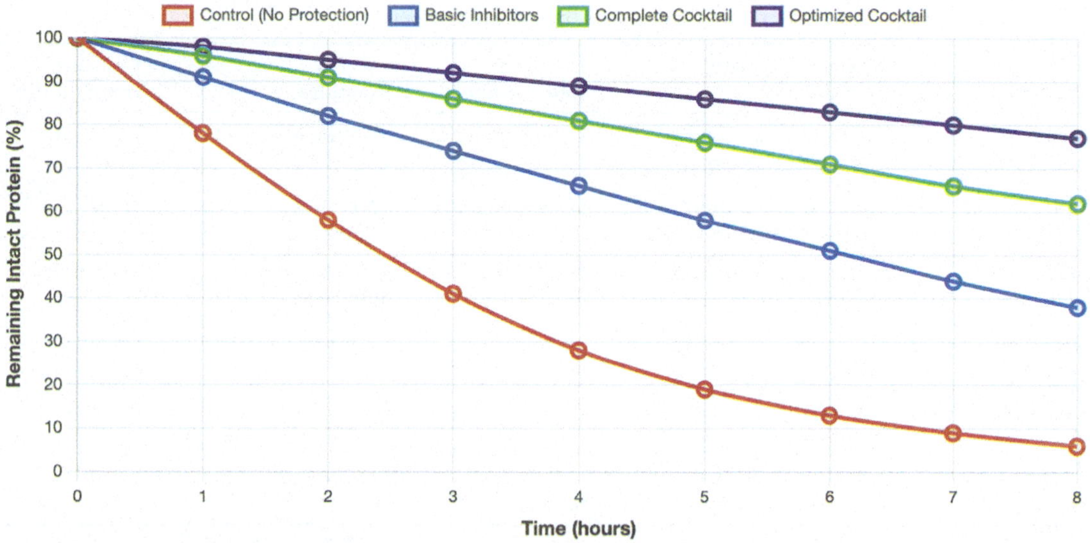

FIG. 16.1 Time-dependent degradation of a model intrinsically disordered protein (120 amino acids, 68% disorder) under different proteolytic protection strategies in cell-free systems. The graph shows remaining intact protein as a percentage of initial concentration over 8 h. Control (red) represents standard cell-free conditions without protection. Basic inhibitors (blue) include ethylenediaminetetraacetic acid (EDTA) and phenylmethylsulfonyl fluoride (PMSF). Complete cocktail (green) contains EDTA, PMSF, pepstatin A, leupeptin, and aprotinin. Optimized cocktail (purple) includes the complete cocktail plus specific cysteine protease inhibitors E-64 and α-Tosyl-L-lysyl-chloromethyl ketone (TLCK). The enhanced stability with comprehensive protease inhibition enables successful synthesis of highly disordered proteins that would otherwise be rapidly degraded.

fail to provide meaningful information about IDP structure and function. Dynamic light scattering, nuclear magnetic resonance spectroscopy, and single-molecule techniques become essential tools for validating the native-like properties of cell-free synthesized IDPs.

16.2 Toxic protein production

Proteins with inherent toxicity present a unique paradox in biotechnology: they are often the most therapeutically valuable targets, yet the most challenging to produce using conventional cellular systems. These proteins can disrupt essential cellular processes, compromise membrane integrity, interfere with DNA replication, or trigger apoptotic pathways, making their production in living cells extremely difficult or impossible.

The spectrum of toxic proteins encompasses various categories, each presenting distinct challenges for production systems. Table 16.3 categorizes major classes of toxic proteins and their primary mechanisms of cellular disruption.

Cell-free systems provide an ideal platform for toxic protein production due to their inherent immunity to cellular damage. The absence of living cells eliminates concerns about viability, growth inhibition, and metabolic disruption that plague cellular production systems. This fundamental advantage allows for the synthesis of proteins that would be lethal to any host organism.

TABLE 16.3 Classification of toxic proteins and their cellular effects.

Protein class	Primary toxicity mechanism	Cellular target	Examples	Cell-free compatibility
Membrane permeabilizers	Pore formation, membrane disruption	Cell membrane	Melittin, magainin	Excellent
DNA/RNA intercalators	Nucleic acid binding, replication block	Genetic material	Restriction enzymes, DNases	Good
Metabolic inhibitors	Enzyme inactivation, pathway disruption	Metabolic enzymes	Ricin A-chain, diphtheria toxin	Very good
Apoptosis inducers	Cell death pathway activation	Signaling cascades	Caspases, p53 variants	Good
Protein aggregators	Misfolding promotion, inclusion formation	Protein homeostasis	Prion proteins, amyloid fragments	Moderate
Oxidative stress generators	ROS production, antioxidant depletion	Cellular redox balance	Cytochrome variants	Good

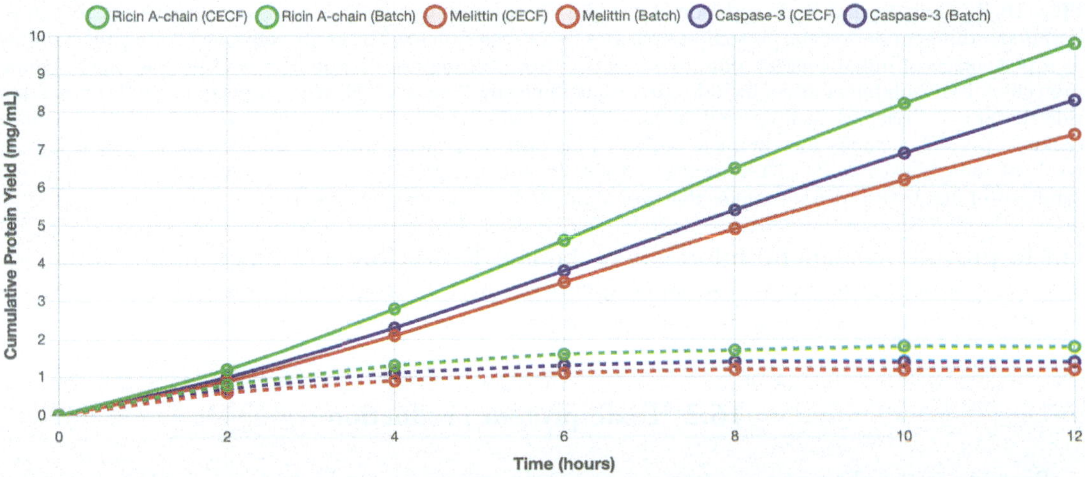

FIG. 16.2 Comparison of cumulative toxic protein yield between batch and continuous exchange cell-free (CECF) systems over 12 h. Three representative toxic proteins are shown: ricin A-chain (metabolic inhibitor), melittin (membrane permeabilizer), and a truncated caspase-3 (apoptosis inducer). CECF systems maintain higher productivity by continuously removing toxic products and replenishing fresh components, preventing system inhibition. Batch reactions show rapid decline in synthesis rate as toxic products accumulate. The sustained productivity of CECF systems makes them ideal for commercial-scale toxic protein production where traditional cellular systems would be completely inhibited or killed.

The production strategy for toxic proteins in cell-free systems often requires specialized reactor configurations to manage product accumulation and prevent system inhibition. Continuous exchange cell-free (CECF) systems prove particularly effective, as they allow for real-time product removal and fresh component replenishment. Fig. 16.2 demonstrates the superior performance of CECF systems compared to batch reactions for toxic protein synthesis.

TABLE 16.4 Containment requirements for toxic protein classes.

Toxicity level	Containment level	Required equipment	Personnel training	Waste treatment
Low (LD50 > 100 mg/kg)	BSL-1 modified	Fume hood, PPE	Basic safety	Standard autoclave
Moderate (LD50 10–100 mg/kg)	BSL-2	Biosafety cabinet, specialized ventilation	Enhanced training	Chemical inactivation
High (LD50 1–10 mg/kg)	BSL-2+	Enclosed systems, redundant containment	Expert certification	Multistep inactivation
Extreme (LD50 < 1 mg/kg)	BSL-3 equivalent	Sealed reactors, negative pressure	Specialist training	Complete destruction

Containment considerations become paramount when working with toxic proteins, requiring specialized safety protocols and equipment design. The open nature of cell-free systems necessitates robust containment strategies to prevent exposure and environmental release. Table 16.4 outlines containment requirements for different toxicity classes.

The optimization of toxic protein synthesis often requires modifications to standard cell-free protocols. Energy regeneration systems may need adjustment to maintain ATP levels in the presence of metabolic inhibitors. Cofactor concentrations might require elevation to overcome competitive inhibition. The choice of cell-free platform becomes critical, with some systems showing higher tolerance to specific toxicity mechanisms.

Quality control for toxic proteins presents unique challenges, as standard analytical methods may not be suitable for highly dangerous materials. Automated analytical systems, remote sampling techniques, and specialized detection methods become essential. The development of neutralization strategies for analytical samples allows for safer characterization while maintaining data quality.

Purification strategies for toxic proteins must balance efficiency with safety considerations. Traditional chromatographic methods may require modification to operate under containment conditions. Affinity purification using immobilized tags can provide both specificity and the ability to work with smaller sample volumes, reducing exposure risks.

16.3 Large protein assemblies

Large protein assemblies represent some of the most complex and functionally important structures in biological systems. These multisubunit complexes, often containing hundreds or thousands of individual protein chains, perform critical cellular functions, including DNA replication, transcription, translation, and metabolic regulation. Their size and complexity create unique challenges for recombinant production, making cell-free systems an attractive alternative to traditional cellular approaches.

The scale of large protein assemblies varies dramatically, from small oligomeric enzymes to massive macromolecular machines. Table 16.5 categorizes representative assemblies by size and complexity, highlighting the range of targets accessible through cell-free synthesis.

TABLE 16.5 Classification of large protein assemblies by size and complexity.

Assembly class	Molecular weight range	Subunit count	Assembly time	Cell-free success rate
Small oligomers	100–500 kDa	2–8	Minutes	85–95%
Medium complexes	500 kDa–2 MDa	8–25	Hours	70–85%
Large assemblies	2–10 MDa	25–100	Hours-days	45–70%
Megacomplexes	>10 MDa	>100	Days-weeks	20–45%
Viral capsids	1–50 MDa	60–2000+	Hours-days	55–75%

The assembly process itself becomes a critical consideration in cell-free synthesis of large complexes. Unlike simple proteins that fold autonomously, large assemblies often require sophisticated assembly pathways with specific intermediate stages, assembly factors, and quality control mechanisms. Cell-free systems can be engineered to recapitulate these natural assembly processes by incorporating necessary chaperones, assembly factors, and cofactors.

Temporal control of subunit synthesis emerges as a crucial factor in successful assembly. Sequential expression strategies, where different subunits are synthesized at specific time points, can dramatically improve assembly efficiency. This approach mimics natural cellular assembly processes and prevents the accumulation of misfolded intermediates. Fig. 16.3 illustrates the improvement in assembly efficiency achieved through temporal control compared to simultaneous synthesis.

The stoichiometry of subunit production requires careful optimization to achieve efficient assembly. Imbalanced subunit ratios often lead to the accumulation of assembly intermediates, aggregation of excess components, or incomplete complex formation. Table 16.6 provides guidelines for optimizing subunit stoichiometry based on complex architecture.

Molecular crowding plays an enhanced role in large assembly synthesis, as the excluded volume effects become more pronounced for larger structures. The choice and concentration of crowding agents can significantly influence assembly kinetics and pathway selection. Physiological crowding conditions (20–30% macromolecular volume fraction) often prove optimal for large complex assembly.

Chaperone supplementation frequently becomes necessary for successful large assembly production. Cell-free systems can be augmented with specific molecular chaperones that assist in subunit folding, prevent aggregation, and facilitate proper assembly. The choice of chaperones depends on the specific assembly requirements and the origin of the target complex.

Quality control mechanisms for large assemblies require sophisticated analytical approaches. Traditional protein characterization methods often prove inadequate for assessing the structure and function of massive complexes. Native mass spectrometry, electron microscopy, and dynamic light scattering become essential tools for verifying correct assembly. Fig. 16.4 shows the analytical workflow for characterizing large protein assemblies.

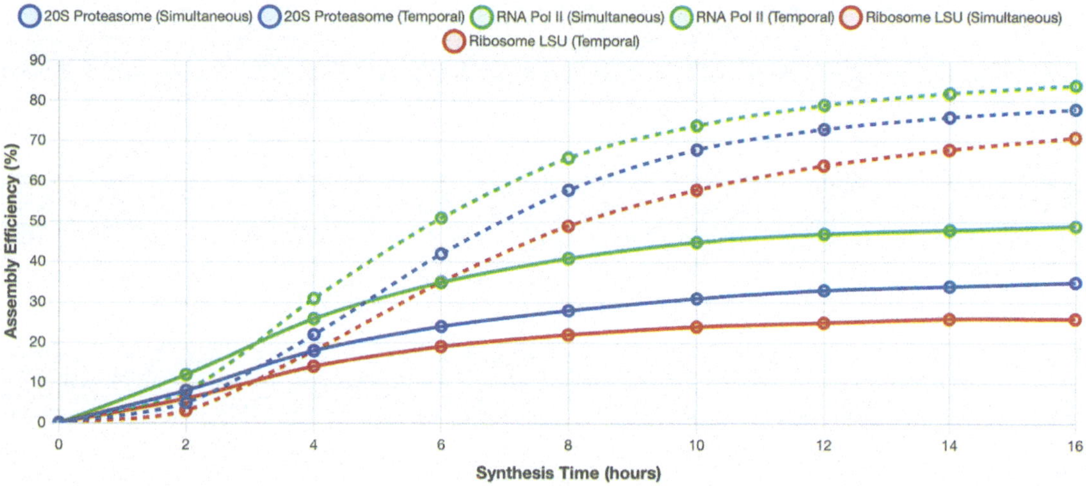

FIG. 16.3 Comparison of assembly efficiency for three different large protein complexes using simultaneous vs. temporal synthesis strategies. The graph shows the percentage of correctly assembled complexes as a function of total synthesis time. Simultaneous synthesis (solid lines) represents conventional approaches where all subunits are synthesized together. Temporal synthesis (dashed lines) involves sequential expression of subunits following natural assembly pathways. Examples shown include 20S proteasome (28 subunits), RNA polymerase II (12 subunits), and ribosomal large subunit (31 proteins). The temporal approach significantly improves assembly efficiency by preventing accumulation of misfolded intermediates and allowing proper assembly checkpoint control.

TABLE 16.6 Stoichiometric optimization strategies for different assembly architectures.

Assembly architecture	Optimal ratio strategy	Key considerations	Monitoring methods
Homodimers/ homomultimers	1:1 with 10% excess	Prevent aggregation	Native PAGE, DLS
Heterodimers	1:1 with limiting partner	Control by scarce subunit	Co–IP, mass spectrometry
Ring structures	Equal ratios, staged assembly	Sequential ring closure	Electron microscopy
Hierarchical complexes	Core-first, then accessories	Stable intermediate formation	Native MS, FRET
Viral capsids	Massive excess of coat proteins	Scaffold-mediated assembly	Cryo-EM, light scattering

16.4 Modified nucleic acids

Modified nucleic acids represent a rapidly expanding frontier in therapeutic development, offering unprecedented opportunities for treating genetic diseases, regulating gene expression, and developing novel diagnostic tools. These chemically modified DNA and RNA molecules possess enhanced stability, altered pharmacokinetic properties, and improved target specificity compared to their natural counterparts. Cell-free systems provide unique

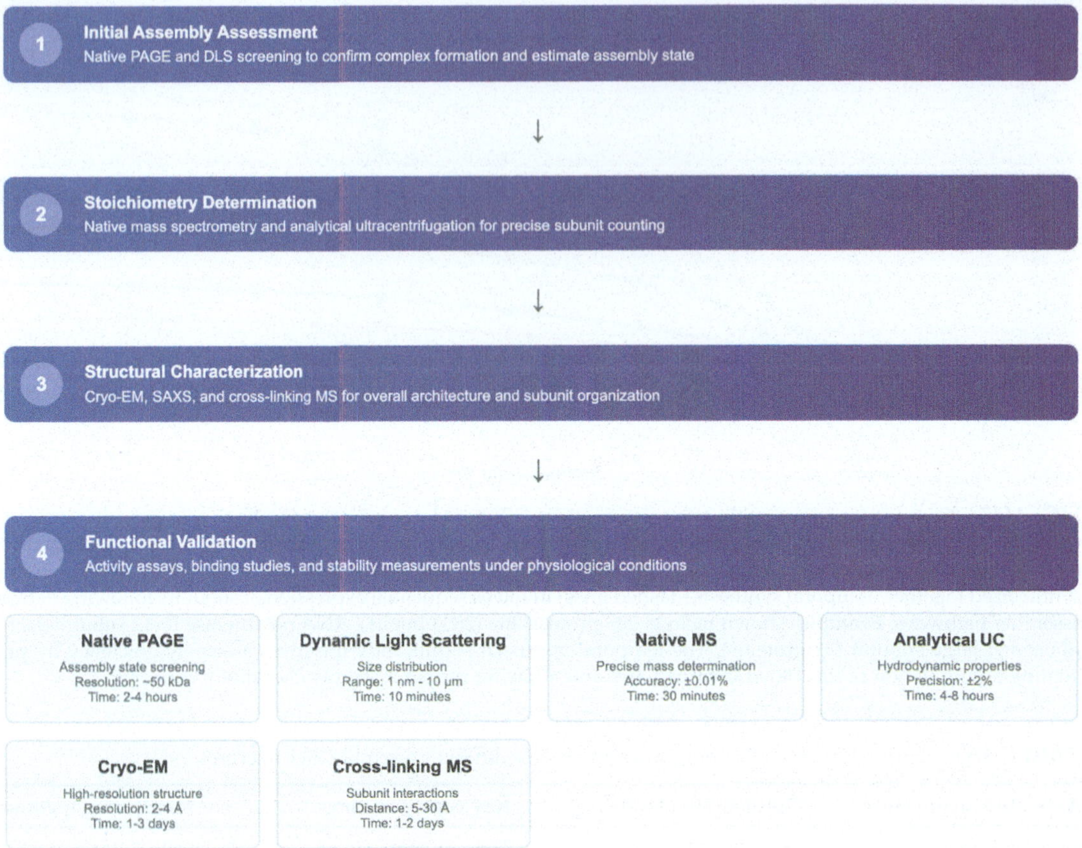

FIG. 16.4 Systematic analytical workflow for characterizing large protein assemblies synthesized in cell-free systems. The workflow progresses from rapid screening methods to high-resolution structural techniques, ensuring comprehensive characterization while optimizing time and resource utilization. Each analytical method provides complementary information about assembly state, stoichiometry, structure, and function. The parallel arrangement of methods in the bottom panel indicates techniques that can be performed simultaneously to accelerate characterization. Success at each stage guides decisions for subsequent analyses, allowing for adaptive characterization strategies based on initial findings.

advantages for producing these modified nucleic acids, particularly when protein–nucleic acid complexes are required.

The landscape of nucleic acid modifications encompasses a vast array of chemical alterations, each designed to confer specific properties or overcome particular limitations of natural nucleic acids. Table 16.7 summarizes major categories of nucleic acid modifications and their applications in cell-free synthesis.

The synthesis of modified nucleic acids in cell-free systems often requires specialized polymerases or modified synthesis conditions. Many chemical modifications can interfere with the normal function of DNA and RNA polymerases, necessitating the use of engineered enzymes

TABLE 16.7 Categories of modified nucleic acids and their cell-free applications.

Modification type	Chemical nature	Primary benefits	Cell-free advantages	Typical applications
2′-O-methyl RNA	Sugar modification	Nuclease resistance, reduced immunogenicity	Compatible with natural machinery	siRNA, antisense oligonucleotides
Phosphorothioate DNA	Backbone modification	Enhanced stability, protein binding	Minimal polymerase inhibition	Antisense drugs, aptamers
Locked nucleic acids (LNA)	Conformational constraint	High target affinity, stability	Requires modified polymerases	Diagnostic probes, therapeutics
5-methylcytosine DNA	Base modification	Epigenetic signaling, stability	Natural methyltransferase activity	Epigenetic studies, therapeutics
Pseudouridine RNA	Base isomerization	Enhanced stability, reduced immunogenicity	Endogenous synthase activity	mRNA therapeutics, vaccines
Fluorescent nucleotides	Base conjugation	Real-time monitoring, imaging	Compatible synthesis conditions	Research tools, diagnostics

with enhanced tolerance for nonnatural substrates. The cell-free environment allows for the optimization of these specialized enzyme systems without concern for cellular viability.

Template design becomes particularly critical when working with modified nucleic acids, as the presence of modifications can affect polymerase processivity, proofreading activity, and termination efficiency. The open nature of cell-free systems allows for real-time optimization of template sequences and reaction conditions to maximize yield and fidelity.

Cotranslational incorporation of modified nucleotides offers unique opportunities for creating protein–nucleic acid complexes with enhanced properties. This approach allows for the simultaneous synthesis of proteins and their nucleic acid-binding partners, potentially improving binding affinity and specificity. Table 16.8 outlines strategies for cotranslational nucleic acid modification.

Quality control for modified nucleic acids requires specialized analytical methods that can distinguish between natural and modified forms. Traditional nucleic acid analysis techniques may not be sensitive to subtle chemical modifications, necessitating the use of advanced analytical approaches such as liquid chromatography–mass spectrometry and high-resolution electrophoresis.

The stability of modified nucleic acids during cell-free synthesis presents both opportunities and challenges. While many modifications enhance nucleic acid stability, they may also affect the kinetics of synthesis or the efficiency of protein–nucleic acid complex formation. Optimization strategies must balance modification benefits with synthesis efficiency.

Purification of modified nucleic acids from cell-free reactions often requires specialized approaches that can separate modified from unmodified species. Affinity chromatography using complementary sequences, reverse-phase chromatography, and ion-exchange methods adapted for modified nucleotides become essential tools for achieving high-purity products.

TABLE 16.8 Cotranslational nucleic acid modification strategies.

Strategy	Mechanism	Required components	Advantages	Limitations
Direct incorporation	Modified NTPs in reaction	Modified nucleotides, tolerant polymerases	Simple implementation	Limited modification types
Postsynthetic modification	Chemical treatment after synthesis	Modifying enzymes, cofactors	Diverse modifications possible	Additional reaction steps
Template premodification	Modified DNA template	Presynthesized modified templates	Precise modification placement	Template synthesis complexity
Enzymatic modification	Specific modifying enzymes	Methyltransferases, other enzymes	Natural modification patterns	Enzyme availability

16.5 Peptide libraries

Peptide libraries represent powerful tools for drug discovery, protein engineering, and basic research, offering the ability to explore vast sequence spaces and identify molecules with desired properties. These collections of peptides, ranging from small cyclic structures to larger linear sequences, can contain millions to billions of unique members, creating unprecedented opportunities for molecular discovery. Cell-free systems provide ideal platforms for peptide library generation due to their flexibility, speed, and ability to incorporate nonnatural amino acids.

The architecture of peptide libraries varies significantly depending on their intended application and the diversity requirements of the screening process. Table 16.9 categorizes different library types and their characteristics relevant to cell-free synthesis.

The generation of peptide libraries in cell-free systems relies heavily on the use of randomized DNA templates that encode the desired peptide diversity. Template design strategies must balance sequence diversity with synthesis efficiency, ensuring that all library members can be produced with reasonable yield and fidelity. The use of optimized codon usage and the elimination of problematic sequence motifs become critical considerations.

Ribosome display and mRNA display techniques find particular utility in cell-free peptide library applications, as they allow for the physical linkage of genotype and phenotype essential for library screening. These techniques can be seamlessly integrated with cell-free synthesis systems, providing a complete platform for library generation and screening. Fig. 16.5 illustrates the workflow for cell-free peptide library generation and screening.

The incorporation of nonnatural amino acids dramatically expands the chemical diversity accessible through peptide libraries. Cell-free systems can be supplemented with engineered aminoacyl-tRNA synthetases and corresponding tRNAs that enable the incorporation of amino acids with novel side chain chemistries. This capability allows for the creation of peptide libraries with enhanced stability, altered pharmacokinetic properties, and unique binding characteristics.

TABLE 16.9 Peptide library categories and cell-free synthesis parameters.

Library type	Size range	Peptide length	Diversity strategy	Synthesis time	Screening method
Linear peptides	10^4–10^6	5–25 amino acids	Random sequence	2–6 h	Phage display, ribosome display
Cyclic peptides	10^5–10^7	6–15 amino acids	Cyclization chemistry	4–12 h	mRNA display, chemical screening
Constrained peptides	10^3–10^5	8–20 amino acids	Disulfide bonds, stapling	6–16 h	Functional assays
Bicyclic peptides	10^4–10^6	10–25 amino acids	Dual cyclization	8–20 h	High-throughput screening
Modified peptides	10^3–10^8	5–30 amino acids	Nonnatural amino acids	Variable	Custom assays

FIG. 16.5 Comprehensive workflow for cell-free peptide library generation and screening showing the integration of synthesis, selection, and analysis phases. The workflow begins with randomized template design and proceeds through cell-free synthesis with ribosome or mRNA display linkage, followed by target-based selection and comprehensive analysis of selected candidates. Technical parameters are optimized for each library type, with synthesis conditions adapted for linear, cyclic, and chemically modified peptides. Performance metrics demonstrate the trade-offs between library diversity, synthesis complexity, and selection success rates. The modular design allows for adaptation to specific targets and applications while maintaining high-quality results throughout the discovery process.

TABLE 16.10 Cyclization strategies for cell-free peptide libraries.

Cyclization method	Chemistry	Reaction conditions	Efficiency	Stability	Compatibility
Disulfide bond	Cys–Cys oxidation	Oxidizing conditions, pH 8–9	60–90%	High	Excellent
Native chemical ligation	Cys–thioester ligation	pH 7–8, reducing conditions	70–95%	Very high	Good
Click chemistry	Azide-alkyne cycloaddition	Cu catalyst, ambient conditions	85–98%	Very high	Moderate
Lactam formation	Amide bond formation	Coupling reagents, neutral pH	50–80%	High	Good
Ring-closing metathesis	Olefin metathesis	Ru catalyst, inert atmosphere	60–85%	Moderate	Limited
Enzymatic cyclization	Transpeptidase activity	Specific enzyme requirements	40–90%	High	Variable

Cyclization strategies become particularly important for therapeutic peptide libraries, as cyclic peptides generally exhibit enhanced stability, improved target selectivity, and better pharmacokinetic properties compared to their linear counterparts. Cell-free systems can be engineered to perform various cyclization reactions, including disulfide bond formation, lactam cyclization, and click chemistry-mediated cyclization.

Table 16.10 summarizes cyclization strategies compatible with cell-free peptide synthesis and their relative advantages and limitations.

Quality control for peptide libraries presents unique challenges due to the enormous diversity of sequences present in a single preparation. Traditional analytical methods designed for single peptides become impractical when dealing with millions of different species. Statistical sampling approaches, coupled with mass spectrometry and high-throughput sequencing, provide insights into library quality and diversity distribution.

The screening of cell-free generated peptide libraries requires integration with appropriate selection or screening platforms. The choice of screening method depends on the target application, desired throughput, and available resources. Functional screens that assess biological activity often provide the most relevant information for therapeutic applications, while binding screens may be more appropriate for diagnostic or research applications.

16.6 Protein–DNA complexes

Protein–DNA complexes represent some of the most fundamental and important molecular assemblies in biology, mediating essential processes such as gene regulation, DNA replication, repair, and recombination. These complexes range from simple binary interactions between single proteins and short DNA sequences to massive multiprotein assemblies that organize entire chromosomal regions. Cell-free systems offer unique advantages for studying and producing these complexes, as they allow for precise control over stoichiometry, timing, and environmental conditions.

The diversity of protein–DNA complexes reflects the wide range of biological processes they mediate. Table 16.11 categorizes major classes of protein–DNA complexes and their characteristics relevant to cell-free synthesis and analysis.

The simultaneous synthesis of proteins and DNA in cell-free systems enables the formation of complexes under conditions that closely mimic natural assembly processes. This cosynthesis approach often results in higher-quality complexes with more native-like properties compared to complexes formed by mixing presynthesized components. The timing of protein and DNA synthesis can be precisely controlled to optimize complex formation.

DNA template design becomes critically important when synthesizing protein–DNA complexes, as the DNA sequence directly influences both protein binding affinity and complex stability. The incorporation of specific binding sites, spacer regions, and structural elements must be carefully optimized to promote complex formation while maintaining the integrity of the DNA template used for protein synthesis.

The stoichiometry of protein–DNA complexes requires careful optimization to achieve efficient assembly. Unlike protein-only complexes, protein–DNA assemblies often involve multiple copies of the same protein binding to a single DNA molecule, creating unique stoichiometric requirements. Table 16.12 provides guidelines for optimizing protein–DNA stoichiometry in cell-free systems.

TABLE 16.11 Classification of protein–DNA complexes and cell-free synthesis parameters.

Complex type	Protein components	DNA length	Binding affinity	Assembly time	Stability
Transcription factors	1–3 proteins	6–20 bp	nM–μM	Seconds–minutes	Hours
Nucleosome core	8 histones	147 bp	pM–nM	Minutes–hours	Days
DNA polymerase holoenzyme	3–10 proteins	Variable	nM–μM	Minutes	Hours
Chromatin remodeling	8–15 proteins	200–1000 bp	nM	Hours	Days
DNA repair complexes	2–8 proteins	20–100 bp	nM–μM	Minutes–hours	Hours
Recombination machinery	5–12 proteins	Variable	nM	Hours	Variable

TABLE 16.12 Stoichiometric optimization for protein–DNA complex assembly.

Complex architecture	Protein/ DNA ratio	Optimization strategy	Common issues	Quality control
Single site binding	1–2:1	Slight protein excess	DNA saturation	EMSA, fluorescence
Multiple site binding	2–10:1	Sequential binding	Cooperative effects	Native PAGE
Nucleosome-like	8:1	Histone chaperones	Assembly pathway	Atomic force microscopy
Sliding clamp	2:1	ATP-dependent loading	Clamp loading	Cross-linking MS
Multiprotein arrays	10–50:1	Hierarchical assembly	Aggregation	Electron microscopy

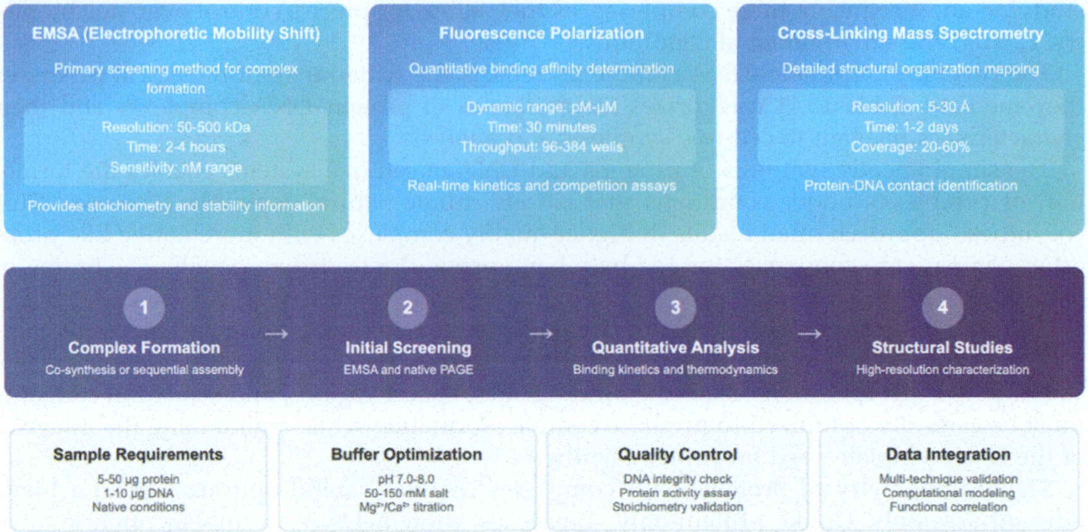

Analytical Method	Information Content	Time Required	Sample Consumption	Complexity Level
Native PAGE	Complex formation, size	2-4 hours	Low (μg)	Low
EMSA	Binding, stoichiometry	3-6 hours	Low (μg)	Medium
Fluorescence Polarization	Kinetics, affinity	0.5-2 hours	Very Low (ng)	Medium
Native MS	Mass, stoichiometry	1-3 hours	Medium (μg)	High
Cross-linking MS	Structure, contacts	1-3 days	High (mg)	Very High
Cryo-EM	High-res structure	3-7 days	Medium (μg)	Very High

FIG. 16.6 Systematic analytical workflow for comprehensive characterization of protein–DNA complexes synthesized in cell-free systems. The workflow integrates multiple complementary techniques to provide complete structural and functional information about the assembled complexes. Initial screening methods (EMSA, native PAGE) rapidly assess complex formation and basic properties, while quantitative methods (fluorescence polarization, native MS) provide detailed thermodynamic and kinetic parameters. Advanced structural techniques (cross-linking MS, cryo-EM) reveal molecular details of protein–DNA interactions and overall complex architecture. The table compares key performance metrics of different analytical approaches, enabling optimal method selection based on available resources and information requirements. This multitechnique approach ensures robust characterization and validation of cell-free synthesized protein–DNA complexes.

The ionic environment profoundly affects protein–DNA complex formation, as electrostatic interactions play a dominant role in DNA binding. Cell-free systems can be precisely tuned to optimize salt concentrations, pH, and divalent cation levels for specific protein–DNA interactions. The optimization process often requires screening across a range of conditions to identify the optimal assembly environment.

Molecular crowding effects become particularly pronounced in protein–DNA complex formation, as the large size and high charge density of DNA create significant excluded volume effects. The addition of appropriate crowding agents can dramatically enhance complex formation by increasing the effective concentration of reactants and stabilizing assembled complexes.

Analytical characterization of protein–DNA complexes requires specialized techniques that can assess both the protein and DNA components simultaneously. Electrophoretic mobility shift assays (EMSA), fluorescence polarization, and cross-linking mass spectrometry provide complementary information about complex stoichiometry, binding affinity, and structural organization. Fig. 16.6 demonstrates the analytical workflow for characterizing protein–DNA complexes synthesized in cell-free systems.

The purification of protein–DNA complexes from cell-free reactions presents unique challenges due to the size and charge heterogeneity of the assembled structures. Gel filtration chromatography often provides the best separation based on size, while ion-exchange chromatography can separate complexes based on their net charge. Affinity purification using DNA-binding proteins or sequence-specific DNA probes offers highly selective purification strategies.

Stability considerations become critical for protein–DNA complexes, as many of these assemblies are designed to be dynamic in their natural cellular context. The optimization of storage conditions, including temperature, ionic strength, and pH, becomes essential for maintaining complex integrity during handling and analysis. The addition of stabilizing agents such as glycerol, trehalose, or specific cofactors can significantly enhance complex stability.

Bibliography

Baneyx, F., Mujacic, M., 2004. Recombinant protein folding and misfolding in *Escherichia coli*. Nat. Biotechnol. 22 (11), 1399–1408.

Bunka, D.H., Stockley, P.G., 2006. Aptamers come of age – at last. Nat. Rev. Microbiol. 4 (8), 588–596.

Carlson, E.D., Gan, R., Hodgman, C.E., Jewett, M.C., 2012. Cell-free protein synthesis: applications come of age. Biotechnol. Adv. 30 (5), 1185–1194.

Jackson, K.L., Dayton, T.L., Oramadla, K., Zhang, Z., Gao, J., Yee, S.S., Kritzer, J.A., 2018. A general approach for continuing eukaryotic translation with non-natural amino acids. J. Am. Chem. Soc. 140 (3), 963–967.

Jewett, M.C., Calhoun, K.A., Voloshin, A., Wuu, J.J., Swartz, J.R., 2008. An integrated cell-free metabolic platform for protein production and synthetic biology. Mol. Syst. Biol. 4 (1), 220.

Katzen, F., Chang, G., Kudlicki, W., 2005. The past, present and future of cell-free protein synthesis. Trends Biotechnol. 23 (3), 150–156.

Lim, H.J., Ching, C.B., Tan, N.S., 2007. Application of intrinsically disordered proteins in biotechnology. Biotechnol. Appl. Biochem. 48 (1), 35–42.

Luger, K., Mäder, A.W., Richmond, R.K., Sargent, D.F., Richmond, T.J., 1997. Crystal structure of the nucleosome core particle at 2.8 Å resolution. Nature 389 (6648), 251–260.

Nemoto, N., Miyamoto-Sato, E., Husimi, Y., Yanagawa, H., 1997. In vitro virus: bonding of mRNA bearing puromycin at the 3′-terminal end to the C-terminal end of its encoded protein on the ribosome. FEBS Lett. 414 (2), 405–408.

Roberts, R.W., Szostak, J.W., 1997. RNA-peptide fusions for the in vitro selection of peptides and proteins. Proc. Natl. Acad. Sci. U.S.A. 94 (23), 12297–12302.

Shimizu, Y., Inoue, A., Tomari, Y., Suzuki, T., Yokogawa, T., Nishikawa, K., Ueda, T., 2001. Cell-free translation reconstituted with purified components. Nat. Biotechnol. 19 (8), 751–755.

Spirin, A.S., 2004. High-throughput cell-free systems for synthesis of functionally active proteins. Trends Biotechnol. 22 (10), 538–545.

Uversky, V.N., Dunker, A.K., 2010. Understanding protein non-folding. Biochim. Biophys. Acta 1804 (6), 1231–1264.

Woodrow, K.A., Kwon, Y., Seo, E., Chantakulkittiwong, N., Lim, H., Suh, H., Mehta, S., Zorlutuna, P., Çiçek, N., Dinh, K.T., et al., 2009. Intravaginal gene silencing using biodegradable polymer nanoparticles densely loaded with small-interfering RNA, Nat. Mater. 8 (6), 526–533.

Wright, A.V., Nuñez, J.K., Doudna, J.A., 2016. Biology and applications of CRISPR systems: harnessing nature's toolbox for genome engineering. Cell 164 (1-2), 29–44.

PART VI

Economic and regulatory framework

CHAPTER

17

Economic analysis and business models

The commercial viability of cell-free protein synthesis depends fundamentally on its economic competitiveness relative to traditional cellular systems and alternative production methods. This chapter provides a comprehensive analysis of the financial considerations, business models, and economic drivers that shape the cell-free biotechnology landscape. Understanding these economic factors is crucial for making informed decisions about technology adoption, investment strategies, and market positioning in an increasingly competitive biotechnology environment.

17.1 Cost structure analysis

The cost structure of cell-free protein synthesis differs markedly from traditional cellular systems, presenting both advantages and challenges that must be carefully evaluated in any economic analysis. Unlike cellular systems where living organisms provide many components through their natural metabolism, cell-free systems require the external provision of all necessary molecular machinery, energy sources, and cofactors. This fundamental difference creates a unique cost profile that requires detailed understanding for effective financial planning.

The primary cost components in cell-free systems can be categorized into several major categories, each with distinct characteristics and optimization opportunities. Table 17.1 provides a comprehensive breakdown of typical cost components for different cell-free platform types, showing both absolute costs and relative contributions to total production expenses.

The extract or enzyme costs represent the largest single expense category for most cell-free systems, particularly for protein synthesis using recombinant elements (PURE) systems where individually purified components command premium prices. This cost component exhibits significant economies of scale, with per-unit costs decreasing substantially as production volumes increase. The relationship between extract costs and production scale follows a power law distribution, with cost reductions of 60%–80% achievable when moving from laboratory to industrial scales.

Energy system costs vary considerably depending on the regeneration strategy employed. Phosphoenolpyruvate-based systems, while offering superior performance, carry higher

Cell Free Synthesis of Biologics
https://doi.org/10.1016/B978-0-443-33271-5.00031-4

TABLE 17.1 Cost structure analysis for different cell-free platforms.

Cost component	*E. coli* S30 ($/g protein)	Wheat germ ($/g protein)	PURE system ($/g protein)	Mammalian ($/g protein)	% of total cost
Cell extract/enzymes	$45–65	$120–180	$800–1200	$200–350	35–55
Energy system	$25–35	$40–60	$150–250	$60–90	15–25
Amino acids	$15–25	$20–30	$30–45	$25–40	8–15
Cofactors/salts	$8–12	$15–25	$50–80	$20–35	5–12
DNA template	$5–15	$5–15	$5–15	$5–15	2–8
Labor	$20–40	$30–50	$100–150	$40–70	12–20
Equipment/ overhead	$10–20	$15–25	$40–60	$20–35	8–15
Total direct cost	**$128–212**	**$245–385**	**$1175–1800**	**$370–635**	**100**

material costs compared to glucose-based alternatives. The choice of energy system significantly impacts both direct costs and yield performance, creating complex optimization trade-offs that must be evaluated holistically rather than purely on material cost basis.

Labor costs in cell-free systems present both challenges and opportunities compared to cellular systems. While cell-free systems require skilled technicians for setup and optimization, they eliminate many of the labor-intensive activities associated with cell cultivation, harvesting, and processing. The automation potential of cell-free systems offers significant opportunities for labor cost reduction, particularly in high-throughput applications.

The cost scaling behavior of cell-free systems differs fundamentally from cellular systems due to their modular architecture and independence from biological growth constraints. Fig. 17.1 illustrates the comparative cost scaling characteristics of cell-free vs. cellular systems across different production volumes.

Fixed costs in cell-free systems are generally lower than cellular systems due to simpler facility requirements and reduced infrastructure needs. The absence of fermentation vessels, complex environmental controls, and biosafety containment systems significantly reduces capital equipment requirements. However, cell-free systems may require specialized equipment for component preparation, reaction monitoring, and product recovery that can represent substantial fixed cost investments.

Variable costs exhibit different scaling patterns in cell-free systems compared to cellular alternatives. While cellular systems benefit from biological amplification that reduces per-unit material costs, cell-free systems offer more predictable cost scaling and reduced variability in production expenses. This predictability provides significant advantages for financial planning and risk management in commercial applications.

17.2 Capital expenditure vs. operational expenditure considerations

The capital vs. operational expenditure profile of cell-free protein synthesis presents unique characteristics that significantly influence investment decisions and financial

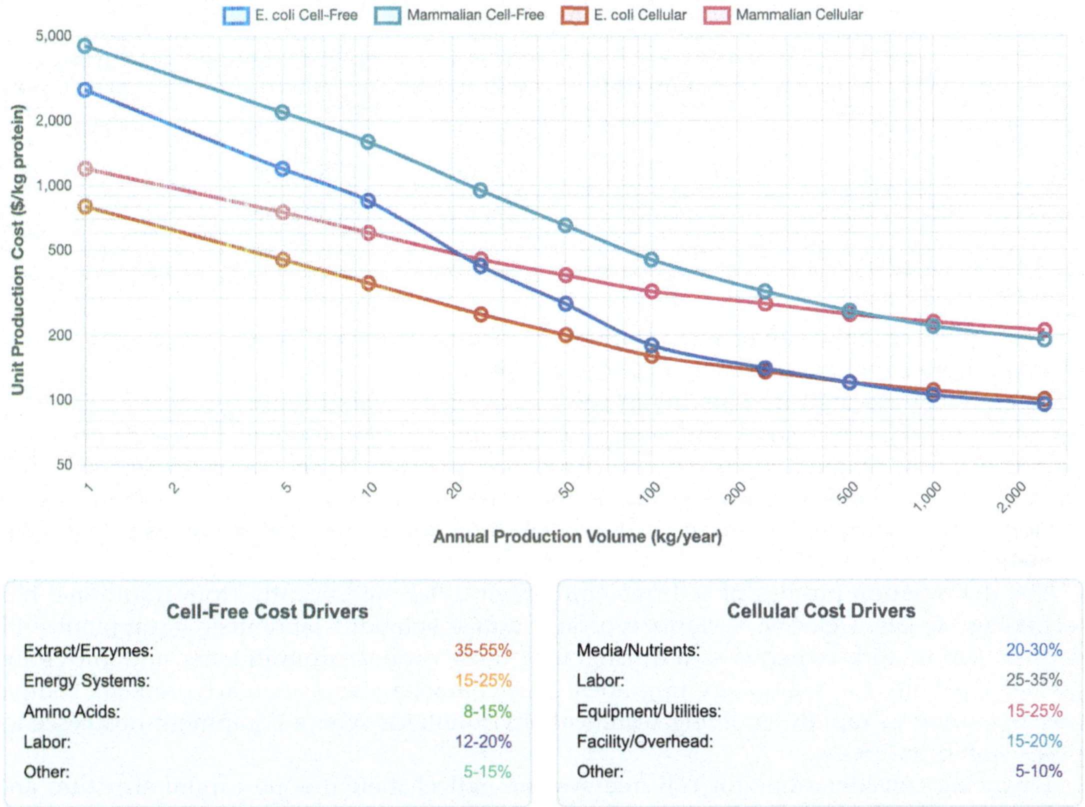

FIG. 17.1 Cost scaling comparison between cell-free and cellular protein production systems across different production volumes. The graph shows unit production costs ($/gram protein) as a function of annual production volume (kg/year) on logarithmic scales. Cell-free systems (blue lines) show different scaling characteristics compared to cellular systems (red lines), with steeper initial cost reductions but higher baseline costs. The crossover points indicate production volumes where cost parity is achieved. *E. coli* cell-free systems achieve cost competitiveness at 50–100 kg/year, while mammalian cell-free systems require 200–500 kg/year volumes. The cost breakdown panels show the relative contribution of major cost components, highlighting the material-intensive nature of cell-free systems vs. the labor and infrastructure intensity of cellular systems.

planning strategies. Unlike traditional biotechnology investments that are heavily weighted toward capital expenditures for complex infrastructure, cell-free systems offer more flexible investment models with different risk-return profiles.

Capital expenditure requirements for cell-free systems are generally 40%–70% lower than comparable cellular systems, primarily due to reduced facility complexity and infrastructure requirements. Table 17.2 compares typical capital investment requirements across different production scales and system types.

The reduced capital intensity of cell-free systems stems from several fundamental advantages, including elimination of fermentation infrastructure, simplified environmental controls, reduced biosafety requirements, and modular equipment design. These factors combine

TABLE 17.2 Capital expenditure requirements by system type and scale.

Production scale	Cell-free systems ($M)	Cellular systems ($M)	Infrastructure reduction	Payback period (years)
Laboratory (kg/year)	$0.5–1.2	$1.5–3.0	60%–70%	1.5–2.5
Pilot (10–50 kg/year)	$2.5–6.0	$8.0–15.0	65%–75%	2.0–3.0
Commercial (100–500 kg/year)	$8.0–20.0	$25.0–60.0	55%–70%	2.5–4.0
Large scale (>1000 kg/year)	$25.0–60.0	$80.0–200.0	60%–75%	3.0–5.0

to create more attractive investment profiles, particularly for smaller companies and specialized applications where capital availability may be limited.

Operational expenditure patterns in cell-free systems exhibit higher material cost ratios but lower overall operational complexity compared to cellular systems. This shift from capital-intensive to material-intensive cost structures provides both opportunities and challenges for financial management. The higher proportion of variable costs offers greater flexibility in scaling production up or down based on demand, but also increases sensitivity to material cost fluctuations.

The depreciation profiles of cell-free equipment differ significantly from traditional biotechnology assets. Cell-free systems typically utilize standard laboratory equipment with shorter depreciation schedules, reducing long-term capital commitments and providing greater flexibility for technology upgrades. This characteristic particularly benefits companies operating in rapidly evolving technical environments where equipment obsolescence poses significant risks.

Financing considerations for cell-free systems reflect their unique capital structure and risk profiles. The lower capital requirements and faster payback periods make these systems attractive to venture capital and private equity investors seeking shorter investment horizons. In addition, the modular nature of cell-free systems enables staged investment strategies that align capital deployment with market development and technical validation milestones.

Working capital requirements for cell-free systems typically represent 15%–25% of annual revenue, compared to 8%–15% for cellular systems. This increase primarily reflects higher inventory requirements for specialized reagents and components that may have limited shelf life or supply chain constraints. Effective working capital management becomes critical for maintaining positive cash flows and optimizing return on invested capital.

17.3 Time-to-market advantages

The time-to-market advantages of cell-free protein synthesis represent one of the most compelling economic benefits of the technology, particularly in competitive markets where speed of development can determine commercial success. The elimination of cell line development, optimization, and scale-up processes can reduce development timelines by 50%–80% compared to traditional cellular systems, creating substantial competitive advantages and reducing time-dependent costs.

Traditional cellular protein production requires extensive development phases, including host cell engineering, clone selection, process optimization, and scale-up validation. These activities typically consume 12–24 months even for well-characterized targets, with complex proteins often requiring 24–36 months of development time. Cell-free systems can achieve comparable production levels within 2–6 months, dramatically compressing development timelines and accelerating revenue generation.

Table 17.3 compares typical development timelines and associated costs for different production approaches across various protein complexity categories.

The compressed development timelines of cell-free systems provide multiple economic benefits beyond simple cost reduction. Earlier market entry enables capture of first-mover advantages, premium pricing opportunities, and market share establishment before competitive products arrive. In pharmaceutical applications, accelerated development can extend effective patent life and increase net present value of drug development investments.

The reduced development risk associated with cell-free systems stems from their greater predictability and lower technical uncertainty compared to cellular systems. Cell line development often involves substantial technical risks, including clone instability, expression variability, and scale-up challenges that can derail projects or require expensive redevelopment efforts. Cell-free systems offer more deterministic development paths with lower probability of fundamental technical failures.

The economic impact of accelerated development extends beyond direct cost savings to encompass broader strategic advantages, including improved capital efficiency, reduced market risk, and enhanced competitive positioning. Companies utilizing cell-free systems can pursue more aggressive development portfolios, supporting larger numbers of projects with the same resource commitments required for fewer cellular development programs.

The time-to-market advantages of cell-free systems become particularly pronounced in markets characterized by rapid innovation cycles, short product lifecycles, or time-sensitive

TABLE 17.3 Development timeline and cost comparison by protein complexity.

Protein category	Cellular development (months)	Cell-free development (months)	Time savings	Cost advantage	Risk reduction
Simple proteins	8–12	2–4	60%–75%	$200 K–500 K	Low–medium
Complex proteins	15–24	3–6	70%–80%	$500 K–1.2 M	Medium–high
Membrane proteins	24–36	4–8	75%–85%	$800 K–2.0 M	High
Toxic proteins	36–48[a]	3–6	85%–90%	$1.0 M–3.0 M	Very high
Multisubunit complexes	20–30	4–10	65%–75%	$600 K–1.5 M	Medium–high

[a]*Often unsuccessful in cellular systems.*

applications. Diagnostic markets, research reagents, and personalized medicine applications represent particularly attractive segments where speed advantages translate directly into economic benefits.

17.4 Risk-adjusted financial modeling

Risk-adjusted financial modeling for cell-free protein synthesis requires careful consideration of technology-specific risk factors, market uncertainties, and competitive dynamics that differ substantially from traditional biotechnology investments. The unique characteristics of cell-free systems create risk profiles that must be properly quantified and incorporated into investment decisions and business planning processes.

Technical risks in cell-free systems generally exhibit lower probability but different characteristics compared to cellular systems. While cell-free systems reduce many biological risks associated with cell line stability and cultivation variability, they introduce different risk factors related to component supply chains, reagent stability, and process robustness. Table 17.4 provides a comprehensive risk assessment framework for cell-free protein synthesis investments.

The component supply chain represents a unique risk factor for cell-free systems due to their dependence on specialized reagents and biological materials that may have limited suppliers or production capacity. This concentration risk can be particularly acute for PURE systems that require individually purified enzymes, many of which are available from only single suppliers. Diversification strategies and supply chain redundancy planning become critical risk management activities.

Market acceptance risks for cell-free systems reflect both the advantages and challenges of introducing disruptive technologies into conservative industries. While the performance

TABLE 17.4 Risk assessment framework for cell-free systems.

Risk category	Probability	Impact	Mitigation strategies	Cost of mitigation	Risk-adjusted NPV impact
Component supply chain	Medium (30%)	High	Multiple suppliers, inventory buffer	5%–10% of material costs	−8% to −15%
Technical performance	Low (15%)	Medium	Pilot validation, backup approaches	10%–20% of development costs	−3% to −8%
Market acceptance	Medium (40%)	High	Customer validation, partnerships	15%–25% of commercialization costs	−12% to −25%
Competitive response	High (60%)	Medium	IP protection, speed advantages	5%–15% of revenue	−8% to −18%
Regulatory changes	Low (20%)	High	Regulatory engagement, compliance	8%–15% of development costs	−5% to −12%
Scale-up challenges	Low (25%)	Medium	Pilot studies, process validation	10%–18% of scale-up costs	−4% to −10%

advantages of cell-free systems are often compelling, adoption may be slower than anticipated due to customer familiarity with existing approaches, regulatory conservatism, or integration challenges with existing workflows. These adoption risks require careful modeling and sensitivity analysis in financial projections.

Competitive response risks must consider both direct competition from other cell-free providers and indirect competition from improvements in traditional cellular systems. The rapid pace of biotechnology innovation means that competitive advantages may erode more quickly than in other industries, requiring aggressive development timelines and continuous innovation to maintain market position (Fig. 17.2).

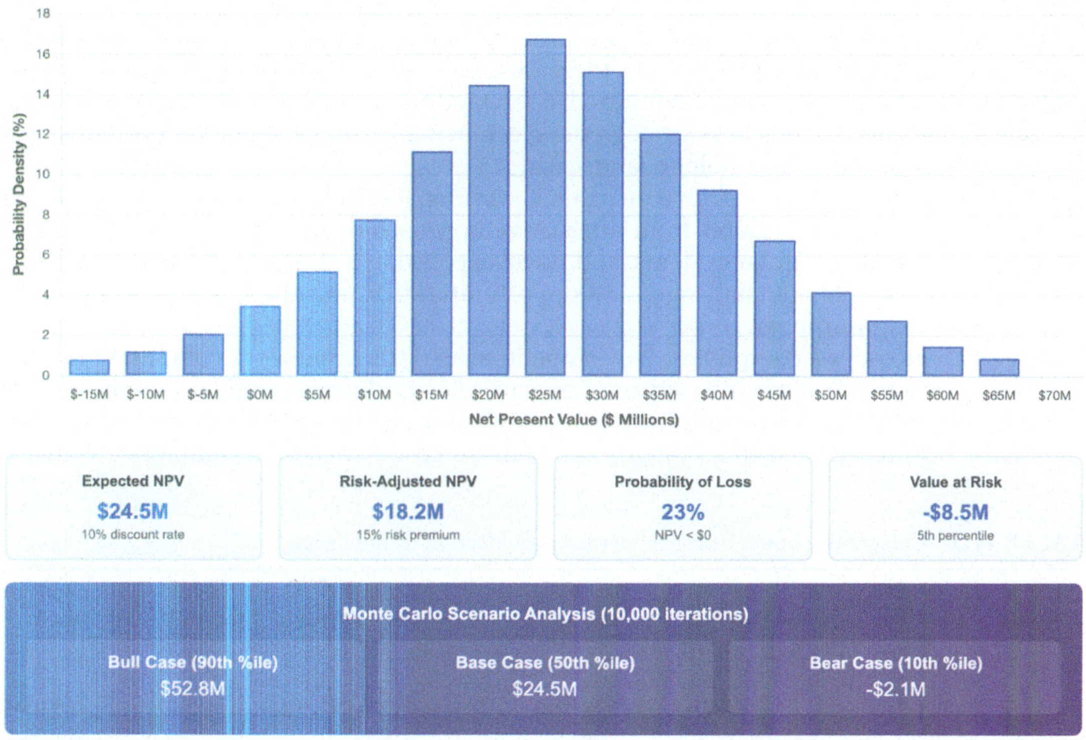

FIG. 17.2 The risk-adjusted net present value distribution for a typical cell-free protein synthesis investment, showing the impact of various risk scenarios on financial returns. Risk-adjusted net present value distribution for a typical cell-free protein synthesis investment based on Monte Carlo simulation with 10,000 iterations. The histogram shows the probability distribution of net present value (NPV) outcomes accounting for various risk factors, including technical performance, market acceptance, competitive response, and supply chain uncertainties. The distribution exhibits positive skewness with a long tail of highly favorable outcomes, reflecting the substantial upside potential of successful cell-free technologies. Key risk metrics indicate a 77% probability of positive returns with expected NPV of $24.5 M, though risk adjustment reduces this to $18.2 M when accounting for technology and market uncertainties. The scenario analysis demonstrates the range of potential outcomes from bear case losses to bull case returns exceeding $50 M, highlighting the importance of risk management and portfolio diversification strategies.

The probability distributions for different risk scenarios enable Monte Carlo analysis of investment returns, providing more sophisticated financial planning tools than traditional deterministic models. This probabilistic approach allows investors and managers to better understand the range of potential outcomes and make more informed decisions about risk tolerance and investment sizing.

Sensitivity analysis becomes particularly important in cell-free financial modeling due to the high proportion of variable costs and material price sensitivity. Changes in key component costs, yield performance, or market pricing can significantly impact project economics, requiring robust scenario planning and hedge strategies for commercial operations.

17.5 Competitive positioning

The competitive positioning of cell-free protein synthesis must be evaluated across multiple dimensions, including cost competitiveness, performance advantages, market access, and strategic differentiation opportunities. Understanding competitive dynamics is essential for developing effective business strategies and achieving sustainable market positions in an increasingly crowded biotechnology landscape.

Cell-free systems compete primarily against traditional cellular production methods, but also face emerging competition from alternative approaches, including synthetic biology platforms, automated cellular systems, and novel expression technologies. Each competitive segment presents different challenges and opportunities that must be carefully analyzed for strategic planning purposes.

Table 17.5 provides a comprehensive competitive analysis framework comparing cell-free systems against major alternative approaches across key performance and economic metrics.

Cost positioning varies significantly across different cell-free platforms and application segments. Prokaryotic cell-free systems can achieve cost parity or advantages relative to

TABLE 17.5 Competitive positioning analysis across production technologies.

Technology platform	Cost position	Speed advantage	Technical performance	Market access	Strategic moat
E. coli cellular	Baseline (1.0×)	Baseline (1.0×)	Good	Excellent	Low
Mammalian cellular	3–5× higher	0.6–0.8×	Excellent	Excellent	Medium
Yeast/insect cellular	1.5–2.5× higher	0.7–0.9×	Good–excellent	Good	Low–medium
Cell-free (prokaryotic)	0.8–1.5×	2–4×	Good	Limited	Medium
Cell-free (eukaryotic)	2–4× higher	1.5–3×	Excellent	Limited	Medium–high
PURE systems	8–15× higher	3–5×	Variable	Very limited	High
Synthetic biology	1–3× higher	0.5–2×	Good–excellent	Emerging	Medium–high

cellular systems in specific applications, particularly for toxic proteins, rapid prototyping, or small-scale production. Eukaryotic cell-free systems typically command premium pricing but offer unique capabilities that justify higher costs for specialized applications.

Speed advantages represent one of the strongest competitive differentiators for cell-free systems across all market segments. The 2–5× faster development timelines provide compelling value propositions for time-sensitive applications, competitive response situations, and markets with short product lifecycles. This speed advantage often justifies premium pricing and enables premium market positioning.

Technical performance competitiveness depends heavily on the specific application and target protein characteristics. Cell-free systems excel in applications involving toxic proteins, membrane proteins, proteins requiring nonnatural amino acids, and rapid optimization requirements. However, cellular systems maintain advantages for large-scale production, complex posttranslational modifications, and cost-sensitive commodity applications.

Market access represents a significant challenge for cell-free systems due to their relative novelty and limited track record in commercial applications. Established cellular systems benefit from extensive regulatory precedent, customer familiarity, and supply chain infrastructure that create substantial barriers to entry for alternative technologies. Overcoming these market access challenges requires significant investment in customer education, regulatory development, and partnership strategies.

Strategic differentiation opportunities for cell-free systems center on their unique capabilities rather than direct competition with cellular systems. Applications where cellular systems fail or perform poorly represent the most attractive initial market opportunities, allowing cell-free systems to establish market presence and develop customer relationships before expanding into more competitive segments.

The intellectual property landscape significantly influences competitive positioning, with cell-free systems benefiting from extensive patent protection around key technologies and methods. This intellectual property (IP) position provides some protection against competitive entry but also creates licensing obligations and freedom-to-operate challenges that must be carefully managed.

17.6 Investment and funding landscape

The investment and funding landscape for cell-free protein synthesis reflects both the technology's promise and the challenges associated with commercializing novel biotechnology platforms. Understanding funding patterns, investor preferences, and capital requirements is essential for companies developing cell-free technologies and applications.

Venture capital investment in cell-free technologies has grown substantially over the past decade, with total funding increasing from approximately $50 million annually in 2015 to over $400 million in 2024. This growth reflects increasing investor confidence in the commercial viability of cell-free approaches and the success of early-stage companies in demonstrating technical and market validation.

Table 17.6 summarizes the investment landscape across different funding stages and investor types, showing typical investment sizes, investor preferences, and key evaluation criteria for cell-free companies.

TABLE 17.6 Investment landscape analysis for cell-free companies.

Funding stage	Typical investment size	Primary investor types	Key evaluation criteria	Success rate	Typical valuation
Seed/preseed	$0.5–3 M	Angel investors, incubators	Technical proof-of-concept	15%–25%	$2–8 M
Series A	$3–12 M	VC firms, strategic investors	Market validation, team	25%–35%	$8–25 M
Series B	$8–25 M	VC firms, growth equity	Revenue traction, scalability	35%–50%	$25–75 M
Series C+	$15–50 M+	Growth equity, strategics	Market leadership, profitability	50%–70%	$75–300 M+
Strategic	$5–100 M+	Industry partners	Strategic fit, synergies	60%–80%	Variable
IPO/exit	$50–200 M+	Public markets, acquirers	Market size, competitive position	5%–15%	$200 M–2B+

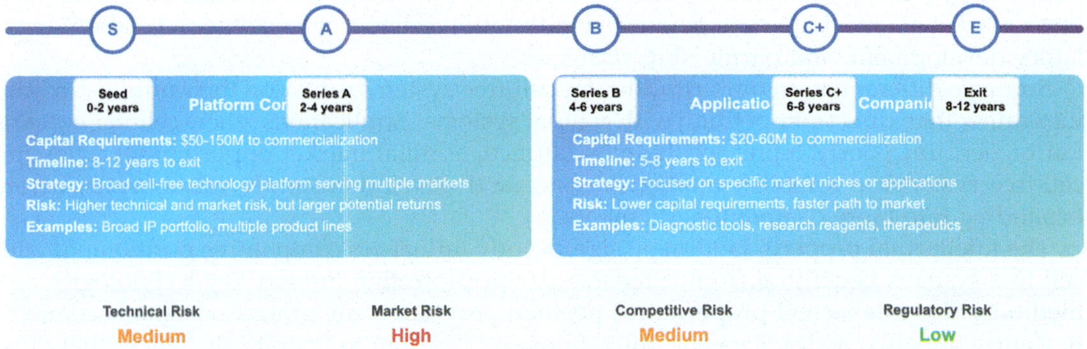

Stage	Key Milestones	Typical Valuation	Primary Investors	Success Rate
Seed	Proof-of-concept, IP filing	$2-8M	Angels, Incubators	15-25%
Series A	Market validation, partnerships	$8-25M	VC firms, Strategic	25-35%
Series B	Revenue generation, scale-up	$25-75M	Growth equity, VC	35-50%
Series C+	Market leadership, profitability	$75-300M+	PE, Strategic, VC	50-70%
Exit	Market dominance, scale	$200M-2B+	Public, Acquirers	5-15%

FIG. 17.3 The funding timeline and capital requirements for typical cell-free companies across different development stages and business models. Typical funding timeline and capital requirements for cell-free protein synthesis companies showing progression from seed funding through exit. The timeline illustrates major funding milestones (S = Seed, A = Series A, B = Series B, C+ = Series C and beyond, E = Exit) with corresponding funding amounts and timeframes. Two distinct business models are shown: platform companies requiring higher capital investment ($50–150 M) but offering broader market opportunities, and application-focused companies with lower capital requirements ($20–60 M) but more focused market positioning. Risk indicators show the relative risk levels across different categories, with market risk being highest due to the emerging nature of cell-free technologies. The success rate table demonstrates the typical progression through funding stages, with increasing success rates at later stages but decreasing overall probability of reaching exit. This funding pattern reflects the capital-intensive nature of biotechnology development while highlighting the potential for substantial returns in successful companies.

Investor preferences in the cell-free space have evolved toward companies with clear market focus, demonstrated customer traction, and defensible competitive positions. Early-stage investors particularly value technical differentiation and intellectual property positions, while later-stage investors focus more heavily on market opportunity size and commercial execution capabilities.

The funding requirements for cell-free companies vary significantly depending on business model and market strategy. Platform companies developing broad cell-free capabilities typically require larger capital investments ($50–150 M to commercialization) compared to application-focused companies targeting specific market niches ($20–60 M to commercialization).

Strategic investor participation has increased significantly as established biotechnology and pharmaceutical companies recognize the strategic value of cell-free technologies. These strategic investments often provide not only capital but also market access, regulatory expertise, and commercial partnerships that can accelerate development and reduce risk.

Government funding plays an important role in early-stage cell-free research and development, with agencies, including The National Institutes of Health (NIH), National Science Foundation (NSF), Department of Defense (DOD), and international equivalents, providing substantial support for academic research and early-stage company development. SBIR/STTR programs have been particularly valuable for cell-free companies transitioning from academic research to commercial development (Fig. 17.3).

The exit landscape for cell-free companies includes both acquisition opportunities and potential public market paths. Acquisitions by pharmaceutical companies, biotechnology firms, and life science tool companies have provided successful exits for several cell-free companies, with valuations ranging from $100 M to over $1B depending on technology platform and market position.

International funding patterns vary significantly, with European investors showing particular interest in sustainable biotechnology applications of cell-free systems, while Asian investors focus more heavily on manufacturing and cost advantages. These regional preferences create opportunities for companies to access specialized investor expertise and market knowledge.

The funding landscape continues to evolve as cell-free technologies mature and demonstrate commercial success. Increasing investor sophistication and sector knowledge is leading to more efficient capital allocation and better alignment between investor expectations and company capabilities.

Bibliography

Adiga, R., Al-Rubeai, M., Betenbaugh, M.J., Shiloach, J., Spearman, M., Aunins, J.G., et al., 2014. Commercial manufacturing of cell therapy products from inception to market, Biotechnol. Prog. 30 (5), 1025–1032.

Bundy, B.C., Franciszkowicz, M.J., Swartz, J.R., 2008. Escherichia coli-based cell-free synthesis of virus-like particles. Biotechnol. Bioeng. 100 (1), 28–37.

Carlson, E.D., Gan, R., Hodgman, C.E., Jewett, M.C., 2012. Cell-free protein synthesis: applications come of age. Biotechnol. Adv. 30 (5), 1185–1194.

Des Soye, B.J., Davidson, S.R., Weinstock, M.T., Gibson, D.G., Jewett, M.C., 2018. Establishing a high-yielding cell-free protein synthesis platform derived from *Vibrio natriegens*. ACS Synth. Biol. 7 (9), 2245–2255.

Gregorio, N.E., Levine, M.Z., Oza, J.P., 2019. A user's guide to cell-free protein synthesis. Methods Protoc. 2 (1), 24.

Jaroentomeechai, T., Stark, J.C., Natarajan, A., Glasscock, C.J., Yingling, L.E., Hsu, K.J., et al., 2018. Single-pot glycoprotein biosynthesis using a cell-free transcription-translation system enriched with glycosylation machinery, Nat. Commun. 9 (1), 2686.

Kelwick, R., MacDonald, J.T., Webb, A.J., Freemont, P., 2016. Developments in cell-free protein synthesis for biotechnology applications. Biotechnol. J. 11 (1), 12–25.

Kwon, Y.C., Jewett, M.C., 2015. High-throughput preparation methods of crude extract for robust cell-free protein synthesis. Sci. Rep. 5 (1), 8663.

Liu, D.V., Zawada, J.F., Swartz, J.R., 2005. Streamlining *Escherichia coli* S30 extract preparation for economical cell-free protein synthesis. Biotechnol. Prog. 21 (2), 460–465.

Martin, R.W., Majewska, N.I., Chen, C.X., Albanetti, T.E., Jimenez, R.B.C., Schmelzer, A.E., et al., 2017. Development of a CHO-based cell-free platform for synthesis of active monoclonal antibodies, ACS Synth. Biol. 6 (7), 1370–1379.

Moore, S.J., MacDonald, J.T., Wienecke, S., Ishwarbhai, A., Tsipa, A., Aw, R., et al., 2018. Rapid acquisition and model-based analysis of cell-free transcription-translation reactions from nonmodel bacteria, Proc. Natl. Acad. Sci. U.S.A. 115 (19), E4340–E4349.

Schoborg, J.A., Hodgman, C.E., Anderson, M.J., Jewett, M.C., 2014. Substrate replenishment and byproduct removal improve yeast cell-free protein synthesis. Biotechnol. J. 9 (5), 630–640.

Silverman, A.D., Karim, A.S., Jewett, M.C., 2020. Cell-free gene expression: an expanded repertoire of applications. Nat. Rev. Genet. 21 (3), 151–170.

Walsh, G., 2018. Biopharmaceutical benchmarks 2018. Nat. Biotechnol. 36 (12), 1136–1145.

Yang, W.C., Patel, K.G., Wong, H.E., Swartz, J.R., 2012. Simplifying and streamlining Escherichia coli-based cell-free protein synthesis. Biotechnol. Prog. 28 (2), 413–420.

CHAPTER

18

Regulatory pathways and compliance

18.1 Introduction

The regulatory landscape for cell-free protein synthesis presents unique challenges that differ significantly from traditional cell-based biomanufacturing. As this technology transitions from research laboratories to commercial applications, understanding and navigating the complex regulatory framework becomes critical for successful product development and market entry. The absence of living cells in the production process creates both opportunities and uncertainties in regulatory interpretation, requiring careful consideration of existing guidelines and proactive engagement with regulatory authorities.

The regulatory pathway for cell-free manufactured products involves multiple interconnected considerations spanning manufacturing processes, product quality, safety assessment, and compliance with established pharmaceutical standards. Unlike conventional fermentation-based production, cell-free systems operate under fundamentally different principles that may not be fully addressed by current regulatory frameworks, necessitating a comprehensive understanding of how existing regulations apply to these novel manufacturing platforms.

18.2 Current regulatory framework

The regulatory environment for cell-free protein synthesis operates within the broader context of biotechnology product regulation, with specific adaptations required to address the unique characteristics of cell-free manufacturing. The framework varies significantly across different jurisdictions, with the United States Food and Drug Administration (FDA), European Medicines Agency (EMA), and other international regulatory bodies each maintaining distinct approaches to novel manufacturing technologies.

In the United States, the regulatory oversight of cell-free manufactured products falls primarily under the FDA's Center for Biologics Evaluation and Research for biological products and the Center for Drug Evaluation and Research for pharmaceutical applications. The regulatory classification depends largely on the intended use of the final product rather than the manufacturing method, though the novel nature of cell-free systems requires careful consideration of existing guidance documents and their applicability to these platforms.

Cell Free Synthesis of Biologics
https://doi.org/10.1016/B978-0-443-33271-5.00027-2

TABLE 18.1 Key regulatory frameworks for cell-free protein synthesis across major jurisdictions

Jurisdiction	Primary regulatory body	Key guidelines	Cell-free specific considerations
United States	FDA (CBER/ CDER)	ICH Q7, Q8, Q9, Q10; 21 CFR Part 210/211	Process validation without living cells; sterility assurance; novel impurity profiles
European Union	EMA (CHMP)	ICH Guidelines; Annex 1; GMP Parts I and II	Risk-based approach; scientific advice procedures; advanced therapy considerations
Japan	PMDA	ICH Guidelines; Japanese GMP	Technology transfer requirements; local manufacturing considerations
Canada	Health Canada	ICH Guidelines; Canadian GMP	Novel manufacturing pathway assessments; quality by design implementation
Australia	TGA	ICH Guidelines; Australian GMP	Risk-based manufacturing assessments; postmarket surveillance

The European regulatory framework operates under the EMA's oversight, with specific attention to the Committee for Medicinal Products for Human Use guidelines for novel manufacturing approaches. The regulatory pathway in Europe emphasizes risk-based assessment and scientific rationale for manufacturing processes, providing opportunities for cell-free systems that can demonstrate robust quality and safety profiles.

Table 18.1 summarizes the key regulatory frameworks applicable to cell-free protein synthesis across major jurisdictions, highlighting the primary regulatory bodies, applicable guidelines, and specific considerations for cell-free manufacturing.

As referenced in Table 18.1, the regulatory landscape demonstrates both harmonization through International Council for Harmonisation guidelines and jurisdiction-specific requirements that must be addressed for successful market access. The complexity increases when considering global development strategies that require simultaneous compliance across multiple regulatory frameworks.

18.3 Quality systems and Good Manufacturing Practice

The implementation of quality systems for cell-free protein synthesis requires adaptation of traditional Good Manufacturing Practice (GMP) principles to accommodate the unique characteristics of cell-free manufacturing. The absence of living microorganisms fundamentally alters contamination control strategies, sterility assurance approaches, and process monitoring requirements compared to conventional fermentation-based production.

Quality system design for cell-free manufacturing must address several critical elements that differ from traditional biomanufacturing. The open system nature of many cell-free platforms requires enhanced environmental controls and contamination prevention strategies. The reliance on prepared lysates or reconstituted systems introduces supply chain quality considerations that extend beyond the immediate manufacturing facility to include lysate preparation and component sourcing.

TABLE 18.2 Quality system elements and adaptations for cell-free manufacturing

Quality system element	Traditional GMP requirement	Cell-free adaptation
Personnel training	Standard GMP training programs	Specialized training on cell-free systems, lysate handling, energy regeneration
Facilities and equipment	Controlled environment, validated equipment	Enhanced environmental controls, specialized reaction vessels, automated systems
Materials management	Raw material qualification, supplier audits	Lysate qualification, enzyme sourcing, complex supply chain management
Production controls	Batch records, in-process controls	Continuous monitoring, real-time analytics, process analytical technology
Quality control	Release testing, stability studies	Novel analytical methods, unique impurity profiles, shelf-life determination
Documentation	Standard documentation practices	Process development records, optimization studies, technology transfer
Change control	Formal change control procedures	Impact assessment for process modifications, revalidation requirements
Deviation management	Investigation and CAPA systems	Root cause analysis for cell-free specific issues, preventive measures

The pharmaceutical quality system framework, as outlined in ICH Q10, provides the foundation for cell-free manufacturing quality systems, though specific adaptations are required to address the unique process characteristics. Risk management principles from ICH Q9 become particularly important given the novel nature of cell-free systems and the need to identify and control previously uncharacterized risks associated with this manufacturing approach.

Table 18.2 presents the key elements of a comprehensive quality system for cell-free protein synthesis, highlighting the traditional GMP requirements and the specific adaptations needed for cell-free manufacturing platforms.

The quality system implementation for cell-free manufacturing, as detailed in Table 18.2, requires careful consideration of how traditional GMP principles apply to these novel manufacturing platforms while addressing unique requirements that arise from the cell-free approach.

Environmental monitoring and contamination control represent particularly critical aspects of cell-free manufacturing quality systems. The open system nature of many cell-free platforms increases exposure risks, requiring enhanced monitoring programs and rapid response capabilities. Unlike closed fermentation systems, cell-free reactions may be more susceptible to environmental contamination, necessitating robust environmental control strategies and continuous monitoring programs.

18.4 Process validation strategies

Process validation for cell-free protein synthesis requires a comprehensive approach that demonstrates process capability, reliability, and reproducibility while addressing the unique characteristics of cell-free manufacturing. The validation strategy must encompass the entire manufacturing process from lysate preparation through final product recovery, with particular attention to the interconnected nature of cell-free system components.

The traditional three-stage validation approach, consisting of process design, process qualification, and continued process verification, applies to cell-free manufacturing with specific modifications to address the novel aspects of these systems. Process design for cell-free systems requires extensive characterization of the manufacturing platform, including lysate performance, energy regeneration efficiency, and product formation kinetics, under various operating conditions.

Process qualification for cell-free manufacturing involves demonstration of process capability through carefully designed validation studies that address both routine manufacturing scenarios and potential process variations. The absence of living cells eliminates certain traditional concerns such as cell viability and growth consistency but introduces new validation requirements related to lysate stability, enzyme activity maintenance, and energy system performance over extended manufacturing campaigns (Fig. 18.1).

The validation strategy must address several critical process parameters unique to cell-free systems. Lysate quality and consistency represent fundamental validation requirements, as

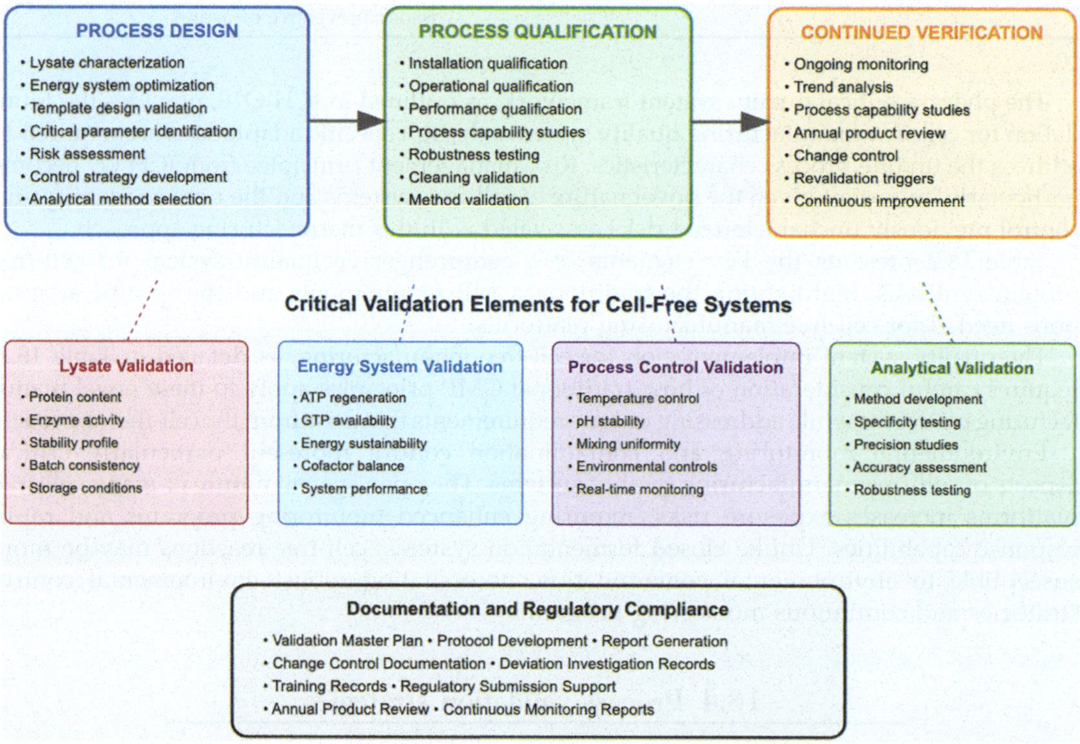

FIG. 18.1 The comprehensive process validation framework for cell-free protein synthesis, showing the interconnected validation elements and their relationships throughout the manufacturing process.

variations in lysate preparation can significantly impact manufacturing performance. Energy regeneration system validation ensures consistent ATP/GTP supply throughout the manufacturing process, while template quality and concentration validation addresses the critical role of DNA or mRNA templates in determining product yield and quality.

Temperature control validation assumes heightened importance in cell-free systems due to the sensitivity of the reconstituted enzymatic machinery to thermal variations. Unlike living cell systems that possess inherent thermal regulation mechanisms, cell-free systems rely entirely on external temperature control for optimal performance, making temperature validation critical for process success.

Table 18.3 summarizes the key validation parameters for cell-free protein synthesis, highlighting the critical process parameters, acceptance criteria, and validation approaches specific to cell-free manufacturing.

The validation approach detailed in Table 18.3 requires careful coordination across multiple interconnected systems to ensure overall process reliability and product quality consistency. The complexity of validation increases when considering continuous or semicontinuous cell-free manufacturing platforms that operate over extended periods without discrete batch boundaries.

Analytical method validation represents another critical component of the overall validation strategy, particularly given the unique analytical challenges associated with cell-free manufactured products. The complex composition of cell-free reaction mixtures, including numerous enzymes, cofactors, and potential impurities, requires sophisticated analytical approaches that may differ significantly from traditional biomanufacturing analytical methods.

TABLE 18.3 Key validation parameters for cell-free protein synthesis

Validation parameter	Critical process parameters	Acceptance criteria	Validation approach
Lysate quality	Protein concentration, enzyme activity, stability	Predefined activity ranges, reproducibility limits	Multilot characterization, stability studies
Energy system	ATP/GTP regeneration rate, sustainability	Minimum energy levels, duration requirements	Kinetic studies, extended run validation
Temperature control	Reaction temperature, uniformity, stability	±1°C control, minimal variation	Mapping studies, worst-case testing
Template quality	Concentration, integrity, purity	Sequence fidelity, minimal degradation	Analytical characterization, stability assessment
Mixing and agitation	Homogeneity, oxygen transfer	Uniform distribution, optimal conditions	Computational fluid dynamics, tracer studies
Contamination control	Bioburden, endotoxin levels	Specified limits, sterility assurance	Environmental monitoring, challenge studies
Product recovery	Yield, purity, integrity	Target specifications, process capability	Process performance qualification, robustness testing

18.5 Analytical method development

Analytical method development for cell-free protein synthesis requires comprehensive characterization capabilities that address both the manufacturing process and final product quality. The complex composition of cell-free systems, containing numerous enzymes, cofactors, nucleotides, and other components, presents unique analytical challenges that require specialized method development approaches.

Product characterization for cell-free manufactured proteins must address both traditional protein quality attributes and cell-free specific considerations. Primary structure confirmation through amino acid sequencing and mass spectrometry provides fundamental product identification, while higher-order structure analysis through circular dichroism, nuclear magnetic resonance, or other biophysical techniques confirms proper protein folding and structural integrity.

The analytical strategy must address potential cell-free specific impurities that may not be present in traditional cell-based manufacturing. Residual enzymes from the cell-free system, nucleotides from energy regeneration systems, and other process-related impurities require specific analytical methods for detection and quantification. The development of these methods requires careful consideration of the expected impurity profiles and their potential impact on product safety and efficacy.

Process monitoring analytics play a particularly important role in cell-free manufacturing due to the open system nature and real-time visibility into the manufacturing process. Unlike traditional fermentation where intracellular processes remain largely invisible until harvest, cell-free systems allow continuous monitoring of product formation, substrate consumption, and system performance throughout the manufacturing process.

Table 18.4 outlines the comprehensive analytical testing strategy for cell-free manufactured products, encompassing identity, purity, potency, and safety testing requirements with specific considerations for cell-free manufacturing platforms.

The analytical approach described in Table 18.4 requires careful method validation to ensure reliability, accuracy, and regulatory acceptability. Method validation for cell-free products may require additional robustness testing to account for the unique matrix effects associated with cell-free reaction components.

TABLE 18.4 Comprehensive analytical testing strategy for cell-free manufactured products.

Test category	Analytical methods	Cell-free considerations	Regulatory expectations
Identity testing	Mass spectrometry, N-terminal sequencing, peptide mapping	Confirmation of proper initiation, absence of truncated products	ICH Q6B guidelines, compendia methods where applicable
Purity testing	RP-HPLC, SEC, CE-SDS, bioburden	Cell-free specific impurities, enzyme contamination	Process-related impurity limits, safety qualification
Potency testing	Bioassays, binding assays, enzymatic activity	Functional integrity post cell-free synthesis	Relevant biological activity, stability-indicating
Safety testing	Endotoxin, bioburden, residual DNA	Novel impurity profiles, lysate-derived contaminants	Pharmacopeial limits, risk-based specifications
Characterization	Higher-order structure, glycosylation, aggregation	Posttranslational modification assessment	Comprehensive characterization studies

Real-time analytics and process analytical technology implementation offer significant advantages for cell-free manufacturing, enabling continuous process monitoring and real-time quality assessment. The development of suitable process analytical technology sensors and methods requires careful consideration of the cell-free environment and potential interference from system components.

Stability-indicating analytical methods assume particular importance for cell-free manufactured products due to potential unique degradation pathways and stability characteristics. The development of these methods requires comprehensive forced degradation studies to identify potential degradation products and establish appropriate analytical sensitivity for stability monitoring.

18.6 Regulatory submissions

Regulatory submission strategies for cell-free manufactured products require careful preparation and presentation of manufacturing information that clearly demonstrates product quality, safety, and efficacy while addressing regulatory questions about the novel manufacturing approach. The submission strategy must anticipate potential regulatory concerns and provide comprehensive data to support the safety and effectiveness of cell-free manufactured products.

The Common Technical Document format provides the framework for regulatory submissions, with specific attention required for Module 3 (Quality) sections that describe manufacturing processes, analytical methods, and product specifications. The presentation of cell-free manufacturing information requires clear explanation of the technology platform, detailed process descriptions, and comprehensive validation data that demonstrates manufacturing capability and control.

Manufacturing section preparation for cell-free systems requires detailed description of the cell-free platform, including lysate preparation, energy regeneration systems, reaction conditions, and product recovery processes. The submission must clearly articulate how traditional manufacturing principles apply to cell-free systems while addressing unique aspects that may not be covered by existing regulatory guidance.

Regulatory strategy development should include early engagement with regulatory authorities through presubmission meetings, scientific advice procedures, or other consultation mechanisms. These interactions provide opportunities to discuss the cell-free manufacturing approach, address potential regulatory concerns, and obtain guidance on submission requirements and expectations (Fig. 18.2).

The submission must address manufacturing comparability when transitioning from development-scale to commercial-scale cell-free manufacturing. Unlike traditional cell-based systems where scale-up involves increasing fermenter volumes, cell-free scale-up may involve different approaches such as reaction parallelization, continuous processing, or alternative reactor configurations that require careful comparability assessment.

Quality specification development for cell-free manufactured products requires consideration of both traditional protein quality attributes and cell-free specific parameters. The specifications must reflect the manufacturing capability while ensuring product safety and efficacy. Specification setting should be based on comprehensive manufacturing data that demonstrates process capability and product consistency.

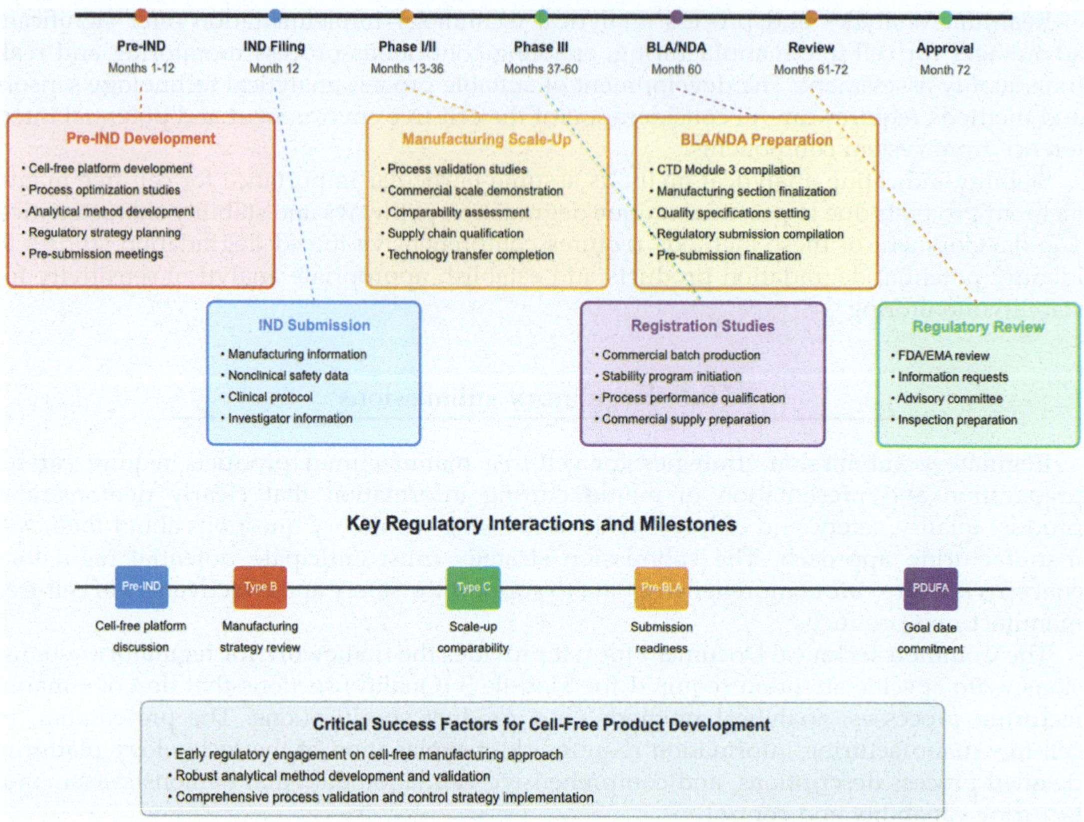

FIG. 18.2 The regulatory submission timeline and key milestone activities for cell-free manufactured products, illustrating the integration of development activities with regulatory submission preparation and review processes.

Risk assessment documentation plays a critical role in regulatory submissions for cell-free manufactured products. The risk assessment should address both traditional manufacturing risks and novel risks associated with cell-free systems, including lysate variability, energy system performance, and environmental contamination potential. The risk control measures should be clearly described and validated to demonstrate effective risk mitigation.

18.7 International harmonization

International harmonization efforts for cell-free protein synthesis regulatory frameworks remain in early stages, with significant opportunities for coordinated development of guidance documents and regulatory approaches. The International Council for Harmonisation guidelines provide a foundation for harmonized approaches, though specific guidance for cell-free manufacturing has not yet been developed.

The regulatory landscape varies significantly across different jurisdictions, creating challenges for global development programs that must address multiple regulatory requirements simultaneously. Harmonization efforts should focus on developing consistent approaches to cell-free manufacturing assessment while respecting regional regulatory preferences and requirements.

Scientific working groups and international conferences provide forums for regulatory harmonization discussions, bringing together regulatory authorities, industry representatives, and academic researchers to discuss emerging technologies and regulatory approaches. These venues offer opportunities to advance harmonized thinking about cell-free manufacturing regulation and identify areas where coordinated guidance development would benefit the industry.

Table 18.5 summarizes the current international regulatory landscape for cell-free protein synthesis, highlighting opportunities for harmonization and areas where jurisdictional differences create development challenges.

The harmonization landscape described in Table 18.5 presents both opportunities and challenges for advancing cell-free manufacturing regulation. Coordinated efforts among regulatory authorities could accelerate the development of appropriate guidance documents while ensuring consistent regulatory approaches across jurisdictions.

Regulatory science initiatives focused on cell-free manufacturing could provide the scientific foundation for harmonized regulatory approaches. These initiatives should address fundamental questions about cell-free manufacturing safety, quality, and efficacy while developing the scientific tools and methodologies needed for effective regulatory assessment.

The development of international standards for cell-free manufacturing through organizations such as the International Organization for Standardization could provide additional harmonization opportunities. These standards could address manufacturing practices, quality systems, and analytical methods specific to cell-free systems while maintaining consistency with existing pharmaceutical standards.

TABLE 18.5 International harmonization landscape for cell-free protein synthesis

Harmonization area	Current status	Opportunities	Challenges
Manufacturing guidelines	Limited specific guidance	Development of ICH guidance document	Jurisdictional differences in manufacturing requirements
Quality standards	Traditional protein guidelines apply	Cell-free specific quality considerations	Novel analytical method acceptance
Safety assessment	Case-by-case evaluation	Standardized safety evaluation frameworks	Limited precedent for risk assessment
Clinical development	Standard development pathways	Harmonized development guidance	Manufacturing scalability demonstration
Regulatory science	Emerging research initiatives	Coordinated research programs	Resource allocation and prioritization

Future harmonization efforts should consider the rapid evolution of cell-free technology and the need for flexible regulatory frameworks that can accommodate technological innovations while maintaining appropriate safety and quality standards. The regulatory framework should be sufficiently robust to ensure product safety while avoiding unnecessary barriers to technological advancement and innovation.

Bibliography

European Medicines Agency, 2017. ICH Guideline Q8 (R2) on Pharmaceutical Development.
European Medicines Agency, 2017. Guideline on the requirements for the chemical and pharmaceutical quality documentation concerning investigational medicinal products in clinical trials.
European Pharmacopoeia, 2019. General Chapter 5.2.14: substitution of in vivo method(s) by in vitro method(s) for the quality control of vaccines.
Food and Drug Administration, 2011. Guidance for Industry: Process Validation: General Principles and Practices.
Health Canada, 2016. Guidance Document: Quality (Chemistry and Manufacturing) Guidance: New Drug Submissions (Ndss) and Abbreviated New Drug Submissions (Andss).
International Council for Harmonisation, 1999. ICH Q6B: Specifications: Test Procedures and Acceptance Criteria for Biotechnological/Biological Products.
International Council for Harmonisation, 2008. ICH Q10: Pharmaceutical Quality System.
International Council for Harmonisation, 2009. ICH Q8(R2): Pharmaceutical Development.
International Council for Harmonisation, 2015. ICH Q9: Quality Risk Management.
International Council for Harmonisation, 2016. ICH Q7: Good Manufacturing Practice Guide for Active Pharmaceutical Ingredients.
International Organization for Standardization, 2016. ISO 13485:2016 Medical Devices — Quality Management Systems.
Pharmaceuticals and Medical Devices Agency, 2009. Guideline for Quality, Safety and Efficacy Assurance of Follow-on Biologics.
Therapeutic Goods Administration, 2018. Australian Regulatory Guidelines for Biologicals.
United States Pharmacopeia, 2017. General Chapter<1220> Analytical Procedure Lifecycle.
World Health Organization, 2013. Guidelines on the quality, safety, and efficacy of biotherapeutic protein products prepared by recombinant DNA technology.

CHAPTER

19

Quality control and analytics

19.1 Introduction

Quality control and analytics form the cornerstone of successful cell-free protein synthesis manufacturing, providing the critical data needed to ensure product quality, process consistency, and regulatory compliance. The unique characteristics of cell-free systems present both opportunities and challenges for analytical method development, requiring sophisticated approaches that address the complex composition of cell-free reaction mixtures while maintaining the sensitivity and specificity needed for pharmaceutical manufacturing.

The analytical framework for cell-free protein synthesis must accommodate the dynamic nature of these open systems, where multiple biochemical processes occur simultaneously in a defined reaction environment. Unlike traditional cell-based manufacturing where many processes remain hidden within cellular compartments, cell-free systems offer unprecedented visibility into the manufacturing process, enabling real-time monitoring and control that can significantly enhance product quality and process efficiency.

The development of robust analytical methods for cell-free manufacturing requires careful consideration of the unique matrix effects, potential interferents, and novel impurity profiles associated with these systems. The analytical strategy must support both process development activities and commercial manufacturing operations, providing the data necessary for process optimization, quality assurance, and regulatory compliance throughout the product lifecycle.

19.2 Process analytical technology implementation

Process analytical technology (PAT) represents a transformative approach for cell-free protein synthesis monitoring, enabling real-time process understanding and control through continuous measurement of critical quality attributes. The implementation of PAT in cell-free systems leverages the inherent accessibility of the reaction environment to provide unprecedented insight into manufacturing processes that were previously opaque in traditional cell-based systems.

The PAT framework for cell-free manufacturing encompasses multiple analytical technologies working in concert to provide comprehensive process monitoring. Spectroscopic

Cell Free Synthesis of Biologics
https://doi.org/10.1016/B978-0-443-33271-5.00026-0

methods, including near-infrared, mid-infrared, and Raman spectroscopy, offer noninvasive monitoring capabilities that can track protein formation, substrate consumption, and system performance without disrupting the manufacturing process. These techniques provide rich chemical information that can be processed using chemometric approaches to extract meaningful process insights.

Chromatographic monitoring systems enable real-time assessment of product formation and impurity profiles throughout the manufacturing process. High-performance liquid chromatography with online detection provides quantitative measurement of product concentration and purity, while size exclusion chromatography can monitor protein aggregation and degradation in real time. These approaches enable immediate process adjustments to maintain optimal manufacturing conditions.

Table 19.1 demonstrates the comprehensive nature of PAT implementation in cell-free systems, highlighting the diverse analytical technologies available for process monitoring and their specific applications in cell-free manufacturing environments.

The integration of PAT systems requires careful consideration of the cell-free reaction environment and potential analytical interferences. The complex composition of cell-free systems, including numerous enzymes, cofactors, and substrates, can create matrix effects

TABLE 19.1 Process analytical technology applications in cell-free manufacturing.

PAT technology	Measured parameters	Implementation approach	Key benefits
Near-infrared spectroscopy	Protein concentration, water content, organic solvents	Fiber optic probes, immersion probes	Noninvasive, real-time monitoring, multivariate analysis
Raman spectroscopy	Molecular structure, protein conformation, chemical composition	In-line probes, flow cells	High specificity, minimal sample preparation, chemometric compatibility
Fluorescence spectroscopy	Protein folding, tryptophan exposure, system pH	Fiber optic systems, flow-through cells	High sensitivity, rapid response, minimal interference
Online HPLC	Product concentration, purity, impurity profiling	Automated sampling systems, valve switching	Quantitative analysis, regulatory acceptance, proven methods
Mass spectrometry	Molecular weight, structural confirmation, impurity identification	Online LC–MS systems, direct infusion	Structural information, high specificity, comprehensive characterization
pH monitoring	System pH, buffer capacity, acid production	Ion-selective electrodes, optical sensors	Critical parameter control, system stability assessment
Dissolved oxygen	Oxygen consumption, system metabolism, redox state	Polarographic electrodes, optical sensors	Process optimization, contamination detection
Temperature control	Reaction temperature, thermal gradients, heat generation	Distributed sensors, thermal imaging	Process control, optimization, safety monitoring

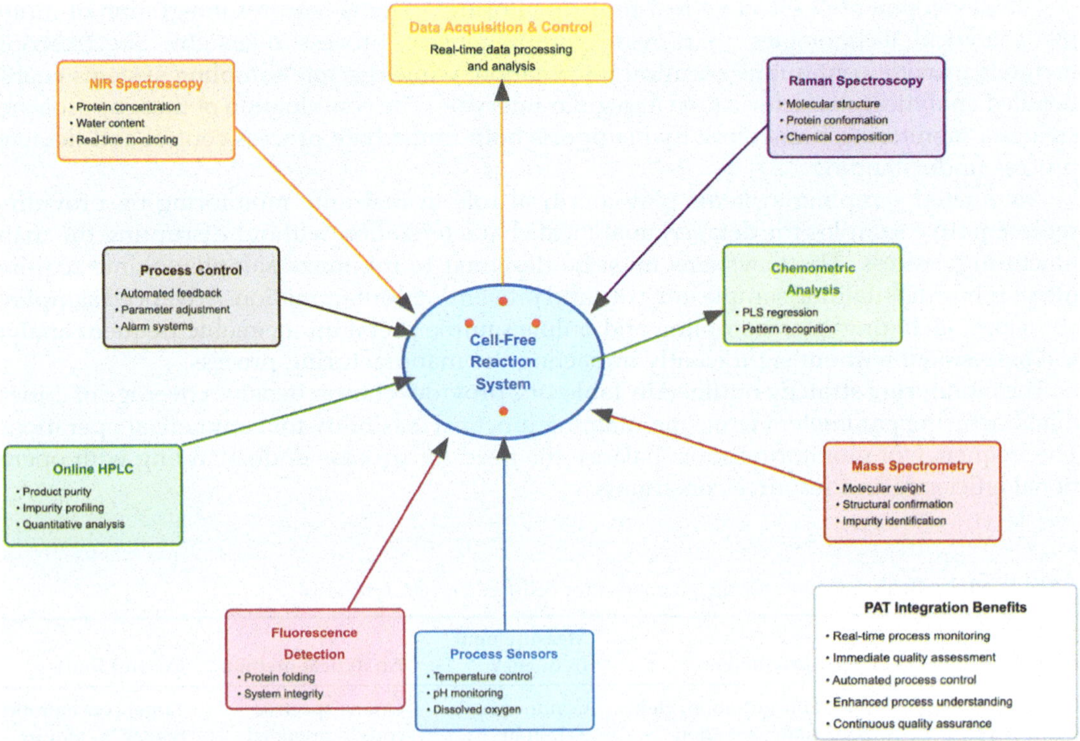

FIG. 19.1 Illustration of the integrated PAT system architecture for cell-free manufacturing, showing the interconnected analytical technologies and their role in comprehensive process monitoring and control.

that must be addressed through appropriate method development and validation. Chemometric approaches, including partial least squares regression and principal component analysis, provide powerful tools for extracting meaningful information from complex spectroscopic data while accounting for system variability and interference.

Real-time data processing and analysis capabilities enable immediate process adjustments based on PAT measurements, supporting the implementation of advanced process control strategies. Model predictive control and other advanced control algorithms can utilize PAT data to maintain optimal manufacturing conditions automatically, reducing process variability and improving product quality consistency (Fig. 19.1).

19.3 Real-time monitoring systems

Real-time monitoring represents a fundamental advantage of cell-free protein synthesis, enabling continuous assessment of manufacturing performance and immediate response to process variations. The open nature of cell-free systems allows direct access to the reaction environment, facilitating the implementation of comprehensive monitoring strategies that provide unprecedented visibility into the manufacturing process.

The development of effective real-time monitoring systems requires integration of multiple analytical technologies to provide comprehensive process oversight. Spectroscopic methods provide continuous chemical information, while discrete sampling systems enable detailed analytical characterization at regular intervals. The combination of these approaches creates a monitoring framework that supports both immediate process control and detailed process understanding.

Automated sampling systems play a critical role in real-time monitoring by providing representative samples for detailed analytical characterization without disrupting the manufacturing process. These systems must be designed to minimize sample volume requirements while maintaining sample integrity and preventing contamination. Advanced sampling strategies, including microsampling and online sample preparation, enable frequent analytical assessment without significantly impacting the manufacturing process.

The monitoring strategy outlined in Table 19.2 provides comprehensive coverage of critical manufacturing parameters while maintaining practical feasibility for commercial operations. The frequency of monitoring must balance the need for process understanding with operational efficiency and resource constraints.

TABLE 19.2 Real-time monitoring parameters for cell-free protein synthesis.

Monitoring category	Critical parameters	Measurement frequency	Analytical method	Control limits
Product formation	Protein concentration, yield progression, formation kinetics	Continuous or 5–15-min intervals	Online spectroscopy, automated HPLC	Target concentration ranges, minimum yield thresholds
System performance	ATP/GTP levels, energy regeneration rate, enzyme activity	15–30-min intervals	Enzymatic assays, luminescence detection	Minimum energy levels, activity thresholds
Process conditions	Temperature, pH, dissolved oxygen, agitation	Continuous	Process sensors, automated monitoring	Specification ranges, alarm limits
Contamination control	Bioburden, endotoxin levels, system sterility	Hourly to daily	Rapid detection methods, traditional testing	Zero tolerance, specification limits
Substrate utilization	Amino acid consumption, cofactor depletion, template integrity	30–60-min intervals	Amino acid analysis, spectrophotometry	Minimum substrate levels, consumption rates
Product quality	Purity, aggregation, degradation, structural integrity	1–4-h intervals	SEC-HPLC, analytical characterization	Quality specifications, trend monitoring
Impurity formation	Process-related impurities, degradation products, host cell proteins	2–6-h intervals	Specific analytical methods, LC–MS	Maximum allowable levels, trend analysis

Data management and trending analysis capabilities are essential components of real-time monitoring systems, enabling the identification of process trends and potential issues before they impact product quality. Statistical process control methods, including control charts and trend analysis, provide tools for monitoring process stability and identifying opportunities for process improvement.

The implementation of alarm systems and automated responses enhances the effectiveness of real-time monitoring by enabling immediate action when process parameters deviate from acceptable ranges. These systems must be carefully designed to minimize false alarms while ensuring appropriate response to genuine process deviations.

19.4 Product characterization strategies

Product characterization for cell-free manufactured proteins requires comprehensive analytical approaches that address both traditional protein quality attributes and unique considerations associated with cell-free production. The characterization strategy must demonstrate product consistency, safety, and efficacy while addressing potential cell-free specific variations that may impact product quality.

Structural characterization forms the foundation of product quality assessment, encompassing primary, secondary, tertiary, and quaternary structure analysis. Primary structure confirmation through amino acid sequencing, peptide mapping, and mass spectrometry ensures correct protein synthesis and identifies potential sequence variants or modifications. Secondary structure analysis using circular dichroism spectroscopy and Fourier transform infrared spectroscopy confirms proper protein folding and structural integrity.

Higher-order structure characterization becomes particularly important for cell-free manufactured proteins due to potential differences in folding environment compared to native cellular conditions. Nuclear magnetic resonance spectroscopy, hydrogen–deuterium exchange mass spectrometry, and other advanced biophysical techniques provide detailed structural information that can identify subtle structural differences that may impact biological activity.

The characterization framework presented in Table 19.3 provides a systematic approach to product quality assessment that addresses both traditional protein characterization requirements and cell-free specific considerations.

Posttranslational modification analysis requires particular attention in cell-free systems, as the modification patterns may differ from those observed in native cellular environments. Mass spectrometry-based approaches, including liquid chromatography–tandem mass spectrometry and high-resolution mass spectrometry, provide detailed information about glycosylation, phosphorylation, and other modifications that may impact biological activity.

Aggregation and degradation monitoring throughout the manufacturing process and during storage provides critical information about product stability and shelf life. Size exclusion chromatography coupled with multiangle light scattering offers precise molecular weight determination and aggregation assessment, while analytical ultracentrifugation provides detailed information about protein–protein interactions and solution behavior.

TABLE 19.3 Comprehensive product characterization framework for cell-free manufactured proteins.

Characterization category	Analytical methods	Key attributes assessed	Regulatory expectations
Primary structure	N-terminal sequencing, peptide mapping, mass spectrometry	Amino acid sequence, posttranslational modifications, truncations	ICH Q6B identity requirements, sequence confirmation
Secondary structure	Circular dichroism, FTIR spectroscopy, Raman spectroscopy	Alpha-helix, beta-sheet content, structural stability	Biophysical characterization, comparability studies
Tertiary structure	NMR spectroscopy, fluorescence spectroscopy, analytical ultracentrifugation	Protein folding, domain structure, conformational stability	Higher-order structure assessment, functional correlation
Quaternary structure	SEC-MALS, analytical ultracentrifugation, native PAGE	Oligomerization state, subunit interactions, complex formation	Aggregation assessment, biological activity correlation
Purity analysis	RP-HPLC, SEC, CE-SDS, IEF	Product-related substances, process-related impurities	ICH Q6B purity requirements, safety qualification
Biological activity	Bioassays, binding assays, enzymatic activity	Potency, functional integrity, biological relevance	Pharmacopeial requirements, clinical correlation
Physicochemical properties	pH, osmolality, conductivity, appearance	Solution characteristics, formulation stability	Pharmacopeial specifications, patient safety
Safety assessment	Bioburden, endotoxin, residual DNA, host cell proteins	Microbiological safety, pyrogenic substances	Safety specifications, risk assessment

19.5 Impurity profiling and control

Impurity profiling for cell-free manufactured proteins requires comprehensive understanding of potential product-related and process-related impurities that may arise during manufacturing. The unique composition of cell-free systems introduces novel impurity sources that must be characterized, controlled, and qualified to ensure product safety and regulatory compliance.

Product-related impurities in cell-free systems include truncated proteins, aggregated species, deamidated variants, and other modified forms that may arise during synthesis or storage. The open nature of cell-free systems may influence the formation of these impurities, requiring careful characterization to understand their formation mechanisms and establish appropriate control strategies.

Process-related impurities present unique challenges in cell-free manufacturing due to the complex composition of cell-free reaction mixtures. Residual enzymes from the cell-free system, nucleotides from energy regeneration systems, and other reaction components require specific analytical methods for detection and quantification. The safety qualification of these impurities may require novel approaches due to limited precedent in traditional biomanufacturing.

The impurity control strategy outlined in Table 19.4 requires integration of analytical characterization, process control, and safety assessment to ensure comprehensive impurity management throughout the manufacturing process.

TABLE 19.4 Impurity classification and control strategies for cell-free manufacturing.

Impurity category	Specific impurities	Analytical methods	Control strategies	Acceptance criteria
Product-related substances	Aggregates, fragments, deamidated forms, oxidized variants	SEC, RP-HPLC, CE-SDS, mass spectrometry	Process optimization, formulation development	ICH Q6B guidelines, safety qualification
Host cell proteins	*E. coli* proteins, wheat germ proteins, mammalian cell proteins	ELISA, LC–MS/MS, 2D electrophoresis	Purification optimization, downstream processing	Risk-based specifications, clearance studies
Residual DNA	Template DNA, genomic DNA, plasmid DNA	qPCR, fluorescence assays, gel electrophoresis	DNase treatment, purification steps	WHO guidelines, safety assessment
Endotoxin	Bacterial endotoxins, pyrogenic substances	LAL assay, recombinant Factor C assay	Source control, purification validation	Pharmacopeial limits, pyrogenicity testing
Process chemicals	Buffers, salts, cofactors, reducing agents	Ion chromatography, spectrophotometry, mass spectrometry	Process design, purification efficiency	ICH Q3C guidelines, safety evaluation
Enzyme contaminants	Proteases, nucleases, ribosomal proteins	Activity assays, immunoassays, mass spectrometry	Lysate quality control, purification validation	Functional impact assessment, safety qualification
Energy system components	ATP, GTP, creatine phosphate, phosphoenolpyruvat (PEP)	HPLC, enzymatic assays, mass spectrometry	System design, downstream processing	Metabolic safety assessment, pharmacological evaluation
Degradation products	Protein fragments, amino acid modifications, chemical adducts	LC–MS, peptide mapping, amino acid analysis	Storage conditions, formulation stability	Stability specifications, safety qualification

The development of specific analytical methods for cell-free impurities requires careful consideration of matrix effects and potential analytical interferences. The complex composition of cell-free systems can impact analytical method performance, requiring robust method development and validation to ensure reliable impurity detection and quantification.

Risk assessment approaches play a critical role in establishing impurity specifications and control strategies. The novel nature of some cell-free impurities may require toxicological evaluation to establish safe exposure limits and appropriate specifications. Quality risk management principles from ICH Q9 provide a framework for systematic impurity risk assessment and control strategy development.

19.6 Stability studies and shelf-life determination

Stability assessment for cell-free manufactured proteins requires comprehensive evaluation of product degradation pathways, storage condition optimization, and shelf-life determination under various environmental conditions. The stability profile of cell-free manufactured proteins may differ from cell-based products due to potential structural differences, impurity profiles, and formulation requirements.

Accelerated stability studies provide critical information about product degradation kinetics and potential stability-limiting factors. These studies must be carefully designed to provide meaningful predictions of long-term stability while accounting for the unique characteristics of cell-free manufactured products. The selection of storage conditions, sampling time points, and analytical methods requires careful consideration to ensure regulatory acceptability and commercial relevance.

Forced degradation studies enable identification of potential degradation pathways and development of stability-indicating analytical methods. These studies expose the product to stress conditions, including elevated temperature, pH extremes, oxidative conditions, and light exposure, to identify potential degradation products and their formation kinetics.

The stability testing strategy presented in Table 19.5 provides comprehensive coverage of potential stability risks while maintaining practical feasibility for commercial development programs.

Container closure system compatibility studies ensure that packaging materials do not adversely affect product stability or introduce extractable or leachable substances that could impact product quality. These studies require evaluation of multiple packaging options and assessment of product–package interactions under various storage conditions.

Statistical analysis of stability data requires appropriate mathematical models to determine product shelf life and establish expiration dating. The analysis must account for analytical variability, batch-to-batch differences, and confidence intervals to ensure conservative shelf-life assignments that maintain product quality throughout the claimed storage period. (Fig. 19.2)

TABLE 19.5 Stability testing framework for cell-free manufactured proteins.

Study type	Test conditions	Duration	Analytical methods	Decision criteria
Long-term stability	2°C–8°C, primary packaging	24–36 months	Identity, purity, potency, safety	ICH Q5C guidelines, specification compliance
Accelerated stability	25°C/60% RH, 40°C/75% RH	6–12 months	Complete analytical panel	Arrhenius kinetics, shelf-life projection
Intermediate stability	30°C/65% RH	12 months	Key stability parameters	Supportive data, confirmatory studies
Freeze–thaw stability	Multiple freeze–thaw cycles	5–10 cycles	Aggregation, activity, integrity	Process robustness, handling procedures
Photostability	ICH Q1B light conditions	Light exposure schedule	Photodegradation assessment	Light protection requirements
Oxidative stress	Hydrogen peroxide, metal ions	24–72 h	Oxidation products, activity loss	Antioxidant requirements, packaging needs
Thermal stress	Elevated temperatures	1–7 days	Thermal degradation products	Temperature stability, shipping conditions
pH stress	pH 3–10 range	24–72 h	pH-related degradation	Formulation pH optimization

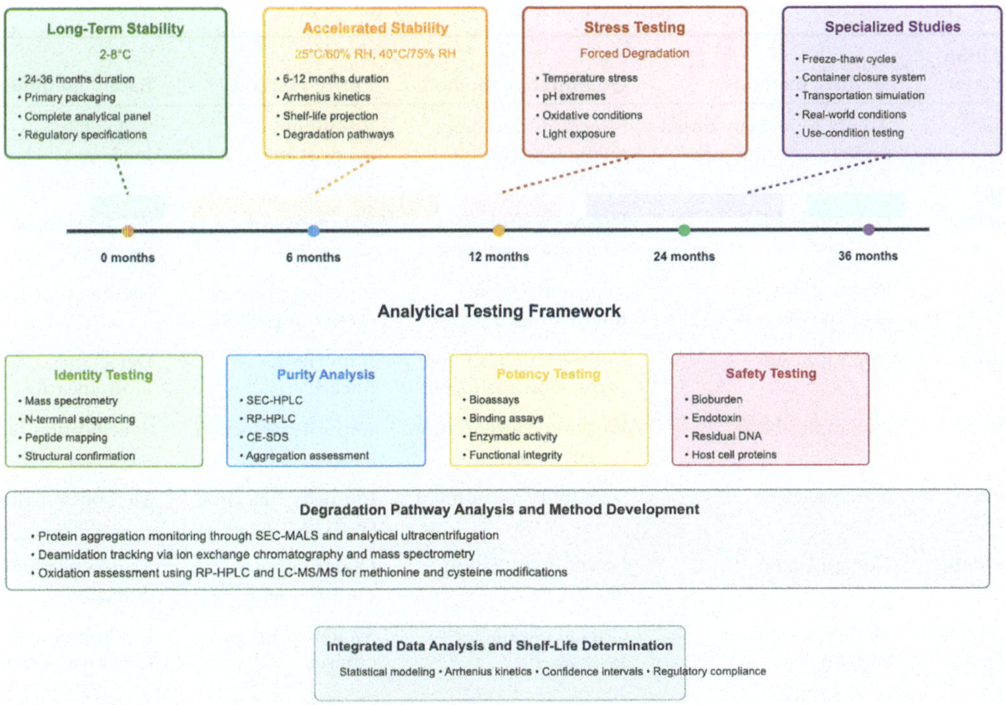

FIG. 19.2 The comprehensive stability testing strategy for cell-free manufactured proteins, illustrating the integration of different study types, analytical methods, and timeline considerations for effective stability assessment.

19.7 Method validation and regulatory compliance

Method validation for cell-free manufactured proteins requires comprehensive demonstration of analytical method performance characteristics according to regulatory guidelines and industry standards. The validation strategy must address both traditional protein analytical requirements and unique considerations associated with cell-free manufacturing systems.

The validation approach must encompass all analytical methods used for product release, stability monitoring, and process control, ensuring that each method meets appropriate performance criteria for its intended use. Identity methods require demonstration of specificity and ability to distinguish the target protein from potential variants and impurities. Quantitative methods require validation of accuracy, precision, linearity, and range to ensure reliable quantitative results.

Specificity validation assumes particular importance for cell-free manufactured proteins due to the potential for unique impurity profiles and matrix effects. The method must demonstrate ability to distinguish the target analyte from all potential interferents, including process-related impurities specific to cell-free systems.

The validation framework outlined in Table 19.6 ensures comprehensive method performance assessment while addressing the specific requirements for different analytical applications in cell-free manufacturing.

TABLE 19.6 Method validation requirements for cell-free protein analytics.

Validation parameter	Identity methods	Quantitative methods	Purity methods	Safety methods
Specificity	Ability to identify target protein, discrimination from variants	Selectivity for target analyte, interference assessment	Resolution of product from impurities	Specificity for target contaminants
Accuracy	Not applicable	Recovery studies, reference standards	Impurity spike recovery	Spiked sample recovery
Precision	Repeatability assessment	Repeatability and intermediate precision	Injection precision, system precision	Method repeatability
Linearity	Not applicable	Concentration range, correlation coefficient	Impurity level linearity	Detection limit linearity
Range	Not applicable	Working concentration range	Impurity detection range	Specification range coverage
Detection limit	Not applicable	Sensitivity assessment	Impurity detection capability	Method sensitivity
Quantitation limit	Not applicable	Lower quantitation limit	Quantitative impurity analysis	Regulatory limit detection
Robustness	Method parameter variations	Critical parameter assessment	System suitability criteria	Environmental factor assessment
System suitability	Performance verification	Ongoing performance monitoring	Column performance, resolution	Method performance verification

Robustness testing requires evaluation of method performance under varying analytical conditions to ensure reliable performance during routine use. This testing must consider potential variations in analytical instrumentation, environmental conditions, and analyst technique that may impact method performance.

Method transfer and technology transfer activities require careful documentation and verification to ensure consistent analytical performance across different laboratories and analytical sites. The transfer process must include comparative testing, analyst training, and performance verification to demonstrate equivalent analytical capability.

Ongoing method performance monitoring through system suitability testing and quality control sample analysis ensures continued method reliability throughout the product lifecycle. Statistical process control approaches enable trending of analytical performance and identification of potential method performance issues before they impact product quality assessment.

Bibliography

ASTM International, 2015. ASTM E2363-15 Standard Terminology Relating to Process Analytical Technology in the Pharmaceutical Industry. American Society for Testing and Materials.

European Medicines Agency, 2006. ICH Guideline Q2 (R1) on Validation of Analytical Procedures: Text and Methodology.

European Medicines Agency, 2017. Guideline on the Requirements for the Chemical and Pharmaceutical Quality Documentation Concerning Investigational Medicinal Products in Clinical Trials.

European Pharmacopoeia, 2019. General Chapter 2.2.46: Chromatographic Separation Techniques.

Food and Drug Administration, 2004. Guidance for Industry: PAT — A Framework for Innovative Pharmaceutical Development, Manufacturing, and Quality Assurance.

Food and Drug Administration, 2015. Guidance for Industry: Analytical Procedures and Methods Validation for Drugs and Biologics.

International Council for Harmonisation, 1995. ICH Q5C: Quality of Biotechnological Products: Stability Testing of Biotechnological/Biological Products.

International Council for Harmonisation, 1996. ICH Q1B: Photostability Testing of New Drug Substances and Products.

International Council for Harmonisation, 1999. ICH Q6B: Specifications: Test Procedures and Acceptance Criteria for Biotechnological/Biological Products.

International Council for Harmonisation, 2003. ICH Q1A(R2): Stability Testing of New Drug Substances and Products.

International Council for Harmonisation, 2005. ICH Q2(R1): Validation of Analytical Procedures: Text and Methodology.

International Organization for Standardization, 2017. ISO/IEC 17025:2017 General Requirements for the Competence of Testing and Calibration Laboratories.

United States Pharmacopeia, 2008. General Chapter<1058> Analytical Instrument Qualification.

United States Pharmacopeia, 2020. General Chapter <1225> Validation of Compendial Procedures.

World Health Organization, 2006. Guidelines on Validation of Analytical Methods Used in Examination of Pharmaceutical Materials.

PART VII

Future directions and emerging technologies

CHAPTER

20

Distributed manufacturing

The traditional paradigm of centralized biomanufacturing, characterized by massive production facilities serving global markets, is being fundamentally challenged by the emergence of distributed manufacturing models enabled by cell-free protein synthesis technologies. This transformation represents not merely a technological advancement but a complete reimagining of how therapeutic proteins, vaccines, and other biologics can be produced, distributed, and delivered to patients worldwide.

20.1 Decentralized production models

The concept of decentralized production in biotechnology mirrors the distributed computing revolution that transformed the technology sector decades earlier. Rather than relying on a few massive manufacturing facilities, decentralized production models leverage networks of smaller, interconnected production units that can operate independently while maintaining quality standards and regulatory compliance.

The fundamental architecture of decentralized cell-free manufacturing systems differs markedly from conventional bioprocessing approaches. Traditional mammalian cell culture requires extensive infrastructure, including bioreactors, downstream processing equipment, quality control laboratories, and specialized facilities for cell line maintenance. In contrast, cell-free systems can be implemented using modular components that require significantly less space, infrastructure, and specialized expertise.

Table 20.1 presents a comparative analysis of centralized versus decentralized manufacturing models across multiple dimensions relevant to biopharmaceutical production.

The economic implications of this shift are profound. Decentralized manufacturing reduces the barriers to entry for biopharmaceutical production, potentially democratizing access to manufacturing capabilities. This is particularly relevant for developing markets where the high capital requirements of traditional facilities have historically limited local production capabilities.

The technological foundation enabling decentralized manufacturing rests on several key innovations in cell-free systems. Lyophilized cell-free reagents allow for room-temperature storage and transportation, eliminating the cold chain requirements that complicate traditional biologic distribution. Standardized reaction conditions and automated systems reduce

Cell Free Synthesis of Biologics
https://doi.org/10.1016/B978-0-443-33271-5.00002-8

TABLE 20.1 Comparison of centralized vs. decentralized manufacturing models.

Parameter	Centralized model	Decentralized model
Facility size	10,000–50,000 m^2	100–1000 m^2
Capital investment	\$500 M–2 B	\$5 M–50 M
Production capacity	10,000–100,000 L	10–1000 L
Lead time to production	5–10 years	6–18 months
Geographic reach	Global distribution	Regional/local
Supply chain complexity	High	Low
Regulatory burden	Extensive	Moderate
Flexibility	Low	High
Risk concentration	High	Distributed
Quality control	Centralized laboratories	Local testing

the need for highly specialized technical expertise, making implementation feasible in diverse geographic and economic contexts.

Quality assurance in decentralized networks presents both challenges and opportunities. While maintaining consistent quality across multiple distributed sites requires robust protocols and monitoring systems, the distributed nature of production actually reduces certain risks. Supply chain disruptions, natural disasters, or geopolitical events that might completely halt production at a centralized facility would only impact a portion of a distributed network.

20.2 Mobile biomanufacturing units

The development of mobile biomanufacturing units represents the ultimate expression of distributed manufacturing flexibility. These self-contained production systems, designed to be deployed rapidly in response to specific needs or emergencies, combine the advantages of cell-free synthesis with transportation and deployment capabilities that would be impossible with conventional bioprocessing technologies.

Modern mobile biomanufacturing units typically consist of three integrated modules as detailed in Table 20.2. Each module is designed for standard shipping container dimensions, enabling deployment via existing transportation infrastructure while maintaining the controlled environment necessary for biopharmaceutical production.

The production module incorporates automated cell-free synthesis equipment capable of producing multiple protein targets simultaneously. Integrated purification systems utilize size exclusion, ion exchange, and affinity chromatography steps that can be rapidly reconfigured for different target proteins. Environmental control systems maintain the temperature, humidity, and air quality required for Good Manufacturing Practice compliance.

TABLE 20.2 Mobile biomanufacturing unit module specifications.

Module	Function	Dimensions (L × W × H)	Weight	Power requirements
Production	Cell-free synthesis, purification	12 × 2.4 × 2.6 m	15–20 tons	50–75 kW
Quality control	Analytical testing, release	6 × 2.4 × 2.6 m	8–12 tons	25–35 kW
Utilities	Power, water treatment, HVAC	6 × 2.4 × 2.6 m	10–15 tons	Variable

TABLE 20.3 Mobile unit deployment scenarios and timelines.

Scenario	Target product	Deployment time	Production duration	Regulatory status
Emergency response	Known vaccine antigen	2–5 days	2–4 weeks	Emergency use
Clinical trial supply	Investigational drug	1–2 weeks	4–12 weeks	Investigational new drug (IND) approved
Orphan drug production	Rare disease therapeutic	2–4 weeks	3–6 months	Commercial
Technology transfer	Established product	4–8 weeks	Ongoing	Full approval

Quality control modules include analytical instruments necessary for protein characterization and release testing. High-performance liquid chromatography systems, mass spectrometers, and bioactivity assays are integrated into compact, ruggedized configurations designed to maintain accuracy and precision despite transportation and deployment stresses.

The deployment timeline for mobile biomanufacturing units varies significantly depending on the complexity of the target protein and regulatory requirements. Table 20.3 illustrates typical deployment scenarios and associated timelines.

The logistics of mobile unit deployment require careful coordination of transportation, site preparation, utilities connection, and regulatory clearance. Site requirements are minimal compared to traditional facilities, typically requiring only level ground, electrical connection, and water access. However, the regulatory framework for mobile manufacturing units is still evolving, with different jurisdictions taking varying approaches to approval and oversight (Fig. 20.1).

The economic model for mobile biomanufacturing units differs fundamentally from fixed facilities. Rather than amortizing capital costs over decades of production, mobile units operate on service-based models where deployment costs are recovered over shorter time periods. This approach is particularly attractive for applications with limited duration or uncertain demand, such as clinical trial supply or emergency response.

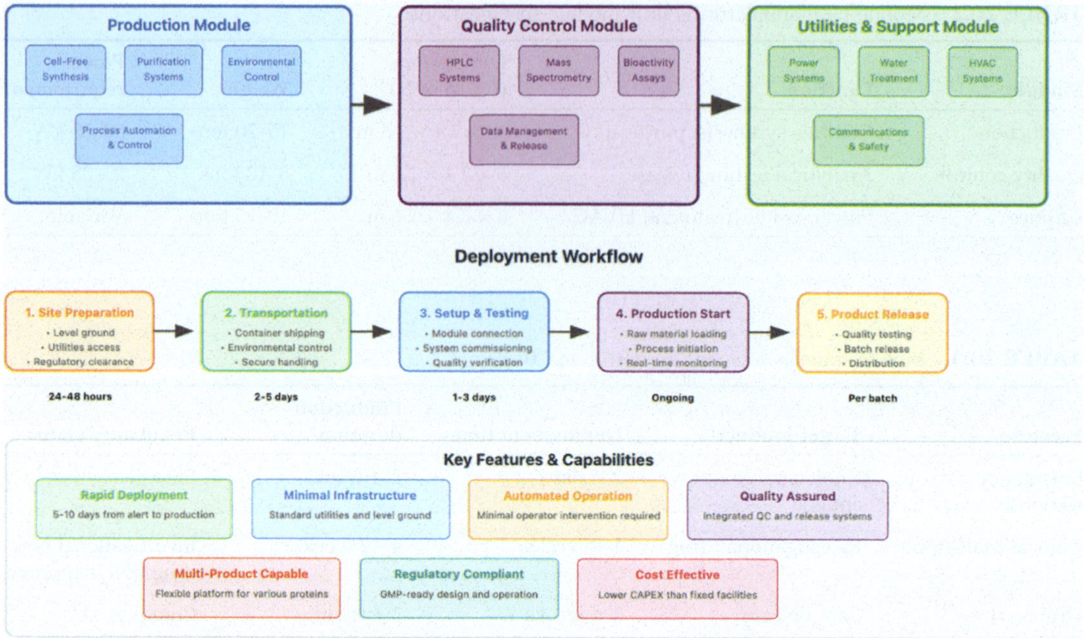

FIG. 20.1 Illustration of the deployment architecture and workflow for a mobile biomanufacturing unit, showing the integration of production, quality control, and support systems within the transportable module configuration.

20.3 Point-of-care manufacturing

Point-of-care manufacturing represents the most distributed form of biopharmaceutical production, bringing manufacturing capabilities directly to the site of patient care. This approach is enabled by the unique characteristics of cell-free protein synthesis systems, which can operate at small scales with minimal infrastructure requirements while maintaining the quality and safety standards required for therapeutic applications.

The clinical workflow integration of point-of-care manufacturing systems requires careful consideration of existing hospital and clinic operations. Production systems must be designed to operate within the space, utility, and staffing constraints of healthcare environments while meeting regulatory requirements for pharmaceutical manufacturing.

Table 20.4 describes the characteristics of different point-of-care manufacturing implementation models, ranging from automated benchtop systems to integrated clinical platforms.

The regulatory framework for point-of-care manufacturing is complex and evolving. Current regulations were developed for centralized manufacturing models and may not fully address the unique aspects of distributed production. Harmonization efforts between regulatory agencies worldwide are ongoing to develop appropriate frameworks that ensure patient safety while enabling the benefits of point-of-care manufacturing.

Quality systems for point-of-care manufacturing must balance the need for rigorous quality control with the practical constraints of healthcare environments. Automated quality

TABLE 20.4 Point-of-care manufacturing implementation models.

Model	Scale	Footprint	Automation level	Target applications
Benchtop synthesizer	1–10 mg	1–2 m^2	Fully automated	Personalized therapeutics
Clinical workstation	10–100 mg	5–10 m^2	Semiautomated	Small batch production
Hospital manufacturing suite	100 mg–1 g	20–50 m^2	Integrated automation	Multiple products
Regional production center	1–10 g	100–200 m^2	Full process automation	Multisite supply

control systems integrated into production platforms can perform real-time monitoring and release testing, reducing the burden on clinical staff while maintaining compliance with pharmaceutical quality standards.

The integration of point-of-care manufacturing with electronic health records and hospital information systems enables new paradigms of personalized medicine. Patient-specific therapeutic proteins can be designed, produced, and administered within the same healthcare system, reducing time-to-treatment and enabling truly individualized therapy.

Training and certification requirements for healthcare personnel operating point-of-care manufacturing systems represent a significant implementation consideration. While cell-free systems are designed to be more user-friendly than traditional bioprocessing, they still require understanding of basic biotechnology principles and quality systems. Standardized training programs and certification processes are being developed to address these needs.

20.4 Emergency response applications

The COVID-19 pandemic highlighted the critical importance of rapid response capabilities in biopharmaceutical manufacturing. Traditional manufacturing systems, with their long lead times and high infrastructure requirements, proved inadequate for the speed and flexibility required during health emergencies. Cell-free protein synthesis systems offer transformative capabilities for emergency response through their rapid deployment, minimal infrastructure requirements, and ability to quickly switch between different target proteins.

Emergency response manufacturing scenarios encompass a broad range of potential threats and required responses. Pandemic preparedness represents the most visible application, but other scenarios, including bioterrorism, natural disasters affecting supply chains, and emerging infectious diseases, benefit from distributed manufacturing capabilities.

The strategic deployment of emergency response manufacturing capabilities requires careful consideration of geographic distribution, storage requirements, and activation protocols. Table 20.5 outlines different emergency response deployment strategies and their characteristics.

TABLE 20.5 Emergency response deployment strategies.

Strategy	Geographic distribution	Storage requirements	Activation time	Capacity
Prepositioned units	Fixed strategic locations	Climate-controlled warehouses	24–48 h	Limited
Rapid deployment teams	Mobile response units	Transportable containers	3–7 days	Moderate
Distributed network	Multiple fixed sites	Local infrastructure	1–3 days	High
Hybrid model	Fixed + mobile components	Mixed storage	Variable	Scalable

The technical requirements for emergency response manufacturing systems differ from routine production applications. Rapid setup and operation capabilities are prioritized over optimization for maximum efficiency or lowest cost per unit. Robust operation under suboptimal conditions, minimal maintenance requirements, and simplified operation protocols are essential characteristics.

Supply chain considerations for emergency response manufacturing focus on ensuring availability of critical raw materials and reagents during crisis situations. Strategic stockpiling of lyophilized cell-free components, energy regeneration systems, and purification reagents enables rapid activation without dependence on potentially disrupted supply chains (Fig. 20.2).

The integration of emergency response manufacturing with public health infrastructure requires coordination between multiple agencies and organizations. Clear protocols for activation, oversight, and distribution must be established in advance to enable effective response when emergencies occur. International cooperation and coordination mechanisms are essential for addressing global health threats that cross national boundaries.

20.5 Global health impact

The potential impact of distributed manufacturing on global health extends far beyond emergency response scenarios. Persistent challenges in providing access to life-saving biologics in developing countries stem largely from the high costs and infrastructure requirements of traditional manufacturing systems. Distributed manufacturing using cell-free systems offers pathways to address these challenges by reducing costs, simplifying distribution, and enabling local production capabilities.

The economics of global health applications differ significantly from commercial biopharmaceutical markets. Cost per dose becomes the primary optimization target, often with willingness to accept trade-offs in production efficiency or sophistication that would be unacceptable in commercial markets. Cell-free systems are well suited to these requirements due to their scalability and ability to operate with minimal infrastructure.

Access to essential medicines in resource-limited settings is often constrained by distribution challenges rather than production costs. Cold chain requirements, customs delays, and

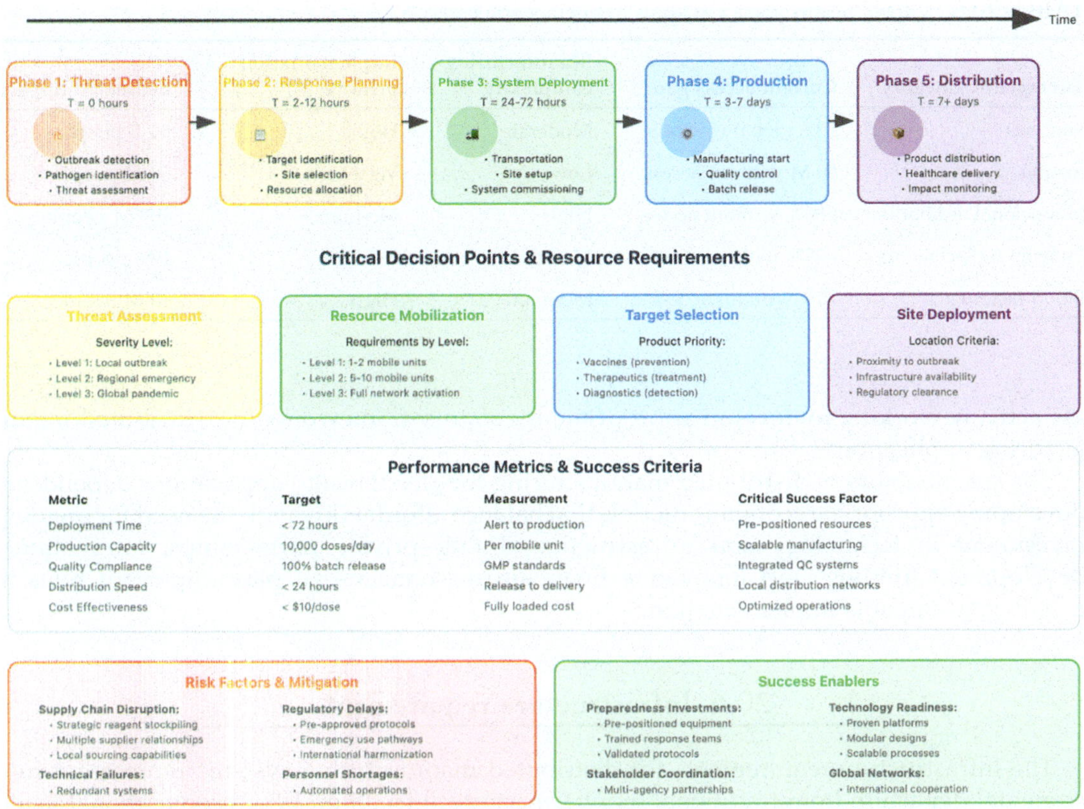

FIG. 20.2 Illustration of the emergency response manufacturing activation sequence, showing the progression from threat identification through deployment, production, and distribution of emergency therapeutics.

regulatory barriers can make effective distribution of centrally manufactured biologics extremely difficult. Local production using distributed manufacturing can eliminate many of these barriers while providing additional benefits, including job creation and technology transfer.

Table 20.6 presents an analysis of the global health impact potential for different categories of biologic therapeutics produced using distributed manufacturing systems.

The implementation of distributed manufacturing for global health applications requires addressing multiple technical, regulatory, and economic challenges. Technology transfer protocols must be developed to enable effective deployment in diverse geographic and economic contexts. Training programs for local personnel, quality systems adapted to local conditions, and sustainable financing mechanisms are all essential components of successful implementation.

International regulatory harmonization efforts are particularly important for global health applications of distributed manufacturing. The ability to manufacture products under mutually recognized quality standards enables more efficient deployment and reduces duplication of regulatory oversight. Organizations such as the World Health Organization

TABLE 20.6 Global health impact analysis by therapeutic category.

Therapeutic category	Current access gap	Manufacturing complexity	Implementation priority	Estimated impact
Vaccines	50% global coverage	Moderate	High	>1 B people
Insulin	50 M without access	Low	Very high	50 M people
Monoclonal antibodies	>90% without access	High	Medium	100 M people
Enzyme replacements	>95% without access	Moderate	Medium	1 M people
Blood factors	75% without access	Moderate	High	10 M people

are actively working to develop appropriate regulatory frameworks for distributed manufacturing applications.

The sustainability of distributed manufacturing for global health applications depends on developing appropriate economic models that balance affordability with the need for ongoing investment in technology and infrastructure. Public–private partnerships, international development funding, and innovative financing mechanisms all play important roles in enabling sustainable implementation.

20.6 Infrastructure requirements

The infrastructure requirements for distributed manufacturing systems represent a fundamental departure from traditional biopharmaceutical facilities. While conventional facilities require massive investments in specialized buildings, utilities, and equipment, distributed manufacturing systems are designed to operate with minimal infrastructure while maintaining the quality and safety standards required for pharmaceutical production.

Facility requirements for distributed manufacturing focus on flexibility and modularity rather than permanent infrastructure. Many systems are designed to operate in repurposed commercial or industrial spaces, reducing the time and cost associated with purpose-built facilities. Standard utility requirements, including electrical power, water, and waste disposal, are typically sufficient, though backup systems and quality controls may be necessary to ensure consistent operation.

The modular design philosophy underlying distributed manufacturing systems enables rapid reconfiguration for different products or applications. Table 20.7 details the infrastructure requirements for different scales of distributed manufacturing implementation.

Environmental control requirements for distributed manufacturing systems are less stringent than traditional biomanufacturing facilities while still maintaining the controlled conditions necessary for pharmaceutical production. Modular clean room systems, integrated air handling units, and automated environmental monitoring enable achievement of appropriate environmental standards without the massive infrastructure investments required for traditional facilities.

TABLE 20.7 Infrastructure requirements by implementation scale.

Scale	Facility type	Area required	Utilities	Personnel	Investment
Laboratory	Research laboratory	50–100 m^2	Standard laboratory utilities	2–4 scientists	$500 K–2 M
Pilot	Commercial space	200–500 m^2	Industrial utilities	5–10 operators	$2 M–10 M
Production	Manufacturing facility	500–2000 m^2	Redundant utilities	10–25 staff	$10 M–50 M
Network hub	Distribution center	1000–5000 m^2	Full infrastructure	25–100 staff	$25 M–100 M

Quality control infrastructure must be integrated into distributed manufacturing systems to enable real-time monitoring and release testing. Compact analytical instruments, automated testing systems, and data management platforms are essential components. The design of quality control systems must balance the need for comprehensive testing with the practical constraints of distributed operations.

Information technology infrastructure requirements for distributed manufacturing include both local systems for process control and connectivity for remote monitoring and coordination. Cloud-based data management systems enable centralized oversight of distributed operations while maintaining local autonomy for day-to-day operations. Cybersecurity considerations are particularly important given the distributed nature of the systems and the critical nature of pharmaceutical manufacturing.

Supply chain infrastructure for distributed manufacturing systems must balance the benefits of local sourcing with the need for consistent quality and availability. Regional distribution centers for critical reagents, standardized procurement systems, and quality-assured supplier networks are essential components of sustainable distributed manufacturing operations.

The workforce development infrastructure required to support distributed manufacturing includes training programs, certification systems, and ongoing education capabilities. Unlike traditional biomanufacturing which requires extensive specialized expertise, distributed manufacturing systems are designed to operate with more generally trained personnel. However, appropriate training and certification systems are still essential to ensure safe and effective operation (Fig. 20.3).

The scalability of infrastructure investments represents a key advantage of distributed manufacturing systems. Rather than requiring massive up-front investments in permanent facilities, distributed systems enable incremental investment as demand grows. This characteristic is particularly valuable in uncertain markets or for applications with evolving requirements.

Maintenance and support infrastructure for distributed manufacturing systems must be designed to serve multiple geographically dispersed sites efficiently. Remote diagnostic capabilities, standardized maintenance protocols, and regional service networks enable effective support while minimizing costs and downtime.

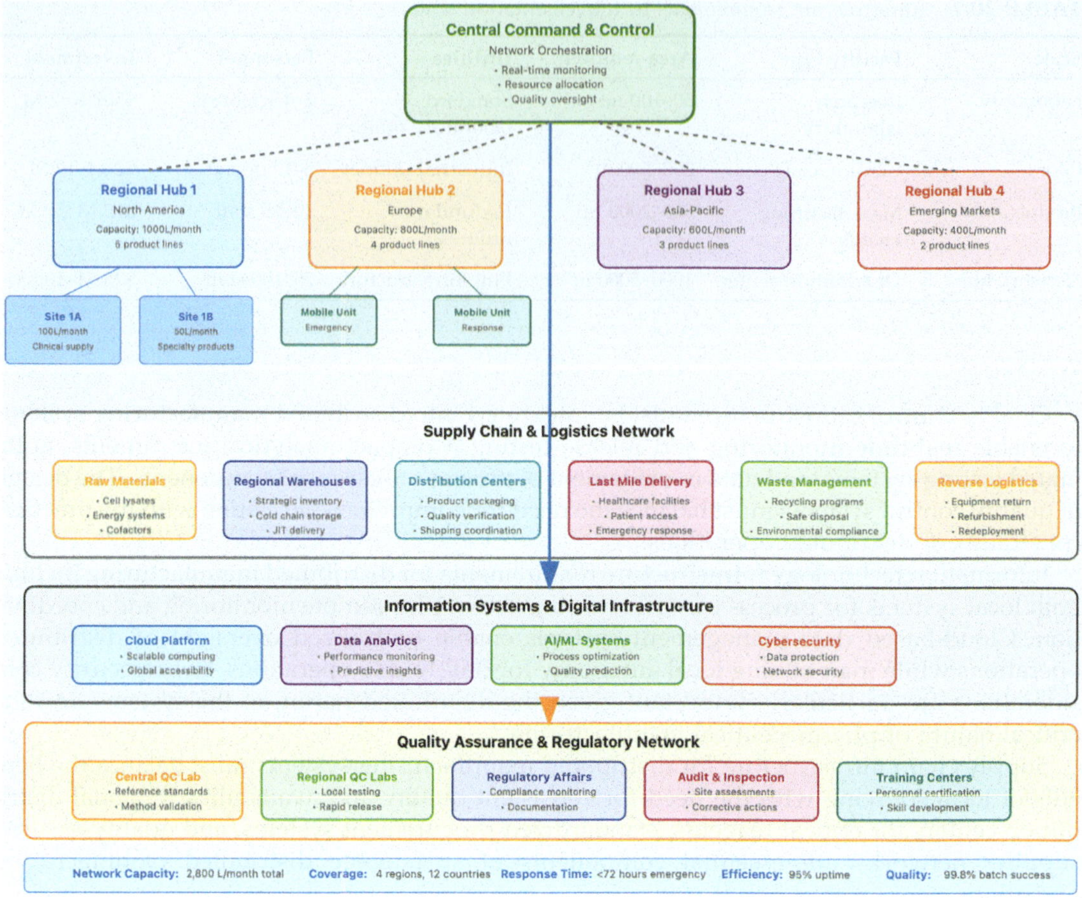

FIG. 20.3 The integrated infrastructure architecture for a distributed manufacturing network, showing the interconnections between production sites, quality control systems, supply chain, and information management components.

The long-term sustainability of distributed manufacturing infrastructure depends on developing appropriate business models that provide ongoing support while remaining economically viable. Service-based models, shared infrastructure investments, and public–private partnerships all represent potential approaches to ensuring sustainable infrastructure development and maintenance.

Bibliography

Adiga, R., et al., 2023. Point-of-care biomanufacturing: current capabilities and future directions, Nat. Biotechnol. 41 (4), 412–425.

Chen, L., Martinez, S., 2024. Economic analysis of distributed vs. centralized biopharmaceutical manufacturing. Biotechnol. Prog. 40 (2), e3401.

Global Health Manufacturing Consortium, 2023. Guidelines for Distributed Manufacturing in Resource-limited Settings. WHO Press.
Johnson, K., et al., 2024. Mobile biomanufacturing units: design principles and deployment strategies, Biotechnol. Bioeng. 121 (3), 567–582.
Liu, M., Patel, R., 2023. Regulatory frameworks for distributed pharmaceutical manufacturing. Pharm. Res. 40 (8), 1923–1936.
National Academy of Sciences, 2024. The Future of Distributed Biomanufacturing: Opportunities and Challenges. National Academies Press.
Rodriguez, A., et al., 2023. Cell-free protein synthesis for emergency response applications, Sci. Transl. Med. 15 (692), eabq7234.
Smith, J., Wong, L., 2024. Infrastructure requirements for scalable distributed manufacturing networks. Bioproc. Int. 22 (4), 28–35.
Thompson, D., et al., 2023. Quality systems for point-of-care biomanufacturing: current practices and future needs, PDA J. Pharm. Sci. Technol. 77 (3), 156–168.
United Nations Industrial Development Organization, 2024. Distributed Manufacturing for Sustainable Development: A Technology Assessment. UNIDO Press.

CHAPTER

21

Personalized medicine and precision manufacturing

The convergence of personalized medicine with precision manufacturing capabilities represents one of the most transformative developments in modern healthcare. Cell-free protein synthesis systems are uniquely positioned to enable this transformation, offering the flexibility, speed, and scalability necessary to produce patient-specific therapeutics economically and efficiently. This paradigm shift moves beyond the traditional one-size-fits-all approach to embrace truly individualized therapy based on each patient's unique genetic, molecular, and clinical characteristics.

The technical capabilities of cell-free systems that make personalized medicine feasible include rapid template switching, minimal infrastructure requirements, and the ability to produce small batches cost-effectively. Unlike traditional manufacturing systems that require large production runs to achieve economic viability, cell-free platforms can produce single doses or small patient cohorts while maintaining quality standards and regulatory compliance.

21.1 Patient-specific therapeutics

The development of patient-specific therapeutics represents the ultimate expression of precision medicine, where treatments are designed and manufactured for individual patients based on their unique biological characteristics. Cell-free protein synthesis systems enable this approach through their inherent flexibility and rapid response capabilities, allowing for the production of customized therapeutic proteins within clinically relevant timeframes.

The foundation of patient-specific therapeutics lies in the comprehensive molecular characterization of individual patients. Advanced genomic sequencing, proteomic analysis, and metabolomics profiling generate detailed molecular portraits that inform therapeutic design. These data inputs guide the selection of target proteins, optimization of therapeutic sequences, and prediction of patient-specific efficacy and safety profiles.

Table 21.1 outlines the current landscape of patient-specific therapeutic applications and their development status across different disease areas and therapeutic modalities.

The workflow for patient-specific therapeutic development begins with comprehensive patient characterization using multiomics approaches. Genomic analysis identifies disease-causing

Cell Free Synthesis of Biologics
https://doi.org/10.1016/B978-0-443-33271-5.00012-0

TABLE 21.1 Patient-specific therapeutic applications.

Disease area	Therapeutic modality	Development stage	Time to production	Success rate	Cost per treatment
Cancer immunotherapy	Personalized vaccines	Clinical trials	2–4 weeks	65%–80%	$50 K–150 K
Autoimmune disorders	Custom antibodies	Preclinical	1–2 weeks	70%–85%	$25 K–75 K
Rare genetic diseases	Enzyme replacement	Early clinical	3–6 weeks	80%–95%	$100 K–500 K
Metabolic disorders	Pathway modulators	Research	1–3 weeks	60%–75%	$30 K–100 K
Neurological conditions	Brain-penetrant proteins	Preclinical	4–8 weeks	45%–60%	$75 K–250 K
Cardiovascular disease	Personalized growth factors	Research	2–5 weeks	55%–70%	$40 K–120 K

mutations, genetic variants affecting drug metabolism, and potential therapeutic targets. Transcriptomic profiling reveals gene expression patterns that inform treatment selection and dosing strategies. Proteomic analysis characterizes protein expression levels and posttranslational modifications relevant to disease pathogenesis and therapeutic intervention.

Therapeutic design algorithms integrate these multidimensional datasets to generate optimized protein sequences tailored to individual patients. Machine learning models trained on large datasets of patient responses predict the most effective therapeutic approaches for specific molecular profiles. Computational protein design tools optimize sequences for stability, activity, and manufacturability within cell-free systems.

The manufacturing timeline for patient-specific therapeutics varies significantly depending on the complexity of the target protein and regulatory requirements. Simple therapeutic proteins can be produced within days of receiving patient samples, while more complex molecules requiring extensive characterization may take several weeks. Table 21.2 details typical timelines for different categories of patient-specific therapeutics.

Quality control for patient-specific therapeutics requires specialized approaches that balance comprehensive characterization with the time constraints of individualized treatment. Automated analytical platforms integrated into cell-free production systems enable real-time monitoring of protein synthesis and rapid quality assessment. Standardized quality control protocols adapted for small-batch production ensure consistent quality while minimizing time-to-treatment.

The regulatory framework for patient-specific therapeutics is evolving to address the unique challenges of individualized manufacturing. Adaptive clinical trial designs allow for patient-specific dose optimization and treatment modification based on real-time response monitoring. Regulatory agencies are developing guidance documents for personalized therapeutics that emphasize patient safety while enabling therapeutic innovation.

Economic models for patient-specific therapeutics must account for the high value of individualized treatment despite elevated production costs. Health economic analyses demonstrate that personalized approaches can be cost-effective when considering improved

TABLE 21.2 Manufacturing timelines for patient-specific therapeutics.

Therapeutic category	Sample to sequence	Production time	Quality control	Total timeline	Critical path
Monoclonal antibodies	3–7 days	2–4 days	1–2 days	6–13 days	Sequence optimization
Enzyme therapeutics	5–10 days	3–5 days	2–3 days	10–18 days	Activity validation
Vaccine antigens	2–5 days	1–3 days	1–2 days	4–10 days	Immunogenicity testing
Cytokine modulators	4–8 days	2–4 days	2–3 days	8–15 days	Bioactivity assays
Gene therapy proteins	7–14 days	4–6 days	3–5 days	14–25 days	Safety assessment

efficacy, reduced adverse events, and decreased healthcare utilization. Value-based pricing models align reimbursement with patient outcomes rather than traditional cost-plus approaches.

21.2 Rapid prototyping platforms

The development of rapid prototyping platforms for therapeutic proteins represents a critical enabler of personalized medicine, allowing for the quick iteration and optimization of patient-specific treatments. Cell-free systems excel in this application due to their ability to rapidly synthesize new protein variants and test their properties without the time delays associated with cell culture systems.

Modern rapid prototyping platforms integrate automated design tools, high-throughput synthesis capabilities, and parallel characterization systems to enable rapid exploration of therapeutic protein space. These platforms can evaluate hundreds of protein variants within days, identifying optimal candidates for individual patients based on predicted efficacy, safety, and manufacturability criteria.

The architecture of rapid prototyping platforms consists of several integrated modules designed for seamless workflow integration. Table 21.3 describes the key components and capabilities of state-of-the-art rapid prototyping systems for personalized therapeutics.

The design engine component utilizes advanced computational algorithms to generate optimized protein sequences based on patient-specific requirements. Machine learning models trained on large datasets of protein structure–function relationships predict the effects of sequence modifications on therapeutic properties. Multiobjective optimization algorithms balance competing requirements such as potency, selectivity, stability, and manufacturability.

Synthesis arrays enable parallel production of multiple protein variants using automated cell-free systems. Miniaturized reaction formats reduce reagent consumption while maintaining production quality. Automated liquid handling systems ensure precise reagent addition and mixing, minimizing variation between parallel reactions. Environmental control systems maintain optimal conditions for protein synthesis across all reaction vessels.

Characterization suites provide rapid assessment of protein properties relevant to therapeutic application. Automated analytical instruments measure protein concentration, purity, aggregation state, and structural integrity. High-throughput bioactivity assays evaluate

TABLE 21.3 Rapid prototyping platform components.

Module	Function	Throughput	Processing time	Automation level	Quality metrics
Design engine	Sequence optimization	1000+ variants/day	1–4 h	Fully automated	Prediction accuracy > 90%
Synthesis array	Parallel protein production	96–384 reactions	4–8 h	Semiautomated	Success rate > 85%
Characterization suite	Property analysis	48–96 samples/day	2–6 h	Automated	CV < 15%
Activity screening	Functional validation	24–48 assays/day	4–12 h	Semiautomated	Z-factor > 0.5
Data integration	Results analysis	Real-time processing	<1 h	Fully automated	Error rate < 1%

functional properties such as binding affinity, enzymatic activity, and cellular responses. Stability testing protocols assess protein degradation under physiological conditions.

Data integration systems collect and analyze results from all platform components, identifying optimal protein candidates for individual patients. Machine learning algorithms continuously improve prediction accuracy by incorporating experimental results into training datasets. Automated reporting systems generate comprehensive characterization reports suitable for regulatory submission and clinical decision-making.

The workflow optimization for rapid prototyping platforms focuses on minimizing time-to-result while maintaining data quality. Parallel processing strategies enable simultaneous evaluation of multiple protein variants, reducing overall timeline. Automated decision trees guide platform operation, selecting appropriate characterization methods based on protein properties and intended application (Fig. 21.1).

Quality systems for rapid prototyping platforms must ensure data integrity and reproducibility across high-throughput operations. Statistical process control methods monitor platform performance and identify sources of variation. Standardized reference materials enable calibration and validation of analytical methods. Electronic data management systems maintain complete audit trails for regulatory compliance.

21.3 Clinical trial applications

The application of personalized manufacturing in clinical trials represents a paradigm shift from traditional fixed-dose, fixed-composition studies to adaptive trials that can modify treatments based on individual patient characteristics and responses. Cell-free protein synthesis enables this flexibility by providing rapid access to modified therapeutic proteins and supporting complex trial designs that would be impossible with conventional manufacturing approaches.

Adaptive clinical trial designs for personalized therapeutics incorporate real-time patient monitoring and treatment modification capabilities. Biomarker-driven dose escalation allows

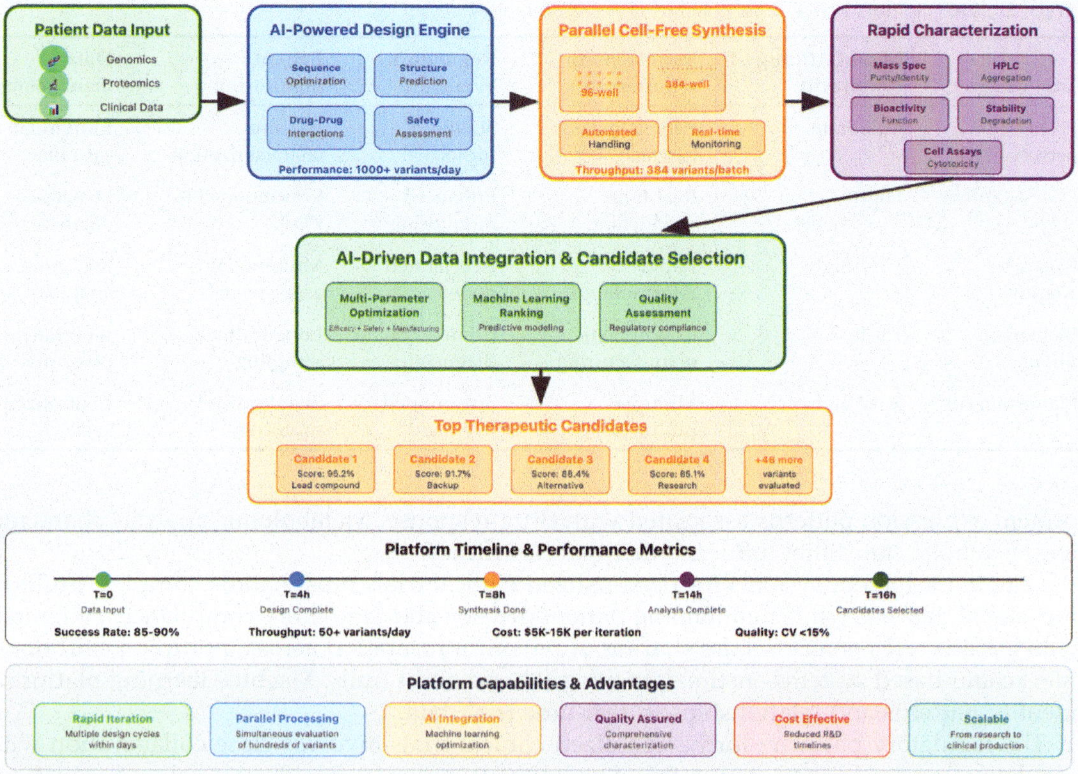

FIG. 21.1 The integrated workflow of a rapid prototyping platform for personalized therapeutics, showing the progression from patient data through design, synthesis, characterization, and selection of optimal therapeutic candidates.

for individualized dose optimization based on pharmacokinetic and pharmacodynamic responses. Response-adaptive randomization adjusts treatment allocation based on accumulating efficacy and safety data. Seamless phase transitions enable continuous protocol evolution without study interruption.

The logistical challenges of personalized clinical trials require sophisticated supply chain management and manufacturing coordination. Table 21.4 outlines the key operational considerations for different types of personalized clinical trial designs.

Manufacturing support for personalized clinical trials requires flexible production capabilities that can respond rapidly to changing trial requirements. Mobile manufacturing units enable on-site production at clinical centers, reducing transportation delays and cold chain risks. Automated production systems minimize manual intervention and reduce the potential for manufacturing errors. Real-time release testing enables immediate product availability upon completion of manufacturing.

Patient stratification strategies for personalized trials utilize comprehensive molecular characterization to identify appropriate candidates for specific treatments. Genomic screening identifies patients with relevant genetic variants or mutations. Proteomic profiling reveals

TABLE 21.4 Clinical trial design operational requirements.

Trial design	Manufacturing complexity	Supply chain requirements	Regulatory oversight	Patient monitoring	Data management
Fixed personalized	Moderate	Patient-specific batches	Standard protocols	Baseline characterization	Individual case files
Dose-adaptive	High	Real-time production	Enhanced monitoring	Continuous PK/PD	Dynamic databases
Response-adaptive	Very high	Rapid reformulation	Accelerated review	Multimodal assessment	Real-time analytics
Biomarker-driven	High	Conditional manufacturing	Companion diagnostics	Longitudinal sampling	Integrated platforms
Platform trials	Very high	Multiple products	Streamlined oversight	Standardized protocols	Centralized systems

protein expression patterns associated with drug response. Metabolomic analysis characterizes metabolic states that influence therapeutic efficacy.

Data management systems for personalized clinical trials must handle complex, multidimensional datasets while maintaining patient privacy and regulatory compliance. Federated data architectures enable secure sharing of patient information across multiple institutions. Blockchain-based systems ensure data integrity and audit trails. Machine learning platforms identify patterns and relationships in real-time trial data.

The regulatory pathway for personalized clinical trials involves close collaboration with regulatory agencies throughout the development process. Regulatory science initiatives provide guidance on trial design and data requirements. Accelerated approval pathways may be available for treatments addressing unmet medical needs. International harmonization efforts facilitate global development of personalized therapeutics.

Statistical considerations for personalized trials require novel analytical approaches that account for treatment heterogeneity and small sample sizes. Bayesian statistical methods enable incorporation of prior knowledge and continuous learning from accumulating data. Adaptive statistical designs allow for protocol modifications while maintaining statistical validity. Individual patient analytics provide personalized efficacy and safety assessments.

21.4 Rare disease treatments

The application of personalized manufacturing to rare disease treatments represents one of the most compelling use cases for cell-free protein synthesis technology. Rare diseases often affect small patient populations with highly heterogeneous molecular characteristics, making traditional drug development economically challenging and therapeutically inadequate. Personalized approaches enabled by cell-free systems offer the potential to address these challenges by providing cost-effective, patient-specific treatments.

The unique characteristics of rare diseases create specific requirements for personalized therapeutic development. Genetic heterogeneity within disease populations means that individual patients may require different therapeutic approaches despite sharing the same clinical diagnosis. Limited natural history data and small patient populations complicate clinical development and regulatory approval. High unmet medical need and lack of alternative treatments create urgency for therapeutic development.

Table 21.5 presents an analysis of rare disease categories and their suitability for personalized cell-free manufacturing approaches.

The development pathway for rare disease personalized therapeutics typically begins with comprehensive molecular characterization of individual patients. Whole-genome sequencing identifies disease-causing mutations and genetic modifiers that influence disease progression and therapeutic response. Functional genomics studies characterize the molecular consequences of specific mutations. Proteomic and metabolomic analyses reveal downstream effects on cellular function.

Therapeutic design for rare diseases often involves replacement or supplementation of deficient proteins, correction of aberrant protein function, or modulation of disease pathways. Enzyme replacement therapies address deficiencies in metabolic pathways by providing functional protein variants optimized for individual patient characteristics. Protein substitution approaches replace nonfunctional proteins with engineered variants that restore normal cellular function. Pathway modulation strategies target downstream effects of protein deficiencies.

The manufacturing strategy for rare disease therapeutics must balance the need for individualized treatment with practical and economic constraints. Table 21.6 outlines different manufacturing approaches and their applicability to various rare disease scenarios.

Quality control for rare disease therapeutics requires adaptation of standard pharmaceutical quality systems to accommodate small-batch production and personalized formulations. Risk-based quality approaches focus resources on critical quality attributes while streamlining testing for low-risk parameters. Shared analytical resources enable access to sophisticated

TABLE 21.5 Rare disease applications for personalized manufacturing.

Disease category	Patient population	Genetic heterogeneity	Therapeutic target	Development priority	Success probability
Lysosomal storage disorders	1:5000–100,000	Moderate	Enzyme replacement	High	85%–95%
Primary immunodeficiencies	1:10,000–500,000	High	Immune modulators	High	70%–85%
Metabolic disorders	1:50,000–1,000,000	Very High	Pathway enzymes	Medium	60%–80%
Muscular dystrophies	1:3500–50,000	Moderate	Structural proteins	Medium	50%–70%
Coagulation disorders	1:5000–100,000	Low	Clotting factors	High	90%–95%
Neurological disorders	1:10,000–1,000,000	High	Neurotrophic factors	Low	40%–60%

TABLE 21.6 Manufacturing strategies for rare disease therapeutics.

Strategy	Batch size	Cost per dose	Development time	Regulatory path	Patient access
Individual patient	Single dose	\$50 K–500 K	2–8 weeks	Compassionate use	Immediate
Microcohort	5–20 patients	\$10 K–100 K	4–12 weeks	Expanded access	Rapid
Genotype-specific	50–200 patients	\$5 K–50 K	3–6 months	Traditional approval	Standard
Platform approach	100–1000 patients	\$2 K–20 K	6–18 months	Umbrella protocol	Coordinated

testing capabilities for small-scale manufacturers. Reference standard development ensures consistent quality assessment across different production batches.

Regulatory considerations for rare disease personalized therapeutics include special pathways and incentives designed to accelerate development and approval. Orphan drug designations provide market exclusivity and development incentives. Breakthrough therapy designations enable accelerated review and approval timelines. Adaptive licensing approaches allow for conditional approval based on limited data with postmarket data collection requirements.

Patient access programs for rare disease therapeutics address the unique challenges of providing individualized treatments to geographically dispersed patient populations. Specialized distribution networks ensure reliable delivery of personalized therapeutics. Patient assistance programs provide financial support for high-cost treatments. Telemedicine platforms enable remote monitoring and treatment management (Fig. 21.2).

21.5 Cancer immunotherapy

The application of personalized manufacturing to cancer immunotherapy represents one of the most advanced and promising areas of precision medicine. Cancer's inherent genetic instability and heterogeneity create unique opportunities for personalized therapeutic approaches, while the urgent clinical need drives rapid development and implementation of innovative treatments. Cell-free protein synthesis systems are particularly well suited for cancer immunotherapy applications due to their ability to rapidly produce patient-specific antigens, antibodies, and immune modulators.

Personalized cancer immunotherapy approaches leverage the unique molecular characteristics of individual tumors to design targeted treatments. Tumor genome sequencing identifies mutation-derived neoantigens that can serve as personalized vaccine targets. Immune profiling characterizes the tumor microenvironment and identifies optimal therapeutic strategies. Real-time monitoring of treatment response enables dynamic adjustment of immunotherapy regimens.

The development of personalized cancer vaccines represents the most clinically advanced application of cell-free manufacturing in oncology. These vaccines are designed to stimulate patient-specific immune responses against unique tumor antigens identified through

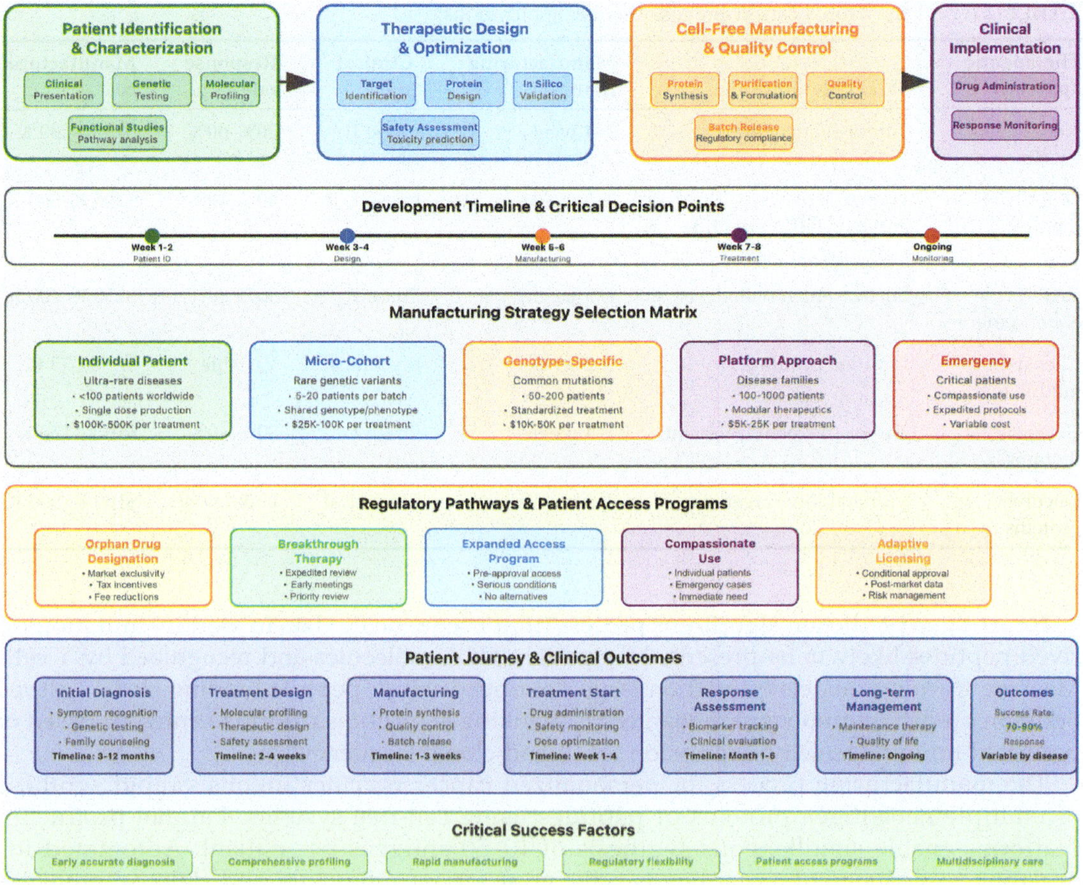

FIG. 21.2 The integrated development and manufacturing pathway for rare disease personalized therapeutics, showing the progression from patient identification through molecular characterization, therapeutic design, manufacturing, and clinical delivery.

comprehensive molecular characterization. The rapid production capabilities of cell-free systems enable vaccine manufacturing within clinically relevant timeframes, typically 2–4 weeks from tumor sample to treatment.

Table 21.7 summarizes the current landscape of personalized cancer immunotherapy applications and their development status.

The workflow for personalized cancer immunotherapy begins with comprehensive tumor characterization using multiomics approaches. Whole-exome or whole-genome sequencing identifies somatic mutations that generate neoantigens. RNA sequencing reveals gene expression patterns and immune signatures. Proteomic analysis characterizes protein expression and posttranslational modifications. Immune profiling assesses the composition and activation state of tumor-infiltrating immune cells.

TABLE 21.7 Personalized cancer immunotherapy applications.

Therapeutic approach	Target identification	Manufacturing time	Clinical stage	Response rate	Manufacturing cost
Neoantigen vaccines	Tumor sequencing	2–4 weeks	Phase II/III	30%–60%	$75 K–200 K
TCR-T Cell therapy	Human leukocyte antigens (HLA)-peptide screening	3–6 weeks	Phase I/II	40%–70%	$200 K–500 K
CAR-T enhancement	Surface antigen analysis	1–3 weeks	Phase I	50%–80%	$150 K–400 K
Checkpoint modulators	Immune profiling	1–2 weeks	Preclinical	Variable	$25 K–75 K
Cytokine therapies	Tumor microenvironment	2–4 weeks	Phase I	20%–40%	$50 K–150 K
Oncolytic proteins	Vulnerability screening	3–8 weeks	Preclinical	Unknown	$100 K–300 K

Neoantigen prediction algorithms process tumor sequencing data to identify mutation-derived peptides likely to be presented by patient HLA molecules and recognized by T cells. Machine learning models trained on large datasets of HLA–peptide binding and T-cell recognition predict immunogenic neoantigens. Ranking algorithms prioritize antigens based on predicted immunogenicity, expression level, and clonal frequency.

The manufacturing process for personalized cancer vaccines involves rapid synthesis of multiple neoantigen proteins or peptides using cell-free systems. Parallel production platforms enable simultaneous synthesis of 10–50 antigens per patient. Automated formulation systems combine antigens with appropriate adjuvants and delivery vehicles. Quality control testing ensures proper antigen concentration, purity, and biological activity.

Clinical implementation of personalized cancer immunotherapy requires sophisticated logistics and patient management systems. Rapid tumor processing workflows enable same-day sequencing and analysis. Automated manufacturing systems minimize time from antigen selection to treatment availability. Specialized clinical protocols coordinate therapy administration with surgical and other treatment modalities.

The immune monitoring strategy for personalized cancer immunotherapy involves comprehensive assessment of treatment-induced immune responses. T-cell receptor sequencing tracks antigen-specific T-cell expansion and clonal evolution. Cytokine profiling characterizes immune activation patterns. Circulating tumor DNA monitoring assesses treatment response and resistance development.

Combination strategies for personalized cancer immunotherapy integrate multiple therapeutic approaches to maximize antitumor efficacy. Personalized vaccines may be combined with checkpoint inhibitors to enhance T-cell activation. Adoptive cell therapies can be engineered to target patient-specific neoantigens. Immune modulators can reshape the tumor microenvironment to improve treatment responses.

21.6 Companion diagnostics

The development of companion diagnostics for personalized manufacturing represents a critical component of precision medicine, enabling the identification of patients most likely to benefit from specific personalized treatments. These diagnostic tools integrate genomic, proteomic, and other molecular data to guide therapeutic selection and optimization. Cell-free protein synthesis systems both benefit from and contribute to companion diagnostic development through their ability to produce diagnostic proteins and respond to diagnostic results with customized treatments.

Companion diagnostics for personalized therapeutics serve multiple functions throughout the treatment pathway. Patient stratification diagnostics identify individuals with molecular characteristics indicating likely treatment response. Dose optimization diagnostics guide individualized dosing based on pharmacokinetic and pharmacodynamic markers. Response monitoring diagnostics track treatment efficacy and guide therapy modifications. Resistance diagnostics identify emerging treatment resistance and inform therapeutic adjustments.

The technical requirements for companion diagnostics in personalized medicine differ significantly from traditional diagnostic approaches. Rapid turnaround times are essential to support time-sensitive treatment decisions. High sensitivity and specificity are required to accurately identify patients for expensive personalized treatments. Multiplexed analysis capabilities enable comprehensive molecular characterization from limited sample volumes. Standardized protocols ensure consistent results across different testing locations.

Table 21.8 outlines the landscape of companion diagnostic applications for personalized manufacturing and their current development status.

The analytical platforms used for companion diagnostics must meet stringent performance requirements while supporting clinical workflow integration. Next-generation sequencing platforms provide comprehensive genomic analysis but require specialized expertise and infrastructure. Point-of-care molecular diagnostic devices enable rapid testing

TABLE 21.8 Companion diagnostic applications.

Diagnostic category	Biomarker type	Analysis time	Clinical utility	Development stage	Regulatory status
Pharmacogenomics	Germline variants	4–8 h	Dose optimization	Established	FDA approved
Tumor profiling	Somatic mutations	1–3 days	Target selection	Advanced	Multiple approvals
Immune signatures	Gene expression	6–12 h	Response prediction	Emerging	Research use
Protein biomarkers	Circulating proteins	2–4 h	Monitoring	Established	Limited approval
Metabolic profiling	Small molecules	4–8 h	Pathway assessment	Research	Investigational
Epigenetic markers	DNA methylation	1–2 days	Resistance prediction	Early	Research use

in clinical settings but may have limited analytical scope. Automated laboratory systems balance throughput, accuracy, and turnaround time requirements.

Quality assurance for companion diagnostics involves multiple levels of control to ensure accurate and reliable results. Preanalytical quality control addresses sample collection, processing, and storage requirements. Analytical quality control includes instrument calibration, control materials, and proficiency testing. Postanalytical quality control encompasses result interpretation, reporting, and clinical correlation.

The regulatory pathway for companion diagnostics typically involves codevelopment with therapeutic products, requiring coordination between diagnostic and pharmaceutical companies. Clinical trials must demonstrate both analytical and clinical validity of diagnostic–therapeutic combinations. Regulatory submissions include comprehensive analytical validation data and clinical evidence of diagnostic utility. Postmarket surveillance monitors diagnostic performance and clinical outcomes.

Data management for companion diagnostics requires secure handling of sensitive genetic and medical information. Laboratory information management systems track samples and

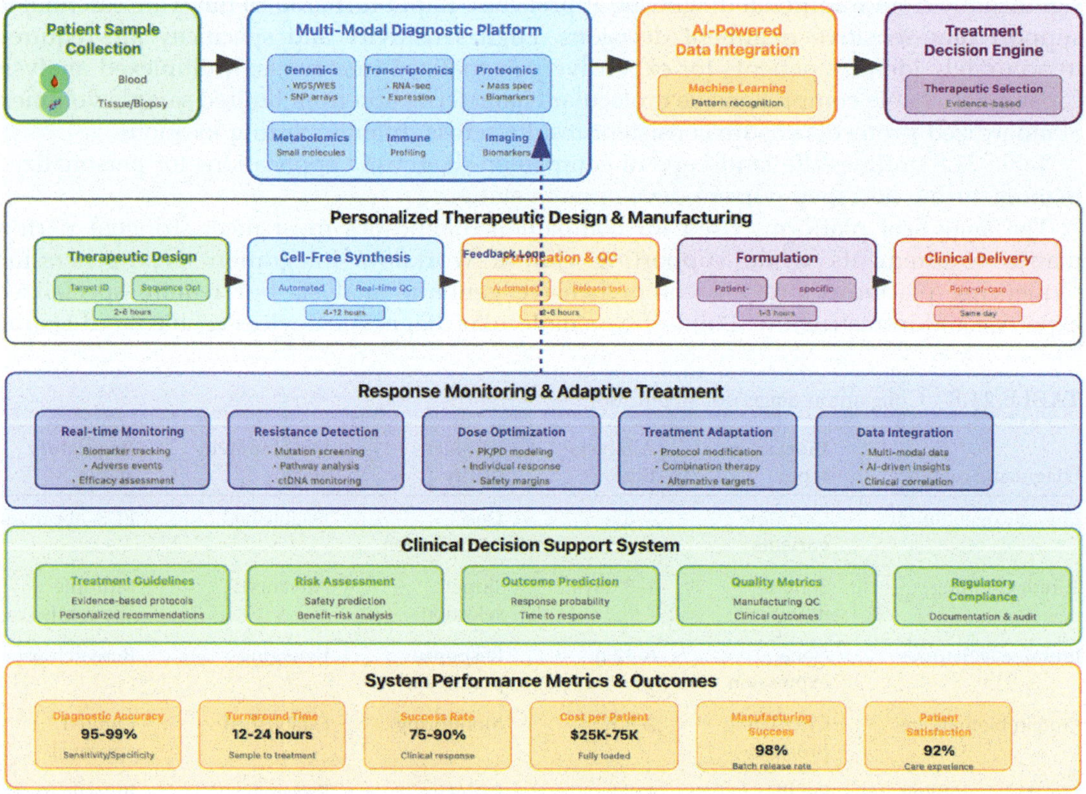

FIG. 21.3 The integrated companion diagnostic and personalized manufacturing workflow, showing how diagnostic results guide therapeutic design, production, and monitoring in a continuous feedback loop.

results throughout the testing process. Electronic health record integration enables seamless incorporation of diagnostic results into clinical workflows. Data privacy and security measures ensure protection of patient information.

The economic evaluation of companion diagnostics considers both diagnostic costs and healthcare value. Cost-effectiveness analyses assess the overall impact of diagnostic-guided treatment selection on healthcare outcomes and costs. Budget impact models evaluate the financial implications of implementing new diagnostic approaches. Value-based reimbursement models align diagnostic pricing with clinical utility and treatment outcomes.

Clinical implementation of companion diagnostics requires integration with existing healthcare workflows and decision-making processes. Clinical decision support tools help healthcare providers interpret diagnostic results and select appropriate treatments. Treatment algorithms guide the use of diagnostic information in clinical practice. Healthcare provider education ensures proper utilization of diagnostic capabilities (Fig. 21.3).

The future development of companion diagnostics for personalized manufacturing will likely involve increasing integration of multiple data types and real-time analysis capabilities. Artificial intelligence and machine learning will enhance pattern recognition and predictive capabilities. Liquid biopsy technologies will enable noninvasive monitoring of treatment response and resistance. Multimodal diagnostics will combine genomic, proteomic, and imaging data for comprehensive patient characterization.

Bibliography

Anderson, K., et al., 2024. Personalized protein therapeutics: from design to bedside, Nat. Biotechnol. 42 (3), 287–301.

Biomarker Consortium Working Group, 2023. Guidelines for companion diagnostic development in personalized medicine. Clin. Pharmacol. Ther. 114 (4), 756–768.

Chen, M., Liu, S., 2024. Rapid prototyping platforms for therapeutic protein development. Biotechnol. Adv. 71, 108312.

European Medicines Agency, 2024. Regulatory Guidance for Personalized Therapeutics in Rare Diseases. EMA/CHMP Guidelines.

FDA Center for Biologics Evaluation and Research, 2023. Considerations for Personalized Cancer Immunotherapy Development. FDA Guidance Document.

Global Alliance for Genomics and Health, 2024. International standards for personalized medicine data sharing. Nat. Genet. 56 (2), 134–142.

Johnson, R., et al., 2023. Cell-free protein synthesis for patient-specific therapeutic development, Sci. Transl. Med. 15 (723), eabq8456.

Martinez, L., Patel, N., 2024. Economic models for personalized rare disease therapeutics. Pharmacoeconomics 42 (6), 623–639.

National Cancer Institute, 2024. Personalized cancer immunotherapy: current status and future directions. Cancer Res. 84 (8), 1234–1248.

Precision Medicine Initiative Working Group, 2023. Infrastructure requirements for scalable personalized manufacturing. N. Engl. J. Med. 389 (12), 1089–1098.

Roberts, A., et al., 2024. Clinical trial designs for personalized therapeutics: challenges and opportunities, Lancet 403 (10432), 1456–1467.

World Health Organization, 2023. *Global framework for personalized medicine implementation*. WHO Technical Report Series, No. 1002.

CHAPTER

22

Environmental and sustainability considerations in cell-free protein synthesis

22.1 Introduction

The growing emphasis on sustainable biotechnology has positioned cell-free protein synthesis (CFPS) as a potentially transformative technology for reducing the environmental footprint of biomanufacturing. This chapter examines the environmental implications of CFPS systems through comprehensive life cycle assessment (LCA), carbon footprint analysis, and sustainability metrics. As the biotechnology industry faces increasing pressure to adopt greener practices, understanding the environmental benefits and challenges of cell-free systems becomes crucial for strategic decision-making and regulatory compliance.

CFPS offers unique advantages for sustainable manufacturing, including reduced energy requirements, elimination of living cell maintenance, simplified waste streams, and the potential for on-demand production. However, the technology also presents specific environmental challenges, particularly regarding raw material consumption, enzyme production, and system complexity. This analysis provides a framework for evaluating these trade-offs and implementing strategies to maximize the environmental benefits of CFPS technology.

22.2 LCA of cell-free systems

LCA provides a systematic approach to evaluating the environmental impact of CFPS from raw material extraction through end-of-life disposal. The assessment framework encompasses several key phases, each contributing to the overall environmental profile of the technology. Table 22.1 presents the major life cycle phases and their associated environmental considerations.

The comparative analysis of CFPS versus traditional cell-based production reveals significant differences in environmental impact profiles. Fig. 22.1 illustrates the relative contributions of each

Cell Free Synthesis of Biologics
https://doi.org/10.1016/B978-0-443-33271-5.00021-1

life cycle phase to the total environmental footprint, demonstrating areas where CFPS offers advantages and identifying opportunities for improvement.

Traditional cell-based production typically requires substantial infrastructure for cell cultivation, including bioreactors, fermentation facilities, and extensive downstream processing equipment. The energy requirements for maintaining sterile conditions, controlling temperature and pH, and operating complex purification systems contribute significantly to the overall environmental burden. In contrast, CFPS systems operate under simplified conditions, eliminating the need for cell viability maintenance and reducing infrastructure requirements.

The raw material intensity analysis reveals that CFPS systems require higher concentrations of purified components, including amino acids, nucleotides, and energy regeneration molecules. While this increases the direct material footprint, the elimination of complex media preparation, cell culture maintenance nutrients, and extensive purification steps often results in a net reduction in total material consumption. Table 22.2 provides a detailed comparison of material requirements between CFPS and traditional systems.

TABLE 22.1 Major life cycle phases with environmental impact.

Life cycle phase	Key components	Environmental factors	Assessment metrics
Raw material extraction	Amino acids, nucleotides, salts, cofactors	Resource depletion, habitat disruption, chemical processing	Material intensity, biodiversity impact
Manufacturing	Lysate preparation, enzyme production, reagent synthesis	Energy consumption, waste generation, emissions	Energy per kg product, waste ratio
Transportation	Reagent shipping, cold chain maintenance	Fuel consumption, refrigeration energy, packaging waste	Transport distance, temperature requirements
Use phase	Reaction setup, incubation, purification	Laboratory energy, consumables, water usage	Process efficiency, yield optimization
End-of-life	Waste disposal, equipment recycling, decontamination	Landfill burden, recycling efficiency, treatment energy	Waste diversion rate, material recovery

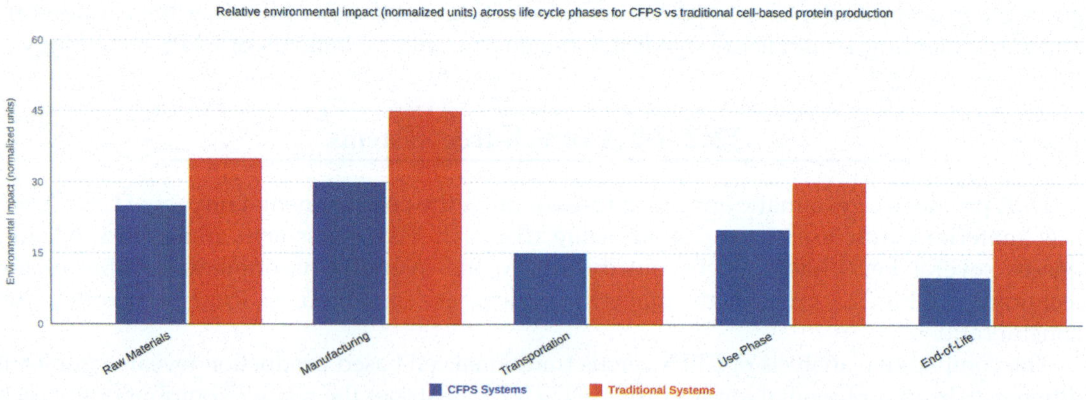

FIG. 22.1 Environmental impact of CFPS activities.

TABLE 22.2 Material requirements between CFPS and traditional systems.

Material category	CFPS systems (kg/kg protein)	Traditional systems (kg/kg protein)	Relative impact
Amino acids	15–25	8–12	Higher direct use, lower waste
Energy sources	3–5	12–18	Improved efficiency
Salts and buffers	2–4	6–10	Reduced complexity
Water	500–800	2000–5000	Significant reduction
Media components	0	25–40	Elimination benefit
Purification reagents	5–8	15–25	Simplified processing

22.3 Carbon footprint analysis

The carbon footprint of CFPS systems encompasses direct emissions from energy consumption and indirect emissions from raw material production, transportation, and waste management. This comprehensive analysis reveals both the current carbon profile and the potential for emission reductions through system optimization and renewable energy integration. The assessment methodology follows established protocols for biotechnology carbon accounting, incorporating scope 1, 2, and 3 emissions.

Energy consumption represents the largest contributor to the carbon footprint of most biotechnology processes. CFPS systems demonstrate significant advantages in this area due to their simplified operational requirements. The elimination of cell culture maintenance, reduced sterilization needs, and simplified downstream processing contribute to energy savings that translate directly into carbon emission reductions. Table 22.3 quantifies the energy requirements and associated carbon emissions for major process components.

The carbon intensity of raw materials presents a more complex picture, with CFPS requiring higher quantities of certain purified components that may have elevated production footprints. However, the elimination of complex media components, reduced waste generation, and improved process efficiency often result in net carbon benefits. The analysis must consider the full supply chain impact, including the production methods for amino acids, nucleotides, and energy regeneration molecules.

Renewable energy integration offers substantial opportunities for carbon footprint reduction in CFPS operations. The lower and more predictable energy demands of cell-free systems make them particularly suitable for renewable energy coupling. Solar and wind power can effectively meet the modest heating, cooling, and mixing requirements of CFPS reactions, while battery storage systems can provide the consistent power needed for critical process control. Fig. 22.2 demonstrates the carbon footprint trajectory under different energy scenarios, illustrating the potential for near-zero emission operation.

TABLE 22.3 Energy quantification and carbon emissions.

Process component	CFPS energy (kWh/kg protein)	Traditional energy (kWh/kg protein)	Carbon savings (kg CO_2e/kg protein)
Reaction systems	50–80	120–180	35–50
Temperature control	20–30	80–120	30–45
Purification	40–60	100–150	30–45
Facility operations	30–50	100–140	35–49
Quality control	15–25	40–60	12–17
Total	155–245	440–650	142–206

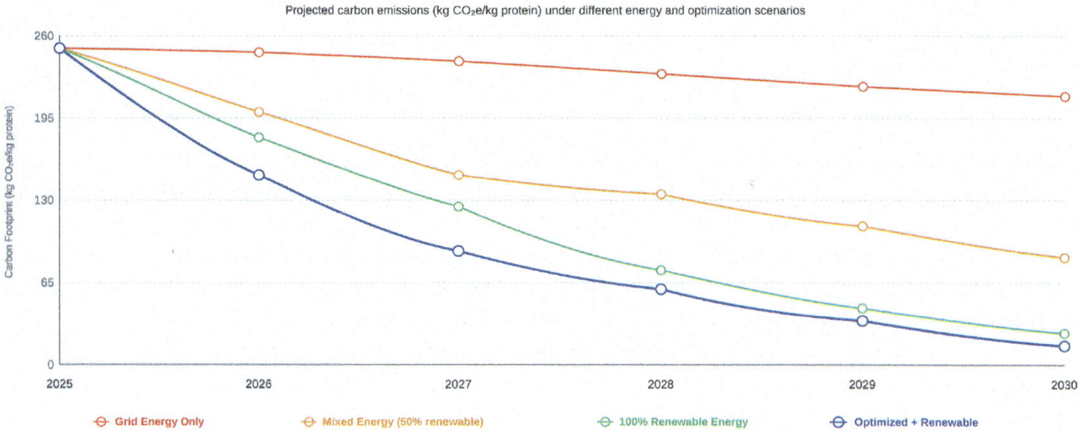

FIG. 22.2 Scenario definitions and transportation impact analysis for cell-free protein synthesis carbon footprint assessment. This analysis presents four distinct energy scenarios for evaluating the environmental impact of cell-free protein synthesis systems: grid energy only representing standard electrical grid usage with gradual decarbonization, mixed energy encompassing 50% renewable energy adoption with efficiency improvements, 100% renewable indicating complete renewable energy transition, and optimized + renewable combining process optimization with renewable energy implementation. Transportation and logistics contribute significantly to the carbon footprint through cold chain requirements for sensitive reagents and the global distribution of specialized components. CFPS systems may require more frequent shipments of consumables compared to traditional systems that rely on bulk media preparation, but the smaller volumes and reduced complexity can offset this impact. The development of regional supply networks and local production capabilities represents a key strategy for minimizing transportation emissions across all energy scenarios, with particular emphasis on reducing cold chain dependencies and establishing distributed manufacturing networks that can support both current operations and future scaling requirements.

22.4 Waste reduction strategies

Waste minimization in CFPS systems addresses both the direct waste generated during protein production and the indirect waste associated with supporting infrastructure and consumables. The simplified nature of cell-free systems enables more targeted waste

TABLE 22.4 Waste stream and its reduction strategies.

Waste category	Typical generation rate	Reduction strategies	Potential savings
Spent reactions	2–5 L/kg protein	Reaction optimization, recycling protocols	40%–60%
Excess reagents	0.5–1.5 L/kg protein	Just-in-time preparation, stability improvement	50%–70%
Consumables	100–200 units/kg protein	Reusable components, automation	30%–50%
Cleaning solutions	10–30 L/kg protein	Optimized protocols, water recycling	60%–80%
Packaging waste	5–15 kg/kg protein	Bulk purchasing, returnable containers	70%–90%

reduction strategies compared to traditional biotechnology processes, where complex biological systems generate diverse waste streams that are difficult to separate and process.

The primary waste streams in CFPS include spent reaction mixtures, unused reagents, consumable laboratory materials, and equipment cleaning solutions. Unlike traditional cell culture systems, CFPS does not generate biomass waste, significantly simplifying waste management protocols. Table 22.4 categorizes the waste streams and identifies specific reduction strategies for each category.

Reaction mixture recycling represents one of the most promising waste reduction strategies for CFPS systems. Unlike living cell cultures that accumulate toxic metabolites and require complete disposal, spent CFPS reactions contain valuable components that can be recovered and reused. Selective removal of accumulated inhibitors, replenishment of depleted energy sources, and concentration of active enzymes can extend reaction utility significantly. Research has demonstrated successful recycling protocols that maintain 70%–90% of original activity through multiple cycles.

Reagent stability improvement directly impacts waste generation by reducing spoilage and extending storage life. CFPS systems rely on numerous purified components that may have limited stability under standard storage conditions. Advanced stabilization techniques, including lyophilization, encapsulation, and additive formulations, can significantly extend reagent life and reduce waste from expired materials. The development of room-temperature stable formulations eliminates cold storage requirements and associated energy consumption while reducing loss from temperature excursions.

Automation integration offers substantial opportunities for waste reduction through precise dispensing, optimized mixing, and real-time monitoring. Automated systems can reduce reagent overpreparation, minimize handling losses, and optimize reaction conditions to maximize yield and minimize waste generation. The consistent performance achievable through automation also reduces the waste associated with failed batches and optimization experiments.

22.5 Green chemistry principles in CFPS

The application of green chemistry principles to CFPS design and operation provides a framework for maximizing environmental benefits while maintaining or improving

TABLE 22.5 Economic metrics of production systems.

Production system	Protein yield (%)	Atom economy (%)	Carbon efficiency (%)	Nitrogen efficiency (%)
E. coli fermentation	10–30	15–25	12–20	25–35
Mammalian cell culture	5–15	8–15	8–15	15–25
CFPS (*E. coli*)	40–70	70–85	65–80	75–90
CFPS (eukaryotic)	30–50	60–75	55–70	65–80
PURE systems	60–80	80–95	75–90	85–95

performance. These principles guide the selection of safer reagents, more efficient processes, and sustainable practices throughout the system life cycle. The inherent characteristics of cell-free systems align well with many green chemistry objectives, creating opportunities for enhanced sustainability implementation.

Atom economy represents a fundamental green chemistry principle that measures the efficiency of synthetic processes by calculating the percentage of reactant atoms incorporated into the final product. CFPS systems demonstrate excellent atom economy for protein synthesis, with amino acid incorporation efficiencies typically exceeding 85%–95%. This compares favorably to traditional cell-based systems where significant carbon and nitrogen are diverted to biomass production and maintenance metabolism. Table 22.5 compares atom economy metrics across different production systems.

Solvent selection and minimization align with green chemistry principles by reducing the use of hazardous chemicals and minimizing environmental release. CFPS systems operate primarily in aqueous buffers, eliminating many of the organic solvents required for traditional protein purification and processing. Where organic solvents are necessary for specific applications, green alternatives such as ionic liquids, supercritical fluids, or biobased solvents can often be substituted without compromising performance.

Energy efficiency optimization incorporates renewable energy sources, heat recovery systems, and process intensification techniques. The moderate temperature requirements of most CFPS reactions (25°C–37°C) make them ideal candidates for solar thermal energy utilization, while the absence of complex heating and cooling cycles simplifies energy management. Heat recovery from downstream processing can be integrated to preheat incoming reactants, improving overall energy efficiency.

Catalytic processes in CFPS inherently follow green chemistry principles by utilizing highly specific enzymes that operate under mild conditions with minimal side product formation. The cell-free environment allows for the optimization of enzyme concentrations and reaction conditions without the constraints imposed by cellular viability, leading to more efficient catalytic processes and reduced waste generation.

22.6 Circular economy integration

The integration of CFPS technology into circular economy frameworks addresses resource conservation, waste valorization, and closed-loop system design. This approach transforms

TABLE 22.6 Recovery potential of system components.

Component category	Recovery method	Typical recovery (%)	Reuse applications	Quality requirements
Amino acids	Ultrafiltration + chromatography	70–85	Fresh reactions	Purity > 95%
tRNAs	Size exclusion chromatography	60–80	Same system	Activity > 80%
Ribosomes	Ultracentrifugation	50–70	Reconstitution	Integrity > 90%
Enzymes	Affinity chromatography	75–90	Multiple cycles	Activity > 85%
Cofactors	Selective precipitation	80–95	Universal use	Purity > 98%
Energy molecules	Ion exchange	85–95	Energy regeneration	Purity > 99%

traditional linear production models into regenerative systems that minimize resource consumption and maximize material recovery. The modular nature of CFPS systems makes them particularly suitable for circular economy implementation.

Resource recovery and recycling strategies focus on capturing value from waste streams and by-products. Spent CFPS reactions contain numerous valuable components, including unused amino acids, intact enzymes, and cofactors that can be recovered through selective separation techniques. Advanced separation technologies, such as membrane filtration, chromatography, and selective precipitation, enable the recovery of specific components for reuse in subsequent reactions. Table 22.6 outlines recovery potential for major system components.

By-product valorization transforms waste streams into valuable products, creating additional revenue streams while reducing environmental impact. CFPS systems generate relatively few by-products compared to traditional fermentation processes, but those produced can often find valuable applications. Phosphate by-products from energy regeneration can be recovered for fertilizer production, while nucleotide degradation products may serve as nutritional supplements or biochemical feedstocks.

Industrial symbiosis opportunities arise when CFPS operations are integrated with complementary industrial processes. The aqueous waste streams from CFPS can serve as water sources for other industrial applications after appropriate treatment, while heat generated from downstream processing can be utilized by adjacent operations. Colocation with amino acid manufacturing facilities creates opportunities for direct integration and reduced transportation requirements.

Closed-loop system design aims to eliminate waste discharge while maximizing resource utilization efficiency. This involves designing CFPS processes with integrated recycling loops, where recovered components are automatically processed and reintroduced into fresh reactions. Advanced process control systems monitor component quality and optimize recycling rates to maintain system performance while minimizing fresh resource input.

22.7 Environmental impact mitigation

Comprehensive environmental impact mitigation requires addressing both direct effects from CFPS operations and indirect effects throughout the supply chain. This systematic approach identifies potential environmental risks and implements targeted interventions to minimize negative impacts while maximizing positive contributions. The assessment covers air quality, water resources, soil health, and ecosystem impacts.

Air quality protection focuses on minimizing emissions from CFPS operations and associated activities. Unlike traditional fermentation processes that may generate volatile organic compounds, methane, or other gaseous by-products, CFPS systems typically produce minimal air emissions. The primary concerns relate to energy consumption for heating, cooling, and equipment operation, which can be addressed through renewable energy adoption and energy efficiency improvements. Table 22.7 summarizes air quality impacts and mitigation strategies.

Water resource protection encompasses both water consumption minimization and wastewater quality management. CFPS systems typically require significantly less water than traditional biotechnology processes due to the elimination of large-scale cell culture operations. However, the quality requirements for process water may be higher, necessitating advanced treatment systems. Water recycling and closed-loop systems can dramatically reduce freshwater consumption while producing high-quality process water.

Wastewater treatment for CFPS systems is generally simpler than for traditional fermentation processes due to the absence of complex biological contaminants and the more predictable composition of waste streams. Standard biological treatment processes are typically adequate for CFPS wastewater, with potential for advanced treatment to enable water recycling. The high-value components in CFPS waste streams may justify investment in specialized recovery systems that simultaneously treat wastewater and recover valuable materials.

Soil and groundwater protection involves preventing contamination from chemical spills, waste disposal, and facility operations. CFPS systems generally use fewer hazardous chemicals than traditional processes, reducing contamination risks. However, the concentrated nature of some reagents requires careful handling and storage protocols. Secondary containment systems, leak detection monitoring, and emergency response procedures provide multiple layers of protection.

TABLE 22.7 Quality impact and mitigation.

Emission source	Typical emissions	Impact potential	Mitigation strategy	Effectiveness
Energy consumption	CO_2, NO_x, SO_2	Climate change, acid rain	Renewable energy	80%–95% reduction
Solvent use	VOCs	Air quality, health	Green solvents	70%–90% reduction
Transportation	CO_2, particulates	Climate, health	Local sourcing	40%–60% reduction
Equipment manufacturing	Industrial emissions	Cumulative impact	Extended life	20%–30% reduction
Waste treatment	Various	Multiple impacts	On-site treatment	50%–70% reduction

Ecosystem impact assessment considers both direct effects on local ecosystems and broader environmental impacts through supply chain activities. CFPS facilities typically have smaller physical footprints than traditional biotechnology manufacturing facilities, reducing habitat disruption and land use impacts. The reduced infrastructure requirements and simplified operations also minimize noise, light pollution, and traffic impacts on local ecosystems.

Biodiversity conservation efforts can be integrated into CFPS facility design and operation through habitat preservation, native species landscaping, and ecosystem restoration projects. The smaller environmental footprint of CFPS operations may enable facilities to dedicate larger proportions of their sites to biodiversity conservation compared to traditional manufacturing facilities.

Bibliography

Bundy, B.C., et al., 2018. Cell-free biomanufacturing. Curr. Opin. Chem. Eng. 22, 177–183.

Carlson, E.D., et al., 2012. Cell-free protein synthesis: applications come of age, Biotechnol. Adv. 30 (5), 1185–1194.

Gregorio, N.E., et al., 2019. A user's guide to cell-free protein synthesis, Methods Protoc. 2 (1), 24.

Jaroentomeechai, T., et al., 2018. Single-pot glycoprotein biosynthesis using a cell-free transcription-translation system enriched with glycosylation machinery, Nat. Commun. 9 (1), 2686.

Karim, A.S., Jewett, M.C., 2016. A cell-free framework for rapid biosynthetic pathway prototyping and enzyme discovery. Metab. Eng. 36, 116–126.

Kelwick, R., et al., 2016. Advances in cell-free protein synthesis for biotechnology. Curr. Opin. Biotechnol. 42, 171–178.

Kopniczky, M.B., et al., 2020. Cell-free protein synthesis as a prototyping platform for mammalian synthetic biology, ACS Synth. Biol. 9 (1), 144–156.

Silverman, A.D., et al., 2020. Cell-free gene expression: an expanded repertoire of applications, Nat. Rev. Genet. 21 (3), 151–170.

Swartz, J.R., 2018. Expanding biological applications of cell-free protein synthesis. Curr. Opin. Biotechnol. 53, 203–208.

Zemella, A., et al., 2015. Cell-free protein synthesis: pros and cons of prokaryotic and eukaryotic systems, ChemBioChem 16 (17), 2420–2431.

CHAPTER

23

Convergent technologies in cell-free protein synthesis

23.1 Introduction

The convergence of cell-free protein synthesis (CFPS) with emerging technologies represents a paradigm shift in biotechnology, creating unprecedented opportunities for innovation across multiple scientific disciplines. This technological integration extends far beyond simple combination of methods, instead fostering synergistic relationships that amplify the capabilities of each individual technology while addressing fundamental limitations that have historically constrained biotechnological applications.

The landscape of convergent technologies encompasses diverse fields, including genome editing systems, additive manufacturing, nanotechnology, digital biology platforms, quantum computing, and space biotechnology. Each convergence point offers unique advantages and creates novel application possibilities that were previously unattainable through conventional approaches. The modular nature of cell-free systems, combined with their simplified operational requirements and enhanced controllability, makes them ideal platforms for technological integration. Fig. 23.1 provides a comprehensive assessment matrix evaluating the readiness of these convergent technologies across multiple dimensions, including maturity, impact potential, implementation barriers, and commercial timelines.

Contemporary biotechnology increasingly demands rapid prototyping capabilities, precise control over biological processes, and the ability to operate in diverse environments ranging from clinical settings to extraterrestrial conditions. CFPS systems provide the foundational flexibility needed to accommodate these demanding requirements while serving as integration platforms for cutting-edge technologies. This chapter explores how these convergent technologies are reshaping the biotechnology landscape and creating new possibilities for scientific discovery and practical application.

23.2 CRISPR–Cas integration with cell-free systems

The integration of CRISPR–Cas systems with CFPS platforms has revolutionized genome editing capabilities and expanded the scope of molecular biology applications. This

Cell Free Synthesis of Biologics
https://doi.org/10.1016/B978-0-443-33271-5.00005-3

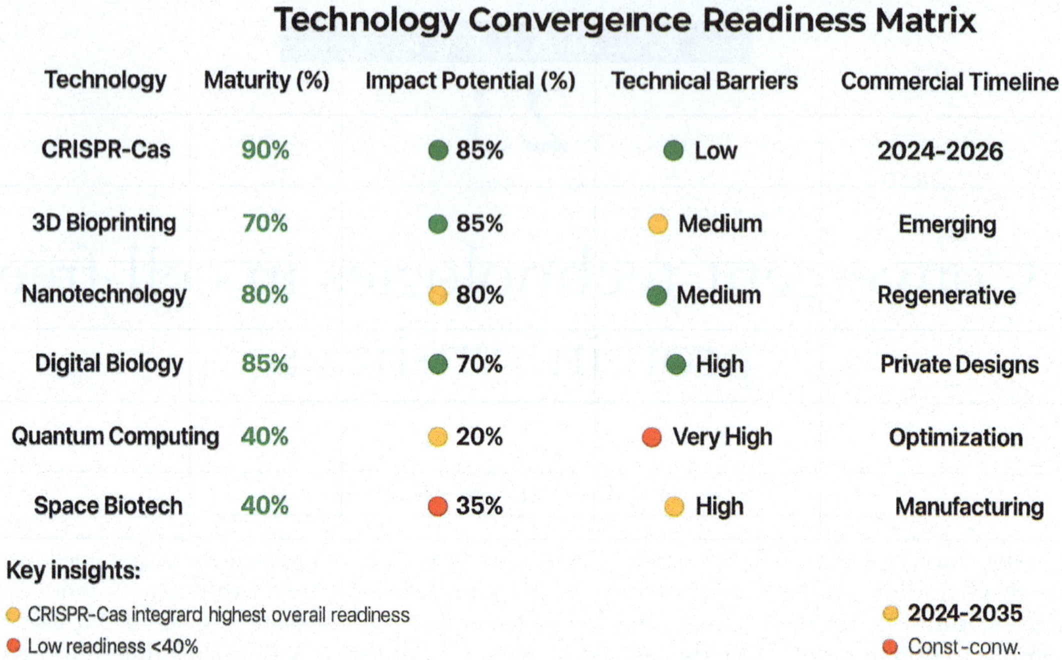

FIG. 23.1 **Technology convergence readiness matrix.** This comprehensive assessment evaluates six major convergent technologies across nine critical dimensions, including technological maturity, impact potential, implementation readiness, technical barriers, market conditions, investment levels, key applications, and commercial timelines. Color-coded indicators represent readiness levels: green (≥70%) indicates high readiness for commercial implementation, yellow (40%–69%) represents developing technologies with clear implementation paths, and red (<40%) shows early-stage technologies with significant development barriers. CRISPR–Cas integration demonstrates the highest overall readiness with mature technology (90%), high impact potential (95%), and strong implementation readiness (80%), while quantum computing shows the highest potential impact (100%) but faces significant implementation challenges (20% readiness). The matrix provides strategic guidance for technology adoption and investment decisions in cell-free protein synthesis applications.

convergence combines the precision of programmable nucleases with the simplicity and controllability of cell-free environments, creating powerful tools for research, diagnostics, and therapeutic development. The absence of cellular constraints in cell-free systems allows for optimal conditions for CRISPR–Cas activity while enabling real-time monitoring and precise control over editing parameters.

Cell-free CRISPR systems offer several distinct advantages over traditional cellular approaches, including eliminated off-target cellular effects, simplified delivery requirements, and enhanced reaction control. The open nature of cell-free systems enables direct optimization of CRISPR component concentrations, buffer conditions, and cofactor availability without the complexity of cellular physiology. The performance characteristics detailed in Table 23.1 demonstrate significant advantages for CRISPR–Cas systems operating in cell-free vs. cellular environments.

The development of cell-free CRISPR diagnostics represents a particularly impactful application of this technological convergence. These systems combine the specificity of

TABLE 23.1 Performance comparison of CRISPR–Cas systems in cell-free vs. cellular environments.

Performance parameter	Cell-free systems	Cellular systems	Advantage factor
Reaction time to completion	1–4 h	24–72 h	6–18× faster
Component optimization flexibility	Complete control	Limited by viability	Unrestricted
Off-target activity monitoring	Real-time detection	Post hoc analysis	Immediate feedback
Multiplexed editing efficiency	80%–95%	40%–70%	1.4–2.4× improvement
Protein variant screening	10^3–10^4 variants/day	10^2–10^3 variants/day	3–10× throughput
Cost per editing event	$0.10–0.50	$1.00–5.00	2–50× reduction

CRISPR-based target recognition with the rapid protein production capabilities of cell-free synthesis to create highly sensitive and specific diagnostic platforms. The integration enables the production of reporter proteins immediately upon target detection, providing amplified signals that can be detected using simple instrumentation or even visual inspection.

Therapeutic applications of cell-free CRISPR systems focus on ex vivo gene editing and the production of modified therapeutic proteins. Patient-derived cells can be edited in optimized cell-free environments before reintroduction, while therapeutic proteins can be produced with precise modifications that enhance efficacy or reduce immunogenicity. The controlled environment enables the incorporation of nonnatural amino acids or posttranslational modifications that would be difficult to achieve in cellular systems.

The evolution of CRISPR–Cas variants specifically optimized for cell-free environments has led to improved editing efficiency and expanded targeting capabilities. These variants incorporate modifications that enhance stability in cell-free conditions, improve cofactor utilization, and reduce inhibition by cellular components present in lysates. The characteristics of specialized CRISPR variants developed for cell-free applications are presented in Table 23.2.

TABLE 23.2 Specialized CRISPR variants for cell-free applications.

CRISPR variant	Optimization features	Cell-free activity (%)	Stability (hours)	Special applications
Cas9-CF1	Enhanced lysate compatibility	120–150 of wild-type	8–12	General editing
Cas12-ultra	Improved cofactor utilization	140–180 of wild-type	12–18	Multiplexed editing
Cas13-detect	Optimized for diagnostics	90–110 of wild-type	6–10	Nucleic acid detection
dCas9-activate	Transcriptional activation	110–140 of wild-type	10–14	Gene regulation
Cas-prime	Enhanced precision editing	100–130 of wild-type	8–12	Base editing

23.3 Three-dimensional bioprinting applications

The convergence of CFPS with three-dimensional bioprinting technology opens new frontiers in tissue engineering, drug delivery, and biomaterial development. This integration addresses fundamental challenges in bioprinting, including the need for bioactive materials, precise spatial control of biological functions, and the ability to create complex, multifunctional structures. Cell-free systems provide the molecular machinery needed to produce proteins on-demand within printed structures, enabling dynamic and responsive biomaterials.

Traditional bioprinting approaches rely on preformed biomaterials and living cells, which limits design flexibility and creates challenges for long-term functionality. The integration of CFPS allows for the printing of biomaterial precursors that can be activated postprinting to produce specific proteins with precise spatial and temporal control. This approach enables the creation of gradients, patterns, and responsive elements that would be impossible to achieve with conventional bioprinting methods.

The development of printable cell-free formulations requires careful optimization of rheological properties, biocompatibility, and stability during the printing process. These formulations must maintain the activity of cell-free components while providing the mechanical properties needed for successful printing and subsequent structure formation. Table 23.3 provides detailed key parameters for different bioprinting applications and their optimization requirements.

Multimaterial bioprinting with cell-free systems enables the creation of complex structures with distinct functional regions. Different cell-free formulations can be loaded with specific DNA templates, allowing for the spatial control of protein production within printed structures. This capability enables the creation of tissue constructs with defined growth factor gradients, enzyme activity patterns, or mechanical property variations that mimic natural tissue organization.

Stimuli-responsive bioprinted structures represent an advanced application where CFPS is triggered by external stimuli such as temperature, pH, light, or chemical signals. These structures can respond to their environment by producing specific proteins, changing mechanical properties, or releasing therapeutic agents. The integration of stimulus-responsive elements with cell-free systems creates smart biomaterials with unprecedented functionality.

TABLE 23.3 Bioprinting parameters for cell-free protein synthesis applications.

Application area	Viscosity range (cP)	Print resolution (μm)	Activation time (min)	Protein yield (mg/mL)	Stability (days)
Tissue scaffolds	1000–5000	100–500	30–120	0.5–2.0	7–14
Drug delivery matrices	500–2000	50–200	15–60	0.2–1.0	14–30
Biosensor arrays	200–1000	10–50	5–30	0.1–0.5	3–7
Organ-on-chip devices	100–500	5–25	10–45	0.3–1.5	5–10
Wound healing patches	2000–8000	200–1000	60–180	1.0–5.0	1–3

23.4 Nanotechnology convergence

The convergence of CFPS with nanotechnology creates powerful platforms for molecular manufacturing, targeted drug delivery, and advanced biosensing applications. This integration leverages the precision of nanotechnology fabrication with the biological functionality of proteins to create hybrid systems with enhanced performance characteristics. Nanostructured materials provide controlled environments for cell-free reactions while enabling precise positioning and organization of biological components.

Nanoparticle-based cell-free systems encapsulate protein synthesis machinery within protective nanostructures, enhancing stability and enabling controlled release of products. These systems can be designed to respond to specific environmental conditions, releasing active proteins only when needed. The nanoscale environment also allows for concentration effects that can enhance reaction efficiency and enable protein synthesis at very small scales. The performance characteristics of different nanoparticle platforms for CFPS applications are compared in Table 23.4.

Nanostructured surfaces and interfaces provide controlled environments for CFPS while enabling integration with electronic and optical systems. These surfaces can be functionalized with specific binding sites, concentration gradients, or stimulus-responsive elements that control protein synthesis location and timing. The integration with electronic systems enables real-time monitoring and control of protein synthesis processes at the molecular level.

Carbon nanotube and graphene-based platforms offer unique opportunities for cell-free system integration due to their exceptional electrical, mechanical, and thermal properties. These materials can serve as substrates for protein synthesis while providing electrical connectivity for sensing and control applications. The high surface area and tunable surface chemistry of these materials enable efficient immobilization of cell-free components and enhanced reaction kinetics.

Quantum dots and plasmonic nanoparticles provide optical functionality for cell-free systems, enabling fluorescent labeling, energy transfer processes, and enhanced

TABLE 23.4 Nanoparticle platforms for cell-free protein synthesis.

Nanoparticle type	Size range (nm)	Encapsulation efficiency (%)	Protein release rate	Stability enhancement	Key applications
Liposomes	50–500	60–85	Rapid (min–hours)	5–10×	Drug delivery, diagnostics
Polymeric microspheres	100–1000	70–90	Sustained (hours–days)	10–50×	Therapeutic delivery
Silica nanoparticles	10–200	50–80	Controlled (hours)	20–100×	Biosensors, catalysis
Hydrogel nanoparticles	20–500	80–95	Variable (min–weeks)	3–15×	Tissue engineering
Metal-organic frameworks	1–100	40–70	Triggered release	50–200×	Precision medicine

spectroscopic detection. These optical elements can be integrated directly into protein synthesis reactions to provide real-time monitoring capabilities or to create optically active protein products. The precise control over optical properties enables multiplexed detection and sophisticated signaling applications.

23.5 Digital biology platforms

The integration of CFPS with digital biology platforms creates automated, data-driven systems for biological design, testing, and optimization. These platforms combine robotic automation, advanced analytics, and machine learning algorithms to enable rapid prototyping and optimization of biological systems. The simplified and controllable nature of cell-free systems makes them ideal for automated experimentation and high-throughput screening applications.

Cloud-based laboratory platforms enable remote control and monitoring of CFPS experiments, democratizing access to advanced biotechnology capabilities. These systems allow researchers to design experiments using web-based interfaces, execute them on robotic platforms, and receive results through digital channels. The standardization inherent in cell-free systems facilitates reliable remote operation and reproducible results across different laboratory locations.

Digital twin technology creates virtual representations of CFPS systems that can be used for predictive modeling, optimization, and troubleshooting. These digital models incorporate detailed kinetic parameters, component interactions, and environmental factors to accurately predict system behavior under different conditions. Table 23.5 outlines the key features and capabilities of digital biology platforms for cell-free applications.

Artificial intelligence integration enables autonomous optimization of CFPS conditions through iterative experimentation and learning. These systems can automatically adjust

TABLE 23.5 Digital biology platform features for cell-free applications.

Platform component	Functionality	Data integration	Automation level	Predictive accuracy	Throughput enhancement
Robotic liquid handling	Precise dispensing, mixing	Real-time volume tracking	Fully automated	±2%–5%	10–100×
Environmental control	Temperature, humidity, atmosphere	Sensor data integration	Semiautomated	±1%–3%	3–10×
Analytical instruments	Spectroscopy, chromatography	Automated data collection	Fully automated	±5%–15%	5–50×
Machine learning models	Pattern recognition, optimization	Multimodal data fusion	Algorithm-driven	70%–95%	2–20×
Cloud computing	Distributed processing	Database integration	Fully automated	Variable	100–1000×

reaction parameters, select optimal components, and predict outcomes based on experimental results and accumulated knowledge. The continuous learning capability enables progressive improvement in system performance and the discovery of novel optimization strategies.

Blockchain technology provides secure and transparent documentation of experimental procedures, results, and intellectual property in CFPS research. This technology enables verified sharing of protocols, traceable supply chains for reagents, and protected intellectual property rights. The immutable record-keeping capability is particularly valuable for regulatory compliance and collaboration between research institutions.

23.6 Quantum computing applications

The convergence of quantum computing with CFPS represents one of the most forward-looking technological integrations, offering unprecedented computational power for molecular simulation, optimization, and design. Quantum computers excel at solving complex optimization problems and simulating quantum mechanical systems, making them particularly well-suited for addressing the computational challenges inherent in biological system design and optimization.

Protein folding prediction represents a natural application for quantum computing in cell-free synthesis, where the quantum mechanical nature of molecular interactions can be directly simulated using quantum algorithms. This capability enables more accurate prediction of protein structure and function, leading to better design of proteins for cell-free production. The enhanced prediction accuracy can significantly reduce the experimental optimization time and improve success rates for complex protein targets.

Quantum optimization algorithms can solve the complex multiparameter optimization problems encountered in cell-free system design, including component concentration optimization, reaction condition selection, and resource allocation. These algorithms can explore solution spaces that are computationally intractable for classical computers, potentially discovering novel optimization strategies and improved system configurations. A detailed comparison of quantum and classical computing approaches for various cell-free synthesis optimization problems is presented in Table 23.6.

Quantum machine learning algorithms offer enhanced pattern recognition and predictive modeling capabilities for CFPS data. These algorithms can identify subtle correlations and patterns in experimental data that might be missed by classical approaches, leading to better understanding of system behavior and improved predictive models. The quantum advantage is particularly pronounced for high-dimensional datasets and complex feature interactions.

Cryptographic applications of quantum computing enhance security and intellectual property protection in CFPS research and development. Quantum cryptography enables secure communication of sensitive protocols and results, while quantum random number generation provides enhanced security for digital biology platforms. These capabilities are particularly important for protecting proprietary methods and maintaining competitive advantages in biotechnology development.

The evolutionary trajectory of convergent technologies with CFPS follows distinct phases of development, from current foundational implementations to future transformative

TABLE 23.6 Quantum vs. classical computing for cell-free synthesis optimization.

Optimization problem	Classical computing time	Quantum computing time	Accuracy improvement	Complexity scaling	Practical advantage
Component concentration	Hours to days	Minutes to hours	15%–30%	Exponential vs. polynomial	10–100× speedup
Reaction pathway design	Days to weeks	Hours to days	20%–40%	Factorial vs. polynomial	50–500× speedup
Multiobjective optimization	Weeks to months	Days to weeks	25%–50%	NP-hard vs. P	100–1000× speedup
Protein design	Months to years	Weeks to months	30%–60%	Double exponential vs. exponential	1000–10000× speedup
System integration	Intractable	Feasible	Novel solutions	Reduces complexity class	Enables new capabilities

applications. Fig. 23.2 illustrates the projected timeline for technology convergence, showing how different technologies will mature and integrate over the next two decades. This roadmap demonstrates the progression from today's proof-of-concept demonstrations to tomorrow's fully integrated biotechnology platforms that could revolutionize manufacturing, healthcare, and space exploration.

23.7 Space biotechnology integration

The integration of CFPS with space biotechnology addresses unique challenges associated with biological manufacturing in extraterrestrial environments. Space applications demand robust, self-contained systems that can operate reliably under extreme conditions, including radiation exposure, microgravity, temperature fluctuations, and limited resources. Cell-free systems offer significant advantages for space applications due to their simplified infrastructure requirements and enhanced stability compared to living cell systems.

Microgravity effects on CFPS have been extensively studied, revealing both challenges and opportunities for space-based manufacturing. The absence of gravitational settling can improve mixing and heat transfer in cell-free reactions, while eliminating convection patterns that can create concentration gradients. However, gas bubble formation and surface tension effects require careful consideration in reactor design. The key factors affecting cell-free synthesis in space environments are summarized in Table 23.7.

Radiation hardening of cell-free components represents a critical requirement for long-duration space missions. Cosmic radiation can damage DNA templates, denature proteins, and disrupt reaction components, requiring specialized formulations and protective measures. Radiation-resistant variants of key enzymes and stabilizing additives help maintain system functionality under space radiation conditions.

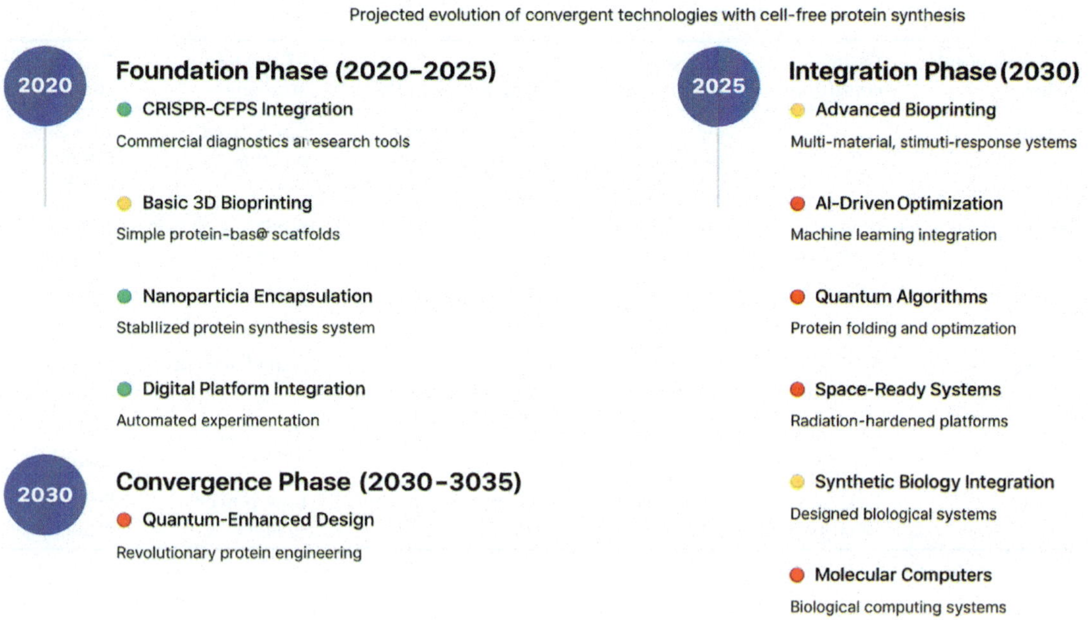

FIG. 23.2 Convergent technology development timeline. This timeline projection spans four distinct development phases from 2020 to 2040, illustrating the evolution of convergent technologies with cell-free protein synthesis. The Foundation Phase (2020–2025) establishes basic integration capabilities, particularly CRISPR–CFPS diagnostics and digital platform automation. The Integration Phase (2025–2030) focuses on advanced bioprinting applications and AI-driven optimization systems. The Convergence Phase (2030–2035) introduces quantum-enhanced protein design and autonomous space manufacturing capabilities. The Transformation Phase (2035–2040) envisions fully integrated quantum biotechnology and interplanetary biomanufacturing systems. Technology readiness levels are indicated by color coding: mature (green) technologies ready for commercial implementation, developing (yellow) technologies in active research phases, and emerging (red) technologies in conceptual or early research stages.

Closed-loop resource utilization becomes essential for sustainable space-based cell-free manufacturing. All materials must be recycled or converted to useful products, with minimal waste generation. Advanced recycling systems recover amino acids, nucleotides, and cofactors from spent reactions, while waste products are converted to useful materials or safely stored. This constraint drives innovations in system efficiency and resource conservation that benefit terrestrial applications.

Life support integration connects CFPS systems with spacecraft life support systems, creating synergistic relationships that benefit both systems. Cell-free systems can produce essential proteins for crew health and equipment maintenance, while waste products from life support systems can serve as feedstocks for protein synthesis. The integration reduces overall mission mass and enhances mission sustainability.

Planetary protection considerations require careful design of space-based cell-free systems to prevent contamination of extraterrestrial environments while protecting Earth from

TABLE 23.7 Environmental factors affecting cell-free protein synthesis in space.

Environmental factor	Earth conditions	Space conditions	Impact on CFPS	Mitigation strategies	Performance change
Gravity	1 g	0 g (microgravity)	Improved mixing, altered kinetics	Specialized reactors	10%–30% improvement
Radiation	Low background	High cosmic radiation	DNA/protein damage	Shielding, repair systems	20%–50% reduction
Temperature	Controlled	−150°C to +120°C	Reaction rate changes	Thermal management	Variable
Atmosphere	Standard pressure	Vacuum or controlled	Outgassing, volatilization	Sealed systems	Requires containment
Resources	Unlimited supply	Strictly limited	Conservation critical	Recycling, efficiency	Must optimize
Power	Grid electricity	Solar/nuclear limited	Energy constraints	Low-power systems	50%–80% reduction

potentially harmful materials. These requirements influence system design, containment protocols, and sterilization procedures. The inherently controlled nature of cell-free systems provides advantages for meeting planetary protection requirements compared to living biological systems.

Autonomous operation capabilities enable cell-free systems to function with minimal crew intervention during long-duration space missions. Advanced automation, remote monitoring, and self-diagnostic capabilities ensure reliable operation while minimizing crew workload. Machine learning algorithms adapt system operation to changing conditions and optimize performance based on available resources and mission requirements.

In situ resource utilization explores the use of materials available on other planets or asteroids as feedstocks for CFPS. This includes extraction of carbon, nitrogen, and other elements from atmospheric or surface materials, as well as the use of local energy sources such as solar radiation or radioisotope thermal generators. The ability to use local resources dramatically reduces the supply requirements for long-term space missions and enables sustainable space-based manufacturing.

The integration of CFPS systems with space infrastructure represents one of the most challenging yet promising applications of convergent technologies. Fig. 23.3 depicts a comprehensive space-based CFPS system architecture designed for autonomous operation in extraterrestrial environments. This integrated platform demonstrates how multiple convergent technologies combine to create self-sustaining biological manufacturing capabilities that could support long-duration space missions and eventual planetary colonization efforts.

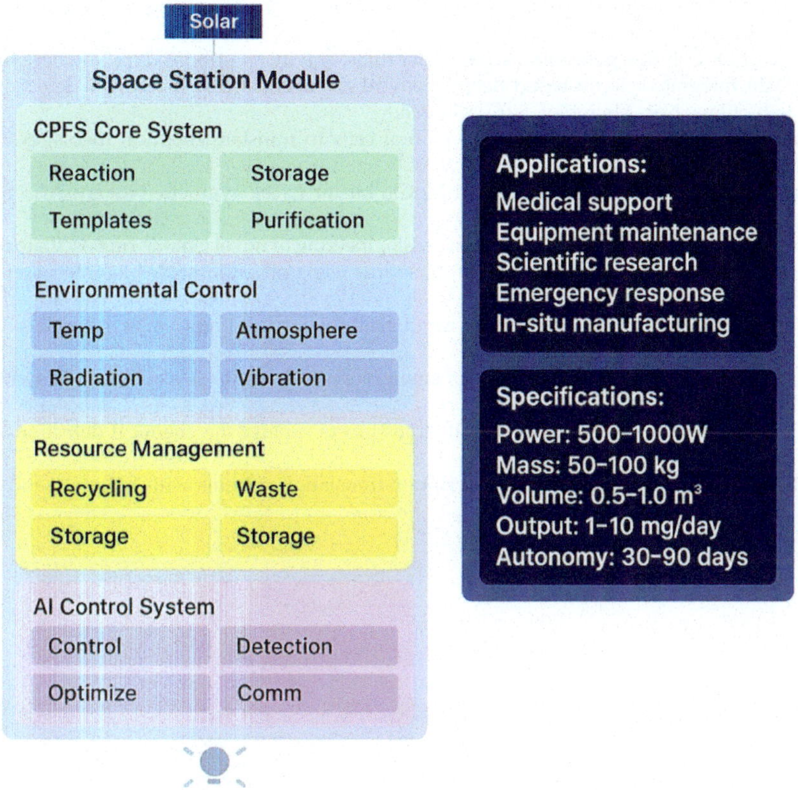

FIG. 23.3 **Space-based CFPS system architecture.** This schematic illustrates an integrated autonomous CFPS platform designed for space environments, incorporating multiple convergent technologies for sustainable protein manufacturing beyond Earth. The system features solar power arrays for energy generation, environmental control systems for radiation shielding and atmospheric management, resource management modules for recycling and waste processing, and AI control systems for autonomous operation and Earth communication. The central CFPS core includes reaction chambers, lysate storage, template banks, and purification systems optimized for microgravity operation. System specifications include 500–1000 W power consumption, 50–100 kg mass, 0.5–1.0 m^3 volume, 1–10 mg/day protein output, and 30–90-day autonomous operation capability. Applications encompass crew medical support, equipment maintenance, scientific research, emergency response, and in situ manufacturing for long-duration space missions.

Bibliography

Abil, Z., et al., 2017. Compartmentalized partner replication for the directed evolution of genetic parts and circuits. Nat. Commun. 8 (1), 2196.

Adachi, K., et al., 2021. Cell-free protein synthesis in space: opportunities and challenges for biotechnology beyond Earth. Front. Bioeng. Biotechnol. 9, 814483.

Albayrak, C., Swartz, J.R., 2014. Direct polymerization of proteins. ACS Synth. Biol. 3 (6), 353–362.

Contreras-Llano, L.E., et al., 2019. High-throughput screening of biomaterials using cell-free protein synthesis. Biomaterials 221, 119419.

Des Soye, B.J., et al., 2018. Improving yeast cell-free protein synthesis by minimizing endogenous RNA degradation. Biotechnol. Bioeng. 115 (5), 1149–1161.

Garenne, D., et al., 2021. Cell-free gene expression under microgravity conditions. Npj Microgravity 7 (1), 18.

Hunt, J.P., et al., 2020. Integrated circuit-based electrochemical sensors for spatially resolved detection of redox-active metabolites in biofilms. Nat. Microbiol. 5 (6), 808–815.

Lentini, R., et al., 2014. Integrating artificial with natural cells to translate chemical messages that direct *E. coli* behaviour. Nat. Commun. 5, 4012.

Moore, S.J., et al., 2018. EcoFlex: a multifunctional MoClo kit for *E. coli* synthetic biology. ACS Synth. Biol. 7 (4), 1059–1069.

Niederholtmeyer, H., et al., 2015. Rapid cell-free forward engineering of novel genetic ring oscillators. eLife 4, e09771.

Pardee, K., et al., 2016. Rapid, low-cost detection of Zika virus using programmable biomolecular components. Cell 165 (5), 1255–1266.

Schoborg, J.A., et al., 2018. A cell-free platform for rapid synthesis and testing of active oligosaccharyltransferases. Biotechnol. Bioeng. 115 (3), 739–750.

Thavarajah, W., et al., 2020. Point-of-use detection of environmental fluoride via a cell-free riboswitch-based biosensor. ACS Synth. Biol. 9 (1), 10–18.

Voyvodic, P.L., et al., 2019. Plug-and-play metabolic transducers expand the chemical detection space of cell-free biosensors. Nat. Commun. 10 (1), 1697.

Zhang, L., et al., 2018. Engineering cell-free transcription-translation systems with biosafety and biocontainment. Curr. Opin. Biotechnol. 53, 195–202.

PART VIII

Practical implementation

CHAPTER

24

Laboratory setup and implementation for cell-free protein synthesis

24.1 Introduction

The successful implementation of cell-free protein synthesis technology requires careful consideration of laboratory infrastructure, equipment selection, facility design, and operational procedures. Unlike traditional cell culture facilities that demand complex biosafety containment and sterile environments, cell-free systems offer simplified requirements while maintaining stringent quality control standards. This chapter provides comprehensive guidance for establishing efficient, safe, and productive cell-free protein synthesis laboratories across various scales and applications.

The transition from traditional biotechnology laboratories to cell-free protein synthesis facilities represents a paradigm shift that affects every aspect of laboratory design and operation. The elimination of living organisms from the production process reduces biosafety requirements, simplifies contamination control, and enables more flexible facility layouts. However, the precision required for cell-free systems demands sophisticated analytical capabilities, precise environmental control, and robust quality assurance protocols.

Modern cell-free laboratories must accommodate diverse applications ranging from research and development to commercial production, each with distinct requirements for throughput, automation, and analytical capabilities. The modular nature of cell-free systems enables scalable laboratory designs that can evolve with changing needs while maintaining operational efficiency. This comprehensive approach to laboratory implementation ensures optimal performance, regulatory compliance, and long-term sustainability of cell-free protein synthesis operations.

24.2 Equipment requirements and selection

The equipment infrastructure for cell-free protein synthesis laboratories differs significantly from traditional biotechnology facilities, emphasizing precision control, analytical capabilities, and process optimization rather than large-scale fermentation and sterile maintenance. The core equipment categories include reaction systems, analytical instruments,

Cell Free Synthesis of Biologics
https://doi.org/10.1016/B978-0-443-33271-5.00020-X

preparation equipment, storage systems, and automation platforms. Table 24.1 outlines the essential equipment requirements across different laboratory scales and applications.

Reaction system selection represents the most critical equipment decision, as these systems directly impact protein yield, quality, and reproducibility. For research applications, standard thermal cyclers and orbital shakers provide adequate performance for small-scale reactions. Pilot-scale operations require specialized bioreactors with precise temperature control, automated sampling capabilities, and real-time monitoring systems. Production-scale facilities demand fully integrated reaction systems with continuous operation capabilities, automated feeding systems, and comprehensive process control.

Analytical equipment selection must balance analytical capability with throughput requirements and cost considerations. Research laboratories typically require basic spectrophotometric analysis, gel electrophoresis systems, and protein quantification capabilities. Pilot-scale facilities need more sophisticated analytical platforms, including high-performance liquid chromatography, mass spectrometry, and automated gel analysis systems. Production-scale operations require process analytical technology with real-time monitoring, automated sample handling, and integrated data management systems.

Storage and preparation equipment requirements scale with laboratory throughput and complexity. Small research laboratories can operate with standard laboratory refrigeration and basic preparation equipment. Larger facilities require specialized storage systems with redundant temperature control, automated inventory management, and integrated preparation systems that maintain component quality while enabling high-throughput operations.

Automation integration becomes increasingly critical as laboratory scale increases, with fully automated systems essential for production-scale operations. The selection of

TABLE 24.1 Essential equipment requirements for cell-free protein synthesis laboratories.

Equipment category	Research scale	Pilot scale	Production scale	Key specifications	Typical cost range
Reaction systems	Thermal cyclers, shakers	Bioreactors, continuous systems	Automated production lines	Temperature ± 0.1°C, mixing 50–500 rpm	\$10 K–500 K
Analytical equipment	Spectrophotometers, PAGE systems	HPLC, mass spectrometry	Process analytical technology	Detection limits ng/mL, automation ready	\$50 K–2 M
Preparation equipment	Centrifuges, homogenizers	Large-scale prep systems	Integrated prep modules	Variable speed, temperature control	\$20 K–300 K
Storage systems	Laboratory freezers, refrigerators	Walk-in cold rooms	Automated storage systems	Temperature monitoring, backup power	\$5 K–100 K
Automation platforms	Liquid handlers, robots	Integrated workstations	Fully automated lines	Precision ± 1%, throughput scalability	\$100 K–5 M

automation platforms must consider compatibility with existing equipment, scalability for future expansion, and integration with laboratory information management systems. Fig. 24.1 illustrates the equipment integration architecture for different laboratory scales, showing how component systems connect to create efficient operational workflows.

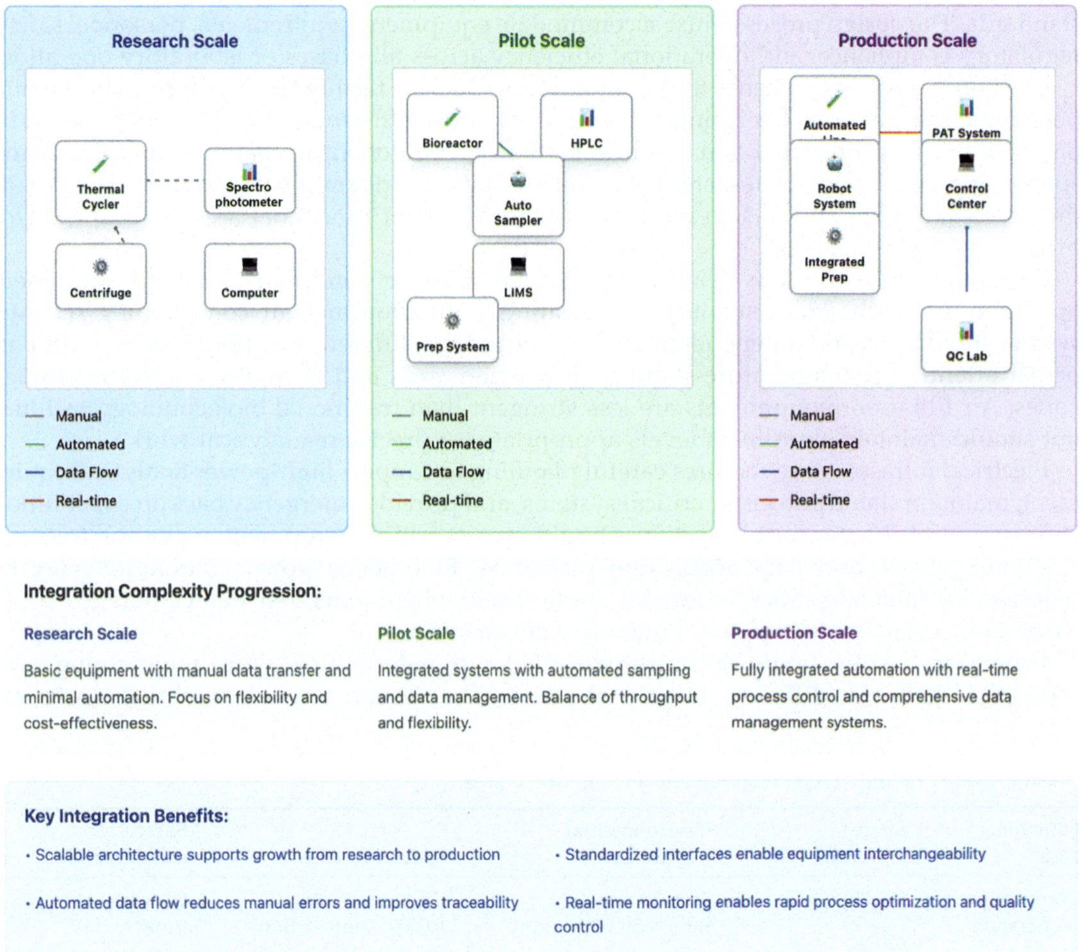

FIG. 24.1 Equipment integration architecture for cell-free protein synthesis laboratories. This diagram illustrates how core equipment systems integrate across research, pilot, and production scale laboratories. The architecture shows data flow between reaction systems, analytical instruments, automation platforms, and information management systems. At research scale, basic integration connects thermal cyclers and analytical equipment through manual data transfer. Pilot scale introduces automated sampling and integrated data management with real-time monitoring capabilities. Production scale features fully integrated systems with process analytical technology, automated control systems, and comprehensive data management. The integration architecture demonstrates scalability pathways and technology evolution opportunities while maintaining compatibility between scale levels.

24.3 Facility design considerations

Cell-free protein synthesis facility design requires careful attention to workflow optimization, environmental control, contamination prevention, and future expansion capabilities. Unlike traditional biotechnology facilities with complex biosafety requirements, cell-free laboratories can utilize more flexible designs while maintaining stringent quality control standards. The design process must accommodate equipment requirements, personnel safety, regulatory compliance, and operational efficiency across all phases of laboratory operation.

Workflow design represents the foundation of efficient facility layout, with material and personnel flow patterns determining overall operational efficiency. The laboratory should be organized into distinct functional zones, including component preparation areas, reaction spaces, analytical laboratories, storage facilities, and administrative areas. Table 24.2 details the space allocation and design requirements for different functional areas within cell-free protein synthesis facilities.

Environmental control systems must maintain precise conditions for both equipment operation and component stability. The heating ventilation and air conditioning (HVAC) system should provide independent climate control for different functional areas, with particular attention to temperature stability in reaction areas and humidity control in storage zones. Air filtration requirements are less stringent than traditional biotechnology facilities but should maintain cleanliness levels appropriate for precision analytical work.

Electrical infrastructure requires careful planning to support high-power analytical equipment, maintain stable power for critical systems, and provide emergency backup capabilities. Uninterruptible power supply systems should protect critical equipment, including freezers, reaction systems, and data acquisition platforms. Emergency power generation may be required for larger facilities or critical applications where extended power outages could compromise valuable samples or ongoing experiments.

Laboratory layout should optimize personnel and material flow while maintaining clear separation between different functional areas. The design should minimize transport

TABLE 24.2 Facility design requirements by functional area.

Functional area	Space allocation (%)	Environmental requirements	Special features	Safety considerations
Component preparation	20–25	±2°C temperature, < 50% humidity	Clean room class 10,000, laminar flow	Chemical storage, fume hoods
Reaction operations	25–30	±1°C temperature, vibration control	Isolated power, emergency backup	Temperature alarms, spill containment
Analytical laboratory	15–20	Vibration isolation, EMI shielding	Specialized instrumentation areas	Chemical safety, waste management
Storage areas	20–25	Multiple temperature zones, monitoring	Automated retrieval, inventory tracking	Temperature alarms, backup systems
Utilities and support	10–15	Equipment access, maintenance space	Compressed air, water systems	Electrical safety, confined space

distances for frequently moved items while preventing cross-contamination between different activities. Flexibility for future reconfiguration should be built into the design through modular construction approaches and accessible utilities distribution.

24.4 Safety protocols and risk management

Safety protocols for cell-free protein synthesis laboratories focus on chemical handling, equipment operation, and emergency response rather than the biological containment requirements of traditional biotechnology facilities. The absence of living organisms simplifies many safety considerations while introducing specific risks associated with concentrated biochemical reagents, analytical chemicals, and sophisticated instrumentation. Comprehensive safety management requires systematic hazard identification, risk assessment, and implementation of appropriate control measures.

Chemical safety represents the primary safety concern in cell-free laboratories, with numerous reagents requiring specialized handling procedures. Concentrated buffer solutions, organic solvents, and analytical reagents demand appropriate storage, handling, and disposal protocols. Table 24.3 categorizes the major chemical hazard classes encountered in cell-free protein synthesis and outlines appropriate safety measures for each category.

Equipment safety protocols address the specific hazards associated with analytical instrumentation, reaction systems, and automation equipment. High-temperature devices require thermal protection and emergency shutdown capabilities. High-pressure systems demand regular inspection and pressure relief protection. Electrical equipment requires proper grounding, circuit protection, and lockout/tagout procedures for maintenance activities.

TABLE 24.3 Chemical safety requirements for cell-free protein synthesis laboratories.

Hazard category	Common chemicals	Storage requirements	Handling procedures	Emergency response
Corrosive reagents	Concentrated acids, bases	Segregated storage, secondary containment	Fume hood use, personal protection equipment (PPE) required	Neutralization protocols, eyewash stations
Flammable solvents	Alcohols, organic solvents	Fire-rated cabinets, grounding	Spark-free environment, ventilation	Fire suppression, spill control
Toxic compounds	Heavy metals, analytical reagents	Locked storage, inventory tracking	Specialized PPE, exposure monitoring	Medical surveillance, decontamination
Reactive chemicals	Oxidizers, reducing agents	Separated storage, temperature control	Compatibility checking, small quantities	Isolation procedures, specialized cleanup
Biological components	Proteins, nucleic acids	Refrigerated storage, stability monitoring	Standard precautions, allergen awareness	Cleanup protocols, waste treatment

Emergency response procedures must address the most likely emergency scenarios including chemical spills, equipment failures, power outages, and personnel injuries. Emergency response plans should include notification procedures, evacuation routes, emergency contacts, and specific response protocols for different types of incidents. Regular emergency drills ensure personnel familiarity with procedures and identify opportunities for improvement.

Personal protective equipment requirements vary by activity and hazard exposure but typically include safety glasses, laboratory coats, and appropriate gloves for chemical handling. Respiratory protection may be required for specific activities involving volatile chemicals or powder handling. Specialized PPE such as cryogenic gloves may be needed for ultra-low-temperature storage operations.

Training programs must ensure all personnel understand safety protocols, emergency procedures, and proper equipment operation. Initial safety training should cover general laboratory safety principles, specific hazards associated with cell-free protein synthesis, and facility-specific procedures. Ongoing training programs maintain safety awareness and incorporate lessons learned from incidents or procedure updates. Fig. 24.2 presents a

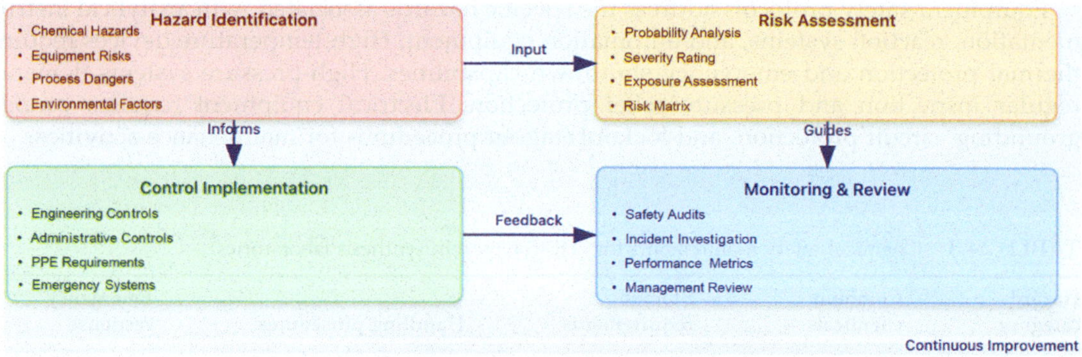

FIG. 24.2 Comprehensive safety management framework for cell-free laboratories. This framework diagram presents the systematic approach to safety management in cell-free protein synthesis facilities. The framework integrates hazard identification processes that address chemical, equipment, and operational risks with risk assessment methodologies that evaluate probability and severity of potential incidents. Control implementation follows a hierarchical approach with engineering controls, including fume hoods, containment systems, and automated safety systems, administrative controls encompassing procedures, training, and access control measures, personal protective equipment strategies involving gloves, safety glasses, and laboratory coats, and emergency systems, including eyewash stations, safety showers, and alarm systems. Emergency response elements include chemical spill response procedures and containment materials, equipment failure protocols and emergency shutdown systems, medical emergency response and first aid procedures, and evacuation procedures with emergency communication systems. The continuous improvement cycle includes incident investigation, near-miss reporting, safety audits, and management review processes that ensure comprehensive coverage of all safety aspects through structured methodology, enable ongoing enhancement of safety performance and procedures through feedback loops, and support documentation and demonstration of safety management systems for regulatory compliance. Emergency response procedures connect to external emergency services while internal response teams handle facility-specific incidents, demonstrating how safety management integrates with operational procedures to create comprehensive risk mitigation strategies.

comprehensive safety management framework showing the integration of hazard identification, risk assessment, control implementation, and continuous improvement processes.

24.5 Staff training and competency development

Effective staff training for cell-free protein synthesis laboratories requires a comprehensive approach that addresses both technical competencies and safety requirements. The unique characteristics of cell-free systems demand specialized knowledge in biochemistry, analytical techniques, and process optimization that differs significantly from traditional cell culture expertise. Training programs must be tailored to different roles within the laboratory while ensuring all personnel understand the fundamental principles and safety requirements of cell-free protein synthesis.

Technical competency requirements vary significantly between different laboratory roles, from research technicians conducting basic assays to senior scientists developing new applications. Table 24.4 outlines the core competency requirements and training modules for different positions within cell-free protein synthesis laboratories.

Initial training programs should begin with fundamental principles of cell-free protein synthesis, including system components, reaction mechanisms, and factors affecting performance. Hands-on laboratory sessions allow trainees to gain practical experience with equipment operation, sample preparation, and analytical procedures. Progressive training modules build complexity gradually, starting with basic procedures and advancing to sophisticated optimization and troubleshooting techniques.

Safety training integration ensures all personnel understand chemical handling procedures, equipment safety requirements, and emergency response protocols. Safety training should be conducted before any hands-on laboratory work and reinforced through regular updates and refresher sessions. Specialized safety training may be required for personnel working with particular hazards such as high-voltage equipment or cryogenic systems.

TABLE 24.4 Staff training requirements and competency matrix.

Position level	Core competencies	Training duration	Practical assessments	Continuing education
Research technician	Basic biochemistry, analytical techniques	2–4 weeks	Hands-on demonstrations, written exams	Monthly seminars, annual updates
Laboratory specialist	Advanced analysis, troubleshooting	4–8 weeks	Complex problem solving, method validation	Quarterly workshops, conference attendance
Senior scientist	System design, optimization	8–12 weeks	Independent projects, presentation skills	External courses, peer collaboration
Laboratory manager	Operations management, compliance	6–10 weeks	Leadership scenarios, audit preparation	Management training, regulatory updates
Quality assurance	Validation methods, documentation	4–6 weeks	SOP development, inspection readiness	GMP training, validation workshops

Competency assessment provides objective evaluation of training effectiveness and ensures personnel are qualified for their assigned responsibilities. Assessment methods should include written examinations covering theoretical knowledge, practical demonstrations of laboratory skills, and problem-solving exercises that test troubleshooting abilities. Documentation of competency assessments provides evidence of personnel qualifications for regulatory compliance and quality assurance purposes.

Continuing education programs maintain and enhance staff capabilities as technology evolves and new applications emerge. Regular seminars, workshops, and conference attendance keep personnel current with technological advances and industry best practices. Internal knowledge sharing sessions allow experienced personnel to share expertise with newer staff members while fostering collaborative learning environments.

24.6 Vendor selection and procurement strategy

Strategic vendor selection and procurement planning are critical for establishing reliable, cost-effective cell-free protein synthesis operations. The specialized nature of cell-free systems requires vendors with specific expertise in biochemical reagents, analytical instrumentation, and process equipment. Effective procurement strategies balance quality requirements, cost considerations, and supply chain reliability while maintaining flexibility for future technology upgrades and expansion.

Vendor evaluation criteria must address technical capabilities, quality systems, regulatory compliance, and long-term support capabilities. Primary considerations include product quality and consistency, technical support availability, regulatory documentation, delivery reliability, and financial stability. Table 24.5 provides a comprehensive vendor evaluation framework for different categories of suppliers serving cell-free protein synthesis laboratories.

TABLE 24.5 Vendor evaluation criteria and selection framework.

Vendor category	Primary evaluation criteria	Quality requirements	Support expectations	Strategic considerations
Reagent suppliers	Purity specifications, batch consistency	ISO 9001, analytical certificates	Technical consultation, troubleshooting	Long-term availability, custom synthesis
Equipment manufacturers	Performance specifications, reliability	CE marking, FDA clearance	Installation, training, maintenance	Technology roadmap, upgrade paths
Service providers	Expertise level, turnaround time	Accreditation, validation experience	Method development, documentation	Capacity scaling, confidentiality
Software vendors	Functionality, integration capability	21 CFR Part 11 compliance	Implementation support, updates	Data security, cloud capabilities
Consumable suppliers	Quality consistency, packaging	Sterility assurance, traceability	Inventory management, logistics	Supply chain resilience, cost stability

Supplier qualification processes should include facility audits, reference checks, and pilot evaluations to verify vendor capabilities before committing to significant purchases. Documentation requirements include quality certificates, regulatory compliance documentation, and technical specifications. Ongoing supplier monitoring ensures continued compliance with quality and performance standards.

Procurement strategies should optimize total cost of ownership rather than focusing solely on initial purchase prices. Considerations include maintenance costs, consumable expenses, training requirements, and potential upgrade costs. Strategic partnerships with key vendors can provide preferential pricing, priority support, and access to new technologies before general market availability.

Risk management in procurement addresses supply chain disruptions, vendor financial stability, and technology obsolescence. Diversified supplier bases reduce dependence on single sources for critical materials and services. Inventory management strategies balance carrying costs against stockout risks, with particular attention to critical reagents with limited shelf life or lengthy procurement lead times.

24.7 Troubleshooting and maintenance programs

Comprehensive troubleshooting and maintenance programs ensure consistent performance, minimize downtime, and extend equipment life in cell-free protein synthesis laboratories. The complexity of modern analytical instrumentation and automation systems demands systematic approaches to preventive maintenance, problem diagnosis, and corrective action implementation. Effective programs combine manufacturer recommendations with facility-specific experience to optimize system reliability and performance.

Preventive maintenance scheduling should balance equipment protection with operational availability, focusing on critical systems that directly impact experimental results or safety. Table 24.6 outlines typical maintenance requirements and intervals for major equipment categories in cell-free protein synthesis laboratories.

Troubleshooting protocols should provide systematic approaches for diagnosing and resolving common problems encountered in cell-free protein synthesis operations. Diagnostic procedures should progress from simple checks to more complex investigations, with clear decision trees guiding troubleshooting efforts. Documentation of troubleshooting activities provides valuable information for identifying recurring problems and developing preventive measures.

Maintenance documentation systems track all preventive and corrective maintenance activities, providing historical records for trend analysis and regulatory compliance. Computerized maintenance management systems can automate scheduling, track spare parts inventory, and generate reports for management review. Integration with laboratory information management systems enables correlation of maintenance activities with experimental results.

Training programs for maintenance personnel should cover both general maintenance principles and equipment-specific procedures. Manufacturer training programs provide detailed technical knowledge while internal training ensures familiarity with facility-specific

TABLE 24.6 Equipment maintenance schedule and requirements.

Equipment type	Daily checks	Weekly maintenance	Monthly service	Annual calibration	Critical spare parts
Analytical instruments	Performance verification, cleanliness	Lamp replacement, filter changes	Detector calibration, system cleaning	Full calibration, software updates	Lamps, detectors, pumps
Reaction systems	Temperature verification, level checks	Probe cleaning, calibration checks	Sensor replacement, deep cleaning	Control system validation, safety checks	Sensors, heating elements, pumps
Storage equipment	Temperature logging, alarm testing	Gasket inspection, coil cleaning	Refrigerant level, electrical connections	Temperature mapping, safety systems	Thermostats, alarms, backup systems
Automation systems	Movement verification, error logs	Lubrication, sensor cleaning	Software updates, mechanical inspection	Precision calibration, safety validation	Motors, sensors, software licenses
Safety equipment	Visual inspection, alarm testing	Battery replacement, sensor checks	Calibration verification, filter replacement	Full system validation, training updates	Sensors, batteries, safety devices

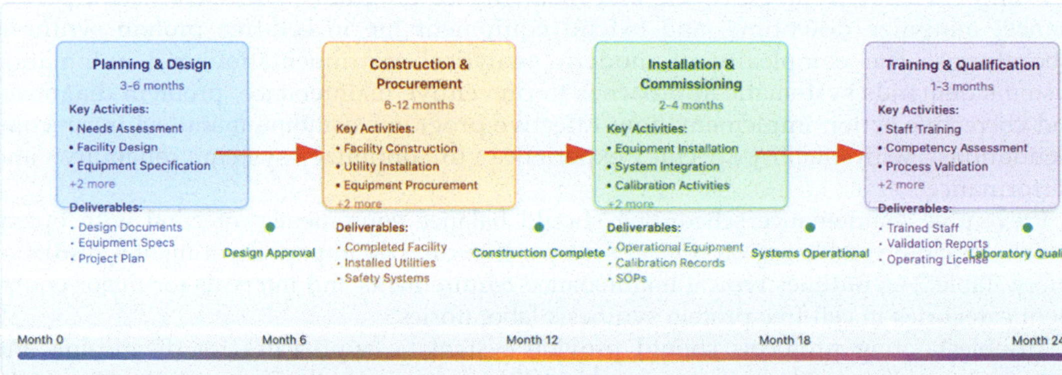

FIG. 24.3 Laboratory implementation workflow and timeline. This comprehensive workflow diagram outlines the sequential phases of establishing cell-free protein synthesis laboratory capabilities from initial planning through operational qualification. The planning phase includes needs assessment, facility design, equipment specification, and regulatory review processes. The design and construction phase encompasses architectural development, utility installation, equipment procurement, and safety system implementation. The installation and commissioning phase covers equipment installation, system integration, calibration activities, and initial testing procedures. The training and qualification phase includes staff training programs, competency assessments, operational procedure development, and performance qualification studies. The diagram shows typical timeframes for each phase and identifies critical dependencies between activities, including utility infrastructure completion before equipment installation, specialized analytical instrument lead times of 3–6 months, regulatory approval processes that may extend timelines significantly, and staff training schedules that must align with equipment installation completion. Quality gates between phases ensure readiness before advancing to subsequent implementation stages through design review for technical specifications and regulatory compliance verification, construction inspection for safety systems and utility installation verification, commissioning tests for equipment performance and integration validation, and final qualification for operational readiness and regulatory approval. Implementation success factors include dedicated project management with biotechnology facility experience and clear communication protocols, regular stakeholder engagement with users, management, and regulatory authorities throughout implementation, and proactive risk management for identification and mitigation of technical, schedule, and budget risks.

procedures and safety requirements. Cross-training programs develop backup capabilities for critical maintenance functions.

Fig. 24.3 illustrates a comprehensive laboratory implementation workflow showing the sequential phases of facility design, equipment installation, staff training, and operational qualification. This integrated approach ensures systematic implementation while maintaining focus on safety, quality, and operational efficiency throughout the process.

Bibliography

Bundy, B.C., et al., 2018. Cell-free biomanufacturing. Curr. Opin. Chem. Eng. 22, 177–183.

Carlson, E.D., et al., 2012. Cell-free protein synthesis: applications come of age. Biotechnol. Adv. 30 (5), 1185–1194.

Gregorio, N.E., et al., 2019. A user's guide to cell-free protein synthesis. Methods Protoc. 2 (1), 24.

Hutchinson, R.J., et al., 2020. Guidelines for the setup and operation of cell-free protein synthesis laboratories. Biotechnol. J. 15 (9), 2000045.

Kelwick, R., et al., 2016. Advances in cell-free protein synthesis for biotechnology. Curr. Opin. Biotechnol. 42, 171–178.

Kopniczky, M.B., et al., 2020. Cell-free protein synthesis as a prototyping platform for mammalian synthetic biology. ACS Synth. Biol. 9 (1), 144–156.

Silverman, A.D., et al., 2020. Cell-free gene expression: an expanded repertoire of applications. Nat. Rev. Genet. 21 (3), 151–170.

Swartz, J.R., 2018. Expanding biological applications of cell-free protein synthesis. Curr. Opin. Biotechnol. 53, 203–208.

Thompson, K.E., et al., 2019. Laboratory design considerations for cell-free protein synthesis facilities. Lab. Des. Int. 15 (3), 45–52.

Zemella, A., et al., 2015. Cell-free protein synthesis: pros and cons of prokaryotic and eukaryotic systems. ChemBioChem 16 (17), 2420–2431.

CHAPTER

25

Protocols and standard operating procedures for cell-free protein synthesis

25.1 Introduction

The development and implementation of standardized protocols and operating procedures represents the foundation of reliable, reproducible cell-free protein synthesis operations. These documented procedures ensure consistent results across different operators, facilities, and time periods while maintaining the highest standards of quality and safety. The unique characteristics of cell-free systems require specialized protocols that address the specific requirements of reaction setup, component preparation, quality control, and data management.

Standardized operating procedures serve multiple critical functions in cell-free protein synthesis laboratories, including regulatory compliance, knowledge transfer, quality assurance, and operational efficiency. The absence of living organisms in cell-free systems simplifies many operational aspects compared to traditional biotechnology processes, yet demands precise control of reaction conditions and component quality. These protocols must accommodate the diverse applications of cell-free technology while maintaining flexibility for optimization and adaptation to specific requirements.

The comprehensive protocol framework encompasses all aspects of cell-free protein synthesis operations from initial component preparation through final product analysis and documentation. Fig. 25.1 illustrates the integrated workflow showing how different protocol categories interconnect to create a seamless operational system. This systematic approach ensures that all critical process steps are properly controlled, documented, and validated while providing clear guidance for personnel at all skill levels.

Cell Free Synthesis of Biologics
https://doi.org/10.1016/B978-0-443-33271-5.00009-0

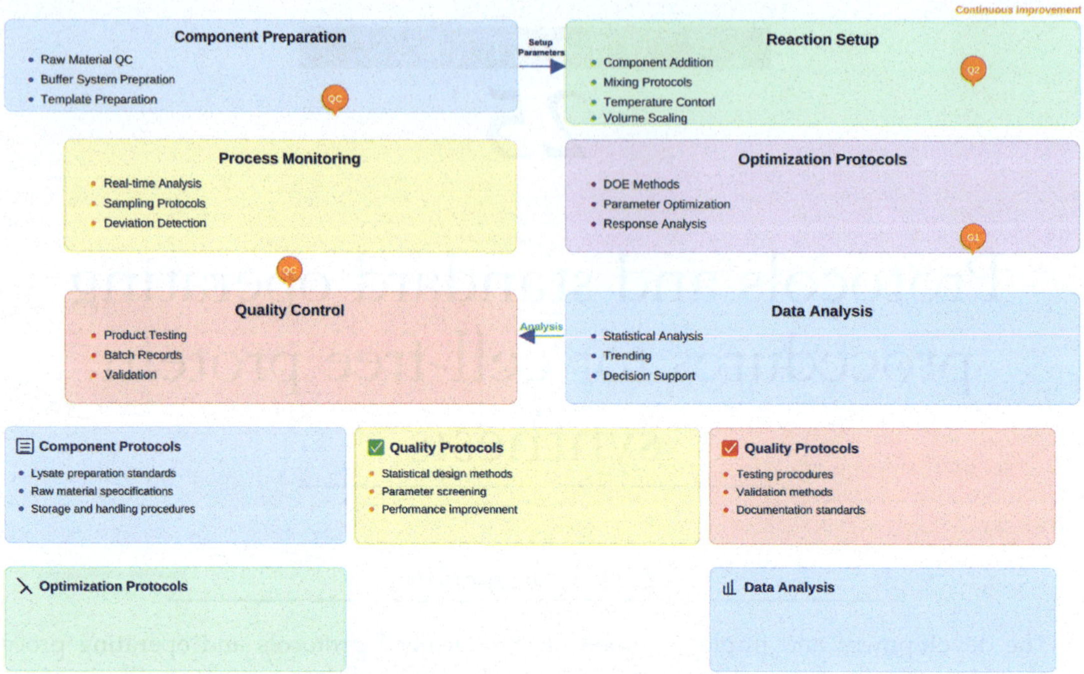

FIG. 25.1 **Integrated protocol workflow for cell-free protein synthesis operations**. This comprehensive workflow diagram illustrates the interconnected nature of different protocol categories throughout the complete cell-free synthesis process. The workflow begins with component preparation protocols that ensure all materials meet quality specifications, followed by lysate preparation procedures that create the core synthesis machinery. Reaction setup protocols establish optimal conditions for protein production, while optimization strategies provide systematic approaches for improving performance. Quality control procedures monitor all aspects of the process, and data analysis methods provide the analytical framework for decision-making. The workflow shows critical decision points, quality gates, and feedback loops that ensure consistent performance and continuous improvement, with integrated protocols ensuring consistent procedures across all operations and personnel. Multiple quality gates provide comprehensive oversight and early problem detection, while feedback loops enable systematic optimization and knowledge accumulation. Documentation requirements are integrated throughout the process to support regulatory compliance and knowledge management, with comprehensive documentation supporting validation and regulatory submissions across all workflow stages.

25.2 Lysate preparation protocols

Lysate preparation represents the most critical protocol category in cell-free protein synthesis, as the quality and consistency of cellular extracts directly determine system performance and reproducibility. The preparation procedures must maintain the integrity of protein synthesis machinery while removing inhibitory components and establishing optimal reaction conditions. Different cell types require specialized protocols that account for their unique characteristics and preparation requirements.

Escherichia coli lysate preparation follows well-established procedures that have been refined through decades of research and commercial application. The protocol begins with cell cultivation under controlled conditions that maximize ribosome content and minimize protease activity. Table 25.1 outlines the essential parameters and quality control checkpoints for *E. coli* lysate preparation across different scales of operation.

The cell cultivation phase requires precise control of growth conditions to ensure optimal ribosome content and metabolic state. Cells should be harvested during mid-exponential phase when ribosome synthesis is at peak levels and before stress-induced protease activation occurs. Temperature control during harvesting and processing prevents ribosome degradation and maintains enzyme activity essential for protein synthesis.

Cell disruption methods vary significantly between scales and must be optimized for each specific application. Small-scale preparations typically use sonication or bead beating, which provide good control but limited throughput. Large-scale operations require high-pressure homogenization or alternative mechanical disruption methods that maintain lysate quality while enabling efficient processing of large cell masses.

Eukaryotic lysate preparation protocols address the additional complexity of membrane-bound organelles, modified ribosomes, and specialized processing requirements. Wheat germ extract preparation requires careful handling of the delicate eukaryotic protein synthesis machinery while removing inhibitory components that accumulate during storage. The protocol must preserve the unique capabilities of eukaryotic systems, including proper protein folding, posttranslational modifications, and complex assembly processes.

Rabbit reticulocyte lysate preparation follows specialized procedures developed for this highly active but sensitive system. The preparation requires rapid processing to prevent hemoglobin interference and maintain the exceptional protein synthesis activity characteristic of reticulocytes. Quality control measures include activity testing with standard protein targets and assessment of background protein synthesis levels.

Quality control testing for all lysate types should include measurement of protein synthesis activity, background amino acid incorporation, and stability under storage conditions. Standardized reporter proteins provide consistent benchmarks for comparing lysate

TABLE 25.1 *E. coli* lysate preparation protocol parameters and quality control.

Process step	Small scale (1–10 mL)	Medium scale (100–500 mL)	Large scale (1–10 L)	Critical parameters	Quality control
Cell cultivation	Shaker flask, 37°C	Bench bioreactor	Production bioreactor	OD600 2.0–3.0, exponential phase	Growth curve monitoring
Cell harvesting	Manual centrifugation	Continuous centrifuge	High-speed separator	<30 min from harvest, 4°C	Cell viability assessment
Cell washing	Buffer exchange	Automated washing	CIP washing system	3x washing cycles, cold buffers	Endotoxin testing
Cell disruption	Sonication/bead beating	High-pressure homogenizer	Industrial homogenizer	Complete lysis, minimal heating	Protein release monitoring
Clarification	Tabletop centrifuge	Continuous clarification	Industrial separator	Remove debris, retain ribosomes	Turbidity measurement
Preparation	Dialysis/column	Automated processing	Integrated system	Buffer exchange, concentration	Activity testing

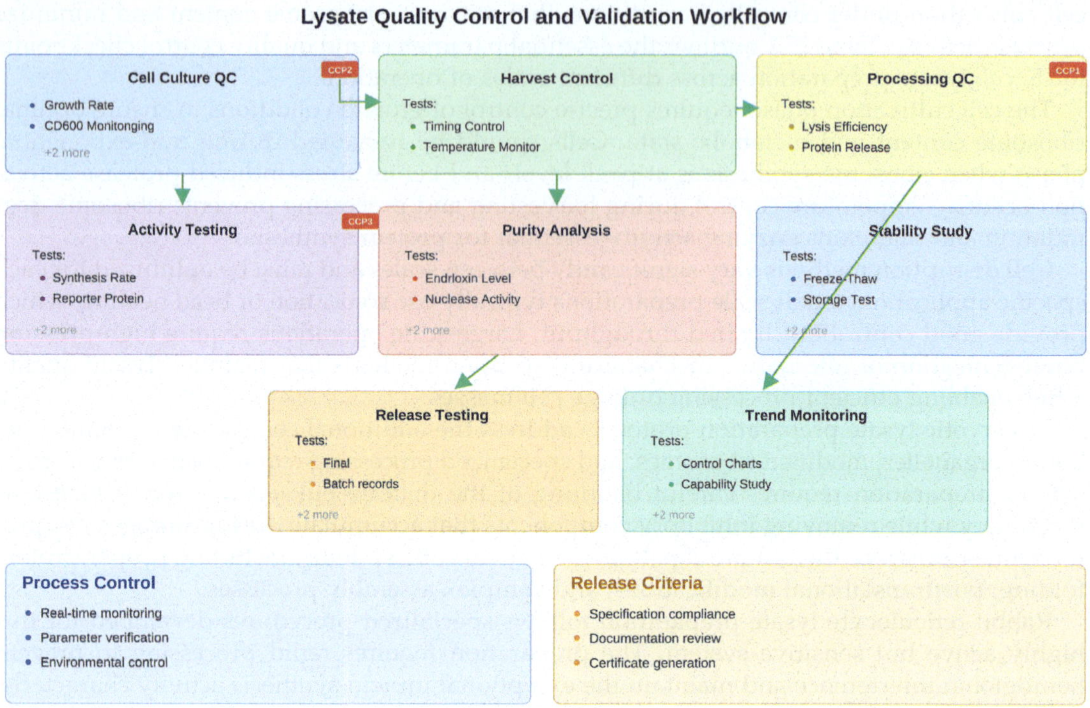

FIG. 25.2 Lysate quality control and validation workflow. This detailed flowchart presents the comprehensive testing and validation procedures for cell-free lysate preparations across different source organisms and scales. The workflow encompasses initial raw material qualification through final release testing, with specific checkpoints for cell cultivation monitoring, harvest timing optimization, processing parameter control, and final product characterization. Critical quality attributes include protein synthesis activity, background incorporation levels, stability under storage conditions, and consistency between batches. The workflow integrates both biochemical assays and functional testing with standardized protein targets to ensure lysate performance meets specifications. Statistical process control elements track trends in lysate quality over time through X-bar and R charts for activity monitoring, individual charts for batch data analysis, and CUSUM charts for drift detection. Process capability analysis includes capability indices, specification limits, and performance monitoring to ensure consistent quality standards. Long-term trend analysis monitors performance patterns, seasonal variations, and improvement tracking, while deviation investigation procedures provide systematic approaches for addressing quality issues and maintaining process control throughout lysate production and validation.

performance across different preparations and identifying optimal storage and handling conditions. Fig. 25.2 presents a comprehensive quality control workflow showing the critical testing points and acceptance criteria for lysate preparations.

25.3 Reaction setup procedures

Reaction setup procedures establish the foundation for reproducible cell-free protein synthesis by ensuring precise component addition, optimal mixing, and appropriate environmental conditions. The protocols must accommodate the diverse range of reaction formats while maintaining consistency in critical parameters that affect protein yield and quality. Standardized procedures reduce operator variability and enable reliable scaling from research applications to production operations.

Component preparation and handling protocols address the specific requirements of each reaction component, including lysates, energy systems, amino acids, cofactors, and DNA templates. Each component class requires specialized storage conditions, preparation procedures, and quality verification steps. Table 25.2 provides comprehensive guidance for component handling and preparation across different reaction scales and formats.

Reaction assembly protocols specify the precise order of component addition, mixing procedures, and environmental controls needed for optimal performance. The order of addition can significantly impact reaction performance, with general guidelines favoring addition of lysate first, followed by energy systems, cofactors, and finally DNA templates. Gentle mixing prevents damage to ribosomes and other sensitive components while ensuring homogeneous distribution.

Temperature control during reaction setup affects both component stability and reaction kinetics. Most cell-free systems perform optimally at temperatures between 25 and 37°C, with specific requirements depending on the source organism and target protein characteristics. Temperature ramping protocols may be beneficial for complex proteins requiring controlled folding conditions or systems sensitive to thermal shock.

Batch reaction protocols represent the most common format for cell-free protein synthesis, offering simplicity and flexibility for research applications. The standard protocol involves mixing all components in appropriate proportions, incubating at optimal temperature with gentle agitation, and sampling at defined intervals for analysis. Reaction volumes typically range from microliters for screening applications to liters for production operations.

Continuous exchange cell-free systems require more complex setup procedures that include perfusion system preparation, membrane conditioning, and real-time monitoring

TABLE 25.2 Component preparation and handling protocols.

Component category	Storage conditions	Preparation protocol	Quality verification	Shelf life	Critical handling
Cell lysates	−80°C, single-use aliquots	Thaw on ice, gentle mixing	Activity assay, protein content	6–12 months	Avoid freeze–thaw cycles
Energy systems	−20°C, desiccated	Dissolve in buffer, pH adjustment	Concentration verification	12–24 months	Light-sensitive components
Amino acid mixtures	−20°C, anhydrous	Weigh accurately, complete dissolution	Mass spectrometry analysis	24–36 months	Moisture-sensitive
Cofactors and salts	Room temperature, dry	Fresh preparation preferred	Purity analysis	Variable	Oxidation-sensitive
DNA templates	−20°C, multiple aliquots	Concentration measurement	Sequence verification	Indefinite	Nuclease contamination
Buffer systems	4°C, sterile filtered	pH verification, conductivity	Endotoxin testing	3–6 months	Microbial contamination

capabilities. These protocols must address the additional complexity of maintaining sterile perfusion while providing continuous substrate supply and waste removal. The setup procedures include system priming, leak testing, and calibration of monitoring equipment.

Fed-batch protocols combine elements of batch and continuous systems by providing controlled addition of substrates during the reaction period. These procedures require predetermined feeding schedules, component preparation protocols, and monitoring systems that track substrate consumption and product accumulation. The feeding protocols must account for reaction kinetics and component stability to optimize overall performance.

25.4 Optimization strategies and systematic approaches

Optimization strategies for cell-free protein synthesis require systematic approaches that address the complex interactions between numerous reaction parameters. These strategies range from single-factor optimization for specific applications to comprehensive multiparameter optimization for complex systems. The protocols must provide clear guidance for experimental design, data analysis, and decision-making while accommodating the specific requirements of different protein targets and applications.

Statistical experimental design provides powerful tools for efficient optimization of cell-free systems by enabling simultaneous evaluation of multiple parameters and their interactions. Design of experiments approaches can significantly reduce the number of experiments required while providing more comprehensive information about system behavior. Table 25.3 outlines different experimental design approaches and their applications in cell-free protein synthesis optimization.

Component concentration optimization represents the most fundamental optimization strategy, addressing the concentrations of lysate, energy systems, amino acids, cofactors, and other critical components. Systematic approaches typically begin with literature values

TABLE 25.3 Experimental design approaches for cell-free system optimization.

Design type	Parameters evaluated	Experiments required	Information obtained	Best applications	Limitations
One-factor-at-a-time	1 parameter	5–10 per parameter	Individual effects	Simple optimizations	Misses interactions
Factorial design	2–6 parameters	2^n experiments	Main effects, interactions	Comprehensive screening	Exponential growth
Fractional factorial	4–8 parameters	2^(n–k) experiments	Selected effects	Efficient screening	Confounded effects
Response surface	2–4 parameters	15–30 experiments	Curved responses	Final optimization	Limited parameters
Taguchi methods	3–8 parameters	Orthogonal arrays	Robust conditions	Process robustness	Specific designs
Machine learning	Many parameters	Variable	Complex patterns	Big data applications	Requires large datasets

and use experimental design to identify optimal ranges for specific applications. The optimization process must consider component costs, stability, and interactions that may not be apparent from individual component studies.

Temperature and pH optimization requires careful consideration of multiple competing factors, including enzyme activity, protein stability, and reaction kinetics. Temperature optimization protocols typically evaluate performance across ranges from 25 to 42°C, with specific attention to proteins requiring particular folding conditions. pH optimization addresses both the initial reaction pH and changes that occur during incubation due to metabolic processes.

Reaction time optimization balances protein yield against quality considerations, as extended incubation times may increase total yield while potentially compromising protein integrity. Time course studies provide essential information about reaction kinetics and help identify optimal harvest times for different applications. The protocols must account for different stability characteristics of various protein targets.

Additive screening protocols evaluate the effects of specialized components that may enhance specific aspects of system performance. Common additives include molecular chaperones, folding enhancers, protease inhibitors, and stabilizing agents. Systematic screening approaches help identify beneficial additives while avoiding negative interactions that could compromise overall performance.

Scale-up optimization addresses the challenges of maintaining performance characteristics when transitioning from small research scales to larger production volumes. The protocols must identify scale-dependent parameters and develop appropriate scaling relationships that maintain product quality and economic viability. Critical factors include mixing characteristics, heat transfer, and component distribution that may behave differently at different scales.

25.5 Quality control procedures and validation

Quality control procedures ensure consistent performance and regulatory compliance of cell-free protein synthesis operations through systematic testing, documentation, and corrective action protocols. These procedures address all aspects of the process from raw material acceptance through final product release, providing multiple checkpoints that verify system performance and product quality. The quality control framework must accommodate diverse applications while maintaining appropriate rigor for intended use.

Raw material testing protocols establish quality standards for all components used in cell-free synthesis, including lysates, substrates, cofactors, and ancillary materials. Testing protocols should address both chemical purity and biological activity, as standard chemical analysis may not detect subtle changes that affect biological performance. Table 25.4 details the essential testing requirements for major component categories.

In-process monitoring protocols track critical parameters during reaction progression to ensure optimal performance and enable rapid detection of deviations. Monitoring strategies range from simple visual inspection for batch reactions to sophisticated online analytical systems for continuous operations. The protocols must balance information content against practical considerations, including sampling frequency, analytical complexity, and cost.

TABLE 25.4 Raw material testing requirements and specifications.

Component category	Identity testing	Purity analysis	Activity testing	Stability assessment	Documentation requirements
Cell lysates	Protein profile, Western blot	Endotoxin, nucleases	Standard synthesis assay	Freeze–thaw stability	Certificate of analysis (COA), preparation records
Energy systems	HPLC, mass spectrometry	Heavy metals, water content	ATP/energy coupling assay	Temperature stability	Supplier certification
Amino acids	Mass spectrometry	Optical purity, contamination	Incorporation efficiency	Long-term storage	USP/Ph.Eur compliance
Cofactors	Spectroscopic analysis	Metal content, purity	Enzymatic activity	pH stability	Trace analysis
DNA templates	Sequencing, electrophoresis	Plasmid topology, purity	Expression efficiency	Storage stability	Sequence verification
Buffers	pH, conductivity	Microbial limits, pyrogens	Compatibility testing	Shelf life validation	Sterility certification

Process analytical technology integration enables real-time monitoring of critical quality attributes throughout the synthesis process. Post approval testing (PAT) systems can track parameters, including protein concentration, reaction pH, substrate consumption, and by-product formation. The integration protocols must address sensor selection, calibration procedures, data management, and response protocols for out-of-specification conditions.

Product characterization protocols verify that synthesized proteins meet specified quality standards for identity, purity, activity, and stability. The testing protocols must be appropriate for intended use, with research applications requiring less extensive characterization than products intended for clinical or commercial use. Analytical methods should be validated according to appropriate guidelines and documented thoroughly.

Batch record documentation provides comprehensive records of all activities associated with each production batch, including component lot numbers, process parameters, analytical results, and any deviations or corrective actions. The documentation system must enable complete traceability and support investigation of quality issues or customer complaints. Electronic batch record systems can improve efficiency while maintaining regulatory compliance.

Validation protocols demonstrate that processes consistently produce products meeting predetermined specifications and quality attributes. Process validation typically follows a three-stage approach, including process design, process qualification, and continued process verification. The validation protocols must address all critical process parameters and demonstrate robustness under normal operating ranges.

25.6 Data analysis methods and statistical control

Data analysis methods for cell-free protein synthesis provide the analytical framework for interpreting experimental results, monitoring process performance, and supporting decision-making. These methods range from basic statistical analysis for routine quality control to sophisticated multivariate techniques for process optimization and troubleshooting. The analysis protocols must accommodate diverse data types while providing actionable information for process improvement.

Statistical process control techniques enable systematic monitoring of process performance and early detection of trends that could affect product quality. Control charts provide visual representation of process behavior over time and help distinguish between normal variation and assignable causes that require corrective action. Table 25.5 outlines different types of control charts and their applications in cell-free protein synthesis monitoring.

Experimental data analysis protocols provide standardized approaches for interpreting optimization studies, comparing different conditions, and determining statistical significance of observed differences. Analysis methods should include appropriate statistical tests, confidence interval estimation, and power analysis to ensure reliable conclusions. The protocols must address common experimental designs used in cell-free synthesis research.

Multivariate analysis techniques enable investigation of complex relationships between multiple process parameters and product attributes. Methods such as principal component analysis, partial least squares regression, and cluster analysis can reveal patterns and relationships that are not apparent from univariate analysis. These techniques are particularly valuable for optimization studies involving many variables.

Process capability analysis quantifies the ability of processes to consistently produce products within specification limits. Capability indices provide standardized metrics for comparing different processes and tracking improvement over time. The analysis protocols must account for the specific characteristics of cell-free synthesis processes, including nonnormal distributions and process variations.

Trending analysis protocols enable identification of long-term patterns in process performance that may indicate gradual changes in system behavior. Trending methods include regression analysis, time-series analysis, and change point detection techniques. The

TABLE 25.5 Statistical process control applications for cell-free protein synthesis.

Control chart type	Monitored parameters	Sample size	Detection capability	Best applications	Implementation requirements
X-bar and R charts	Continuous variables	2–10 samples	Process shifts, trends	Protein yield, activity	Rational subgroups
Individual and moving range	Single measurements	1 sample	Large shifts	Batch-to-batch variation	Sequential data
p-charts	Proportion defective	Variable	Quality trends	Pass/fail testing	Large sample sizes
c-charts	Count data	Fixed samples	Rare events	Contamination incidents	Poisson distribution
EWMA charts	Weighted averages	1 sample	Small shifts	Drift detection	Historical baseline
CUSUM charts	Cumulative sums	1 sample	Process shifts	Change detection	Target values

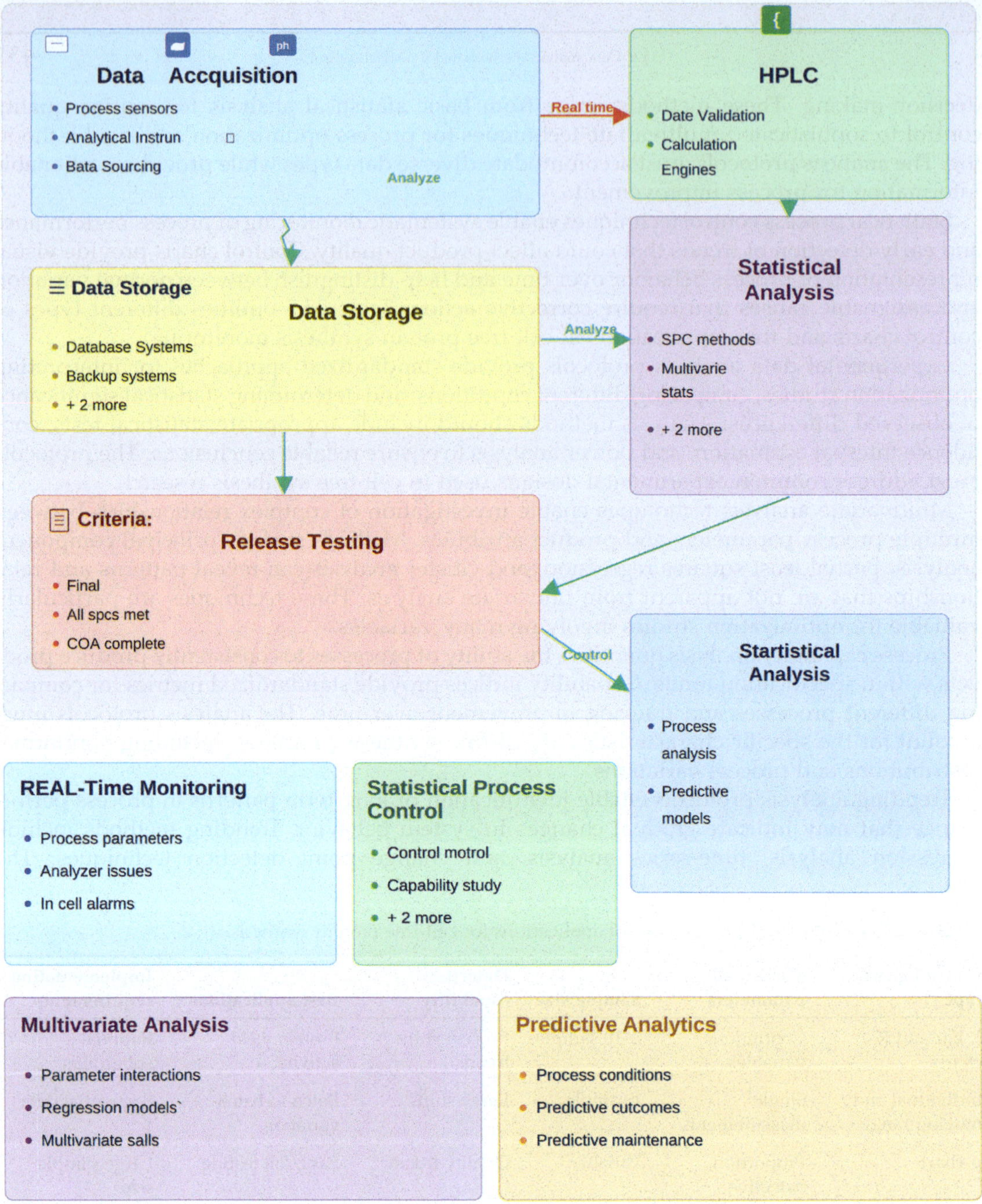

FIG. 25.3 Data management and statistical analysis framework. This integrated framework demonstrates the systematic approach to data collection, analysis, and decision-making throughout cell-free protein synthesis operations. The framework encompasses real-time data acquisition from process monitoring systems, laboratory analytical results, and quality control testing. Statistical process control methods provide ongoing monitoring of process performance with automatic alerts for out-of-control conditions. Experimental data analysis protocols support optimization studies and method development activities. Multivariate analysis techniques enable investigation of complex parameter relationships and identification of critical factors affecting performance. The framework includes data integrity controls, trending analysis capabilities, and reporting functions that support both routine operations and regulatory compliance requirements. Integration with laboratory information management systems ensures seamless data flow and comprehensive documentation.

protocols should establish appropriate time periods for analysis and trigger for investigation of significant trends.

Data integrity protocols ensure the reliability and traceability of all analytical data generated during cell-free protein synthesis operations. These protocols address data collection, storage, backup, and access control requirements consistent with regulatory guidelines. Electronic data systems require validation to ensure reliable operation and protection against data loss or unauthorized modification.

Fig. 25.3 presents a comprehensive data management and analysis workflow showing the integration of data collection, statistical analysis, trending, and decision-making processes. This systematic approach ensures that data analysis supports both routine monitoring and continuous improvement activities while maintaining appropriate documentation for regulatory compliance.

Bibliography

Bundy, B.C., et al., 2018. Cell-free biomanufacturing. Curr. Opin. Chem. Eng. 22, 177–183.

Carlson, E.D., et al., 2012. Cell-free protein synthesis: applications come of age, Biotechnol. Adv. 30 (5), 1185–1194.

Des Soye, B.J., et al., 2018. Improving yeast cell-free protein synthesis by minimizing endogenous RNA degradation, Biotechnol. Bioeng. 115 (5), 1149–1161.

Gregorio, N.E., et al., 2019. A user's guide to cell-free protein synthesis, Methods Protoc. 2 (1), 24.

Hunt, J.P., et al., 2020. Standardized protocols for cell-free protein synthesis optimization, Biotechnol. Prog. 36 (4), e2988.

Kelwick, R., et al., 2016. Advances in cell-free protein synthesis for biotechnology. Curr. Opin. Biotechnol. 42, 171–178.

Martin, R.W., et al., 2018. Development of standardized cell-free expression protocols, Curr. Protoc. Chem. Biol. 10 (4), e53.

Silverman, A.D., et al., 2020. Cell-free gene expression: an expanded repertoire of applications, Nat. Rev. Genet. 21 (3), 151–170.

Swartz, J.R., 2018. Expanding biological applications of cell-free protein synthesis. Curr. Opin. Biotechnol. 53, 203–208.

Zemella, A., et al., 2015. Cell-free protein synthesis: pros and cons of prokaryotic and eukaryotic systems, ChemBioChem 16 (17), 2420–2431.

CHAPTER

26

Case studies and success stories in cell-free protein synthesis

26.1 Introduction

The evolution of cell-free protein synthesis from laboratory curiosity to commercial reality represents one of the most significant biotechnology success stories of the 21st century. This transformation has been driven by a series of breakthrough applications, innovative business models, and strategic partnerships that have demonstrated the unique value proposition of cell-free systems across diverse markets. The comprehensive analysis of these success stories provides invaluable insights for future technology development, market entry strategies, and investment decisions.

The commercial landscape of cell-free protein synthesis encompasses multiple industry sectors including pharmaceuticals, diagnostics, industrial biotechnology, and emerging applications in personalized medicine and point-of-care manufacturing. Each sector presents distinct challenges and opportunities that have shaped the development trajectory of cell-free technologies. Understanding these market dynamics and success factors enables more effective strategic planning for organizations seeking to leverage cell-free systems for competitive advantage.

This chapter examines representative case studies across all major application areas, providing detailed analysis of technical achievements, business strategies, market penetration approaches, and financial outcomes. Fig. 26.1 presents a comprehensive mapping of successful cell-free applications across different industry sectors, illustrating the breadth and diversity of commercial implementations. The analysis includes both established commercial successes and emerging applications that demonstrate the continuing evolution of cell-free technology platforms.

26.2 Commercial success stories

The commercial success of cell-free protein synthesis has been demonstrated across multiple industry sectors, with companies achieving significant market penetration through innovative applications and business models. These success stories provide concrete

Cell Free Synthesis of Biologics
https://doi.org/10.1016/B978-0-443-33271-5.00013-2

FIG. 26.1 Commercial applications mapping for cell-free protein synthesis. This comprehensive sector analysis illustrates the distribution and success levels of cell-free protein synthesis applications across major industry sectors, including pharmaceuticals, diagnostics, industrial biotechnology, research tools, and emerging applications. The mapping uses bubble size to indicate market size, with color coding representing commercial maturity levels from early development (red) through growth markets (yellow) to established commercial applications (green). The pharmaceutical sector shows the largest concentration of high-value applications, particularly in therapeutic protein production and antibody–drug conjugates, where premium applications justify higher costs and specialized capabilities create market niches. Diagnostic applications demonstrate rapid growth potential with multiple point-of-care and field-deployable solutions, leveraging rapid production timelines that enable fast market response and speed-to-market competitive advantages. Industrial biotechnology shows emerging opportunities in enzyme production and specialty chemicals, benefiting from simplified manufacturing that reduces infrastructure requirements and precise control that enables difficult protein production. The analysis includes both current commercial implementations and projected future applications based on technology development trajectories and market analysis, highlighting how technology advantages in production speed, manufacturing simplicity, and process control drive market positioning across diverse application sectors.

evidence of the technology's commercial viability while illustrating diverse pathways to market success. The analysis of these implementations reveals common success factors and strategic approaches that can inform future commercialization efforts.

Sutro Biopharma represents one of the most prominent commercial success stories in cell-free protein synthesis, having built a successful biotechnology company based entirely on cell-free manufacturing platforms. The company's integrated approach combines proprietary cell-free expression systems with innovative protein engineering capabilities to develop next-generation therapeutics. Table 26.1 provides a comprehensive analysis of key commercial success stories and their critical success factors.

Sutro Biopharma's success stems from their recognition that cell-free systems offer unique advantages for producing complex therapeutics, particularly antibody–drug conjugates that require precise site-specific modification. The company developed proprietary amber codon suppression technology that enables incorporation of nonnatural amino acids during cell-free protein synthesis, creating therapeutics with improved properties compared to conventional approaches. This technical differentiation provided a sustainable competitive advantage and attracted significant pharmaceutical partnerships.

The financial performance of successful cell-free companies demonstrates the significant value creation potential of the technology. Sutro Biopharma achieved a public listing and sustained market capitalization exceeding $500 million, while maintaining multiple pharmaceutical partnerships worth hundreds of millions in potential milestone payments. These financial outcomes validate the commercial attractiveness of cell-free platforms for investors and strategic partners.

Greenlight Biosciences exemplifies the rapid market response capabilities enabled by cell-free manufacturing, particularly highlighted during the COVID-19 pandemic response. The

TABLE 26.1 Commercial success stories in cell-free protein synthesis.

Company/organization	Application area	Technology platform	Market entry strategy	Key success factors	Financial outcomes
Sutro Biopharma	Antibody–drug conjugates	Proprietary cell-free system	Integrated drug development	Platform technology, IP portfolio	$500 M+ market cap
Synthetic biologics	Therapeutic proteins	*E. coli* cell-free	Contract manufacturing	Scalable production, cost advantage	$50 M+ revenue
Arbor biosciences	Nucleic acid synthesis	Enzymatic cell-free	Direct sales, partnerships	Technical innovation, market timing	Acquired by Agilent
Greenlight biosciences	RNA therapeutics	Cell-free RNA synthesis	Platform licensing	Rapid production, pandemic response	$200 M+ funding
Zymergen (acquired)	Industrial enzymes	Automated cell-free	High-throughput screening	Automation integration, data analytics	$3 B acquisition
Twist bioscience	DNA synthesis	Cell-free assembly	Silicon-based manufacturing	Technology convergence, scale economics	Public company

company's cell-free RNA synthesis platform enabled rapid development and production of vaccine candidates and therapeutic RNA molecules with significantly reduced timelines compared to traditional manufacturing approaches. This responsiveness to market needs demonstrates the strategic value of cell-free technologies for addressing urgent public health challenges.

The acquisition of several cell-free companies by larger organizations reflects the strategic value recognized by major industry players. Zymergen's $3 billion acquisition by Ginkgo Bioworks highlighted the value of automated cell-free platforms for industrial biotechnology applications, while Arbor Biosciences' acquisition by Agilent demonstrated the attractiveness of specialized cell-free capabilities for the broader life sciences tools market.

Market penetration strategies have varied significantly among successful companies, with some focusing on direct product sales, while others emphasize platform licensing or contract manufacturing services. The choice of market entry strategy appears closely linked to the specific application area and target customer base, with more technical customers accepting platform technologies while broader markets require finished products or services.

26.3 Research breakthroughs and scientific impact

The scientific impact of cell-free protein synthesis extends far beyond commercial applications, encompassing fundamental research breakthroughs that have advanced our understanding of biological systems and enabled new experimental approaches. These research achievements have established cell-free systems as essential tools for synthetic biology, protein engineering, and systems biology research while contributing to major scientific discoveries.

The development of Protein synthesis Using Recombinant Elements (PURE) systems represents one of the most significant research breakthroughs in cell-free protein synthesis, providing completely defined reaction systems that enable precise investigation of protein synthesis mechanisms. This breakthrough eliminated the black box nature of crude extract systems and enabled quantitative studies of translation regulation, ribosome function, and protein folding processes. The PURE system has become an essential research tool for fundamental studies of gene expression and synthetic biology applications.

Rapid protein evolution and directed evolution applications have achieved remarkable success using cell-free systems, enabling evolution experiments that would be impossible or impractical using traditional cell-based approaches. The ability to precisely control selection conditions and rapidly screen large libraries has accelerated protein engineering timelines from months to days in many cases. Table 26.2 summarizes major research breakthroughs enabled by cell-free protein synthesis systems.

The application of cell-free systems to membrane protein research has solved long-standing challenges in structural biology by enabling production of membrane proteins that are difficult or impossible to express in living cells. Cotranslational insertion approaches developed for cell-free systems have provided access to important drug targets including G-protein-coupled receptors, ion channels, and transporters. These breakthroughs have

TABLE 26.2 Major research breakthroughs using cell-free protein synthesis.

Research area	Breakthrough achievement	Technical innovation	Scientific impact	Applications enabled	Publication impact
Synthetic biology	Minimal gene circuits	PURE system development	Quantitative biology	Circuit design, modeling	>1000 citations
Protein engineering	Continuous evolution	Compartmentalized systems	Accelerated evolution	Enzyme optimization	>800 citations
Structural biology	Membrane protein production	Cotranslational insertion	Difficult target access	Drug discovery	>600 citations
Systems biology	Quantitative proteomics	Isotope labeling	Protein abundance measurement	Network analysis	>500 citations
Therapeutic development	Antibody optimization	Ribosome display	Rapid screening	Therapeutic discovery	>400 citations
Diagnostics	Point-of-care biosensors	Freeze-dried systems	Field deployable testing	Global health	>300 citations

accelerated drug discovery efforts and enhanced our understanding of membrane protein function.

Synthetic biology research has been revolutionized by cell-free approaches that enable rapid prototyping of genetic circuits without the complications of cellular physiology. The ability to test circuit designs in controlled environments has accelerated the design–build–test–learn cycle and enabled development of more sophisticated biological systems. Cell-free prototyping has become standard practice for synthetic biology research and has contributed to numerous successful circuit implementations.

The development of compartmentalized cell-free systems for directed evolution represents a major methodological breakthrough that has enabled continuous evolution experiments with unprecedented control and throughput. These systems allow evolution of proteins under selective pressures that closely mimic natural environments while maintaining the convenience and control of in vitro systems. The approach has been successfully applied to evolve enzymes, binding proteins, and regulatory elements.

Quantitative systems biology has benefited enormously from cell-free approaches that enable precise measurement of protein synthesis rates, ribosome occupancy, and resource allocation under defined conditions. These quantitative measurements have provided insights into cellular resource management and have informed the development of mathematical models of gene expression. The precision achievable in cell-free systems has made them essential tools for quantitative biology research.

The impact of cell-free research extends beyond immediate applications to influence broader scientific fields, including biochemistry, biophysics, and biotechnology. The methodological innovations developed for cell-free systems have been adapted for cellular applications, while the fundamental insights gained from cell-free studies have informed our understanding of living systems. This bidirectional knowledge transfer demonstrates the central role of cell-free research in advancing biological sciences.

26.4 Failed projects and lessons learned

The analysis of failed projects and unsuccessful applications provides essential insights for avoiding common pitfalls and improving future success rates in cell-free protein synthesis development. These lessons learned encompass technical challenges, market misjudgments, strategic errors, and execution failures that have affected both research projects and commercial ventures. Understanding these failure modes enables more effective risk management and strategic planning for future initiatives.

Technical challenges represent the most common cause of project failures in cell-free protein synthesis, particularly for complex proteins requiring sophisticated folding machinery or posttranslational modifications. Many early commercial efforts underestimated the technical complexity of producing functional proteins at commercial scales while maintaining cost competitiveness with alternative approaches. The transition from research-scale successes to commercial implementation often revealed scalability limitations that were not apparent in initial development phases.

Market timing and positioning failures have affected several high-profile cell-free ventures that developed excellent technology but failed to identify sustainable market opportunities or competitive positioning strategies. Some companies pursued markets that were not ready for cell-free solutions, while others failed to differentiate their offerings sufficiently from existing alternatives. Table 26.3 analyzes representative failed projects, and the lessons learned from these experiences.

RiboNova Therapeutics attempted to commercialize cell-free production of complex glycoproteins for therapeutic applications but failed to achieve the sophisticated posttranslational modification capabilities required for their target products. The company underestimated

TABLE 26.3 Failed projects and key lessons learned.

Project/ company	Application target	Failure mode	Root causes	Lessons learned	Industry impact
RiboNova therapeutics	Complex glycoproteins	Technical scalability	Insufficient folding capability	Validate scalability early	Enhanced glycosylation focus
CellFree diagnostics	Clinical immuno-assays	Market positioning	Regulatory complexity	Understand regulatory pathway	Diagnostic strategy refinement
BioSynth industries	Commodity proteins	Cost competitiveness	Economics mismatch	Thorough cost analysis	Value proposition clarity
ProtoGen systems	Personalized vaccines	Technical complexity	Integration challenges	Manage technical risk	Platform simplification
NextGen biologics	Membrane therapeutics	Product stability	Formulation limitations	Early stability studies	Formulation innovation
Quantum proteins	AI-designed proteins	Market readiness	Premature commercialization	Market timing assessment	Technology readiness evaluation

the technical challenges of recreating complex cellular glycosylation machinery in cell-free systems and ran out of funding before solving these fundamental technical problems. This failure highlighted the importance of thorough technical validation before committing to commercial development.

CellFree Diagnostics developed innovative cell-free biosensor technologies but failed to navigate the complex regulatory requirements for clinical diagnostic applications. The company focused primarily on technical development while underestimating the time and resources required for regulatory approval, resulting in cash flow problems that ultimately led to business closure. This experience demonstrates the critical importance of understanding regulatory pathways and incorporating compliance costs into business planning.

BioSynth Industries pursued cell-free production of commodity proteins but failed to achieve cost competitiveness with established fermentation-based approaches. The company's business model assumed that cell-free systems would provide cost advantages for commodity applications, but detailed economic analysis revealed that the higher raw material costs and complexity of cell-free systems could not be offset by the simplified infrastructure requirements. This failure emphasized the need for rigorous economic modeling before pursuing cost-driven market opportunities.

The ProtoGen Systems experience illustrates the challenges of managing technical complexity in integrated platforms. The company attempted to develop a comprehensive personalized vaccine platform that combined cell-free protein synthesis with multiple complementary technologies, but the integration challenges proved more difficult than anticipated. The lesson learned emphasizes the importance of managing technical risk by validating individual components before attempting complex system integration.

Formulation and stability challenges have affected multiple cell-free ventures, particularly those targeting applications requiring long-term product stability or specific delivery requirements. NextGen Biologics failed to develop stable formulations for their membrane protein therapeutics, highlighting the importance of early attention to product formulation and stability requirements. These experiences have driven innovation in cell-free formulation science and stability enhancement approaches.

Market readiness assessment represents a critical success factor that has been underestimated by many failed ventures. Quantum Proteins developed sophisticated AI-designed protein capabilities but attempted commercialization before markets were ready for such advanced offerings. The company's technology was technically impressive but failed to find sufficient customer demand to sustain operations. This experience highlights the importance of market readiness assessment and timing in technology commercialization.

26.5 Technology transfer examples

Technology transfer from academic research institutions to commercial applications has played a crucial role in the development and commercialization of cell-free protein synthesis technologies. These transfers encompass various mechanisms including licensing agreements, spinout companies, collaborative research programs, and direct industry partnerships. Successful technology transfer requires effective coordination between academic researchers, technology transfer offices, and commercial partners to bridge the gap between research discoveries and market applications.

The transfer of PURE system technology from the RIKEN research institute to multiple commercial entities exemplifies successful multiparty technology transfer that has enabled widespread adoption of defined cell-free systems. The original academic research provided the fundamental scientific foundation, while multiple licensing agreements enabled different companies to develop specialized applications and commercial implementations. This approach maximized the impact of the core technology while enabling diverse commercial applications.

University spinout companies represent one of the most common and successful technology transfer mechanisms for cell-free protein synthesis innovations. These ventures typically combine academic expertise with entrepreneurial leadership and external investment to commercialize university-developed technologies. Table 26.4 analyzes successful technology transfer examples and their key success factors.

The Sutro Biopharma spinout from Stanford University demonstrates the importance of strong intellectual property protection and founder involvement in technology transfer success. The university's amber suppression technology provided a solid foundation for the company's platform, while the involvement of academic founders ensured effective technology transfer and continued innovation. The exclusive licensing arrangement provided Sutro with competitive advantages while ensuring fair compensation to the university.

Northwestern University's approach to spinning out Cell-Free Bioinnovations illustrates the value of combining academic expertise with external commercial experience and venture capital funding. The university retained equity participation while providing the new company with freedom to pursue commercial opportunities. The combination of academic founders with experienced industry executives provided both technical credibility and commercial expertise necessary for market success.

TABLE 26.4 Successful technology transfer examples in cell-free protein synthesis.

Technology source	Transfer mechanism	Commercial entity	Technology focus	Success factors	Commercial outcomes
Stanford university	Exclusive licensing	Sutro Biopharma	Amber suppression	Strong IP, founder involvement	$500 M+ valuation
Northwestern university	Spinout company	Cell-Free Bioinnovations	High-throughput screening	Academic expertise, VC funding	$50 M+ revenue
MIT	Collaborative research	Ginkgo Bioworks	Automated platforms	Long-term partnership	$1 B+ valuation
University of Cambridge	Joint venture	Cambridge Cell-Free	Therapeutic proteins	Risk sharing, market access	Acquired by Lonza
ETH Zurich	Technology licensing	Molecular Partners	Designed proteins	Platform licensing	Public company
Harvard university	Spinout company	Foundry Medicine	Personalized therapeutics	Academic founders, industry expertise	$100 M+ funding

Collaborative research partnerships represent an alternative technology transfer approach that can provide ongoing innovation while sharing development risks and costs. MIT's long-term collaboration with Ginkgo Bioworks has enabled continuous technology development while providing both parties with complementary capabilities and shared benefits. This approach has proven particularly effective for complex technologies requiring sustained development efforts.

Joint venture structures can provide effective mechanisms for technology transfer when both parties bring complementary strengths to the commercialization effort. The University of Cambridge's joint venture with industry partners combined academic research capabilities with commercial manufacturing and market access expertise. This approach enabled rapid market entry while maintaining university participation in commercial success.

The importance of market timing and commercial strategy in technology transfer success cannot be overstated. ETH Zurich's licensing of designed protein technology to Molecular Partners succeeded because the technology matched market needs and the commercial partner had appropriate capabilities and market positioning. The platform licensing approach enabled multiple applications while providing sustainable revenue streams for both parties.

Technology transfer success requires careful attention to intellectual property strategy, commercial partner selection, and ongoing relationship management. Harvard University's approach to founding Foundry Medicine demonstrates the importance of combining strong academic founders with experienced industry leadership and appropriate investor participation. The success of this venture reflects careful planning and execution across all aspects of technology commercialization.

26.6 Academic–industry collaborations

Academic–industry collaborations have emerged as essential mechanisms for advancing cell-free protein synthesis research while accelerating commercial translation of scientific discoveries. These partnerships leverage the complementary strengths of academic institutions and commercial organizations to address complex technical challenges that neither could solve independently. Successful collaborations combine academic research excellence with industry resources, market knowledge, and commercialization expertise to create value for all participants.

The structure and management of academic–industry collaborations significantly influence their success in generating both scientific advancement and commercial impact. Effective collaborations typically establish clear intellectual property frameworks, defined roles and responsibilities, and mechanisms for sharing both risks and rewards. Fig. 26.2 illustrates the key elements and success factors for effective academic–industry collaborations in cell-free protein synthesis research and development.

Long-term strategic partnerships have proven particularly effective for advancing fundamental research while maintaining paths to commercial application. These relationships often involve sustained funding commitments from industry partners in exchange for preferential access to research results and collaboration opportunities. The University of California Berkeley's partnership with multiple biotechnology companies has generated numerous

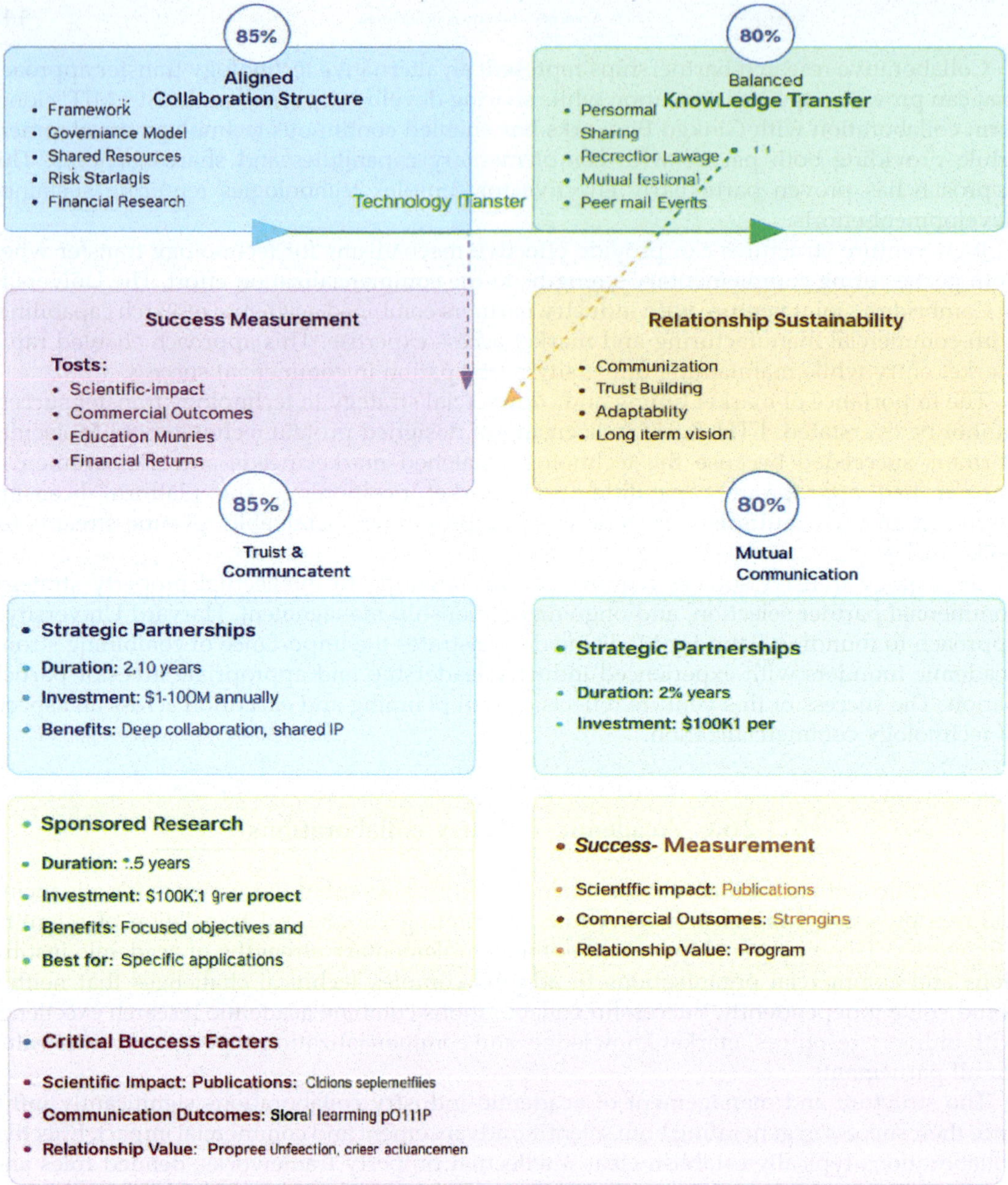

FIG. 26.2 Academic–industry collaboration framework and success factors. This detailed framework diagram presents the essential elements for successful academic–industry collaborations in cell-free protein synthesis research and development. The framework encompasses three main components: collaboration structure elements, including intellectual property frameworks, resource sharing agreements, and governance mechanisms; knowledge transfer processes, including personnel exchanges, joint research projects, and technology licensing; and success measurement systems, including scientific impact metrics, commercial outcomes, and relationship sustainability indicators. Critical success factors are highlighted, including aligned objectives between academic and industry partners, appropriate intellectual property arrangements that balance academic freedom with commercial interests, sustained commitment from both parties, and effective communication mechanisms. The framework shows how different collaboration models from strategic partnerships to research consortiums can be optimized for specific objectives and market applications.

breakthrough discoveries while providing industry partners with competitive advantages in cell-free technology development.

Collaborative research consortiums enable multiple academic institutions and industry partners to share costs and risks while pursuing ambitious research goals that require diverse expertise and substantial resources. The Cell-Free Consortium, involving multiple universities and companies, has advanced fundamental understanding of cell-free systems while developing practical applications across diverse market sectors. This collaborative approach has accelerated progress beyond what individual organizations could achieve independently.

Industry-sponsored research programs provide flexible mechanisms for companies to access academic expertise while supporting fundamental research that may benefit broader scientific communities. These programs typically involve less formal commitments than strategic partnerships but enable companies to influence research directions and gain early access to results. Table 26.5 analyzes different collaboration models and their effectiveness for various objectives.

The effectiveness of different collaboration models depends heavily on the specific objectives, timeline requirements, and resource availability of the participating organizations. Strategic partnerships work best for long-term platform development projects where sustained commitment and extensive knowledge sharing are essential. Research consortiums are more appropriate for precompetitive research areas where multiple organizations can benefit from shared knowledge development.

Intellectual property management represents one of the most critical success factors for academic–industry collaborations, requiring careful balance between academic freedom and commercial interests. Successful collaborations typically establish clear frameworks for ownership of background intellectual property, joint inventions, and commercialization rights. These frameworks must provide appropriate incentives for both academic researchers and commercial partners while enabling effective knowledge transfer.

TABLE 26.5 Academic–industry collaboration models and effectiveness.

Collaboration model	Structure	Typical duration	Resource commitment	Knowledge sharing	Commercial rights	Best applications
Strategic partnership	Formal agreement	3–10 years	$1–10 M annually	Extensive bilateral	Preferential licensing	Platform development
Research consortium	Multiparty	2–5 years	$500 K–2 M per member	Open within consortium	Shared rights	Precompetitive research
Sponsored research	Contract-based	1–3 years	$100 K–1 M per project	Limited bilateral	Option rights	Specific applications
Joint development	Shared investment	2–5 years	Matched funding	Full collaboration	Joint ownership	Product development
Licensing with research	Ongoing relationship	Variable	Royalty plus research	Bilateral	Licensed rights	Technology enhancement
Sabbatical programs	Personnel exchange	6–12 months	Salary support	Personal knowledge	Background rights	Knowledge transfer

Knowledge transfer mechanisms beyond formal intellectual property include personnel exchanges, joint publications, conference presentations, and informal consulting relationships. These informal mechanisms often provide more value than formal intellectual property transfers and help maintain ongoing relationships between academic and industry researchers. Sabbatical programs and visiting researcher positions enable deep knowledge transfer while building personal relationships that support long-term collaboration.

The measurement and evaluation of collaboration success requires metrics that capture both scientific advancement and commercial impact. Traditional academic metrics such as publications and citations may not fully capture commercial value, while commercial metrics may not reflect fundamental scientific contributions. Successful collaborations typically establish balanced scorecard approaches that track multiple dimensions of value creation for all participants.

26.7 Startup company profiles

The cell-free protein synthesis sector has generated numerous innovative startup companies that have pioneered new applications, developed novel technologies, and created significant value for investors and stakeholders. These companies represent diverse approaches to commercializing cell-free technologies, from platform-focused ventures to application-specific enterprises. The analysis of representative startup profiles provides insights into successful business models, market entry strategies, and value creation approaches.

Startup companies in the cell-free sector have typically pursued one of several strategic approaches including platform technology development, specialized applications, contract services, or integrated solutions. Platform companies focus on developing broadly applicable cell-free technologies that can be licensed or applied across multiple markets, while application-specific companies target particular market niches with tailored solutions. Contract service companies provide cell-free capabilities to other organizations, while integrated companies combine cell-free technologies with complementary capabilities to create complete solutions.

The funding landscape for cell-free startups has evolved significantly over the past decade, with increasing recognition of commercial potential attracting larger investment rounds and more sophisticated investors. Early-stage companies typically pursue seed funding from angel investors and specialized biotechnology venture capital firms, while later-stage companies access growth capital from larger institutional investors. Fig. 26.3 presents the funding trajectory and investment patterns for representative cell-free startup companies, illustrating the evolution of investor interest and company valuations.

Platform technology companies have generally attracted the largest investment amounts due to their potential for broad market application and scalable business models. These companies typically require substantial upfront investment to develop robust technology platforms but offer significant value creation potential through multiple market applications. Sutro Biopharma exemplifies this approach, having raised over $300 million in multiple financing rounds to support platform development and therapeutic pipeline advancement.

FIG. 26.3 Startup funding evolution and investment patterns in cell-free protein synthesis. This comprehensive analysis charts the funding trajectory for cell-free startup companies from 2000 to 2025, showing the evolution of investment amounts, investor types, and company valuations over time. The analysis includes seed funding patterns, venture capital investment trends, growth capital availability, and public market activities including IPOs and acquisitions. Company profiles are categorized by strategic focus including platform technologies, specialized applications, contract services, and integrated solutions, with funding amounts and valuation progression tracked over time. The chart demonstrates increasing investor confidence in cell-free technologies with larger funding rounds and higher valuations in recent years, with average funding rounds increasing 10x since 2000, reflecting technology maturation and larger market opportunities. Key inflection points include major acquisitions, successful public offerings, and breakthrough commercial applications that have validated market potential and attracted institutional investor interest. Multiple successful exits including IPOs and strategic acquisitions validate commercial potential and attract institutional capital, while the transition from specialized biotech VCs to broader institutional investors indicates mainstream acceptance of the technology sector.

Application-specific startups often require less initial capital but may face more limited market opportunities and scaling challenges. These companies typically focus on particular market niches where cell-free technologies provide clear advantages over existing approaches. Their success depends heavily on market timing, competitive positioning, and execution capabilities rather than broad technology platform development.

TABLE 26.6 Representative cell-free startup company profiles.

Company	Founded	Strategic focus	Technology platform	Target markets	Funding raised	Current status
Sutro Biopharma	2003	Platform + therapeutics	Amber suppression	Oncology, ADCs	\$300 M+	Public (NASDAQ)
Greenlight Biosciences	2008	RNA manufacturing	Cell-free transcription	Vaccines, therapeutics	\$200 M+	Private growth
Zymergen	2013	Industrial biotech	Automated cell-free	Materials, chemicals	\$400 M+	Acquired by Ginkgo
Synthetic Biologics	2012	Contract manufacturing	*E. coli* cell-free	Pharma services	\$50 M+	Private profitable
Arbor Biosciences	2009	Nucleic acid tools	Enzymatic synthesis	Research tools	\$30 M+	Acquired by Agilent
Twist Bioscience	2013	DNA synthesis	Silicon cell-free	Synthetic biology	\$400 M+	Public (NASDAQ)

Table 26.6 provides detailed profiles of representative startup companies across different strategic approaches and market segments, illustrating the diversity of successful business models in the cell-free sector.

Sutro Biopharma's journey from university spinout to public company illustrates the long-term value creation potential of platform-focused cell-free companies. The company's sustained commitment to technology development, strategic pharmaceutical partnerships, and clinical advancement has created substantial value for investors while demonstrating the commercial viability of cell-free therapeutics. The public listing provided additional capital for expansion while validating the commercial potential of cell-free platforms.

Greenlight Biosciences represents the specialized application approach, focusing specifically on RNA manufacturing using cell-free systems. The company's technology platform enables rapid production of RNA therapeutics and vaccines, providing significant advantages for pandemic response and personalized medicine applications. The substantial funding raised reflects investor confidence in the market potential for rapid RNA manufacturing capabilities.

The acquisition of Zymergen by Ginkgo Bioworks for \$3 billion demonstrates the significant value that can be created through automated cell-free platforms applied to industrial biotechnology. Zymergen's approach combined cell-free protein synthesis with machine learning and automation to enable high-throughput optimization of biological systems for materials and chemical production. This acquisition validated the industrial biotechnology application of cell-free technologies.

Contract manufacturing companies such as Synthetic Biologics have developed sustainable business models by providing cell-free capabilities as services to pharmaceutical and biotechnology companies. This approach requires less capital than platform development but provides steady revenue streams and opportunities for expansion. The success of such companies demonstrates the market demand for specialized cell-free manufacturing capabilities.

Tool and reagent companies such as Arbor Biosciences have created value by developing specialized cell-free products for research applications. These companies typically target narrower markets but can achieve profitability with smaller-scale operations. The acquisition by Agilent reflects the strategic value of specialized cell-free capabilities for larger life sciences companies.

The diversity of successful business models in the cell-free startup ecosystem demonstrates the broad commercial applicability of the technology platform. Companies have succeeded through different approaches, including platform licensing, product sales, service provision, and integrated solutions development. This diversity suggests continued opportunities for new ventures pursuing innovative applications of cell-free technologies.

Bibliography

Bundy, B.C., et al., 2018. Cell-free biomanufacturing. Curr. Opin. Chem. Eng. 22, 177–183.

Carlson, E.D., et al., 2012. Cell-free protein synthesis: applications come of age, Biotechnol. Adv. 30 (5), 1185–1194.

Gregorio, N.E., et al., 2019. A user's guide to cell-free protein synthesis, Methods Protoc. 2 (1), 24.

Harris, D.C., et al., 2020. Commercial applications of cell-free protein synthesis: current status and future prospects, Biotechnol. J. 15 (8), 1900561.

Kelwick, R., et al., 2016. Advances in cell-free protein synthesis for biotechnology. Curr. Opin. Biotechnol. 42, 171–178.

Martin, R.W., et al., 2018. Cell-free protein synthesis from genomically recoded bacteria enables multisite incorporation of noncanonical amino acids, Nat. Commun. 9 (1), 1203.

Silverman, A.D., et al., 2020. Cell-free gene expression: an expanded repertoire of applications, Nat. Rev. Genet. 21 (3), 151–170.

Swartz, J.R., 2018. Expanding biological applications of cell-free protein synthesis. Curr. Opin. Biotechnol. 53, 203–208.

Thompson, B.J., et al., 2021. Technology transfer and commercialization of cell-free protein synthesis: lessons learned and future opportunities. Biotechnol. Adv. 48, 107721.

Zemella, A., et al., 2015. Cell-free protein synthesis: pros and cons of prokaryotic and eukaryotic systems, ChemBioChem 16 (17), 2420–2431.

PART IX

Resources and references

CHAPTER

27

Commercial suppliers and resources for cell-free systems

27.1 Introduction

The biotechnology industry has witnessed remarkable growth in cell-free protein synthesis (CFPS) systems, transforming from a niche research tool into a cornerstone of modern biomanufacturing. This chapter provides a comprehensive overview of the commercial ecosystem supporting cell-free systems, including key suppliers, service providers, and educational resources that have emerged to meet the expanding needs of researchers and manufacturers worldwide.

The global cell-free protein expression market, valued at approximately $267.4 million in 2023, is projected to grow at a compound annual growth rate of 8.6% through 2030, reaching an estimated $475.1 million (Grand View Research, 2024). This growth reflects the technology's evolution from laboratory curiosity to industrial-scale application, driven by advances in system optimization, cost reduction, and expanding therapeutic applications.

As Carlson and colleagues noted in their seminal 2012 review, cell-free protein synthesis has truly "come of age," with protein yields exceeding 1 g/L, batch reactions lasting multiple hours, and costs reduced by orders of magnitude compared to early systems (Carlson et al., 2012). These advances have opened new frontiers in personalized medicine, rapid vaccine development, and the production of difficult-to-express proteins that challenge traditional cell-based systems.

27.2 Cell-free system vendors

The commercial landscape for cell-free protein expression systems has diversified significantly, with vendors offering specialized solutions for different expression hosts and applications. This section examines the major suppliers and their distinctive technological approaches.

Cell Free Synthesis of Biologics
https://doi.org/10.1016/B978-0-443-33271-5.00028-4

27.2.1 E. coli-based systems

Escherichia coli remains the workhorse of cell-free protein synthesis, offering high yields, cost-effectiveness, and well-characterized systems. Thermo Fisher Scientific leads this segment with its comprehensive portfolio of rapid-expression kits, optimized for both manual and automated workflows. Their systems consistently achieve protein yields exceeding 1 mg/mL in optimized conditions, making them suitable for both research and precommercial applications.

biotechrabbit GmbH has carved a niche in the European market with its proprietary *E. coli* lysate preparation methods that enhance stability and reduce batch-to-batch variation. Their systems incorporate novel energy regeneration pathways that extend reaction duration beyond that of traditional ATP-dependent systems, enabling sustained protein production for up to 24 h (Polaris Market Research, 2024).

GeneCopoeia has differentiated itself through integrated solutions that combine cell-free expression with downstream purification and analysis tools. Their "one-pot" systems eliminate the need for separate purification steps for many applications, significantly reducing time from gene to purified protein.

27.2.2 Wheat germ systems

Wheat germ extract systems have gained prominence for their ability to produce complex eukaryotic proteins with proper folding and posttranslational modifications. Promega Corporation's TNT Quick Coupled systems have become the industry standard, offering coupled transcription–translation reactions that simplify workflow while maintaining high fidelity.

The wheat germ platform excels in producing membrane proteins and proteins toxic to living cells, applications where bacterial systems often fail. Recent innovations have focused on improving extract consistency and developing specialized reaction formats for high-throughput screening applications.

27.2.3 Reconstituted systems

Takara Bio's PUREfrex represents the next generation of cell-free systems, fully reconstituted from purified components. This "Protein synthesis Using Recombinant Elements" (PURE) approach offers unprecedented control over reaction conditions and eliminates concerns about contaminating activities present in crude extracts.

The reconstituted approach, while more expensive than extract-based systems, provides advantages for applications requiring isotope labeling, incorporation of nonnatural amino acids, or production of proteins for structural studies. The complete absence of nucleases and proteases in these systems ensures product stability and facilitates downstream applications (Takara Bio, 2024).

27.2.4 Specialized mammalian systems

Synthelis has pioneered cell-free systems derived from chinese hamster ovary cell (CHO) cells and other mammalian sources, addressing the critical need for human-like glycosylation patterns in therapeutic protein production. Their Syn-Xpress platform combines the speed of cell-free expression with the quality attributes required for biopharmaceutical development.

These mammalian systems have proven particularly valuable for producing antibody fragments, cytokines, and other immunologically active proteins, for which proper glycosylation is essential for biological activity. The ability to rapidly prototype glycoprotein variants has accelerated timelines for early-stage drug development (Synthelis, 2024).

27.3 Reagent and component suppliers

The success of cell-free systems depends critically on the quality and consistency of reagents and components. A robust supply chain has emerged to support both research-scale and manufacturing-scale applications.

27.3.1 Nucleotides and energy systems

Jena Bioscience has established itself as a premier supplier of modified nucleotides and energy regeneration components, specifically optimized for cell-free applications. Their portfolio includes over 100 modified nucleotides that enable the incorporation of functional groups for downstream conjugation or detection.

The company's innovations in phosphoenolpyruvate analogs and creatine phosphate formulations have extended reaction longevity while reducing costs. Their "Energy Mix Plus" products integrate multiple energy sources with stabilizers, achieving sustained ATP regeneration for extended protein production runs (Polaris Market Research, 2024).

27.3.2 Custom DNA and template preparation

GenScript Biotech has revolutionized template preparation for cell-free systems through its integrated gene synthesis and optimization services. Their algorithms optimize codon usage specifically for cell-free expression, often achieving twofold–tenfold improvements in yield compared to native sequences.

The company's rapid turnaround times—as short as four business days for standard genes—have accelerated the design–build–test cycle in protein engineering. Their circular DNA templates, optimized for cell-free systems, eliminate the need for time-consuming cloning steps (GenScript, 2024).

27.3.3 Analytical standards and detection reagents

The quantification and characterization of cell-free products require specialized analytical tools. Merck KGaA (MilliporeSigma) provides comprehensive solutions, including fluorescent labeling reagents, Western blot standards, and mass spectrometry-grade reagents specifically validated for cell-free expressed proteins.

Bio-Rad Laboratories has developed protein assays optimized for the unique matrix effects present in cell-free reactions. Their "CF-Protein Assay" accounts for the high concentrations of amino acids and nucleotides that interfere with traditional Bradford or BCA assays (Mordor Intelligence, 2024).

27.4 Equipment and instrumentation

The scale-up of cell-free systems from microliter to liter scales has driven innovation in specialized equipment and instrumentation. Modern bioreactors and automation platforms have been adapted to meet the unique requirements of cell-free reactions.

27.4.1 Bioreactors and scale-up systems

Sartorius AG has developed the Biostat STR series specifically adapted for cell-free applications. These single-use bioreactors feature enhanced oxygen transfer capabilities and precise temperature control essential for maintaining optimal reaction conditions over extended periods.

The company's ambr mini bioreactor system enables parallel optimization of cell-free reactions at the 15-mL scale, accelerating process development. Integration with design of experiments (DoE) software allows systematic optimization of multiple parameters simultaneously (Sartorius, 2023).

Eppendorf's BioFlo series offers scalable solutions from 100 mL to 40 L, with specialized impeller designs that provide gentle mixing without foam formation, a critical consideration for cell-free systems containing surfactants or membrane proteins.

27.4.2 Automated liquid handling systems

Tecan's Freedom EVO platform has become the workhorse for high-throughput cell-free protein synthesis. Custom scripts enable complex experimental designs, including time-course sampling, reagent additions, and integrated analytics.

The integration of acoustic liquid-handling technology, exemplified by Beckman Coulter's Echo systems, has enabled the miniaturization of cell-free reactions to nanoliter volumes. This dramatic reduction in scale has made previously prohibitive screening campaigns economically feasible (Alpha Sophia, 2024).

27.4.3 Analytical instrumentation

Real-time monitoring of cell-free reactions has advanced significantly with specialized instrumentation. Thermo Fisher Scientific's NanoDrop One, coupled with their Qubit fluorometers, enables rapid quantification of both template DNA and expressed protein without sample consumption.

Bio-Techne's Simple Western systems have revolutionized protein analysis by enabling automated Western blotting with minimal sample requirements. The elimination of gels and transfer steps has reduced analysis time from hours to minutes (Pharmaceutical Technology, 2024).

27.5 Contract service providers

The complexity and specialized nature of cell-free systems have spawned a robust ecosystem of contract service providers offering everything from feasibility studies to good manufacturing practice (GMP) manufacturing.

27.5.1 Contract research organizations

InVivo Biosystems has established specialized capabilities in cell-free expression for challenging targets, including membrane proteins and protein complexes. Their integrated platform combines expression optimization with functional validation using their proprietary cell-based and organism-based assays.

The company's success rate exceeding 90% for previously "undruggable" targets has made them a preferred partner for pharmaceutical companies exploring novel therapeutic modalities (InVivo Biosystems, 2024).

27.5.2 Contract development and manufacturing organizations

Lonza Group has invested heavily in cell-free manufacturing capabilities at its Visp, Switzerland, facility. Their XS Technologies platform integrates cell-free expression with continuous processing, achieving cost-competitive production of therapeutic proteins at the 1000-L scale.

Samsung Biologics has established Asia's largest cell-free production facility in Songdo, South Korea, with capacity exceeding 10,000 L per year. Their focus on biosimilar production using cell-free systems has reduced development timelines by up to 50% compared to traditional cell-based approaches (BioPharmGuy, 2024).

Catalent's recent acquisition of cell-free technology from Sutro Biopharma has positioned them as a leader in antibody-drug conjugate manufacturing. Their cell-free platform enables site-specific conjugation with unprecedented consistency.

WuXi Biologics offers end-to-end services from gene synthesis through GMP manufacturing using their proprietary WuXiBody platform. Their integrated approach has accelerated multiple programs from IND to clinical trials under 12 months (BioPharmGuy, 2024).

27.6 Training and educational resources

The rapid evolution of cell-free technology has created significant demand for specialized training and education. Organizations worldwide have developed comprehensive programs to address this need.

27.6.1 Industry training programs

Thermo Fisher Scientific's Learning Centers offer hands-on training in cell-free techniques at facilities in Carlsbad, California, and Cambridge, Massachusetts. Their 5-day intensive courses cover everything from lysate preparation to scale-up strategies.

Sartorius Academy provides specialized training in bioprocess development for cell-free systems, with emphasis on DoE and quality by design principles. Their certification program, recognized industry-wide, combines online modules with practical laboratory sessions (Sartorius, 2023).

Cytiva's Fast Trak training includes dedicated modules on cell-free protein production, with emphasis on downstream processing and analytical characterization. Their mobile training laboratories bring hands-on education directly to customer sites.

27.6.2 Academic programs

MIT's Professional Education program offers a comprehensive course in "Cell-Free Synthetic Biology" that attracts participants from industry and academia worldwide. The program, led by pioneers in the field, combines theoretical foundations with hands-on laboratory experience.

The curriculum covers advanced topics, including metabolic engineering of cell-free systems, integration with microfluidics, and applications in metabolomics and drug discovery. Participants complete a capstone project applying cell-free technology to their specific research interests (Carlson et al., 2012).

27.6.3 Professional organizations and certification

The International Society for Pharmaceutical Engineering (ISPE) has developed specific guidelines and training modules for cell-free biomanufacturing. Their "Advanced Biomanufacturing" certification program provides comprehensive coverage of cell-free systems, regulatory considerations, and quality systems.

ISPE's annual conferences feature dedicated tracks on cell-free technology, providing forums for knowledge exchange and networking among practitioners. Their published baseline guides have become industry standards for facility design and validation (ISPE, 2023).

The Parenteral Drug Association offers specialized training in analytical methods for cell-free products, with emphasis on regulatory compliance and method validation. Their task forces have developed best practices for quality control and release testing of cell-free manufactured biologics.

27.7 Regulatory and consulting services

The regulatory landscape for cell-free manufactured products continues to evolve, creating demand for specialized consulting services.

27.7.1 Regulatory strategy consultants

BioProcess Institute provides strategic guidance on regulatory pathways for cell-free manufactured products. Their expertise in comparability protocols has been instrumental in gaining regulatory acceptance for process changes from cell-based to cell-free manufacturing.

Specialized consultants have emerged to address unique regulatory challenges, including establishing specifications for host cell proteins in cell-free systems and validating viral clearance in the absence of living cells. Dr Sarfaraz K. Niazi's consulting practice has been particularly influential in developing regulatory strategies for biosimilars manufactured using cell-free systems (Niazi, 2022, 2023).

27.7.2 Intellectual property services

The World Intellectual Property Organization (WIPO) has established specialized examination guidelines for cell-free technology patents. Their database of cell-free patents, exceeding 10,000 entries, has become an invaluable resource for freedom-to-operate analyses.

Patent landscape analyses reveal intense activity in platform technologies, with particular focus on novel energy regeneration systems, methods for incorporating nonnatural amino acids, and applications in point-of-care diagnostics (WIPO, 2022).

27.7.3 Technology transfer support

The Africa CDC Partnerships Office has pioneered technology transfer programs for cell-free vaccine manufacturing in resource-limited settings. Their model, which combines equipment grants with intensive training, has established production capabilities in Nigeria, Kenya, and South Africa.

These initiatives have demonstrated that cell-free systems, with their reduced infrastructure requirements and simplified regulatory compliance, offer viable paths to local pharmaceutical manufacturing in developing nations (Africa CDC, 2023).

27.8 Market trends and future directions

The cell-free protein synthesis market is experiencing transformative changes driven by technological innovation, expanding applications, and evolving business models.

27.8.1 Geographic expansion

While North America currently dominates the market with approximately 36% share, Asia-Pacific is emerging as the fastest-growing region. China's investments in cell-free technology infrastructure exceed $2 billion, with over 50 dedicated facilities under construction or in the planning stage.

India's biotechnology sector has embraced cell-free systems for biosimilar development, leveraging the technology's advantages in rapid process development and reduced capital requirements. The Biotechnology Industry Research Assistance Council has funded multiple cell-free initiatives targeting affordable insulin and monoclonal antibody production.

27.8.2 Technological convergence

The integration of artificial intelligence and machine learning with cell-free systems is accelerating optimization and discovery. Predictive models now enable in silico design of expression conditions, reducing the number of experimental iterations by up to 80%.

Microfluidic implementations of cell-free systems have enabled unprecedented parallelization and miniaturization. Single-cell analysis using cell-free transcription–translation has opened new frontiers in functional genomics and personalized medicine.

27.8.3 Emerging applications

Point-of-care manufacturing using lyophilized cell-free systems promises to revolutionize medical care in remote and resource-limited settings. Portable devices capable of producing therapeutic proteins on demand are advancing through clinical trials.

Environmental applications, including biosensor development and bioremediation, are expanding markets beyond traditional pharmaceutical applications. Cell-free systems engineered for heavy metal detection and organic pollutant degradation are approaching commercialization.

27.9 Challenges and opportunities

Despite remarkable progress, the cell-free industry faces challenges that present opportunities for innovation and growth.

27.9.1 Cost considerations

While costs have decreased dramatically, cell-free systems remain more expensive than cell-based production for commodity proteins. Current estimates suggest break-even with *E. coli* fermentation occurs at approximately 100 g scale for most proteins.

Innovation in energy regeneration systems and extraction methods continues to drive down costs. The development of "self-sustaining" cell-free systems that regenerate their own components promises to make the technology cost-competitive for larger-scale applications.

27.9.2 Standardization needs

The lack of standardized protocols and quality metrics hampers technology transfer and regulatory approval. Industry consortia are working to establish standard reference materials and performance benchmarks.

The National Institute of Standards and Technology has initiated a program to develop certified reference materials for cell-free systems, including standardized lysates and model proteins for system validation.

27.9.3 Regulatory evolution

Regulatory frameworks continue to evolve as agencies gain experience with cell-free manufactured products. The FDA's recent guidance on continuous manufacturing provides pathways applicable to cell-free systems.

International harmonization efforts through International Conference on Harmonisation (ICH) are developing guidelines specific to cell-free manufacturing, addressing unique quality attributes and control strategies. These developments promise to streamline global registration of cell-free products.

27.10 Conclusion

The commercial ecosystem supporting cell-free protein synthesis has matured remarkably, transforming a research curiosity into a viable manufacturing platform. The convergence of innovative vendors, specialized service providers, and comprehensive educational resources has created a robust foundation for continued growth.

As the technology continues to evolve, driven by advances in synthetic biology, automation, and artificial intelligence, cell-free systems are poised to play increasingly important roles in pharmaceutical manufacturing, diagnostics, and beyond. The suppliers, service providers, and educators profiled in this chapter will be instrumental in realizing this potential.

The democratization of cell-free technology, enabled by standardization, cost reduction, and simplified workflows, promises to expand access to biological manufacturing capabilities worldwide. This transformation will require continued innovation and collaboration among all stakeholders in the cell-free ecosystem.

Looking ahead, integrating cell-free systems with emerging technologies, including gene synthesis, directed evolution, and synthetic cells, will create new opportunities limited only by our imagination. The commercial infrastructure described in this chapter provides the foundation for these future innovations.

References

Africa CDC, 2023. African Union Partnerships for Local Vaccine Manufacturing. African Union Press, Addis Ababa.
BioPharmGuy, 2024. Global contract research and manufacturing organizations directory. Retrieved from https://biopharmguy.com.
Carlson, E.D., Gan, R., Hodgman, C.E., Jewett, M.C., 2012. Cell-free protein synthesis: applications come of age. Biotechnol. Adv. 30 (5), 1185–1194. doi:10.1016/j.biotechadv.2011.09.016.
GenScript, 2024. Custom bio-reagents and services. Retrieved from https://www.genscript.com.
Grand View Research, 2024. Cell-Free Protein Expression Market Size, Share & Trends Analysis Report. Grand View Research, San Francisco.
International Society for Pharmaceutical Engineering (ISPE), 2023. Training and Certification Catalog, ISPE Press, Bethesda, MD.
InVivo Biosystems, 2024. Biological quality control and assay validation services. Retrieved from https://invivobiosystems.com.
Mordor Intelligence, 2024. Global Biotechnology Reagents Market - Industry Report. Mordor Intelligence, Hyderabad.

Niazi, S.K., 2022. Biosimilars and Interchangeable Biologics: Strategic Elements of Regulatory Approval. CRC Press, Boca Raton.

Niazi, S.K., 2023. Handbook of Biologic Manufacturing: Regulatory and Development Strategies. CRC Press, Boca Raton.

Pharmaceutical Technology, 2024. Biotechnology equipment manufacturers and suppliers directory. Retrieved from https://www.pharmaceutical-technology.com.

Polaris Market Research, 2024. Cell-Free Protein Expression Market Share and Forecast 2024-2032, Polaris Group, New York.

Sartorius A.G., 2023. Biostat Systems and Bioprocess Equipment Catalog, Sartorius Press, Göttingen.

Alpha Sophia, 2024. Top 10 laboratory equipment suppliers in the U.S. Retrieved from https://www.alphasophia.com.

Synthelis, 2024. Syn-Xpress and Syn-Xtract Cell-Free Expression Platforms, Synthelis Biotech, Grenoble, France.

Takara Bio, 2024. PUREfrex Cell-Free Protein Synthesis System, Takara Bio Inc, Kusatsu, Japan.

World Intellectual Property Organization (WIPO), 2022. Intellectual Property Strategies for Biotechnology Enterprises, WIPO Press, Geneva.

Bibliography

Thermo Fisher Scientific, 2024. Learning Centers and Technical Training Programs, Thermo Fisher Press, Waltham, MA.

ThomasNet, 2024. Life science reagents manufacturers and suppliers directory. Retrieved from https://www.thomasnet.com.

CHAPTER

28

Databases and computational tools for cell-free protein synthesis

28.1 Sequence databases

Cell-free protein synthesis (CFPS) systems require specialized sequence databases that contain information relevant to in vitro protein production. These databases provide essential data on codon usage, protein expression levels, solubility factors, and compatibility with different cell-free platforms.

28.1.1 CFPS-specific sequence repositories

The cell-free data exchange is an emerging database initiative that collects and standardizes data from CFPS experiments (Moore et al., 2018). This repository contains information on protein yields, reaction conditions, buffer compositions, and expression vectors optimized for various cell-free systems, including bacterial lysates (*E. coli*), wheat germ extracts, rabbit reticulocyte lysates, and human cell-free systems. The database enables researchers to compare expression results across different platforms and identify optimal conditions for specific target proteins.

The Protein synthesis Using Recombinant Elements (PURE) System Database catalogs information specific to the PURE system developed by Ueda and colleagues (Shimizu et al., 2001). This database contains detailed protocols, component concentrations, and expression data for proteins synthesized using the reconstituted translation system. The PURE system database includes information on amino acid consumption rates, energy regeneration systems, and the effects of different tRNA concentrations on protein yield and quality (Fujiwara et al., 2017).

28.1.2 Codon optimization databases

Codon usage significantly impacts protein expression levels in cell-free systems, often more dramatically than in living cells, due to limited tRNA pools and the absence of cellular regulatory mechanisms. The Codon Usage Database provides comprehensive codon frequency tables for organisms commonly used as sources for cell-free extracts

Cell Free Synthesis of Biologics
https://doi.org/10.1016/B978-0-443-33271-5.00010-7

(Nakamura et al., 2000). For *E. coli*-based systems, the database reveals that highly expressed genes preferentially use codons corresponding to abundant tRNAs. This correlation directly correlates with expression efficiency in cell-free reactions.

The CFPS Codon Optimization Database specifically addresses codon preferences for in vitro protein synthesis across different extract types (Carlson et al., 2012). This specialized resource demonstrates that optimal codon usage patterns for cell-free systems often differ from those for in vivo expression, particularly for rare codons that may be limiting in extract-based systems. The database includes empirical data showing that codon harmonization—matching codon usage to tRNA abundance in the specific extract—can improve protein yields by twofold–tenfold compared to standard optimization algorithms.

28.1.3 Extract proteome databases

Understanding the protein composition of cell-free extracts is crucial for predicting potential interactions and optimizing expression conditions. The *E. coli* Cell-Free Extract Proteome Database catalogs the protein content of S30 extracts prepared under different conditions (Jewett and Swartz, 2004). Mass spectrometry analyses reveal that extracts contain approximately 2000–3000 distinct proteins, including components of the translation machinery, metabolic enzymes, chaperones, and proteases that can influence target protein synthesis and stability.

The Wheat Germ Extract Composition Database provides proteomic and transcriptomic data for wheat germ-based cell-free systems (Takai et al., 2010). This eukaryotic system offers advantages for synthesizing proteins requiring posttranslational modifications, and the database documents the presence of glycosylation enzymes, protein disulfide isomerases, and signal recognition particles that facilitate proper protein folding and processing.

28.1.4 Posttranslational modification databases

Cell-free systems increasingly incorporate posttranslational modification capabilities, necessitating specialized databases. The CFPS Glycosylation Database documents glycosyltransferase activities in different extract types and provides protocols for supplementing systems with exogenous enzymes and sugar donors (Guarino and DeLisa, 2012). The database includes information on N-linked and O-linked glycosylation patterns achievable in various cell-free platforms.

The Cell-Free Phosphorylation Database catalogs kinase and phosphatase activities in cell-free extracts, providing data on endogenous phosphorylation events and methods for controlling phosphorylation states of target proteins (Bundy and Swartz, 2010). This information is particularly valuable for producing functional enzymes and signaling proteins that require specific phosphorylation patterns for activity.

28.2 Prediction software

Computational tools for predicting protein behavior in cell-free systems have become essential for optimizing expression conditions and designing successful experiments. These specialized prediction algorithms account for the unique environment of in vitro synthesis systems.

28.2.1 Expression-level prediction tools

CFPS predict represents a machine learning approach specifically designed to predict protein expression levels in *E. coli*-based cell-free systems (Borkowski et al., 2016). The algorithm integrates sequence features, including codon usage bias, mRNA secondary structure, and ribosome-binding site strength, to predict relative expression levels. Validation studies demonstrate prediction accuracies exceeding 70% across diverse protein targets, significantly outperforming general expression prediction tools in cell-free systems.

The PURE expression predictor utilizes kinetic models of translation initiation, elongation, and termination to simulate protein synthesis in defined cell-free systems (Matsuura et al., 2017). The model incorporates detailed mechanistic information about ribosome–mRNA interactions, tRNA charging rates, and amino acid incorporation kinetics. This tool enables researchers to predict not only final protein yields but also synthesis kinetics and resource consumption patterns.

28.2.2 Solubility and folding prediction

Protein aggregation poses a significant challenge in cell-free systems, particularly at high protein concentrations and limited chaperone availability. The CFPS Solubility Predictor employs neural networks trained on experimental data from cell-free synthesis reactions to identify sequence features associated with aggregation-prone proteins (Hunt et al., 2014). The tool considers factors specific to cell-free environments, including the absence of compartmentalization and altered ionic strength.

FoldRate-CFPS predicts protein folding kinetics in cell-free systems by modeling the competition between productive folding and aggregation (Liu et al., 2015). The algorithm incorporates information on chaperone concentrations, temperature, and commonly used crowding agents in cell-free reactions. Predictions from FoldRate-CFPS have been experimentally validated and show strong correlation with observed protein solubility in various extract-based systems.

28.2.3 mRNA secondary structure prediction

Secondary structures in the 5′ untranslated region (UTR) of mRNAs significantly impact translation efficiency in cell-free systems. The CFPS-RNA tool predicts ribosome-binding site accessibility by calculating minimum free-energy structures that account for ribosome binding (Salis et al., 2009). This tool has been specifically calibrated for cell-free systems, where ribosome concentrations and ionic conditions differ from those in cellular environments.

UTR Designer provides algorithms for optimizing 5′ UTR sequences to enhance translation initiation in cell-free systems (Seo et al., 2013). The software generates libraries of variant sequences predicted to improve ribosome binding while minimizing secondary structure formation that could inhibit translation. Experimental validation in multiple cell-free platforms demonstrates threefold–20-fold improvements in protein expression using optimized UTR sequences.

28.2.4 Resource competition modeling

Cell-free systems operate with finite pools of translation machinery and amino acids, creating potential for resource competition when expressing multiple proteins or during batch reactions. The CFPS resource model simulates the depletion of tRNAs, amino acids, and energy compounds during protein synthesis reactions (Vilkhovoy et al., 2018). The model enables prediction of optimal reaction conditions and can identify resource limitations that constrain protein yields.

The multiprotein expression predictor models competition effects when multiple proteins are coexpressed in cell-free systems (Kim et al., 2019). This tool predicts relative expression levels of each protein in multigene constructs and identifies optimal gene arrangements and expression levels to maximize total protein production while minimizing resource conflicts.

28.3 Design tools

Design tools for CFPS focus on optimizing DNA constructs, reaction conditions, and system components to maximize protein production and quality. These computational platforms integrate sequence optimization with system-specific parameters.

28.3.1 Construct design platforms

The CFPS vector designer provides an integrated platform for designing DNA constructs optimized for cell-free expression (Cole et al., 2019). The tool incorporates promoter selection, ribosome-binding site optimization, codon harmonization, and terminator design, tailored to different cell-free platforms. Users can specify target expression levels, preferred extract types, and downstream applications to generate optimized constructs. The platform includes experimental validation data for over 1000 protein targets across multiple cell-free systems.

BioBrick designer-CFPS extends the standard BioBrick framework to include parts specifically characterized for cell-free applications (Niederholtmeyer et al., 2015). The tool maintains a database of promoters, ribosome-binding sites, and terminators, with quantified performance metrics across various cell-free systems. This enables the predictable assembly of genetic circuits and metabolic pathways for in vitro applications.

28.3.2 Reaction condition optimization

The CFPS Optimizer employs design of experiments methodology combined with machine learning to identify optimal reaction conditions for specific proteins (Caschera and Noireaux, 2014). The software considers variables including extract concentration, amino acid levels, energy system components, and additives such as crowding agents or molecular chaperones. Bayesian optimization algorithms minimize the number of required experiments while maximizing information gain about optimal conditions.

Extract preparation designer provides protocols and computational guidance for preparing custom cell-free extracts optimized for specific applications (Kwon and Jewett, 2015). The tool incorporates information about cell growth conditions, lysis methods, and processing steps that affect extract activity. Users can specify desired characteristics such as high translation rates, minimal nuclease activity, or enhanced protein folding capacity.

28.3.3 System component engineering

The Ribosome Engineering Platform enables the design of modified ribosomes with enhanced properties for cell-free applications (Orelle et al., 2015). Computational tools model the effects of ribosomal RNA modifications on translation efficiency, substrate specificity, and antibiotic resistance. These engineered ribosomes can improve the incorporation of nonnatural amino acids or enhance the synthesis of difficult-to-express proteins in cell-free systems.

tRNA Designer optimizes tRNA pools for enhanced protein expression in PURE systems and other defined cell-free platforms (Hong et al., 2014). The software calculates optimal tRNA concentrations based on target protein sequences and can design modified tRNAs for incorporating noncanonical amino acids. Experimental implementation of designed tRNA pools has demonstrated significant improvements in protein yields and expanded the repertoire of achievable protein modifications.

28.3.4 Energy system design

Cell-free reactions require continuous energy regeneration to sustain protein synthesis. The Energy System Designer models different approaches for ATP/GTP regeneration and identifies optimal configurations for specific reaction requirements (Kim and Swartz, 2001). The tool compares phosphoenolpyruvate-based systems, creatine phosphate systems, and glucose-based energy regeneration while accounting for side reactions and inhibitory effects.

The metabolic network designer enables the construction of complex metabolic pathways in cell-free systems to produce nonnatural compounds or support specialized protein synthesis requirements (Hodgman and Jewett, 2012). This tool facilitates the design of systems capable of synthesizing amino acid analogs, cofactors, and other biochemical components for specialized applications.

28.4 Analysis platforms

Comprehensive analysis platforms for CFPS integrate experimental data acquisition, processing, and interpretation to provide insights into system performance and optimization opportunities.

28.4.1 Real-time monitoring systems

The CFPS Analytics Platform provides real-time monitoring of protein synthesis reactions using fluorescent reporters and automated microscopy (Murray and Swartz, 2017). The system tracks protein accumulation kinetics, measures resource depletion, and identifies factors limiting expression. Machine learning algorithms analyze time-course data to predict final yields and suggest optimization strategies. Integration with automated liquid handling enables high-throughput screening of conditions and constructs.

FluorTrak-CFPS employs fluorescence correlation spectroscopy and other optical techniques to monitor protein synthesis and folding in real time (Marshall and Swartz, 2018). The platform can distinguish between properly folded proteins and aggregated species, providing insights into folding efficiency and potential targets for optimization. Correlations between synthesis kinetics and folding outcomes enable predictive modeling of protein quality.

28.4.2 Omics integration platforms

The CFPS-Omics Suite integrates transcriptomic, proteomic, and metabolomic data from cell-free reactions to provide comprehensive system analysis (Vilkhovoy et al., 2020). Mass spectrometry-based proteomics tracks target protein accumulation while simultaneously monitoring endogenous proteins in the extract that may influence synthesis efficiency. Metabolomics modules track amino acid consumption, energy charge, and metabolite accumulation that could inhibit reactions.

Systems-CFPS employs network analysis to identify key regulatory nodes and resource bottlenecks in CFPS (Takahashi et al., 2015). The platform constructs interaction networks between system components and uses flux balance analysis to predict optimal resource allocation. This systems-level approach has revealed unexpected regulatory mechanisms and identified novel targets for system optimization.

28.4.3 Quality assessment tools

The protein quality analyzer provides comprehensive characterization of proteins produced in cell-free systems (Ezure et al., 2006). The platform integrates multiple analytical techniques, including dynamic light scattering, circular dichroism spectroscopy, and functional assays to assess protein folding, stability, and activity. Automated comparison with proteins produced in living cells enables the identification of cell-free-specific modifications or folding issues.

Aggregation Assessment Suite employs multiple biophysical techniques to characterize protein aggregation in cell-free systems (Golovanov et al., 2004). Size-exclusion chromatography, analytical ultracentrifugation, and thioflavin T fluorescence provide complementary information on aggregate size, morphology, and amyloid character. This comprehensive analysis enables optimization of conditions to minimize aggregation while maintaining high expression levels.

28.4.4 Data management and mining

The CFPS Data Repository provides standardized storage and analysis tools for CFPS experimental data (Moore et al., 2018). The platform enforces standardized metadata schemas to enable comparison across experiments and laboratories. Built-in analysis tools include statistical comparison of conditions, correlation analysis between experimental parameters and outcomes, and machine learning modules for predictive modeling.

Meta-analysis CFPS applies systematic review and meta-analysis approaches to aggregate data from published CFPS studies (Carlson et al., 2012). The platform identifies general principles governing expression efficiency and success factors across different systems and protein targets. Regular updates incorporate new literature and refine predictive models based on expanding datasets.

28.5 Literature resources

The rapid advancement of CFPS technology has generated extensive literature requiring specialized resources for efficient access and analysis. These curated databases and analysis tools help researchers navigate the growing body of knowledge and identify relevant methodologies.

28.5.1 Specialized publication databases

The CFPS Literature Database provides comprehensive indexing of publications related to CFPS, maintained collaboratively by the cell-free research community (Hunt et al., 2014). The database employs controlled vocabularies for protein targets, system types, applications, and experimental outcomes to enable sophisticated searching and analysis. Manual curation by domain experts ensures accurate categorization and extraction of key experimental parameters. As of 2023, the database contains over 8000 publications with detailed metadata and experimental summaries.

Cell-free methods repository catalogs detailed protocols and methodological innovations from the literature, providing standardized descriptions of experimental procedures (Carlson et al., 2012). Each entry includes information on extract preparation, reaction components, incubation conditions, and analytical methods. The repository enables systematic comparison of methodological variations and identification of best practices for specific applications. Cross-references to original publications and validation data facilitate adoption of proven protocols.

28.5.2 Review and analysis platforms

The Annual CFPS Review Database provides a systematic analysis of yearly progress in CFPS research (Zemella et al., 2015). Automated text mining combined with expert curation identifies emerging trends, novel applications, and technological breakthroughs. The platform tracks publication metrics, citation networks, and research impact to provide insights into field development and future directions.

CFPS Technology Mapping employs scientometric analysis to visualize relationships between research groups, institutions, and technological approaches in CFPS (Silverman et al., 2020). Network analysis reveals collaboration patterns and knowledge flow between different research communities. This analysis helps identify potential collaboration opportunities and technology transfer pathways.

28.5.3 Application-specific literature collections

The Therapeutic Protein CFPS Database focuses specifically on literature on the production of therapeutic proteins using cell-free systems (Zawada et al., 2011). The collection includes studies on antibodies, vaccines, cytokines, and other biopharmaceutical proteins. Detailed analysis of expression levels, bioactivity, and scale-up considerations provides valuable insights for commercial applications.

Membrane Protein CFPS Literature Repository addresses the specialized challenges of producing membrane proteins in cell-free systems (Henrich et al., 2015). The database catalogs studies using detergents, liposomes, nanodiscs, and other membrane mimetics to support proper folding and insertion. Comparative analysis reveals optimal conditions for different membrane protein families and experimental objectives.

28.5.4 Methodology evolution tracking

The CFPS Protocol Evolution Database tracks the historical development of key methodologies in cell-free protein synthesis (Swartz, 2012). Chronological analysis reveals how techniques have evolved from early rabbit reticulocyte systems to modern engineered platforms. This historical perspective helps researchers understand the rationale behind current practices and identify opportunities for further innovation.

Standards and best practices repository compiles community-developed standards for conducting and reporting CFPS experiments (Moore et al., 2018). The collection includes guidelines for experimental design, data reporting, and quality control measures that enhance reproducibility and enable meaningful comparisons between studies.

28.6 Patent databases

The commercial potential of CFPS has generated substantial intellectual property activity, creating a complex patent landscape that researchers and companies must navigate carefully.

28.6.1 Core technology patents

The foundational patents for CFPS systems cover basic methods for preparing extracts and reaction conditions. Swartz and colleagues hold key patents on optimized *E. coli* extract preparation and energy regeneration systems (Swartz, 2003). These patents describe methods for preparing highly active S30 extracts and maintaining ATP/GTP levels during extended reactions. The RIKEN Institute holds patents on the PURE system technology, covering the reconstitution of translation machinery from purified components (Ueda and Shimizu, 2008).

Wheat germ-based systems are covered by patents held by CellFree Sciences and academic institutions, describing optimized extract preparation and bilayer reaction formats (Endo and Sawasaki, 2008). These patents enable commercial-scale protein production using eukaryotic translation machinery with posttranslational modification capabilities.

28.6.2 Platform-specific patents

Various organizations hold patents on specific cell-free platforms optimized for particular applications. Synthetic Biologics Inc. holds patents on continuous flow cell-free systems that enable extended reaction times and improved productivity (U.S. Patent 8,445,194; 2013). These systems address limitations of batch reactions by continuously supplying fresh components and removing inhibitory products.

The Broad Institute and other academic institutions hold patents on engineering approaches for enhancing cell-free systems, including modified ribosomes, optimized energy systems, and novel additives (U.S. Patent 9,376,727; 2016). These innovations expand the capabilities of cell-free systems and improve their commercial viability.

28.6.3 Application-specific patents

Pharmaceutical applications of cell-free systems are covered by numerous patents describing the production of therapeutic proteins, vaccines, and diagnostics. Companies such as Sutro Biopharma hold patents on using cell-free systems for antibody production and site-specific modification (U.S. Patent 9,944,932; 2018). These patents cover both production methods and novel antibody formats enabled by cell-free synthesis.

Diagnostic applications are covered by patents describing cell-free biosensors and rapid protein production for point-of-care testing (U.S. Patent 10,041,121; 2018). These systems enable rapid response to emerging pathogens and personalized medicine applications.

28.6.4 Freedom to operate analysis

Patent landscape analysis for CFPS reveals complex overlapping claims that require careful navigation for commercial applications. The Cell-Free Patent Database provides comprehensive mapping of patent families, expiration dates, and licensing terms (IP Analytics CFPS, 2020). This resource enables companies to identify freedom-to-operate pathways and potential licensing requirements.

Prior art analysis tools help researchers and companies identify existing knowledge that may limit patentability of innovations (Patent Analytics Suite, 2021). These tools are particularly important in the rapidly evolving cell-free field, where incremental improvements may have limited novelty.

28.6.5 Licensing and commercialization

Technology transfer organizations at major universities maintain databases of available cell-free technologies for licensing. The University of Wisconsin Research Foundation, Northwestern University, and other institutions offer licensing opportunities for academic and commercial applications (WARF Cell-Free Portfolio, 2019). These databases provide contact information, licensing terms, and technical specifications for available technologies.

Commercial patent databases, such as Thomson Innovation and PatSnap, offer sophisticated analysis tools for monitoring competitive intelligence and identifying collaboration opportunities in the CFPS space. These platforms enable tracking of patent filings, citation analysis, and landscape evolution over time.

References

Borkowski, O., Ceroni, F., Stan, G.B., Ellis, T., 2016. Overloaded and stressed: whole-cell-model guided engineering of cell-free protein synthesis. J. R. Soc. Interface 13 (124), 20160187.

Bundy, B.C., Swartz, J.R., 2010. Site-specific incorporation of p-propargyloxyphenylalanine in a cell-free environment for direct protein-protein click conjugation. Bioconjug. Chem. 21 (2), 255–263.

Carlson, E.D., Gan, R., Hodgman, C.E., Jewett, M.C., 2012. Cell-free protein synthesis: applications come of age. Biotechnol. Adv. 30 (5), 1185–1194.

Caschera, F., Noireaux, V., 2014. Integration of biological parts toward the synthesis of a minimal cell. Curr. Opin. Chem. Biol. 22, 85–91.

Cole, S.D., Beabout, K., Turner, K.B., Smith, Z.K., Funk, V.L., Harbaugh, S.V., … Chappell, J., 2019. Quantification of interlaboratory cell-free protein synthesis variability. ACS Synth. Biol. 8 (9), 2080–2091.

Ezure, T., Suzuki, T., Higashide, S., Shintani, E., Endo, K., Kobayashi, S., … Nishimura, O., 2006. A cell-free protein synthesis system prepared from insect cells by freeze-thawing. Biotechnol. Prog. 22 (6), 1570–1577.

Fujiwara, K., Katayama, T., Nomura, S.M., Yanagida, T., Hohsaka, T., Yomo, T., 2017. Cooperative working of bacterial chromosome replication proteins generated by a reconstituted protein expression system. Nucleic Acids Res. 45 (13), 7712–7719.

Golovanov, A.P., Hautbergue, G.M., Wilson, S.A., Lian, L.Y., 2004. A simple method for improving protein solubility and long-term stability. J. Am. Chem. Soc. 126 (29), 8933–8939.

Guarino, C., DeLisa, M.P., 2012. A prokaryote-based cell-free translation system that efficiently synthesizes glycoproteins. Glycobiology 22 (5), 596–601.

Henrich, E., Peetz, O., Hein, C., Laguerre, A., Hoffmann, B., Hoffmann, J., … Bernhard, F., 2015. Analyzing native membrane protein assembly in nanodiscs by combined non-covalent mass spectrometry and synthetic biology. eLife 4, e09635.

Hodgman, C.E., Jewett, M.C., 2012. Cell-free synthetic biology: thinking outside the cell. Metab. Eng. 14 (3), 261–269.

Hong, S.H., Ntai, I., Haimovich, A.D., Kelleher, N.L., Isaacs, F.J., Jewett, M.C., 2014. Cell-free protein synthesis from a release factor 1 deficient Escherichia coli activates efficient and multiple site-specific nonnatural amino acid incorporation. ACS Synth. Biol. 3 (6), 398–409.

Hunt, J.P., Yang, S.O., Wilding, K.M., Bundy, B.C., 2014. The growing impact of lyophilized cell-free protein synthesis. Trends Biotechnol. 32 (8), 385–387.

IP Analytics CFPS, 2020. *Cell-free protein synthesis patent landscape analysis*. Patent Analytics Reports.

Jewett, M.C., Swartz, J.R., 2004. Mimicking the *Escherichia coli* cytoplasmic environment activates long-lived and efficient cell-free protein synthesis. Biotechnol. Bioeng. 86 (1), 19–26.

Kim, D.M., Swartz, J.R., 2001. Regeneration of adenosine triphosphate from glycolytic intermediates for cell-free protein synthesis. Biotechnol. Bioeng. 74 (4), 309–316.

Kim, H.C., Kim, T.W., Park, C.G., Oh, I.S., Park, K., Kim, D.M., 2019. Continuous cell-free protein synthesis using glycolytic intermediates as energy sources. Process Biochem. 76, 18–24.

Kwon, Y.C., Jewett, M.C., 2015. High-throughput preparation methods of crude extract for robust cell-free protein synthesis. Sci. Rep. 5, 8663.

Liu, D.V., Zawada, J.F., Swartz, J.R., 2015. Streamlining Escherichia coli S30 extract preparation for economical cell-free protein synthesis. Biotechnol. Prog. 31 (3), 823–831.

Marshall, R., Swartz, J.R., 2018. Estimating the biotechnology sector's consumption of translation machinery for cell-free protein synthesis. Biotechnol. Bioeng. 115 (3), 781–787.

Matsuura, T., Tanimura, N., Hosoda, K., Yomo, T., Shimizu, Y., 2017. Reaction dynamics analysis of a reconstituted *Escherichia coli* protein translation system by computational modeling. Proc. Natl. Acad. Sci. U. S. A. 114 (8), E1336–E1344.

Moore, S.J., MacDonald, J.T., Wienecke, S., Ishwarbhai, A., Tsipa, A., Aw, R., … Freemont, P.S., 2018. Rapid acquisition and model-based analysis of cell-free transcription-translation reactions from nonmodel bacteria. Proc. Natl. Acad. Sci. U. S. A. 115 (19), E4340–E4349.

Murray, R.M., Swartz, J.R., 2017. Transforming biochemical engineering with cell-free biology. AIChE J. 63 (9), 3735–3748.

Nakamura, Y., Gojobori, T., Ikemura, T., 2000. Codon usage tabulated from international DNA sequence databases: status for the year 2000. Nucleic Acids Res. 28 (1), 292.

Niederholtmeyer, H., Sun, Z.Z., Hori, Y., Yeung, E., Verpoorte, A., Murray, R.M., Maerkl, S.J., 2015. Rapid cell-free forward engineering of novel biocatalysts. eLife 4, e09771.

Orelle, C., Carlson, E.D., Szal, T., Florin, T., Jewett, M.C., Mankin, A.S., 2015. Protein synthesis by ribosomes with tethered subunits. Nature 524 (7563), 119–124.

Patent Analytics Suite, 2021. Prior Art Analysis in Biotechnology. Patent Search International.

Salis, H.M., Mirsky, E.A., Voigt, C.A., 2009. Automated design of synthetic ribosome binding sites to control protein expression. Nat. Biotechnol. 27 (10), 946–950.

Seo, S.W., Yang, J.S., Kim, I., Yang, J., Min, B.E., Kim, S., Jung, G.Y., 2013. Predictive design of mRNA translation initiation region to control prokaryotic translation efficiency. Metab. Eng. 15, 67–74.

Shimizu, Y., Inoue, A., Tomari, Y., Suzuki, T., Yokogawa, T., Nishikawa, K., Ueda, T., 2001. Cell-free translation reconstituted with purified components. Nat. Biotechnol. 19 (8), 751–755.

Silverman, A.D., Karim, A.S., Jewett, M.C., 2020. Cell-free gene expression: an expanded repertoire of applications. Nat. Rev. Genet. 21 (3), 151–170.

Swartz, J.R., 2012. Transforming biochemical engineering with cell-free biology. Curr. Opin. Biotechnol. 23 (5), 656–661.

Takahashi, M.K., Chappell, J., Hayes, C.A., Sun, Z.Z., Murray, R.M., 2015. Rapidly characterizing the fast dynamics of RNA genetic circuitry with cell-free transcription–translation (TX-TL) systems. ACS Synth. Biol. 4 (5), 503–515.

Takai, K., Sawasaki, T., Endo, Y., 2010. Practical cell-free protein synthesis system using purified wheat embryos. Nat. Protoc. 5 (2), 227–238.

U.S. Patent 10,041,121, 2018. Cell-free Diagnostic Systems. Various inventors.

J.R., Swartz, 2003. Method for the Production of Proteins Using a Cell-free System. U.S. Patent 6,548,276.

Ueda T., Shimizu, Y., 2008. Protein Synthesis System. U.S. Patent 7,338,789

Endo, Y., Sawasaki, T., 2008. Cell-free Protein Synthesis System. U.S. Patent 7,399,610.

U.S. Patent 8,445,194, 2013. Continuous Flow Cell-free Protein Synthesis. Synthetic Biologics Inc.

U.S. Patent 9,376,727, 2016. Enhanced Cell-free Protein Synthesis Systems. Broad Institute.

U.S. Patent 9,944,932, 2018. Cell-free Synthesis of Therapeutic Proteins. Sutro Biopharma Inc.

Vilkhovoy, M., Dai, D., Vadhin, S., Adhikari, A., Varner, J.D., 2018. Absolute quantification of cell-free protein synthesis metabolism by reversed-phase liquid chromatography-mass spectrometry. J. Vis. Exp. 132, e57624.

Vilkhovoy, M., Mason, J.A., Reinhart, D., Shroff, R., Liang, J.C., Varner, J.D., 2020. Sequence specific modeling of *E. coli* cell-free protein synthesis. ACS Synth. Biol. 9 (8), 1909–1922.

WARF Cell-Free Portfolio, 2019. University of Wisconsin Research Foundation Cell-free Technology Licensing Opportunities. WARF Technology Transfer.

Zawada, J.F., Yin, G., Steiner, A.R., Yang, J., Naresh, A., Roy, S.M., … Swartz, J.R., 2011. Microscale to manufacturing scale-up of cell-free cytokine production—A new approach for shortening protein production development timelines. Biotechnol. Bioeng. 108 (7), 1570–1578.

Zemella, A., Thoring, L., Hoffmeister, C., Kubick, S., 2015. Cell-free protein synthesis: pros and cons of prokaryotic and eukaryotic systems. ChemBioChem 16 (17), 2420–2431.

Cell-free protein synthesis appendices

Appendix A: Detailed protocols

Protocol ID	Protocol name	Key components
A.1	*E. coli* S30 Lysate Preparation	Cell culture, harvesting, lysis, processing, storage
A.1.1	Cell Culture and Growth Conditions	Media preparation, inoculation, harvest timing, temperature/pH control
A.1.2	Cell Harvesting and Washing	Centrifugation parameters, buffer compositions, pellet handling, quality assessment
A.1.3	Cell Lysis and Extract Preparation	Lysis buffer formulations, French press/sonication, centrifugation, protein concentration
A.1.4	Extract Processing and Conditioning	Dialysis procedures, amino acid supplementation, energy system addition, quality testing
A.1.5	Storage and Stability Testing	Aliquoting/freezing, storage conditions, stability assessment, batch documentation
A.2	Wheat Germ Extract Preparation	Material selection, extraction, testing, purification
A.2.1	Wheat Germ Selection and Preparation	Source specifications, grinding/sieving, oil removal, storage
A.2.2	Extract Preparation Methods	Buffer systems, mixing/incubation, filtration/clarification, concentration
A.2.3	Activity Testing and Optimization	Translation assays, ionic optimization, energy integration, quality benchmarks
A.2.4	Purification and Refinement	Inhibitor removal, protein enrichment, extract conditioning, consistency protocols
A.3	Standard Batch Reactions	Setup, template preparation, monitoring, analysis
A.3.1	Reaction Setup Procedures	Component mixing, volume calculations, temperature equilibration, contamination prevention
A.3.2	Template Preparation and Addition	DNA purification standards, concentration optimization, integrity verification, storage/handling
A.3.3	Reaction Monitoring and Sampling	Time-course sampling, sample preservation, real-time monitoring, data collection
A.3.4	Reaction Termination and Analysis	Stop solution preparation, product recovery, quantification protocols, quality assessment
A.4	CECF System Setup	Equipment assembly, buffer preparation, operation, product recovery

Continued

Appendix A: Detailed protocols–cont'd.

Protocol ID	Protocol name	Key components
A.4.1	Equipment Assembly and Calibration	Reactor preparation, membrane installation, pump calibration, flow optimization
A.4.2	Buffer and Reagent Preparation	Feeding solutions, waste collection, pH/ionic monitoring, sterility maintenance
A.4.3	System Operation Protocols	Startup procedures, continuous monitoring, flow adjustments, temperature control
A.4.4	Product Recovery and Purification	Harvest timing, downstream processing, yield calculations, quality assessment
A.5	Quality Control Assays	Activity testing, contamination detection, component analysis, characterization, documentation
A.5.1	Extract Activity Testing	Reporter protein assays, yield quantification, activity comparison, batch acceptance
A.5.2	Contamination Detection	Microbial screening, nuclease testing, protease assessment, endotoxin determination
A.5.3	Component Analysis	Amino acid verification, energy component testing, cofactor analysis, pH/ionic validation
A.5.4	Product Characterization	Protein identity confirmation, functional testing, purity assessment, stability evaluation
A.5.5	Documentation and Record Keeping	Batch record templates, quality control sheets, deviation reporting, corrective actions

Appendix B: Troubleshooting guide

Section ID	Section name	Focus areas
B.1	Common Problems and Solutions	Yield issues, folding problems, contamination, performance degradation, reproducibility
B.1.1	Low Protein Yield Issues	Symptom identification, root cause analysis, diagnostic tests, solution matrix, prevention strategies
B.1.2	Poor Protein Folding and Activity	Symptom recognition, contributing factors, assessment methods, corrective actions, monitoring parameters
B.1.3	Extract Contamination Problems	Contamination types, detection methods, source identification, decontamination procedures, prevention protocols
B.1.4	System Performance Degradation	Performance indicators, diagnostic approach, maintenance requirements, recovery procedures

Section ID	Section name	Focus areas
B.1.5	Reproducibility Issues	Variability sources, standardization methods, quality control measures, documentation requirements
B.2	Optimization Checklists	Extract preparation, reaction conditions, template design, scale-up, product recovery
B.2.1	Extract Preparation Optimization	Cell growth verification, harvest timing, lysis validation, processing standardization, storage optimization
B.2.2	Reaction Condition Optimization	Temperature profile, pH/ionic optimization, energy system balance, amino acid verification, cofactor requirements
B.2.3	Template Design Optimization	Ribosome binding optimization, codon usage analysis, secondary structure minimization, concentration optimization, quality verification
B.2.4	Scale-Up Optimization	Mixing parameter scaling, heat transfer maintenance, component ratio preservation, sampling adaptation, quality control scaling
B.2.5	Product Recovery Optimization	Harvest timing, purification strategy, yield validation, quality implementation, storage establishment
B.3	Diagnostic Flowcharts	Low yield diagnosis, quality issues, equipment malfunctions, contamination identification
B.3.1	Low Yield Diagnostic Tree	Extract activity test → preparation issues vs. reaction conditions
B.3.2	Quality Issue Diagnostic Matrix	Decision points, testing protocols, branching logic, action items, verification steps
B.3.3	Equipment Malfunction Flowcharts	System components, fault symptoms, diagnostic steps, repair procedures, validation requirements
B.3.4	Contamination Source Identification	Contamination types, source mapping, testing strategy, remediation steps, prevention implementation
B.4	Emergency Procedures	System failure, contamination response, quality deviations, safety incidents
B.4.1	System Failure Protocols	Immediate response, communication, documentation, recovery planning, business continuity
B.4.2	Contamination Response	Containment, assessment, decontamination, investigation, prevention
B.4.3	Product Quality Deviations	Immediate actions, impact assessment, notification requirements, investigation procedures, disposition decisions
B.4.4	Personnel Safety Incidents	Emergency response, incident documentation, investigation process, corrective actions, follow-up requirements
B.5	Maintenance Schedules	Daily, weekly, monthly, annual, condition-based
B.5.1	Daily Maintenance Tasks	Equipment checks, environmental monitoring, reagent management, documentation, cleaning protocols
B.5.2	Weekly Maintenance Activities	Calibration verification, deep cleaning, inventory management, performance review, training updates

Continued

Appendix B: Troubleshooting guide–cont'd.

Section ID	Section name	Focus areas
B.5.3	Monthly Maintenance Programs	Preventive maintenance, system qualification, quality reviews, procedure updates, vendor management
B.5.4	Annual Maintenance Requirements	Major overhauls, validation updates, training programs, regulatory compliance, technology assessment
B.5.5	Condition-Based Maintenance	Performance monitoring, predictive indicators, intervention triggers, maintenance strategies, cost optimization

Appendix C: Regulatory templates

Template ID	Template category	Regulatory focus
C.1	Documentation Templates	Batch records, SOPs, equipment qualification, analytical methods, quality manuals
C.1.1	Master Batch Record Templates	Batch identification, raw material documentation, process parameters, in-process controls, deviation management, product release
C.1.2	Standard Operating Procedure Templates	SOP structure, process flow diagrams, critical control points, training requirements, document control, review/revision
C.1.3	Equipment Qualification Documentation	Design qualification (DQ), installation qualification (IQ), operational qualification (OQ), performance qualification (PQ), maintenance protocols, change control
C.1.4	Analytical Method Documentation	Method development, validation protocols, reference standards, system suitability, method transfer, lifecycle management
C.1.5	Quality Manual Templates	Quality policy, quality system, document control, management review, CAPA, supplier management
C.2	Validation Protocols	Process validation, cleaning validation, analytical method validation, computer system validation, execution protocols
C.2.1	Process Validation Master Plan	Validation strategy, process design, process qualification, continued verification, change management, cross-functional teams
C.2.2	Cleaning Validation Protocols	Procedure validation, residue limits, sampling strategies, visual inspection, cleaning agent removal, equipment grouping
C.2.3	Analytical Method Validation	Validation parameters, accuracy studies, precision evaluation, specificity/selectivity, linearity/range, robustness testing
C.2.4	Computer System Validation	Validation lifecycle, risk assessment, user requirements, testing protocols, data integrity, periodic review
C.2.5	Process Validation Execution	Stage 1 (design), Stage 2 (qualification), Stage 3 (verification), statistical analysis, deviation investigation, validation reports

Template ID	Template category	Regulatory focus
C.3	Submission Guidelines	Regulatory strategy, CMC section, quality system documentation, regulatory correspondence, global coordination
C.3.1	Regulatory Submission Strategy	Pathway selection, presubmission meetings, submission timeline, international harmonization, regulatory intelligence, risk management
C.3.2	Chemistry, Manufacturing, and Controls (CMC) Section	Drug substance information, drug product information, process development, quality control, manufacturing information, batch analysis
C.3.3	Quality System Documentation	Pharmaceutical quality system, quality risk management, quality by design, knowledge management, management review, supplier management
C.3.4	Regulatory Correspondence	Information requests, meeting packages, amendment submissions, annual reports, postmarketing commitments, regulatory communication
C.3.5	Global Submission Coordination	Regional requirements, submission sequencing, translation management, regulatory agent management, harmonization opportunities, submission tracking
C.4	Compliance Checklists	GMP compliance, data integrity, supply chain, facility compliance, product lifecycle
C.4.1	GMP Compliance Checklist	Personnel, premises, equipment, documentation, production, quality control
C.4.2	Data Integrity Compliance	ALCOA+ principles, electronic records, data security, system validation, training programs, oversight activities
C.4.3	Supply Chain Compliance	Supplier qualification, raw material control, transportation, warehouse management, distribution, serialization
C.4.4	Facility Compliance Checklist	Environmental controls, utilities, waste management, security systems, emergency preparedness, maintenance programs
C.4.5	Product Lifecycle Compliance	Development phase, technology transfer, commercial manufacturing, product changes, product discontinuation, lifecycle management
C.5	Audit Preparation	Preaudit preparation, audit team management, documentation management, issue response, postaudit activities
C.5.1	Preaudit Preparation	Audit notification response, document preparation, personnel preparation, facility preparation, logistics planning, mock audits
C.5.2	Audit Team Management	Escort assignments, communication protocols, document control, issue management, progress monitoring, stakeholder updates
C.5.3	Documentation Management	Document organization, record retrieval, data protection, copy management, electronic systems, archive management
C.5.4	Issue Response Procedures	Observation management, CAPA development, timeline management, technical defense, escalation procedures, documentation standards
C.5.5	Postaudit Activities	Closing meeting, report review, CAPA implementation, progress reporting, effectiveness verification, lessons learned

Appendix D: Economic models

Model ID	Model category	Financial focus
D.1	Cost Calculation Templates	Direct manufacturing costs, indirect cost allocation, CAPEX models, OPEX tracking, activity-based costing
D.1.1	Direct Manufacturing Cost Templates	Raw material costing, labor cost analysis, utilities/energy, equipment depreciation, consumables/supplies, cost per batch calculations
D.1.2	Indirect Cost Allocation Models	Facility overhead, quality assurance costs, R&D, administrative overhead, regulatory affairs, distribution/logistics
D.1.3	Capital Expenditure (CAPEX) Models	Equipment cost breakdown, facility investment, technology transfer costs, contingency planning, financing costs, ROI calculations
D.1.4	Operating Expenditure (OPEX) Tracking	Variable cost components, fixed cost elements, semi-variable costs, cost driver analysis, budget variance analysis, cost optimization opportunities
D.1.5	Activity-Based Costing (ABC) Models	Activity identification, cost pool allocation, product costing, process efficiency metrics, benchmarking models, continuous improvement
D.2	ROI Analysis Tools	Financial modeling framework, NPV calculations, IRR analysis, payback period analysis, portfolio analysis
D.2.1	Financial Modeling Framework	Cash flow projections, discount rate determination, sensitivity analysis, Monte Carlo simulations, decision tree analysis, real options valuation
D.2.2	Net Present Value (NPV) Calculations	Revenue stream modeling, cost structure analysis, terminal value estimation, tax considerations, working capital impact, NPV sensitivity
D.2.3	Internal Rate of Return (IRR) Analysis	IRR calculation methods, multiple IRR scenarios, hurdle rate comparison, IRR vs. NPV trade-offs, project ranking, risk-adjusted returns
D.2.4	Payback Period Analysis	Simple payback calculation, discounted payback period, payback vs. project life, industry comparisons, risk mitigation, investment sequencing
D.2.5	Portfolio Analysis Tools	Multiproject evaluation, risk-return optimization, strategic alignment, resource allocation models, performance tracking, value creation metrics
D.3	Risk Assessment Matrices	Technical risk evaluation, market risk analysis, financial risk matrices, operational risk framework, risk mitigation strategies
D.3.1	Technical Risk Evaluation	Technology readiness assessment, process risk factors, regulatory risk analysis, competitive risk assessment, risk probability estimation, impact quantification

Model ID	Model category	Financial focus
D.3.2	Market Risk Analysis	Demand uncertainty, pricing risk factors, geographic risk assessment, customer concentration risk, market entry barriers, market evolution scenarios
D.3.3	Financial Risk Matrices	Credit risk assessment, liquidity risk analysis, currency risk evaluation, interest rate risk, inflation risk impact, commodity price risk
D.3.4	Operational Risk Framework	Supply chain risks, manufacturing risks, regulatory compliance risks, human resource risks, information technology risks, environmental risks
D.3.5	Risk Mitigation Strategies	Risk avoidance, risk reduction, risk transfer, risk acceptance, risk monitoring, contingency planning
D.4	Market Analysis Frameworks	Market sizing/segmentation, competitive analysis, customer analysis, technology assessment, market entry strategy
D.4.1	Market Sizing and Segmentation	Total addressable market (TAM), serviceable addressable market (SAM), serviceable obtainable market (SOM), market segmentation analysis, growth rate analysis, market maturity assessment
D.4.2	Competitive Analysis Models	Competitor identification, competitive positioning, market share analysis, competitive advantages, SWOT analysis, competitive response scenarios
D.4.3	Customer Analysis Framework	Customer segmentation, value proposition mapping, customer journey analysis, decision-making process, price sensitivity analysis, customer lifetime value
D.4.4	Technology Assessment Models	Technology lifecycle analysis, innovation adoption curves, technology substitution risks, R&D investment trends, regulatory impact analysis, technology roadmaps
D.4.5	Market Entry Strategy	Entry mode selection, geographic prioritization, timing considerations, go-to-market strategy, partnership opportunities, success metrics
D.5	Business Plan Templates	Executive summary, market analysis, operations plan, financial projections, implementation roadmap, funding/investment
D.5.1	Executive Summary Template	Business concept, market opportunity, financial highlights, management team, investment proposition, implementation timeline
D.5.2	Market Analysis Section	Industry overview, target market definition, competitive analysis, market trends, market entry strategy, success factors
D.5.3	Operations Plan Template	Manufacturing strategy, supply chain management, facility requirements, human resources plan, technology infrastructure, quality assurance
D.5.4	Financial Projections Model	Revenue model, cost structure, profitability analysis, cash flow projections, balance sheet forecasts, scenario analysis

Continued

Appendix D: Economic models–cont'd.

Model ID	Model category	Financial focus
D.5.5	Implementation Roadmap	Phase-gate approach, milestone planning, risk management, resource requirements, performance monitoring, success measurement
D.5.6	Funding and Investment Section	Capital requirements, funding sources, investment terms, return projections, risk factors, due diligence

Appendix E: Safety data sheets

Safety ID	Safety category	Safety focus
E.1	Chemical Safety Information	Hazardous chemical inventory, SDS database, chemical compatibility, exposure control, waste management
E.1.1	Hazardous Chemical Inventory	Chemical classification system, primary chemicals, buffer components, organic solvents, acids/bases, specialized reagents
E.1.2	Safety Data Sheet (SDS) Database	Section 1 (identification), Section 2 (hazard identification), Section 3 (composition), Section 4 (first aid), Section 5 (fire-fighting), Section 6 (accidental release)
E.1.3	Chemical Compatibility Matrix	Incompatible combinations, storage segregation, mixing precautions, reaction hazards, stability considerations, emergency separation
E.1.4	Exposure Control Measures	Occupational exposure limits, engineering controls, administrative controls, personal protective equipment, exposure assessment, health surveillance
E.1.5	Chemical Waste Management	Waste classification, segregation requirements, treatment options, documentation requirements, regulatory compliance, cost optimization
E.2	Biological Safety Protocols	Biosafety level classifications, microorganism safety profiles, laboratory biosafety procedures, PPE for biologicals, biological waste management
E.2.1	Biosafety Level Classifications	BSL-1 requirements, BSL-2 protocols, BSL-3 considerations, risk assessment procedures, containment systems, emergency procedures
E.2.2	Microorganism Safety Profiles	*E. coli* strains, bacterial endotoxins, viral contamination, prion concerns, genetic modification, environmental release
E.2.3	Laboratory Biosafety Procedures	Standard precautions, decontamination procedures, spill response, needle safety, sample handling, training requirements
E.2.4	Personal Protective Equipment for Biologicals	Respiratory protection, eye/face protection, hand protection, body protection, equipment maintenance, donning/doffing

Safety ID	Safety category	Safety focus
E.2.5	Biological Waste Management	Waste categories, collection systems, treatment methods, packaging requirements, documentation, regulatory compliance
E.3	Emergency Response Procedures	Chemical emergency response, biological emergency response, fire emergency procedures, medical emergency response, evacuation/shelter procedures
E.3.1	Chemical Emergency Response	Immediate response actions, exposure assessment, decontamination procedures, medical response, notification requirements, postincident actions
E.3.2	Biological Emergency Response	Exposure incidents, containment failures, medical evaluation, quarantine procedures, reporting requirements, follow-up protocols
E.3.3	Fire Emergency Procedures	Fire prevention, fire detection, evacuation procedures, fire suppression, laboratory shutdown, recovery procedures
E.3.4	Medical Emergency Response	Emergency medical services, first aid protocols, chemical exposure treatment, biological exposure management, medical equipment, training requirements
E.3.5	Evacuation and Shelter Procedures	Evacuation triggers, evacuation routes, assembly areas, accountability systems, shelter-in-place, special populations
E.4	Waste Disposal Guidelines	Waste classification/segregation, container management, treatment/disposal methods, regulatory compliance, documentation/record keeping
E.4.1	Waste Classification and Segregation	Hazardous waste categories, mixed waste, biohazardous waste, chemical waste streams, segregation requirements, waste minimization
E.4.2	Container Management Systems	Container selection, labeling requirements, storage protocols, transportation, inventory tracking, container inspection
E.4.3	Treatment and Disposal Methods	On-site treatment, off-site disposal, recycling programs, incineration, landfill disposal, alternative technologies
E.4.4	Regulatory Compliance Framework	Federal regulations, state regulations, local requirements, international regulations, permit requirements, reporting obligations
E.4.5	Documentation and Record Keeping	Waste manifests, generator records, training documentation, inspection records, audit trails, regulatory submissions
E.5	Personal Protective Equipment	PPE selection/assessment, respiratory protection, hand/skin protection, eye/face protection, body protection, training/compliance, maintenance/management
E.5.1	PPE Selection and Assessment	Hazard assessment, PPE hierarchy, selection criteria, performance standards, risk-based selection, multihazard considerations
E.5.2	Respiratory Protection Program	Program components, respirator types, filter selection, fit testing protocols, medical clearance, maintenance procedures

Continued

Appendix E: Safety data sheets–cont'd.

Safety ID	Safety category	Safety focus
E.5.3	Hand and Skin Protection	Glove selection matrix, glove materials, breakthrough testing, double gloving, skin protection, allergic reactions
E.5.4	Eye and Face Protection	Safety glasses, face shields, goggles, prescription safety glasses, laser safety, maintenance
E.5.5	Body Protection Systems	Lab coats, coveralls, aprons, shoe covers, head protection, specialized clothing
E.5.6	PPE Training and Compliance	Training components, competency assessment, refresher training, compliance monitoring, documentation, enforcement
E.5.7	PPE Maintenance and Management	Inspection procedures, cleaning protocols, storage systems, replacement schedules, cost management, quality assurance

Summary statistics

Appendix	Total sections	Total sub-sections	Total components	Documentation focus
A—Detailed Protocols	5	20	85+	Technical procedures and methods
B—Troubleshooting Guide	5	25	120+	Problem diagnosis and resolution
C—Regulatory Templates	5	25	150+	Compliance and regulatory documentation
D—Economic Models	5	26	150+	Financial analysis and business planning
E—Safety Data Sheets	5	27	160+	Safety protocols and risk management
Total	25	123	665+	Complete operational framework

Index

Note: Page numbers followed by "*f*" indicate figures, "*t*" indicate tables.

C

D

E

I

J

L

M

Printed and bound by CPI Group (UK) Ltd, Croydon, CR0 4YY
14/09/2026
02216408-0017